Bioregulators for Pest Control

ACS SYMPOSIUM SERIES **276**

Bioregulators for Pest Control

Paul A. Hedin, EDITOR
U.S. Department of Agriculture

ASSOCIATE EDITORS
**Horace G. Cutler, Bruce D. Hammock,
Julius J. Menn, Donald E. Moreland,
and Jack R. Plimmer**

Based on a symposium sponsored by
the Division of Pesticide Chemistry
at the Division of Pesticide Chemistry
Special Conference II,
Snowbird, Utah,
June 24–29, 1984

American Chemical Society, Washington, D.C. 1985

Library of Congress Cataloging in Publication Data
Bioregulators for pest control.
  (ACS symposium series, ISSN 0097–6156; 276)

  "Based on a symposium sponsored by the Division of
Pesticide Chemistry at the Division of Pesticide
Chemistry Special Conference II, Snowbird, Utah, June
24–29, 1984."

  1. Pesticides—Congresses. 2. Pests—Control—
Congresses.
  I. Hedin, Paul A. (Paul Arthur), 1926–   .
II. American Chemical Society. Division of Pesticide
Chemistry.
SB950.93.B56   1985   632′.95   85–6087
ISBN 0–8412–0910–3

ACS Symposium Series

M. Joan Comstock, *Series Editor*

Advisory Board

FOREWORD

The ACS SYMPOSIUM SERIES was founded in 1974 to provide a medium for publishing symposia quickly in book form. The format of the Series parallels that of the continuing ADVANCES IN CHEMISTRY SERIES except that, in order to save time, the papers are not typeset but are reproduced as they are submitted by the authors in camera-ready form. Papers are reviewed under the supervision of the Editors with the assistance of the Series Advisory Board and are selected to maintain the integrity of the symposia; however, verbatim reproductions of previously published papers are not accepted. Both reviews and reports of research are acceptable, because symposia may embrace both types of presentation.

CONTENTS

ix

PREFACE

PESTICIDES ARE BIOREGULATORS because they regulate various aspects of life processes. We are moving into a new era of regulating growth in a broad spectrum of pests that includes plants, insects, and diseases. Bioregulators can be characterized as being "endogenous," that is, originating within the organism, or "exogenous" where the agent obtained from an outside source acts to induce a desirable response in a treated species. These bioregulators may have a natural or synthetic origin.

Included in the special conference of the Division of Pesticide Chemistry were symposia on "Control of Plant Growth," "Control of Insect Growth," "Control of Pests with Natural Products," and "Molecular Biology and Genetic Engineering." The first three symposia comprise the three major sections of the book. Additionally, a chapter on "The Impact of Biotechnology on Crop Improvement" is included.

The "Control of Plant Growth" symposium, organized by Donald E. Moreland, highlighted the current status of knowledge on the mechanisms of action of herbicides, fungicides, and endogenous plant hormones. Other topics that were discussed included the phloem transport system of plants, development of pathogen resistance to fungicides, strategies for controlling pathogens by manipulation or potentiation of the plant's defense mechanisms, integration of exogenous plant bioregulators into crop production practices, methods for increasing tolerance of crop plants to herbicides, and the effects of allelochemicals on plant growth and development.

The "Control of Insect Growth" symposium, organized by Julius Menn, highlighted recent advances in the biochemistry of regulation of development by insect growth regulators, anti juvenile hormones, and behavior modification governed by antifeedants, pheromones, and defensive secretions.

The "Control of Pests with Natural Products" symposium was organized by Jack R. Plimmer and moderated by Horace G. Cutler in Plimmer's absence. The role of natural products in the control of pests has increased in recent decades as the chemist has acquired more sophisticated tools with which to elucidate complex structures. This, in turn, has led to explosive growth in the understanding of biochemical processes. Knowledge of metabolism, biosynthetic processes, neurochemistry, regulatory mechanisms, and many other aspects of plant, animal, and insect biochemistry has provided a more complete basis for understanding the modes of action of pesticides. The exploitation of biological information with a chemical basis (i.e., biorational approaches) may lead to the synthesis of a new molecule designed to act at a particular site or to block a key step in a biochemical process.

Additionally, a poster session was held during the conference under the direction of Bruce D. Hammock that consisted of 15 presentations. From these, seven manuscripts and two abstracts were submitted, of which six and the abstracts are included in the section on "Control of Insect Growth." The other is included in the section on "Control of Pests with Natural Products." Finally, chapters based on the banquet address by John H. Law and on "Biotechnology in Crop Improvement" by John T. Marvel are also included in the section on "Control of Pests with Natural Products."

It is the hope of the editors that this book will contribute to the elucidation and subsequent adoption of yet additional "New Concepts" including novel chemical compounds for the control of pests. These compounds should be effective at low concentrations, selective in activity against specific pests, of limited toxicity to nontarget organisms, and environmentally nonpersistent, and we should have an understanding of their mechanisms of action. It is the further hope of the editors that this book will serve as a document that identifies unifying themes by which research to control pests can be conducted.

We thank Nicholas A. Mangan and Henry J. Dishburger for their excellent administrative management of the conference and Gerald G. Still for his organization and chairing of the section on "Molecular Biology and Genetic Engineering." We are also grateful to all of the participants for their contributions that added materially to this book. Finally, I thank the Agricultural Research Service of the U.S. Department of Agriculture for granting me permission to organize the conference and compile the book.

PAUL A. HEDIN
U.S. Department of Agriculture
Mississippi State, MS 39762

Associate Editors

Horace G. Cutler
U.S. Department of Agriculture
Athens, GA 30613

Bruce D. Hammock
University of California
Davis, CA 95616

Julius J. Menn
Zoecon Corporation
Palo Alto, CA 94304

Donald E. Moreland
U.S. Department of Agriculture
Raleigh, NC 27695–7620

Jack R. Plimmer
International Atomic Energy Agency
Vienna, Austria

January 17, 1985

CONTROL OF PLANT GROWTH

INTRODUCTION TO CONTROL
OF PLANT GROWTH

THE PLANT GROWTH SECTION of the conference on "New Concepts and Trends in Pesticide Chemistry" highlighted the current status of knowledge and prospects for the future in the general areas of herbicide, fungicide, and growth regulator research. The presentations served as the basis for the eight chapters of this volume that follow.

The application, redistribution, and delivery of xenobiotics to the tissues of interest (sites of action) depend on a basic understanding of the phloem transport system in plants. Giaquinta reviews the structural, biochemical, biophysical, and physiological aspects of the phloem system. On the basis of distribution patterns within the plant, transport of agrichemicals was originally characterized as being either apoplastic or symplastic. However, in recent years, the ambimobility of many comounds has becomie recognized. Giaquinta also discusses the two hypotheses that have been proposed to explain the entry and systemic mobility of chemicals in the phloem, that is, the "weak-acid" and "intermediate-diffusion" hypotheses. Information on the structural and chemical properties of molecules that are required for, and the mechanisms associated with, phloem mobility should provide a basis for the rational development of systemic agrichemicals that are designed to reach specific target sites.

A large percentage of the currently used commercial herbicides inhibits photosynthetic electron transport. Considerable progress is being made in identifying and characterizing the proteins associated with photosystem II to which 3-(3,4-dichlorophenyl)-1,1-dimethylurea and phenolic types of herbicides bind. Oettmeier reports on the contributions made with radiolabeled and photoaffinity-labeled herbicides. Through the use of molecular biology techniques, identification of nucleotide and amino acid sequences has associated the replacement of a serine (in susceptible plants) by glycine (in resistant plants), in the 32–34 kilodalton binding protein, with triazine tolerance in higher plants.

Species-selective herbicides are available for use to control weeds in most of the major crops. However, differential responses between a crop plant and a weed to a given herbicide are frequently quantitative rather than qualitative. In many situations, there is still a need to increase the tolerance of crops to given herbicides. Crop tolerance can be conferred either mechanically, genetically, or chemically. Stephenson and Ezra discuss ways that can be used to improve the selective action of herbicides and to protect agronomic plants from herbicide injury. They present results of some of their recent studies that involve pretreatment of seed or early treatment of seedlings with subtoxic concentrations of herbicides to increase the metabolic tolerance of the plant to higher concentrations of the same herbicide when applied at a later stage in the growth cycle.

Considerable attention continues to be given to incorporation of herbicide resistance into crop plants. This becomes feasible after the molecular mechanism of herbicidal activity is understood as it is with the triazines. The alteration associated with triazine resistance involves a single gene transformation, is maternally inherited, and is encoded on chloroplast DNA. In addition to target site modification associated with triazine resistance, tolerance for some herbicides has been associated with metabolic detoxification or with restricted uptake or translocation. In the latter situations, transferring tolerance will be considerably more complex if the trait should be polygenic or encoded by the nuclear genome. However, in all situations, advantage can be taken of both classical and genetic engineering techniques to develop herbicide-tolerant cultivars of crop plants, when the molecular basis for tolerance has been established.

Ragsdale and Siegel review the types of chemicals that are being used to control plant pathogens, the mechanisms of actions of these fungitoxic chemicals, and the development of pathogen resistance to the fungicides. Fungicides vary in that the older compounds have nonspecific mechanisms of action, whereas some of the newer classes of compounds have rather specific sites of action. Development of pathogen resistance has been more of a problem with the site-specific, newer types of fungicides. Efforts are being made in pathogen control programs to avoid development of resistance by taking advantage of information on mechanisms of action, mechanisms of resistance, and the biochemistry and physiology of host–parasite interactions.

Strategies for controlling pathogens (bacteria, viruses, and fungi) by manipulating or potentiating the plant's defense mechanisms are discussed by Salt and Kuć. Genetic, biotic, chemical, and physical agents can be used as stimuli to increase resistance to pathogenic attack. The authors present results of their studies on immunization of tobacco, cucurbits, and beans. Additionally, they briefly review the work of other investigators on biotically and chemically induced disease resistance. Biotic and chemical sensitization of plants, for example, immunization, offers promise as an alternative or complementary method to our present strategies for protecting plants from diseases and pests.

A concise, up-to-date review of the responses induced by and mechanisms of action, for example, specific molecular interactions, of endogenous plant hormones is provided by Moore. Considered are the auxins, the gibberellins, the cytokinins, abscisic acid, ethylene, and the postulated florigens and anthesins. The plant hormones express multiple physiological and biochemical responses. Consequently, difficulty has been encountered in separating primary from secondary effects. The hormones are considered to

4

bind to specific cellular receptors, but the receptors themselves have not been identified. The varied responses could result from a single primary action of the hormone; that is, the molecular basis through which a hormone acts may be the same although the manifestations of the responses may differ in different types of cells. Some aspect of gene expression has been associated with the action of auxins, gibberellins, cytokinins, and abscisic acid.

In addition to effects on gene expression, some plant hormones have been shown to induce or modulate the action of certain proteins; however, the functions of the proteins remain to be identified. The hormone receptors also need to be identified and characterized together with the identity and role of secondary messengers. The ability to identify and quantify plant hormones is important in studies concerned with hormone physiology and biochemistry. Needed here are techniques that are more sensitive than either GC–MS or HPLC to rapidly identify and quantify the plant hormones. Development of radioimmunoassays for the hormones may be the answer.

As summarized by Jung, most, but not all, of the exogenous plant bioregulators (PBRs) function as homologues, precursors, synergists, antagonists, or biosynthetic inhibitors of the endogenous plant hormones. Presently available PBRs are being used in specific crops and in particular geographic areas to enhance biomass production, induce flowering, hasten maturity, stimulate latex flow, hasten senescence, and prevent lodging. In the past, PBRs have been identified through an empirical approach. However, biorational design of new candidates should become feasible as plant biochemistry and physiology and the regulation of cellular metabolism become defined in molecular terms. In the future, PBRs can be expected to play a much more important integrated role in crop production practices than they do at the present time.

Development of PBRs will probably be associated with advances that are made with the endogenous regulators because most of the PBRs seem to interact in some way with the endogenous regulators. Consequently, rapid developments in the area of plant growth regulation with the application of PBRs can be anticipated when the mechanisms of actions of plant hormones have been identified at the molecular level.

Einhellig reviews the types of chemistry represented by compounds synthesized by higher plants and microorganisms (allelochemicals) that have the capacity to influence plant growth and development. For the most part, the biochemical mechanisms through which allelochemicals interact with hormones or alter the physiology and biochemistry of plants remain to be identified. The possibility exists for a coordinated use of allelochemicals and herbicides to control weeds. This can be achieved by the selection of crop cultivars that have a high allelopathic potential and selection of rotational and/or cover crops that contain appropriate allelochemicals. Investigators are taking a new look at the action of allelochemicals with the advocation of limited-tillage practices as soil conservation measures that involve planting through residues of previous crops. Allelochemicals are also of interest because of their potential use as models for the development of new pesticides and bioregulators.

In the next few years, we can expect to witness a significant impact of recombinant DNA techniques on elucidation of the basic biology of plants and animals. Such an application has already been used to identify the molecular basis for plant tolerance to triazine herbicides. Future contributions can be expected in the elucidation of plant hormone biochemistry and physiology, metabolic regulatory mechanisms, factors associated with allelopathic potential, and tolerances to pathogens, insects, xenobiotics, and environmental stresses. When the molecular bases and factors associated with various tolerances and traits have been identified, a number of approaches can be taken. One approach will be to take advantage of inherent defense mechanisms and to "immunize" crop plants so that resistance or tolerance is strengthened against pathogens, insects, xenobiotics, and environmental stresses. A second approach involves a combination of traditional plant breeding and molecular biology techniques to develop new cultivars that possess the desired alterations.

If the traits involve single gene transformations and are encoded by organelle DNA, the problem may be resolvable with currently available techniques. However, many of the desired traits may be polygenic and/or encoded by the nucleus genome. The problems and procedures associated with transference in such cases will be considerably more complex.

Cultivars that are resistant or tolerant to given factors may have a lower ecological fitness or a lower agronomic value than their counterparts. Ecological and agronomic fitness will need to be evaluated by classical methods; for example, the agronomic properties of the transformed plant will have to be identified. This evaluation alone may take a number of years to produce a cultivar that a farmer can grow with assurance and have confidence in the fact that it has been tested adequately.

We hope that the chapters described in this section will stimulate thinking and research (both basic and applied) in the dynamic areas of plant science that are represented.

DONALD E. MORELAND
Agricultural Research Science
U.S. Department of Agriculture
North Carolina State University
Raleigh, NC 27695–7620

RECEIVED February 17, 1985

Physiological Basis of Phloem Transport of Agrichemicals

ROBERT T. GIAQUINTA

Experimental Station, Central Research and Development Department, E. I. du Pont de Nemours & Co., Inc., Wilmington, DE 19898

Agrichemical transport in the phloem is discussed in terms of the physiological, biochemical, and structural bases of assimilate translocation. Specifically, the cellular pathways and mechanisms of phloem loading in source leaves, long distance transport, and phloem unloading in sinks, are used as a framework for examining the biological basis of the systemic mobility of agrichemicals.

Phloem transport is the process responsible for the systemic mobility of agrichemicals in plants. From a practical viewpoint, knowledge of the structural and chemical properties of molecules that are necessary for phloem mobility should have considerable impact on the rational design of systemic agrichemicals with improved efficacy. Unfortunately, little practical information exists on structure-activity relationships of agrichemicals with respect to phloem mobility. That is, what is it about a molecule that governs its ability to be translocated in the phloem? In general, several factors ultimately determine whether an agrichemical moves to its site of action in the plant. These include: (1) efficient chemical penetration through the cuticle of the leaves and stems; (2) the ability of the chemical to enter the symplast or metabolic compartment of the cell (i.e., crossing the cell membrane); (3) short- and long-distance transport, either cell-to-cell via plasmodesmata or in the xylem, or phloem; (4) metabolism or conjugation of an agrichemical to an inactive form; and (5) immobilization of the agrichemical at non-active sites (e.g., binding at the cell walls, sequestering in the vacuole, or adsorption to cellular protein). No single factor will dictate whether a chemical is translocated in all cases and a complex interrelationship probably exists between all of these. The reader is referred to several recent and comprehensive reviews on various aspects of xenobiotic entry and transport within plants (1-5).

In this review I address the phloem mobility of agrichemicals from the viewpoint of a phloem physiologist. First, I will present an overview of the physiological basis of translocation by using

0097-6156/85/0276-0007$06.00/0

what we know about the in vivo transport of sucrose as a way of
describing the characteristics of the translocation system which are
relevant to agrichemical transport. This is important because if we
desire to rationally design molecules with improved phloem mobility,
we need to be aware of the structural, biochemical, biophysical, and
physiological aspects of the transport process itself. The second
part of this review will highlight the properties of xenobiotic
movement in plants. Although space constraints inevitably limit an
in-depth treatment of this subject, I hope the broad-stroke approach
presented here will generate a better understanding of the biological basis for translocation and, more importantly, spur additional
research in the relatively uncharted area of phloem physiology and
agrichemical transport.

Physiological Basis of Phloem Translocation

The translocation system is usually divided into three structurally
and physiologically distinct regions: (1) the source, usually the
photosynthetic leaves producing sugars; (2) the path, a series of
connecting sieve elements which comprise the conduits for assimilate
flow; and (3) the sink, which is comprised of assimilate-consuming
or target cells (e.g., growing, utilizing, or storage regions).
    In source leaves, phloem loading is the process whereby photo-
synthetically-derived sucrose produced in the mesophyll is accumulated into the minor vein network of the phloem. This sucrose accumulation increases the solute potential of the sieve tubes, causing
water from the surrounding tissues to enter the phloem to produce
hydrostatic pressure ($\underline{6}$, $\underline{7}$). At the sink end, assimilates exit the
sieve tubes by a variety of mechanisms (see below) thereby reducing
the sucrose concentration in this part of the system ($\underline{8}$). Because
of this pressure and concentration difference, water, sucrose, and
any other substance (including agrichemicals) present in the phloem
will move in bulk or mass flow from source to sink. The direction
of this osmotically-driven flow in the phloem is governed solely by
the position of sources and sinks in the plant. However, the position of these sources and sinks can differ at different stages of
leaf development or plant ontogeny ($\underline{7}$).

Phloem Loading. How does sucrose which is produced in the mesophyll
cells of source leaves enter the translocation stream and how does
this sucrose exit from the translocation stream in the sink regions?
More importantly, can this tell us anything about agrichemical
transport? Figure 1A shows an autoradiograph of a source leaf
following the accumulation of ^{14}C-sucrose (^{14}C-label is in white).
The ^{14}C-label is accumulated markedly into the vein network comprised of the minor vein phloem. This is an extensive network -
about 70 cm veins/cm^2 leaf - and thus it represents an efficient
collecting system for both sucrose which is produced in the mesophyll and for chemicals entering the leaf. Figure 1B illustrates
diagrammatically a cross section of a single minor vein traced from
an electron micrograph. The vascular bundle is composed of a single
xylem element and the phloem bundle which contains two centrally-
located sieve tubes surrounded by specialized phloem cells called
either companion cells or transfer cells depending on the presence

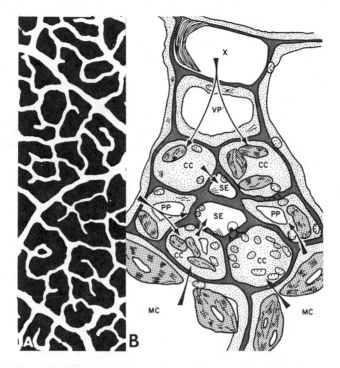

Figure 1.   Source leaf minor vein phloem. (A) Autoradiograph of
leaf tissues following ^{14}C-sucrose accumulation showing radioactivity (white) in veins.   (B) Tracing of an electron micrograph of a cross section of minor vein. x, xylem, vp, vascular
parenchyma; cc, companion cell; se, sieve element; pp, phloem
parenchyma, mc, mesophyll cell. Reproduced with permission from
Ref. 6. Copyright 1983. Annual Reviews.

of cell wall ingrowths (transfer cells have wall ingrowths). The entire bundle is surrounded by mesophyll cells and phloem parenchyma cells (6). Substances can enter the phloem via several routes, for example, from the apoplast, and cell-to-cell (see arrows in Figure 1B). Several lines of evidence show that sugars do not simply diffuse down a concentration gradient from the mesophyll to phloem. Instead there is a marked concentration of sugars in the sieve element-companion cell complex indicating that a concentrating mechanism exists at the mesophyll-phloem interface. The current body of evidence indicates that photosynthetically derived sucrose travels symplastically (via plasmodesmata) to the phloem region, then exits the symplasm into the apoplast where it is then actively accumulated across the phloem cell membranes (8). The current working model for the mechanism of sucrose uptake across the phloem membranes is illustrated in Figure 2. In this model, sucrose which is the free space, interacts with a sucrose-specific carrier on the membrane (6). We know very little about this putative sucrosyl carrier other than that it contains essential sulfhydryl groups and is highly selective for sucrose.

The characteristics of the phloem itself figure prominently in this proposed mechanism. The phloem interior has a low proton concentration (pH 7.5 to 8.0) relative to the apoplast which has a high proton concentration (pH 5.5). Thus, a substantial proton gradient of up to 2-3 pH units exists across the phloem membranes. More correctly, there is an electrochemical potential gradient across the phloem membrane which gives rise to an interior negative membrane potential of about -150 mv. It is believed that this electrochemical potential of protons which exists across the phloem membranes is the driving force for sucrose loading. The gradient is established by a "metabolism-dependent" proton pump, presumably an ATPase enzyme which is located on the phloem membrane. It is envisioned that sucrose uptake is coupled to the co-transport of protons whereby the energetically "downhill" movement of protons into the phloem is coupled to the "secondary" active transport of sucrose into the phloem (6). As discussed below, these chemical and electrical properties of the phloem can influence the ability of agrichemicals to enter the phloem.

The characteristics of the phloem loading system can be summarized as follows. Sucrose loading is: (1) dependent on metabolism; (2) carrier-mediated; (3) selective for sucrose; (4) maintains a high concentration inside the phloem which is the basis for the osmotically-driven mass flow of solutions; and (5) dependent on the factors which control assimilate supply to the loading sites (e.g., photosynthesis, sucrose synthesis, and sucrose movement between leaf cells, and within subcellular compartments such as the cytoplasm and vacuole) (6, 7).

Vascular Anatomy. One aspect of the translocation system that is often overlooked is the influence of the structural features of the plant's vascular system on solute transport. For example, although much of what is known about phloem loading has been derived from a few dicotyledon leaves, all dicotyledon leaves are not similar. A notable example is the soybean leaf. Soybean leaves are specialized in that they have a unique cell type called the paraveinal mesophyll

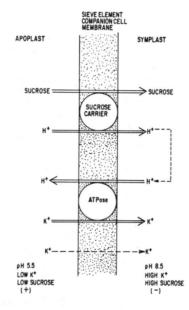

Figure 2. Model for phloem loading of sucrose. See text for details. Reproduced with permission from Ref. 8. Copyright 1980. Academic Press.

(PVM) (9, 10). In cross section the PVM appears as a single, discontinuous layer of cells in the center of the leaf (Figure 3A). The
significance of the cell type to transport, however, is indicated in
a paradermal section (Figure 3B) which shows that the PVM forms an
interconnecting network of cells in the center of the leaf that
connects to the phloem. All assimilates produced in the palisade
and spongy mesophyll appear to pass through the PVM before they
enter the phloem. It is also interesting that the vacuoles of these
cells accumulate substantial amounts of a glycoprotein during certain stages of leaf development (flowering to early pod-fill in soybeans) (9, 10). This protein, which provides a nitrogen reserve
for seed growth, could cause sequestering of agrichemicals at non-
active sites.

    Another example of vascular differences in plants is repre-
sented by monocotyledon grasses such as wheat (Figure 3C). The
vascular bundle in wheat and in many grasses is surrounded by a
mestome sheath which is comprised of an impermeable, suberized layer
of cells (6). This may represent a formidable barrier to foliar-
applied hydrophilic agrichemicals. These structural features may
need to be taken into account when seeking to rationally design crop
specific systemic chemicals. There are even structural differences
among the different grass species. For example, in grasses that
have C-3 photosynthesis, such as barley, oats, wheat, and fescue,
there are 11-15 mesophyll cells between each longitudinal vascular
bundle, with a distance between each vascular bundle of about 0.30
mm. In contrast, grasses with C-4 photosynthesis, like corn, sorghum, sugarcane, foxtail, crabgrass, barnyard grass, have only 2
mesophyll cells between each vascular bundle and the vascular bun-
dles are only 0.1 mm away from each other (11). This shorter route
of sucrose transport from mesophyll to phloem in the C-4 species is
thought to be one of the prime reasons why C-4 plants translocate at
a faster rate than C-3 species. It is not known if this influences
the transport characteristics of agrichemicals, but there are data
which suggest that these structural differences may influence movement. For example, Martin and Edgington (12) found that only 3% of
the total amount of fenarimol that was transported in barley occurred symplastically, whereas in soybeans this value was 43%. Similarly, the percent of oxamyl transport within the symplasm was 4 and
31% in barley and soybean, respectively. The percent of 2,4-dichlo-
rophenoxyacetic acid that was transported symplastically was 65 and
96% in barley and soybean, respectively. The reduced amount of
symplastic transport of these three chemicals in barley compared to
soybean may be related to the differences in vascular anatomy
between these species.

Path and Sink Features. In the translocation path (e.g., stems and
petioles), assimilates and solutes move in mass flow through the
cylindrical sieve tubes which have open sieve pores. The ability of
a chemical to leak across the membrane from the sieve tube during
transit will affect its ability to be transported through the entire
pathway.
    In sink regions, there are essentially three in vivo pathways
by which sucrose exits the sieve tubes (Figure 4). All three are
operating in different types of sinks and all are metabolism depend-

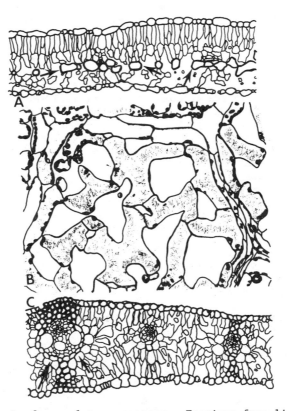

Figure 3.   Leaf vasculature anatomy.   Tracings from light micro-
graphs of: (A) soybean leaf cross section showing PVM (arrows);
(B) paradermal section of soybean leaf showing interconnecting
PVM (shaded cells); and (C) cross section of a wheat leaf showing
mestome sheath cells surrounding vascular tissue.   Tracings
provided by Shiela McKelvey.

ent (8, 13). Sucrose can exit the sieve tube and: (1) enter the
free space where it is hydrolyzed by a cell wall invertase to hexose
prior to uptake (this route occurs in corn kernels and sugarcane
storage stalks; (2) enter the free space and then be accumulated
as the intact molecule (e.g., sugarbeet root, soybean seeds, wheat
seeds); and (3) enter sink cells via plasmodesmata connections
without leaving the symplasm or metabolic space (growing roots and
young leaves). Different pathways exist in different organs and
this should be recognized when considering the target-specific
phloem mobility of agrichemicals.

Thus, based on the characteristics of the transport system,
several factors can be identified that influence the entry of a
compound into cells and its subsequent translocation in the phloem.

First, the binding of compound to the cell wall can prevent
initial entry into the cytoplasm. In general, because of the negative charge of the cell walls, positively charged compounds will
bind more than uncharged or negatively charged ones. Second, permeability barriers, such as the suberized layers surrounding the
vascular system in certain grasses, can prevent the movement of a
hydrophilic chemical directly to phloem. Third, structural features
of the vasculature, such as vein distance or the paraveinal mesophyll in soybeans, may also influence movement of a chemical to the
vein. Fourth, increased lipophilicity is usually a prerequisite for
phloem mobility but, as will be discussed below, highly penetrating
chemicals can show very little systemic movement. Fifth, the chemical cannot be a short-term inhibitor of metabolism because phloem
loading and translocation are highly dependent on metabolic energy.
Compounds such as uncouplers of phosphorylation, photosynthetic
inhibitors, and compounds which increase the permeability of cell
membranes will all inhibit translocation and thus prevent compound
movement. Sixth, the overall direction of assimilate movement is
determined by the relative position of sources and sinks on a plant.
Seventh, environmental factors such as light, temperature, and water
stress affect translocation. These factors should be taken into
account, particularly during application of a chemical in order to
maximize translocation efficiency.

Agrichemical Transport

The systemic mobility of an efficacious agrichemical depends on: (1)
effective penetration of the cuticle; (2) long distance movement
within the plant; (3) metabolic stability; and (4) selective toxicity. There are two components of the plant's structure and volume
that are important to long distance xenobiotic movement: the apoplast and the symplast. The apoplast is essentially the non-
metabolic space residing outside the cell membrane and consists of
the cell walls, xylem, and non-living fibers. It is bounded by the
cuticle on both leaf surfaces. Solute movement in the apoplast is
strongly directional and movement is usually by diffusion or by mass
flow in the transpiration stream. All chemicals enter the plant via
the apoplast. The symplast is defined as the metabolic or cytoplasmic space residing inside the plasmamembrane. It also includes
the phloem. Chemicals enter the symplast by crossing the cell
membrane.

Agrichemicals which travel mainly in the apoplast characteristically accumulate at the leaf tips and margins of mature leaves, whereas compounds that travel in the phloem accumulate at growing regions (i.e., new leaves, buds, root tips, and storage organs).

Based on the overall distribution pattern in plants, chemical transport historically has been characterized as being apoplastic or symplastic. Since the mid-1970's it has been increasingly clear that many compounds are ambimobile (4), in that these chemicals travel in both the apoplast and symplast depending on the physical characteristics of the molecule. In fact, most of the chemicals that were previously characterized as moving only in the apoplast or xylem are now regarded as ambimobile because they penetrate membranes quite readily (4).

The first clues that "apoplastic" or xylem mobile chemicals were not limited to the xylem came from several anomolies. These included the following observations: (1) many apoplastic chemicals have symplastic sites of action (the photosynthesis inhibitors like diuron and atrazine have to transverse not only the cell membrane, but also the double membrane of the chloroplast and the internal thylakoid membrane); (2) many xylem transported insecticides are toxic to aphids which feed exclusively on the phloem; (3) benzimidazole fungicides have cytokinin-like activty suggesting interaction with the symplasm; (4) many "apoplastic" compounds are either metabolized to CO_2 or conjugated with amino acids or sugars; (5) "apoplastic" chemicals that transport across the water impermeable casparian strips in the roots; and (6) the basipetal transport of certain fungicides (4).

Edgington and Peterson (4) have subdivided apoplastic xenobiotics into two classes. Euapoplastic (only transported in the apoplast) and pseudoapoplastic (transport occurs mainly in the xylem but entry into the symplast occurs). Most traditional "apoplastic" chemicals are now known to really be pseudoapoplastic chemicals, e.g., atrazine, diuron, oxamyl, etc. The unresolved question is why don't these pseudoapoplastic chemicals which cross the cell membranes and enter the symplast remain in the symplasm of the phloem? There have been numerous studies focusing on the molecular requirements for phloem mobility (1-5). In general, there is not a good correlation between phloem mobility and water solubility, metabolism of the xenobiotic, or the presence of various substitution groups in a molecule.

"Weak Acid" and "Intermediate-Diffusion" Hypotheses. Two hypotheses (which are not necessarily mutually exclusive) have been proposed for the entry and systemic mobility of chemicals in the phloem: the "weak-acid" hypothesis proposed by Crisp and colleagues (5), and the "intermediate-diffusion" hypothesis proposed by Edgington and Peterson (4). These are illustrated in Figure 5.

The weak-acid hypothesis proposes that compounds which have a free COOH group on the molecule will be in the protonated state in the apoplast because of the low pH of the apoplast (pH 5 to 5.5). The uncharged molecule will cross the phloem membranes because of increased lipid solubility. Once inside the phloem, where the pH is alkaline (pH 8) the COOH group will be ionized. The ionized species will be relatively impermeable to the phloem membrane both

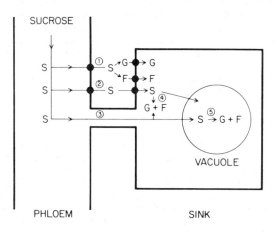

Figure 4. Pathways of phloem unloading in sink regions. See
text for details. Reproduced with permission from Ref. 13.
Copyright 1983. American Society of Plant Physiologists.

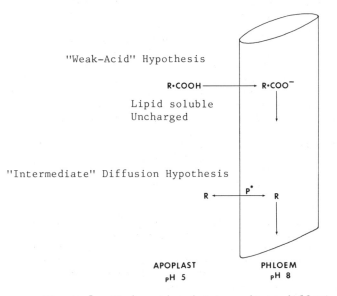

Figure 5. Weak-acid and intermediate diffusion hypotheses for
the entry and systemic mobility of chemicals in the phloem.

because of its charge and its reduced solubility in the membrane.
This ionized species can then move in mass flow with the translocation stream.   Testable features of this hypothesis are that the
agrichemical will tend to accumulate in the phloem above its
external concentration because of this "trapping" mechanism and that
agrichemical uptake will be pH dependent (higher at more acidic
pH).

There is qualified support for the weak-acid hypothesis, particularly for compounds such as 2,4-dichlorophenoxyacetic acid.   Crisp
and Look (5) compared the phloem mobility of several synthetic 4-
chlorophenoxy derivatives.   The carboxyl derivative was loaded and
transported in the phloem, whereas derivatives in which the COOH
group was replaced by an ethyl ester, amide, ketone, alcohol, or
amino group were not translocated.

Although the weak acid hypothesis appears to explain the mobility of compounds such as chlorophenoxy derivatives, there are
several exceptions to the weak-acid hypothesis (4, 14, 15).   For
example, some xenobiotics are phloem mobile but are not weak acids
and do not appear to be converted to a weak acid prior to transport
(e.g., amitrole, oxamyl).   Also, some xenobiotics (e.g., glyphosate)
which have an ionizable COOH group are loaded into the phloem independently of apoplast pH.   These should lose their mobility under
pH conditions which ionize the chemical in the free space.   Furthermore, accumulation of the weak acid glyphosate against a concentration gradient does not occur (14).

The "intermediate diffusion" hypothesis proposes that the
critical determinant of phloem mobility is the optimum permeability
coefficient of a given molecule, P.   P is calculated as:

$$P = \frac{rV}{21} \quad \ln \left[1 - \frac{l}{0.9L} \right]$$

where L is the length of the vascular system; l, the length of the
source or loading region; r, the radius of the sieve tubes; and
V, the average daily translocation velocity.

A compound that is permeable enough to enter the phloem will be
transported as long as the compound is not so permeable to the
phloem membrane that it leaks back out of the phloem into the
adjacent and opposing xylem stream.   The compound has to have a
retention time long enough to be carried in the phloem.   Each
compound and each plant have their own optimum P and Tyree et al.
(15) propose that it is not theoretically possible to devise a
xenobiotic which is optimally ambimobile for plants of all sizes.

In summary, the intent of this review was to examine the
systemic transport of xenobiotics from the viewpoint of a phloem
physiologist in order to highlight certain biochemical, physiological, and structural features of the translocation system that may
be relevant to the future design of phloem mobile chemicals.   I hope
the review has taken a modest step in that direction.

Acknowledgment

The typing of the manuscript by T. Sparre and micrograph tracings
by Shiela McKelvey are greatly appreciated.

Literature Cited

1. Christ, R. A. In "Advances in Pesticide Science"; Geissbühler,
    H., Ed.; Pergammon Press: New York, 1979; Vol. III, p. 420-9.
2. Kirkwood, R. C. In "Herbicides and Fungicides"; McFarlane,
    N. R., Ed.; Burlington House: London, 1977; p. 67-80.
3. Price, C. E. In "Herbicides and Fungicides"; McFarlane, N. R.,
    Ed.; Burlington House: London, 1977; p. 426.
4. Edgington, L. V.; Peterson, C. A. In "Antifungal Compounds";
    Siegal, M. R.; Sisler, H. D.; Eds.; Marcel Dekker, Inc.: New
    York, 1977; Vol. II, Chap. 2, p. 51-89.
5. Crisp, C. E.; Look, M. In "Advances in Pesticide Science";
    Geissbühler, H., Ed.; Pergammon Press: New York, 1979; Vol. III.
    p. 430-7.
6. Giaquinta, R. T., Annu. Rev. Plant Physiol. 1983; 34, 347-87.
7. Geiger, D. R.; Giaquinta, R. T. In "Photosynthesis: CO_2
    Assimilation and Plant Productivity"; Govindjee, Ed.;
    Academic Press: New York, 1982; Vol. II, p. 345-86.
8. Giaquinta, R. T. In "The Biochemistry of plants"; Preiss, J.,
    Ed.; Academic Press: New York, 1980; Vol. III, p. 271-320.
9. Franceschi, V. R.; Giaquinta, R. T. Planta, 1983a, 157, 411-21.
10. Franceschi, V. R.; Giaquinta, R. T. Planta, 1983b, 157, 422-31.
11. Crookston, K. R.; Moss, D. N. Crop Sci. 1974, 14, 123-5.
12. Martin, R. A.; Edgington, L. V. Pest. Biochem. Physiol. 1981,
    16, 87-96.
13. Giaquinta, R. T.; Lin, W.; Sadler, N. L.; Franceschi, V. R.
    Plant Physiol. 1983, 72, 362-7.
14. Gougler, J. A.; Geiger, D. R., Plant Physiol. 1981, 68, 668-72.
15. Tyree, M. T.; Peterson, C. A.; Edgington, L. V. Plant Physiol.
    1979, 63, 367-74

RECEIVED November 15, 1984

Interference by Herbicides with Photosynthetic Electron Transfer

WALTER OETTMEIER

Lehrstuhl Biochemie der Pflanzen, Ruhr-Universität, Postfach 10 21 48, D-4630 Bochum 1, Federal Republic of Germany

Herbicides that inhibit photosynthetic electron flow prevent reduction of plastoquinone by the photosystem II acceptor complex. The properties of the photosystem II herbicide receptor proteins have been investigated by binding and displacement studies with radiolabeled herbicides. The herbicide receptor proteins have been identified with herbicide-derived photoaffinity labels. Herbicides, similar in their mode of action to 3-(3,4-dichlorophenyl)-1,1-dimethylurea (DCMU) bind to a 34 kDa protein, whereas phenolic herbicides bind to the 43-51 kDa photosystem II reaction center proteins. At these receptor proteins, plastoquinone/herbicide interactions and plastoquinone binding sites have been studied, the latter by means of a plastoquinone-derived photoaffinity label. For the 34 kDa herbicide binding protein, whose amino acid sequence is known, herbicide and plastoquinone binding are discussed at the molecular level.

Plastoquinone is one of the most important components of the photosynthetic electron transport chain. It shuttles both electrons and protons across the photosynthetic membrane system of the thylakoid. In photosynthetic electron flow, plastoquinone is reduced at the acceptor side of photosystem II and reoxidized by the cytochrome b_6/f-complex. Herbicides that interfere with photosynthesis have been shown to specifically and effectively block plastoquinone reduction. However, the mechanisms of action of these herbicides, i. e., how inhibition of plastoquinone reduction is brought about, has not been established. Recent developments have brought a substantial increase to our knowledge in this field and one objective of this article will be to summarize the recent progress.

It was originally assumed that the herbicides bind to a protein component of photosystem II (named "B" or "R") (1,2). This protein component was assumed to contain a special bound plastoquinone whose midpoint potential is lowered due to herbicide binding. Consequently, electron flow is interrupted (1,2). The photosystem II

0097–6156/85/0276–0019$06.00/0

herbicide binding protein component was later established to function as a "proteinaceous shield" for photosystem II by Renger (3).
The "proteinaceous shield" can be removed by treatment with the proteolytic enzyme trypsin and, subsequently, DCMU-sensitivity of photosynthetic electron transport is lost (3,4).

In 1979, the concept of a photosystem II herbicide binding protein with different but overlapping binding sites for the various
photosystem II herbicides was simultaneously established by Trebst
and Draber (5) and Pfister and Arntzen (6). This idea of a herbicide receptor protein proved to be extremely fruitful because the
techniques of receptor biochemistry were now applicable. Tischer and
Strotmann (7) were the first investigators to study binding of
radiolabeled herbicides in isolated thylakoids.

Herbicide Binding Experiments

Typical results obtained in a binding experiment for two different
photosystem II herbicides are presented in Figure 1 for the tria-
zinone [^{14}C] metribuzin (right formula), a so-called "DCMU-type"
herbicide, and the phenolic herbicide [^{3}H] 2-iodo-4-nitro-6-isobutyl-
phenol (8) (left formula). The term "DCMU-type" herbicide does not
denote a chemical definition, but is a functional definition because
DCMU (diuron) is the most widely used photosynthesis inhibitor. In
binding experiments metribuzin seems to saturate at relatively low
concentrations (Figure 1). However, the binding of metribuzin is in
fact biphasic: it has a so-called specific (high affinity) binding
and an unspecific (low affinity) binding (7). The latter shows a
linear dependency on the concentration. This (extrapolated) unspecific binding of the phenolic herbicide is much higher, as compared
to that of metribuzin (Figure 1, upper dashed line). This is just
one of the many differences that can be found between "DCMU-type"
and phenolic herbicides and which justifies to view them as two different classes of herbicides (for review, see (9)).

The binding curves of the herbicides in Figure 1, especially
the one for metribuzin, look very much like Michaelis-Menten enzyme
kinetics. Indeed, herbicide binding can be treated in the same way
(7). Figure 2 presents a Lineweaver-Burk plot of the binding data
for 2-iodo-4-nitro-6-isobutylphenol. Clearly, the two types of binding, specific and unspecific binding, can be recognized. This is
even more evident in the Scatchard plot of the binding data (inset
Figure 2). Furthermore, from these plots, binding parameters, such
as the binding constant K_b, and number of binding sites, x_t, can be
obtained (7). These are also listed in Figure 2. According to
Tischer and Strotmann (7), the binding constant K_b corresponds to
the inhibition constant, i. e. the I_{50} value (the concentration
necessary for 50% inhibition of photosynthetic electron transport),
provided the I_{50} value is extrapolated to zero chlorophyll concentration. The value of 527 molecules of chlorophyll per molecule of
bound inhibitor indicates that roughly one molecule of herbicide
binds per electron transport chain, because about 400-600 molecules
of chlorophyll are considered to be associated with each electron
transport chain.

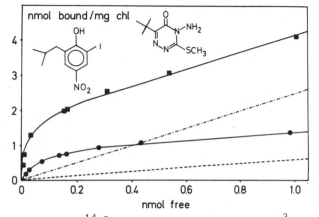

Figure 1. Binding of $[^{14}C]$ metribuzin (●————●) and $[^3H]$ 2-iodo-
4-nitro-6-isobutylphenol (■————■) to isolated spinach thylakoids.

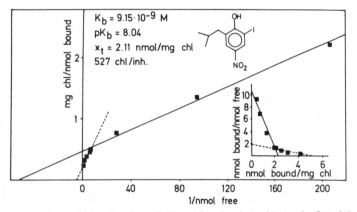

Figure 2. Lineweaver-Burk and Scatchard plot (inset) for binding
of $[^3H]$ 2-iodo-4-nitro-6-isobutylphenol to isolated spinach thylakoids.

Herbicide-Resistant Weeds

The importance of binding experiments with radiolabeled herbicides
became immediately evident in the case of herbicide-resistant weeds.
The use of certain s-triazine herbicides like atrazine for more than
two decades had led to biotypes that are resistant to normally applied doses. Binding experiments with atrazine in thylakoids isolated from resistant weed plants demonstrated that the specific binding
of atrazine was completely absent and only some unspecific binding
was left (10,11). Thus, the resistance is due to decreased binding
of the herbicide in the thylakoid of the resistant plant, which does
not inhibit photosynthetic electron transport at concentrations that
are lethal to the susceptible type. Similarily, the specific binding
of metribuzin also is completely lost, whereas the binding of urea
and biscarbamate herbicides is only slightly affected (11). In contrast, the binding of phenolic herbicides, in general, is more pronounced in thylakoids of resistant plants than in those of the susceptible types (11).

Herbicide Displacement Experiments

The photosystem II herbicides bind reversibly and non-covalently to
their binding site. Consequently, a radiolabeled herbicide can be
displaced from the binding site by another herbicide or inhibitor,
provided it has an identical binding site. Even a displacement from
a different binding site is feasible, if both binding sites interact
with each other. A typical displacement experiment is shown in Figure 3. Evidently, [^{14}C] metribuzin is easily displaced from the thylakoid membrane by DCMU. Since the pI_{50} values (negative logarithm
of concentration achieving 50% inhibition) of both compounds are
in the same order of magnitude , about 50% of the bound metribuzin
is removed from the membrane at the isomolar point, i. e. when the
concentrations of both compounds are identical. For the phenolic
herbicide dinoseb (2,4-dinitro-6-sec.-butylphenol), a much higher
concentration, 7 x 10^{-6} M, is necessary to obtain 50% removal. This
is a consequence of the lower pI_{50} value of dinoseb of 5.5 (12).
Thus, the concentration necessary for 50% displacement roughly corresponds to the pI_{50} value. It is possible, therefore, to assay the
pI_{50} value of a new compound just by examination of its displacement
behaviour. It is no longer necessary to determine the pI_{50} value by
testing the inhibition of a light-driven photoreduction. Another very
potent inhibitor of photosynthetic electron transport, DBMIB (2,5-
dibromo-3-methyl-6-isopropyl-1,4-benzoquinone) (13), almost completely fails to displace metribuzin from the membrane (Figure 3). This
is due to the fact that DBMIB has a completely different site of action as compared to the photosystem II herbicides, i. e. it inhibits
plastohydroquinone oxidation by acting at the cytochrome b_6/f-complex
(13).

Photoaffinity Labeling of the Herbicide Binding Proteins

As already stressed, photosystem II herbicides bind reversibly to
their binding site. Altough radiolabeled herbicides are available,
it is impossible to identify the herbicide receptor protein without
a chemical modification of the herbicide that allows for covalent

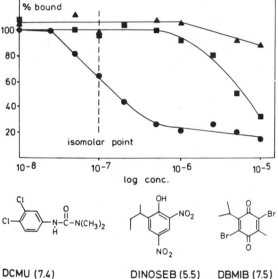

DCMU (7.4)   DINOSEB (5.5)   DBMIB (7.5)

Figure 3. Displacement of $[^{14}C]$ metribuzin from the thylakoid membrane by DCMU (●————●), Dinoseb (■————■), and DBMIB (▲————▲). The numbers in paranthesis below the structural formulas of the compounds correspond to their pI_{50} values.

attachment of the herbicide. This modification is achieved by introduction of an azidofunction into the herbicide molecule. An organic
azide upon illumination with visible or UV light readily splits-off
molecular nitrogen and forms a nitrene. Nitrenes are extremely electrophilic compounds and react immediately with any nucleophilic
groups in their environment. If the azidoderivative of the herbicide
is as good an inhibitor as its parent compound, its specific binding
should exclusively occur at its receptor protein. Consequently, the
nitrene should form a covalent bond to the receptor protein. Since
the azidoderivative of the herbicide is radiolabeled, the receptor
protein can be easily identified because it becomes radioactive by
the attachment of the nitrene. The common procedure for identification includes disruption of the thylakoid membrane system by deter-
gent treatment, separation of the thylakoid proteins by polyacrylamide gel electrophoresis, and assaying for radioactivity either by
cutting the gel into pieces, which are solubilized and counted in a
liquid scintillation counter, or by exposure of the gel on X-ray
film.
     So far, three different photoaffinity labels of photosystem II
herbicides are available (Figure 4): azidodinoseb (phenolic) (14),
azidoatrazine (15), and azidotriazinone (16) (both "DCMU-type" herbicides). Azidoatrazine in isolated thylakoids from spinach, and the
alga Chlamydomonas reinhardtii as well labels a protein with an apparent molecular weight of 34 kDa (15,17). Furthermore, azidoatrazine in the weed Amaranthus binds to the 34 kDa protein only in thylakoids from atrazine-susceptible and not to thylakoids from atrazine-resistant plants (18). It was concluded, therefore, that the
34 kDa protein is the photosystem II herbicide binding protein for
"DCMU-type" herbicides. This 34 kDa herbicide binding protein is
identical to the "photogene" or "rapidly turning over" 34 kDa protein
that stands out amongst all of the thylakoid proteins due to its
rapid destruction and de novo biosynthesis (19).
     The idea of the 34 kDa herbicide binding protein has met some
criticism by Gressel (20). This criticism is due to the fact that
photoaffinity labeling experiments, in general, are not unambiguous.
It is feasible, that a photoaffinity label does not bind to the target protein, but to a neighbouring protein instead. Specifically,
Gressel's criticism is based on the fact that in the azidoatrazine
molecule, the azido group and the structural element generally recognized for herbicidal activity lie on opposite parts of the molecule. Therefore, there is a possibility that the 34 kDa protein that
is tagged by azidoatrazine is not the real herbicide binding protein.
To clarify this question, we recently synthesized another "DCMU-type"
photoaffinity label: azidotriazinone (Figure 4) (16). Figure 5 shows
results of a labeling experiment with [^{14}C] azidotriazinone. Only
one protein is heavily labeled. From the position of the marker proteins, a molecular weight of 34 kDa can be estimated. If samples of
the thylakoid labeled by [^{14}C] azidoatrazine or [^{14}C] azidotriazinone
are run in adjacent lanes of a gel, radioactivity in both cases is
found in exactly the same position. Furthermore, prelabeling with
inactive [^{12}C] azidotriazinone prevents labeling of the 34 kDa protein by [^{14}C] azidoatrazine (16). Azidotriazinone is completely different in its chemical structure from azidoatrazine. However, both
photoaffinity labels bind to an identical 34 kDa protein. It has to
be concluded that Gressel's suggestion that the 34 kDa protein is
not the herbicide binding protein is not valid.

[³H]Azido-dinoseb   [¹⁴C]Azido-atrazine   [¹⁴C]Azido-triazinone

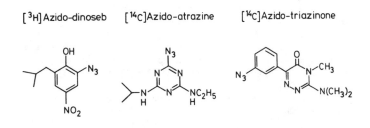

Figure 4. Structural formulas of herbicidal photoaffinity labels.

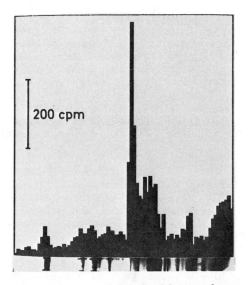

Figure 5. Photograph of a Li-dodecylsulfate polyacrylamide electrophoresis gel (10-15%) and radioactivity distribution therein of spinach thylakoids isolated by 20 nmol/mg chlorophyll [¹⁴C] azidotriazinone.

     The phenolic photoaffinity label azidodinoseb (Figure 4) binds
less specifically than either azidoatrazine or azidotriazinone (14).
In addition to other proteins, it labels predominantly the photosystem II reaction center proteins (spinach: 43 and 47 kDa; Chlamydomonas: 47 and 51 kDa) (17). Because of the unspecific binding of
azidodinoseb, this can best be seen in photosystem II preparations
(17). Thus, the phenolic herbicides bind predominantly to the photosystem II reaction center, which might explain many of the differences observed between "DCMU-type" and phenolic herbicides (9). The
photosystem II reaction center proteins and the 34 kDa herbicide
binding protein must be located closely to and interact with each
other in order to explain the mutual displacement of both types of
herbicides (8,12,21). Furthermore, it should be noted that for phenolic herbicides, some effects at the donor side of photosystem II
(22) and on carotenoid oxidation in the photosystem II reaction center have been found (23).
     The herbicidal photoaffinity labels are also useful tools for
elucidation of herbicide binding properties in various photosynthetic preparations. In photosystem II preparations with an intact water
splitting enzyme system both, azidoatrazine and azidodinoseb bind to
their respective proteins (9). In contrast, in a photosystem II particle without the water splitting enzyme complex, azidoatrazine does
not bind, whereas azidodinoseb still does (17).This does not indicate, however, that the 34 kDa protein is not present in this photosystem II particle. As already stressed, the 34 kDa herbicide binding protein has a high turnover rate (19). If Chlamydomonas cells
are grown in a medium containing [^{14}C] acetate and photosystem II
particles are prepared from these algae, the maximum of the radioactivity in the gel is found exactly at that position where the 34
kDa herbicide binding protein migrates (24). No binding of azidoatrazine is observed in n-hexane extracted thylakoids, whereas azidodinoseb binding is unaffected by this procedure (24). These results
indicate that the herbicide binding properties of the 34 kDa protein
are very sensitive to changes in its protein or lipid environment.

Herbicide/Quinone Interactions

Recent developments have led to a revision of the original idea that
the "B" or "R" protein (see above) which is identical to the 34 kDa
herbicide binding protein contains a bound plastoquinone. Velthuys
(25) from flash-induced absorbance changes of plastosemiquinone in
the presence of various inhibitors, and Lavergne (26) from fluores-
cence experiments, inferred that there is an electron-dependent direct competition between plastoquinone and herbicide. A plastoquinone
molecule from the plastoquinone pool gets bound to the acceptor complex of photosystem II to become Q_B. Q_B, in turn, gets reduced by Q_A
via the semiquinone anion radical Q_B^-, the primary acceptor of photosystem II, again another special plastoquinone molecule. Q_B^- sta-
bilizes upon binding to Q_A. In a subsequent second step, Q_B^- gets
reduced to plastohydroquinone which is exchanged with another plastoquinone from the pool. Photosystem II herbicides compete with plastoquinone for binding to the herbicide/quinone environment. Urbach et
al. (27) recently demonstrated a flash-induced binary oscillation of
herbicide binding. Herbicide binding is higher in the dark or at an

even number of flashes, i. e. when Q_B is oxidized, than at an odd number, when Q_B is singly reduced. Therefore, herbicide binding can only take place when the binding site is vacant, i. e. not occupied by Q_B^-.

It is worthy of special interest to study directly the displacement of a herbicide by plastoquinone or its analogues. In normal thylakoids, almost no displacement of DCMU even by a million-fold excess of the short-chain plastoquinone analogue plastoquinone-1 can be observed (28). This may be due to the high endogenous plastoquinone content of the thylakoid membrane. If the thylakoids are depleted of plastoquinone by means of n-hexane extraction, a competitive displacement of DCMU by plastoquinone-1 is observed (28). This result establishes a direct interaction between herbicide and plastoquinone, though not necessarily at an identical binding site. From the displacement experiments, a binding constant for plastoquinone-1 of 51± 19 µM in plastoquinone-depleted thylakoids can be calculated (28). As compared to DCMU (binding constant 34 nM (24)) the affinity of plastoquinone-1 is more than three orders of magnitude less. In a similar displacement experiment of bromoxynil by plastoquinone-1 in triazine-resistant thylakoids, Vermaas et al. (29) found a plastoquinone-1 binding constant of 20 µM which is in the same order of magnitude as our value (28).

In an attempt to learn more about the nature of the plastoquinone binding site, we have analyzed the displacement behaviour of 25 different 1,4-benzoquinones to DCMU. A quantitative structure activity relationship revealed that the displacing activity of a quinone toward DCMU is governed by the redox potential and the geometrical conformation of the quinone (30).

An Azidoplastoquinone Photoaffinity Label

To identify plastoquinone binding proteins, we have recently synthesized an azidoplastoquinone photoaffinity label (31). Figure 6 shows a typical labeling pattern of a spinach photosystem II preparation. Only one major protein in the 32-34 kDa molecular weight range is heavily tagged. A similar picture is obtained, if normal thylakoids are used (32). Labeling of the 32-34 kDa protein is prevented, if the samples are preincubated either with DCMU, the phenolic herbicide 2-iodo-4-nitro-6-isobutylphenol, or the photosystem II inhibitor tetraiodo-1,4-benzoquinone (33). The question arises as to whether the 32 - 34 kDa protein, as labeled by azidoplastoquinone, and the 34 kDa protein, as labeled by azidoatrazine or azidotriazinone, are identical. If samples labeled by either azidoplastoquinone or azidoatrazine are run in adjacent lanes of a gel, the R_f-values of the spots with the maximum amount of radioactivity differ by 0.05 (24). This experiment has been repeated several times. If 4 M urea is included in the gel, the maxima of radioactivity coincide (24). There are two possible explanations. One is that the proteins labeled by the two different photoaffinity labels are identical. The differences in the R_f-values in one gel system may be due to the attachment of the two different moieties originating from the label to the protein. The second possibility would be that the two proteins are really different, i. e. herbicide and plastoquinone binding sites are on different proteins, but the two proteins have very similar molecular weights. Most recently, evidence is accumulating that in addition to

the 34 kDa herbicide binding protein and a 33 kDa lysine-rich protein, which is part of photosystem II, but is associated with the
water splitting enzyme complex, yet another unidentified 32-34 kDa
protein may play a role in herbicide and plastoquinone binding (34,
35).

The 34 kDa Herbicide Binding Protein: The Molecular Level

The 34 kDa herbicide binding protein is a plastid encoded protein.
In the process of sequencing the plastid genome, the DNA sequence of
the gene coding for the 34 kDa herbicide binding protein became
known and, hence, also its amino acid sequence (36). One great surprise arising from the amino acid sequence is the complete lack of
lysine residues. Except for comparative studies, the knowledge of
the amino acid sequence of the 34 kDa herbicide binding protein does
not seem to yield very much additional information. However, recently Kyte and Doolittle (37) and Argos et al. (38) have devised a program that progressively evaluates the hydrophilicity and hydrophobicity of a protein along its amino acid sequence. This approach is
very important for membrane bound proteins like the 34 kDa herbicide
binding protein because it allows one to predict regions of the protein that may be embedded within the membrane system.

     In the Kyte and Doolittle process (37), each amino acid is as-
signed a hydropathy ("strong feeling about water") value which ranges
from 4.5 for isoleucine (as t.e most hydrophobic amino acid) to -4.5
for arginine (as the most hydrophilic amino acid). Now a certain
"window", i. e. a certain length of the amino acid sequence is selected. Assuming a window of 11 amino acids, the sum of the hydropathy values of the amino acid sequence for amino acids 1 to 11 is
calculated. Next the window is moved one amino acid ahead within the
sequence, and the sum of the hydropathy values for amino acids 2 to
12 is calculated. This process is repeated for sequence 3-13, 4-14
etc. until the end of the sequence is reached. The sum of the hydropathy values over the number of the amino acid residue is plotted.
Such a hydropathy plot for the 34 kDa herbicide binding protein is
presented in Figure 7. (It should be noted that for the hydropathy
plot, the original amino acid sequence as reported by Zurawski et al.
(36) was not used. Instead, a sequence shorter by 36 amino acids
which starts at the second Met was used (39)). The positive (shaded)
areas in Figure 7 correspond to regions of high hydropathy, the negative areas to regions of low hydropathy. Only regions of high hydropathy of the protein are thought to be buried within the membrane
system, provided they are approximately 20 amino acids long to account for a helical span through the lipid bilayer of the membrane.
Based on this assumption, seven helical spans are predicted for the
34 kDa herbicide binding protein (Figure 7). A very similar hydropathy plot for the 34 kDa herbicide binding protein was presented
by Argos's group (40) calculated according to their method (38). It
differs from that in Figure 7 by the assignments of spans V and VI
(41).
     A schematic picture of the 34 kDa herbicide binding protein as
it is thought to be located in the membrane is given in Figure 8.
The seven helical spans through the membrane are indicated. Furthermore, Figure 8 provides three additional relevant facts or sugges-
tions. The first one deals with the possible binding site of azido-

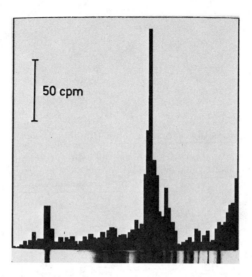

Figure 6. Photograph of a Li-dodecylsulfate polyacrylamide electrophoresis gel (10-15%) and radioactivity distribution therein of a spinach photosystem II preparation labeled by 2 nmol/mg chlorophyll [^{3}H] azidoplastoquinone.

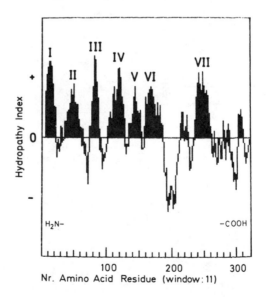

Figure 7. Hydropathy plot of the 34 kDa herbicide binding protein.

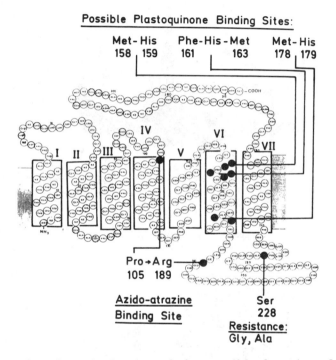

Figure 8. Schematic drawing of the possible location of the 34 kDa herbicide binding protein within the thylakoid membrane.

atrazine on the 34 kDa herbicide binding protein. From their recent work on proteolytic digestion after azidoatrazine labeling, Wolber and Steinback (42) have concluded that the azidoatrazine binding site presumably is located in the region Pro 105 to Arg 189 of the amino acid sequence (spans IV to VI, Figure 8). The second suggestion relates to herbicide resistance. By sequencing the DNA of the 34 kDa herbicide binding protein from atrazine-resistant Amaranthus, Hirschberg and McIntosh (39) have found 4 nucleotide differences as compared to the susceptible type. Only one of the differences leads to a change in the amino acid sequence: Ser in position 228 is replaced by Gly. Similarily, in an atrazine-resistant mutant from Chlamydomonas the same Ser in position 228 is exchanged, but this time against Ala (43). The third suggestion concerns possible plastoquinone binding sites in the 34 kDa herbicide binding protein, if there are any. According to a proposal by Hearst and Sauer (44), the sequence Met-His is a possible quinone binding site, probably with an additional Phe. There are three consecutive Met-His sequences in the 34 kDa herbicide binding protein (Figure 8). It should be pointed out, however, that the arginine residues are also possible candidates for quinone binding (45). Indeed, the arginine-modifying reagent phenylglyoxal was found to decrease atrazine binding (46).

In conclusion, observations made in the last few years, especially the binding studies with radiolabeled herbicides, the photoaffinity labeling technique, and the advances of molecular biology have substantially added to our knowledge of the mechanism of action of photosynthetic herbicides. However, many questions also remain to be answered.

Acknowledgements

This work was supported by Deutsche Forschungsgemeinschaft. I am indebted to Eva Neumann, Dr. Udo Johanningmeier, Klaus Masson, and Hans-Joachim Soll for their help in the experiments.

Literature Cited

1. Bouges-Bocquet, B. Biochim. Biophys. Acta 1973, 314, 250-6.
2. Velthuys, B.R.; Amesz, J. Biochim. Biophys. Acta 1974, 333, 85-94.
3. Renger, G. Biochim. Biophys. Acta 1976, 440, 287-300.
4. Regitz, G; Ohad, I. J. Biol. Chem. 1976, 251, 247-52.
5. Trebst, A; Draber, W. In "Advances in Pesticide Science"; Geissbühler, H., Ed.; Pergamon Press: Oxford, New York, 1979; Part 2, p. 223.
6. Pfister, K; Arntzen, C.J. Z. Naturforsch. 1979, 34c, 996-1009.
7. Tischer, W; Strotmann, H. Biochim. Biophys. Acta 1977, 460, 113-25.
8. Oettmeier, W.; Masson, K.; Johanningmeier, U. Biochim. Biophys. Acta 1982, 679, 376-83.
9. Oettmeier, W.; Trebst, A. In "The Oxygen Evolving System of Photosynthesis"; Inoue, Y., et al., Eds.; Academic Press: Tokyo, 1983, p. 411.
10. Pfister, K.; Radosevich, S.R.; Arntzen, C.J. Plant Physiol. 1979, 64, 995-9.

11. Oettmeier, W.; Masson, K.; Fedtke, C.; Konze, J.; Schmidt, R.R. Pestic. Biochem. Physiol. 1982, 18, 357-67.
12. Oettmeier, W.; Masson, K. Pestic. Biochem. Physiol. 1980, 14, 86-97.
13. Trebst, A; Harth, E.; Draber, W. Z. Naturforsch. 1970, 25b, 1157-9.
14. Oettmeier, W.; Masson, K.; Johanningmeier, U. FEBS Lett.1980, 118, 267-70.
15. Gardner, G. Science 1981, 211, 937-40.
16. Oettmeier, W.; Masson, K.; Soll, H.J.; Draber, W. Biochim. Biophys. Acta 1984, in press.
17. Johanningmeier, U.; Neumann, E.; Oettmeier, W. J. Bioenerg. Biomembr. 1983, 15, 43-66.
18. Pfister, K.; Steinback, K.E.; Gardner, G.; Arntzen, C.J. Proc. Natl. Acad. Sci.USA 1981, 78, 981-5.
19. Mattoo, A.K.; Pick, U.; Hoffmann-Falk, H.; Edelman, M. Proc. Natl. Acad. Sci. USA 1981, 78, 1572-6.
20. Gressel, J. Plant Sci. Lett. 1982, 25, 99-106.
21. Laasch, H.; Pfister, K.; Urbach, W. Z. Naturforsch. 1982, 37c, 620-31.
22. Pfister, K.; Schreiber, U. Z. Naturforsch. 1984, 39c, 389-92.
23. Mathis, P.; Rutherford, A.W. Biochim. Biophys. Acta 1984, in press.
24. Oettmeier, W.; Soll, H.J.; Neumann, E. Z. Naturforsch. 1984, 39c, 393-6.
25. Velthuys, B.R. FEBS Lett. 1981, 126, 277-81.
26. Lavergne, J. Biochim. Biophys. Acta, 1982, 682, 345-53.
27. Urbach, W.; Laasch, H.; Schreiber, U. Z. Naturforsch. 1984, 39c, 397-401.
28. Oettmeier, W.; Soll, H.J. Biochim. Biophys. Acta 1983, 724, 287-90.
29. Vermaas, W.F.J.; Renger, G.; Arntzen, C.J. Z. Naturforsch. 1984, 39c, 368-73.
30. Soll, H.J.; Oettmeier, W. In "Advances in Photosynthesis Research"; Sybesma, C., Ed.; Martinus Nijhoff/Dr. W. Junk Publishers: The Hague, 1984, Vol. 4, p. 5.
31. Oettmeier, W.; Masson, K.; Soll, H.J.; Hurt, E.; Hauska, G. FEBS Lett. 1982, 144, 313-7.
32. Oettmeier, W.; Masson, K.; Soll, H.J.; Olschewski, E. In "Advances in Photosynthesis Research"; Sybesma, C., Ed.; Martinus Nijhoff/ Dr. W. Junk Publishers: The Hague, 1984, Vol. 1, p.469.
33. Oettmeier, W.; Soll, H.J. in preparation.
34. Satoh, K.; Nakatani, H.Y.; Steinback, K.E.; Watson, J.; Arntzen, C.J. Biochim. Biophys. Acta 1983, 724, 142-50.
35. Renger, G.; Hagemann, R.; Vermaas, W.F.J. Z. Naturforsch. 1984, 39c, 362-7.
36. Zurawski, G.; Bohnert, H.J.; Whitfield, P.R.; Bottomley, W. Proc. Natl. Acad. Sci. USA 1982, 79, 7699-703.
37. Kyte, J; Doolitlle, R.F. J. Mol. Biol. 1982, 157, 105-32.
38. Argos, P.; Rao, J.K.M.; Hargrave, P.A. Eur. J. Biochem. 1982, 128, 565-75.
39. Hirschberg, J; McIntosh, L. Science 1983, 222, 1346-9.
40. Rao, J.K.M.; Hargrave, P.A.; Argos, P. FEBS Lett. 1983, 156, 165-9.

41. Trebst, A. In "Proceedings of the Southern Section, American Society of Plant Physiology, 1984, in press.
42. Wolber, P.K.; Steinback, K.E. Z. Naturforsch. 1984, 39c, 425-9.
43. Erickson, J.M; Rahire, M.; Bennoun, P.; Delepelaire, P.; Diner, B.; Rochaix, J.D. Proc. Natl. Acad. Sci. USA 1984, 81, 3617-21.
44. Hearst, J.E.; Sauer, K. Z. Naturforsch. 1984, 39c, 421-4.
45. Shipman, L.L. J. Theor. Biol. 1981, 90, 123-48.
46. Gardner, G.; Allan, C.D.; Paterson, D.R. FEBS Lett. 1983, 164, 191-4.

RECEIVED November 8, 1984

New Approaches to Chemical Control of Plant Pathogens

NANCY N. RAGSDALE[1] and MALCOLM R. SIEGEL[2]

[1] Cooperative State Research Service, U.S. Department of Agriculture, Washington, DC 20251
[2] Plant Pathology Department, University of Kentucky, Lexington, KY 40546

New approaches to chemical control of plant pathogens
emphasize selective use of fungicides based on an
understanding of the physiological, biochemical, and
molecular levels of the modes of action and mechanisms
of resistance. Antifungal compounds currently available
for disease control may have either non-specific or
specific biochemical modes of action. Nearly all
systemic fungicides are of the latter type. Compounds
with specific modes of action are prone to a loss of
efficacy through selection pressure for resistant
pathogens. Evidence suggests that chemicals with
certain modes of action may be less likely to encounter
resistance problems. The possibility of circumventing
resistance by chemically affecting metabolic pathways or
through stereoselectivity are discussed. In addition,
promising methods for plant disease control with
chemicals that activate plant defense mechanisms or
interfere with pathogenesis are considered.

Most of the pests that cause infectious plant diseases may be
classified as fungi, bacteria, nematodes, or viruses. Chemicals
frequently are either unavailable or impractical to use in
controlling viruses and most bacteria. Efforts to achieve nematode
control usually involve compounds associated with insect control.
Thus, this article will be devoted to the use of chemicals to control
fungi that cause plant diseases, a far greater use than for other
disease agents. These fungicides, or antifungal agents as they are
more correctly called, are but a part of disease control strategy.
Plant breeding programs and cultural practices are also part of plant
disease management practices.
  The traditional approach in the development of fungicides has
involved large scale laboratory and greenhouse screening tests
followed by closer examination of structurally related chemicals to
optimize activity. More recently scientists have become aware that
this type of assay system has limitations. There is a need to
examine vulnerabilities in both the pest and the host, and to design

0097–6156/85/0276–0035$06.00/0

antifungal materials that have modes of action based on these
principals. This will identify chemicals that have both direct and
indirect actions. New approaches emphasize an understanding at the
physiological, molecular, and biochemical levels of modes of action
as well as the mechanisms for pathogen resistance. Some of the more
recent antifungal compounds resulted from this type of approach. Not
only will these approaches provide knowledge for the rational
development of new antipathogenic compounds, they will also provide
strategies for the use of established chemicals, especially in
relation to reducing or preventing the development of resistant
pathogens.

This chapter will endeavor to condense a rather wide ranging
subject into some basic principles regarding the types of chemicals
used to control plant pathogens; the strategies behind these; and
some thoughts on new approaches to the development and use of
chemicals to control fungal diseases of plants.

First Generation Fungicides

The first generation fungicides are primarily a group referred to as
surface protectants. These compounds are not taken up to any degree
by the plant, have a broad antifungal spectrum, and demonstrate
multiple sites of action (1). The function of these pesticides is
based on their being present when the fungus arrives and before
infection occurs. They exhibit a differential toxicity; thus a
specific dose will kill the pathogen and not injure the host plant.
Differential toxicity is based primarily on fungi rapidly
accumulating toxic concentrations of the fungicide while plant
tissues do not (2). The surface protectants include inorganic
chemicals such as elemental sulfur, Bordeaux mixture and copper oxide
as well as organic fungicides such as dialkyldithiocarbamates
(thiram, ferbam), ethylenebisdithiocarbamates (maneb, zineb),
phthalimides (captan, folpet), and chlorothalonil. Because these
first generation fungicides are multisite inhibitors, they are not
likely to lose effectiveness because of fungal resistance. Pathogen
resistance has not been a problem with this group and their continued
use as an important part of disease management programs is highly
likely.

Second Generation Fungicides

During the mid 1960's, the systemic toxicants, or second generation
fungicides as they are called, entered the market. Although most of
the older materials are still in use, current efforts have shifted
toward developing systemic compounds which move in the symplast,
apoplast, or are ambimoble (3). Almost all the commercially
available systemic fungicides move only in the apoplast and are
therefore dependent on transpiration for their movement and
accumulation in plant tissue. The systemic fungicides have a number
of advantages over surface protectants. Foremost is their ability to
penetrate host tissue and control or eradicate an established
infection. In order for these compounds to provide internal therapy,
they must have certain qualities. They or their active derivatives
must be relatively stable within the host tissue with either a

mechanism specific for fungal cells or a concentration so as not to
produce phytotoxic effects (4). In contrast to first generation
fungicides, they are generally effective against a narrower range of
pathogens, and systemics have a specific site or a limited number of
sites of action, making them more subject to loss of efficacy due to
fungal resistance.

Following determination of their modes of action, tremendous
contributions have been made through the use of these chemicals to
elucidate various biochemical pathways and cellular functions. We
will look at a number of examples of these second generation
systemics and briefly discuss their modes of action.

Cyclohexamide. Cycloheximide, an antifungal antibiotic produced by
Streptomyces griseus, inhibits protein synthesis not only in fungi,
but in eukaryotic organisms in general (5). Due to phytotoxicity
problems, cycloheximide can only be used successfully for disease
control when low concentrations are effective or on plants that are
relatively insensitive (6). Although cycloheximide has been reported
to inhibit a number of cellular processes including respiration and
nucleic acid synthesis, these are regarded as secondary effects that
result from the primary site of action, i.e., ribosomal protein
synthesizing systems (7). Resistance to cycloheximide has been
correlated with changes in ribosomal sensitivity or in cellular
permeability (1).

Benomyl. Benomyl is the best known of several benzimidazole
fungicides. This compound is effective against a large number of
fungi and plays an important role in disease management. Benomyl
interferes with tubulin polymerization in fungi by binding to the
beta tubulin subunit (8). As a result of the failure to form
microtubules, mitosis is blocked (9) as well as other
microtubular-dependent processes (10). It has been shown that a
single gene mutation alters the affinity of the tubulin binding site
for the inhibitor resulting in fungal resistance (8). The nature of
this highly specific mode of action and the control of toxicity
through a single gene has resulted, as one might expect, in the
development of problems in the field with fungal resistance.

Metalaxyl. Since its introduction the late 1970's for the control of
pathogens such as potato late blight, downy mildews and soil-borne
Pythium and Phytophthora spp., metalaxyl, an acylalanine fungicide,
and other related phenylamide fungicides have seen increasing usage.
Although the precise mechanism of inhibition by metalaxyl has not
been determined, RNA synthesis appears to be the most sensitive
pathway (11). Field resistance to metalaxyl has developed rapidly in
situations that involved exclusive and repeated use of the compound
(11). This has prompted the development of use strategies such as
that of combining chemicals which have different modes of action.

Carboxin. Carboxin is used primarily to control Basidiomycetes such
as rusts, bunts, and smuts. It is used primarily for seed treatment,
but also has soil and foliar applications. The mode of action of
this and related compounds is probably the best understood of any
fungitoxic mechanism. Carboxin blocks the transfer of electrons from

succinic dehydrogenase to coenzyme Q in the complex II region of the
mitochondrial electron transport chain (12, 13). Laboratory studies
have indicated that resistance is single-gene based and several loci
are involved (11). Based on the highly specific mode of action and
studies on laboratory developed resistant mutants, it is surprising
that in spite of quite a few years' use, resistance in the field has
not been a problem. One would speculate that natural mutants must be
less fit or do not have the pathogenic capabilities of the wild type.
A more likely explanation is related to the pattern of use. Because
most carboxin is used for seed treatment, the selection pressure is
rather low.

Polyoxins. The polyoxin antibiotics are effective against fungi that
contain chitin in their cell walls. The resistance to polyoxins in
some fungi of this type appears to be related to a permeability
factor rather than to a change in the site of action (14). Polyoxins
competitively inhibit chitin synthetase and thus prevent the
incorporation of uridinediphospho-N-acetylglucosamine into the chitin
polymer (14).

Edifenphos. Edifenphos, an organophosphorus compound, has a rather
limited use because of its highly selective action against the rice
blast fungus. Some ambiguity exists regarding the mode of action.
Although the compound has been shown to inhibit chitin synthetase
thus blocking cell wall synthesis, it also inhibits the synthesis of
phosphatidylcholine, a phospholipid and important membrane component
(11). The latter inhibition, which has been suggested as the primary
mechanism of action, results from blocking the activity of
phospholipid-N-methyltransferase, which is necessary in the
conversion of phosphatidylethanolamine to phosphatidylcholine (11).
The effects on chitin synthesis probably result from altered membrane
properties. Resistance against edifenphose in the field has been
slow to develop.

Polyenes. Although amphotericin B and nystatin, both polyene
antibiotics, are very active against a wide variety of fungi, they
are impractical as agricultural fungicides due to their rapid
photodecomposition (14). These compounds affect membrane
permeability through their interactions with sterols (15). These
interactions cause membrane changes that result in leakage of small
molecular materials. Resistance may generally be attributed to a
change in sterol quality or growth conditions.

Dicarboximides and Aromatic Hydrocarbons. The dicarboximide
fungicides, procymidone, vinclozolin and iprodione, and aromatic
hydrocarbons such as chloroneb, dichloran and quintozene, show a
great deal of similarity in their modes of toxicity (16). Although
the effects of these compounds on a number of biosynthetic processes
have been investigated, the exact mode of action has not been
resolved (4, 11). Interference with nuclear function, membrane
damage, and interference with cellular motile functions, such as
flagellar movement and cytoplasmic streaming, have been suggested as
possibilities. A better understanding of the mode of action may
explain why resistance has presented few problems in the practical
use of these fungicides.

Ergosterol Biosynthesis Inhibitors (EBI). Since the determination
that triarimol owed its antifungal activity to the inhibition of
ergosterol biosynthesis (17), a large number of structurally diverse
compounds that block ergosterol biosynthesis have become available
for plant disease control. This group consists of pyrimidines,
imidazoles, triazoles, and miscellaneous compounds in the pyridine,
morpholine and piperazine classes. These compounds control a wide
range of plant diseases including smuts, rusts, and powdery mildews
(4). Currently, this is the largest and most important group of
systemic compounds available for controlling fungal diseases in
plants and animals.

While these compounds are conventionally referred to as EBI
fungicides, they can block sterol biosynthesis in organisms that
synthesize sterols other than ergosterol (18). All of the EBI's
interfere with steps in the biosynthetic pathway following the
cyclization of squalene. A recently developed class of antifungal
agents for animal use the allylamines, developed by modification of
naftifine, block the pathway prior to cyclization by inhibiting
squalene epoxidase (19). Most of the EBI fungicides act primarily by
blocking the cytochrome P-450 dependent sterol C-14 demethylation
reaction (18). Although tridemorph produces effects in fungi similar
to those produced by compounds that inhibit sterol C-14
demethylation, it does not interfere with this reaction but inhibits
the isomerization of the sterol C-8(9) double bond to C-7(8) (20) or
the reduction of the sterol C-14(15) double bond (21).

It is interesting that many structurally diverse compounds all
inhibit sterol C-14 demethylation. However, with the exception of
triforine, these EBI's have in common a nitrogen-containing
heterocycle substituted with one large lipophilic group. It has been
proposed that a nitrogen atom of the heterocycle interacts with the
protohaem iron atom of the cytochrome P-450 enzyme(s) involved in
sterol C-14 demethylation and that the lipophilic substituent
increases binding affinity through an interaction with a nearby
region of the enzyme (22, 23). Evidently, there can be a wide
variation in the lipophilic substituent; however, variation in the
structure controls specificity as indicated by the fact that
compounds which effectively inhibit fungal sterol demethylase systems
may be relatively ineffective against the demethylase system from
mammalian cells (23).

The EBI's produce the following typical secondary effects (18):
- No inhibition of initial cell growth
- Morphological abnormalities
- No immediate effect on respiration
- No immediate effect on RNA, DNA or protein syntheses
- Sterol intermediates accumulate
- Free fatty acids accumulate

The morphological abnormalities resulting from treatment with
these compounds are quite striking (18). Sporidia of Ustilago become
multicelled and often branched. Germinating conidia of some fungi
show swelling, leakage, and membrane rupture. In addition, effects
on membranes apparently lead to abnormalities in synthesis of the
cell wall as well as that of other cellular components.

Major lipid fractions other than sterols are not initially
inhibited. However, a delayed accumulation of free fatty acids

characteristically results after sterol inhibition. These are
thought to result from the breakdown of membranes and triglycerides
together with a disproportionate synthesis and utilization of fatty
acids (18).

Growth retardation in host plants has been associated with the
use of a number of these EBI's. This effect results from the
inhibition of a reaction in the gibberellin biosynthetic pathway that
involves cytochrome P-450 enzymes (24).

Although the inhibition of ergosterol biosynthesis is regarded
as the primary mode of action, it is possible that the success of
these compounds as antifungal agents is related to other effects as
well. Not a great deal is known about fungal hormones. A recent
article concerning human pathogenic fungi indicated that fungal
steroids may be associated with the pathogenicity of these organisms
(25). Since the steroids are synthesized from sterol precursors, an
inhibition of steroid biosynthesis in addition to the effects on
precursor sterols may be responsible for the effectiveness of EBI's
in controlling plant pathogenic fungi.

Resistance has not thus far been a serious problem for the
EBI's. Although numerous resistant mutants have been developed in
laboratory studies, resistance problems in the field have not
occurred. However, recent investigations related to the use of a
triazole fungicide on barley indicate that resistant strains of
pathogenic fungi may be increasing (26). The fact that there have
not been major problems may be associated with the nature of the mode
of action and the fact that mutants appear to have reduced fitness
(11, 18). In other words, the genetic alterations conducive to
survival in the presence of these compounds may also reduce fitness.

Third Generation Fungicides

The third generation fungicides consist of chemicals designed to take
advantage of principles underlying host and/or pathogen
vulnerabilities. These compounds typically are highly specific,
often targeted for a specific disease on a selected host. They
characteristically show little or no toxicity to fungi in vitro.
They act as antipathogenic agents and thus affect the process of
pathogenesis. They may act on the host through the induction of
plant resistance mechanisms such as stimulation of lignification or
enhancement of phytoalexin production. (Please refer to the chapter
by Salt and Kuc in this volume for further discussion of this type of
compound.) They may act on the pathogen to accentuate elicitor
release or to prevent infection (host penetration), colonization
(inhibition of phytotoxin synthesis, extracellular enzyme production
and action, or phytoalexin degradation) or reproduction.

Melanin Biosynthesis Inhibitors. The melanin biosynthesis
inhibitors, which are used to control rice blast disease, are a good
example of compounds that prevent infection. Included in this group
are tricyclazole, pyroquilon, fthalide, and chlorbenthiazone.
Tricyclazole, which was the first of these compounds identified as
being effective through the inhibition of melanin synthesis (27),
produces no effects on fungal growth at concentrations of the
chemical that are effective for disease control. Treated fungi

appear brown rather than grayish-black. Tricyclazole blocks the
conversion of 1,3,8-trihydroxynaphthalene to vermelone and the
conversion of 1,3,6,8-tetrahydroxynaphthalene to scytalone in the
poklyketide pathway of melanin biosynthesis (28). The failure to
synthesize melanin was later tied to failure of the fungus to
penetrate the host epidermis (29). Apparently melanin or an oxidized
melanin precursor is involved in the fungal cell wall architecture in
such a way as to provide the rigidity necessary for host penetration
(4).

Compounds that affect Host Reactions. Examples of compounds which
enhance or induce host reactions to pathogens include
2,2-dichloro-3,3-dimethyl-cyclopropane carboxylic acid (DDCC),
probenazole, and fosetyl-Al (4). Although these chemicals do not
stop the fungus from penetrating the plant, they are quite effective
at preventing colonization through the enhancement of the host's
resistance mechanisms. Further studies are needed to elucidate how
these resistance mechanisms are triggered.

Resistance Mechanisms

With the use of second generation antifungal chemicals there has been
an increasing awareness of the need to understand not only the modes
of action but also the mechanisms of fungal resistance. It has long
been recognized that plants may be bred for two types of disease
resistance, race specific and general. Highly specific disease
resistance is very effective for a limited time. Usually the
selection pressure on the disease-producing organism is such that a
new strain, which circumvents the host's resistance, emerges. Plants
with moderate resistance generally retain this property for long
periods. A parallel exists with fungicides (30). Non-specific
types, such as the previously mentioned first generation fungicides,
have been effective for many years whereas the site specific second
and third generation compounds are more likely to encounter
resistance problems.
    Among the biochemical mechanisms of fungicide resistance are
reduced permeability, metabolism (increased detoxification or
decreased conversion to the toxic material), and reduced affinity of
the target site for the toxin.
    Efforts are now being made to avoid resistance problems through
strategies of use in conjunction with knowledge on modes of action
and mechanisms of resistance. Application strategies include timing
of use, reduction of use, combinations of protectants and eradicants
that have non-specific and specific modes of action, and combinations
of eradicants with different specific modes of action. Knowledge on
modes of action and resistance mechanisms offer some approaches that
can be developed to circumvent resistance problems. Many of these
ideas were discussed at a recent symposium addressing novel
approaches for research on agricultural chemicals (31).

Selective Action. Evidence suggests that attention should be given
to the development and use of chemicals with modes of action that
select for resistant strains which do not cause disease problems. In
the section on EBI fungicides, it was pointed out that resistant

mutants have been developed in the laboratory, but that there have
been no cases of disease control failure attributed to resistance in
the field. Perhaps this particular mechanism is such that mutations
to overcome it impart characteristics in a fungus that reduce
pathogenicity (18).

Circumvention of Site Modification. As indicated earlier, carboxin
inhibits succinic dehydrogenase activity. In studies using resistant
laboratory mutants and a large number of analogues of the parent
compound, data indicated that for any mutation in the loci affecting
carboxin sensitivity, a specific structural change in the parent
compound could alleviate or reverse the failure to control the fungus
(32). In screening studies, a number of analogues were identified
that showed in vitro binding affinity for succinic
dehydrogenase–mitochondrial complex II. Some analogues were
effective against wild type and some against resistant mutants. A
logical strategy would be to use mixtures of these chemicals or to
use them separately in an alternating schedule. Unfortunately, this
scheme would not work in this particular case because most of the
active analogues were not systemic. However, modification of the
chemical structure is one approach to control the development of
resistance.

Consideration of Activation and Stereoselectivity. Studies on
triadimefon, one of the EBI compounds, provide thoughts to consider
in developing new compounds as well as addressing resistance
problems. In evaluating the effectiveness of triadimefon applied to
plants for pest control, the following must be taken into
consideration (33):
  * Extent of conversion to triadimenol by host and pathogen
  * Inherent activities of the chemical present
    (stereoselectivity)
  * Time required for an effective dose to reach the specific
    intracellular site of action
  * Further metabolism of the chemical by the pathogen or host
    plant to less toxic compounds
While all of these factors must be taken into account, the latter two
are dependent on the first two, and we will focus on the first two as
points of consideration for resistance as well as strategies for the
future effective use of this compound.
     Fungi sensitive to triadimefon convert the parent compound
through biochemical reduction to triadimenol at a high rate, whereas
resistant species demonstrate little or no conversion (34). This
investigation also demonstrated that plants enhance fungitoxicity by
this conversion. Thus an activation process in both pathogen and
host could possibly be altered to potentiate effectiveness. To add
to this somewhat complex situation, two diastereoisomers of
triadimenol are formed in the reduction process, and one is more
active than the other (35). A high rate of conversion to triadimenol
in addition to a high ratio of the more active diastereomer to the
less active can be assumed to be present in triadimefon–sensitive
fungi whereas the opposite situation may be found in relatively
insensitive fungi. The conversion process in plants affords the
opportunity to control insensitive fungi which lack the conversion
mechanism.

A number of factors must be considered in the area of stereoselectivity (31). Distinct differences are noted between stereoisomers in absorption/ uptake, activity at the target or receptor site, pathways and rate of bioconversion, and pharmacokinetics. All of these factors are applicable in the case of triadimefon and are examples of processes that may be manipulated to enhance disease control effectiveness.

Inhibition of a Resistance Mechanism. Another interesting case serves as an example of the importance of understanding resistance mechanisms. In this situation resistance is not related to the mode of action. Observations indicate that in a wild type strain of Aspergillus nidulans, uptake of the EBI fungicide, fenarimol, is a rapid, passive influx which induces an energy dependent efflux resulting in an energy dependent state of equilibrium (36). Toxic action occurs prior to sufficient efflux to reduce levels below toxic concentrations. This same study found that the efflux system in resistant strains is constitutive resulting in a lower initial uptake of the fungicide. Other investigations show that a similar mechanism is involved in A. nidulans strains resistant to imazalil, another EBI fungicide (37).

Efflux is blocked by compounds such as respiratory inhibitors that decrease levels of ATP. Apparently the influx-efflux mechanism is dependent on intracellular ATP, which is utilized by the plasma membrane ATPase and thus drives the efflux process (36). The synergistic effects of respiratory inhibitors could serve as the basis for using a combination of inhibitors. This also brings up the possibility that through the use of respiratory inhibitors, fungi that are normally insensitive or only slightly sensitive to fenarimol or imazalil due to efflux systems could be controlled.

Conclusions

New approaches for the chemical control of plant pathogens emphasize the need to understand the physiological, biochemical, and molecular levels of the modes of action and the mechanisms of pathogen resistance. These approaches involve the use of chemicals in a way that will reduce or prevent the development of resistant pathogens. They also take advantage of compounds that serve as antipathogenic agents to prevent disease development, an approach which requires an understanding of host-parasite physiology and pathogenic processes.

The antipathogenic agents are highly specific and usually targeted for a designated pathogen on a given plant. In looking to the future, one must be practical and take into account the fact that such chemicals are likely to be commercially available only for major diseases on major crops. The current costs of development, registration, and marketing put restrictions on widespread use of these third generation fungicides. Thus, one must take advantage of first and second generation type compounds. The information we have presented on these materials indicates that a reasonable level of knowledge exists on how they work. We have seen examples of the broad range of cellular processes that second generation fungicides affect. Additional data will undoubtedly enhance this information base. New approaches will take advantage of existing knowledge and

emphasize the importance of additional basic information on the
mechanisms of action and resistance as well as host-pathogen
relationships.

Literature Cited

1. Sisler, H. D.; Ragsdale, N.N. in "Handbook of Pest Management in
   Agriculture"; Pimental, D., Ed.; CRC Press: Boca Raton,
   Florida, 1981; pp. 17-47.
2. Siegel, M. R. in "Pesticide Selectivity"; Street, J. C., Ed.;
   Marcel Dekker: New York, 1975; pp. 21-46.
3. Edgington, L. V.; Martin, R. A.; Bruin, G. C.; Parsons, I.M.
   Plant Dis. 1980, 64, 19-23.
4. Sisler, H. D.; Ragsdale, N. N. in "Agricultural Chemicals of the
   Future"; Hilton, J. L., Ed.; Rowman and Allanheld Publishers:
   Totown, New Jersey, in press.
5. Sisler, H. D. Annu. Rev. Phytopathology. 1969, 7, 311-30.
6. Dekker, J. in "Fungicides", Vol. 2; Torgeson, D. C., Ed.;
   Academic Press: New York, 1969; pp. 580-635.
7. Siegel, M. R. in "Antifungal Compounds", Vol. 2; Siegel, M. R.;
   Sisler, H. D. , Eds.; Marcel Dekker: New York, 1977; pp.
   399-438.
8. Davidse, L. C. in "Fungicide Resistance in Crop Protection";
   Dekker, J.; Georgopoulos, S. G., Eds.; Pudoc: Wageningen, The
   Netherlands, 1982; pp. 60-70.
9. Hammerschlag, R. S.; Sisler, H. D. Pestic. Biochem. Physiol.
   1973, 3, 42-54.
10. Howard, R. J.; Aist, J. R. Protoplasma 1977, 92, 195-210.
11. Davidse, L. C.; deWaard, M. A. Adv. Plant Pathology. 1984, 2,
    191-257.
12. Mathre, D. E. Pestic. Biochem. Physiol. 1971, 1, 216-24.
13. White, G. A. Biochem. Biophys. Res. Commun. 1971, 44, 1210-12.
14. Misato, T.; Kakiki, K. in "Antifungal Compounds", Vol. 2;
    Siegel, M. R.; Sisler, H. D., Eds.; Marcel Dekker: New York,
    1977; pp. 277-300.
15. Ragsdale, N. N. in "Antifungal Compounds, Vol. 2; Siegel, M. R.;
    Sisler, H. D., Eds.; Marcel Dekker; New York, 1977; pp. 333-63.
16. Kaars Sypesteijn, A. in "Fungicide Resistance in Crop
    Protection"; Dekker, J.; Georgopoulos, S. G., Eds.; Pudoc:
    Wageningen, The Netherlands, 1982; pp. 32-45.
17. Ragsdale, N. N.; Sisler, H. D. Biochem. Biophys. Res. Commun.
    1972, 46, 2048-53.
18. Sisler, H. D.; Ragsdale, N. N. in "Mode of Action of Antifungal
    Agents"; Trinci, A. P.J.; Ryley, J. F., Eds.; Cambridge
    University Press: London, 1984; pp. 257-82.
19. Petranyi, G.; Ryder, N. S.; Stütz, A. Science 1984, 224,
    1239-41.
20. Kato, T.; Shoami, M.; Kawase, Y. J. Pestic. Sci. 1980, 5,
    69-79.
21. Kerkenaar, A.; Uchiyama, M.; Versluis, G. G. Pestic. Biochem.
    Physiol. 1981, 16, 97-104.
22. Henry, M. J.; Sisler, H. D. Pestic. Biochem. Physiol. 1984, 22,
    262-75.
23. Gadher, P.; Mercer, E. I.; Baldwin, B. C.; Wiggins, T. E.
    Pestic. Biochem. Physiol. 1983, 19, 1-10.

24. Coolbaugh, R. C.; Hirano, S. S.; West, C. A. Plant Physiol. 1978, 62, 571-6.
25. Kolata, G. Science 1984, 225. 913-4.
26. Wolfe, M. S. personal communication.
27. Chrysayi Tokousbalides, M.; Sisler, H. D. Pestic. Biochem. Physiol. 1978, 8, 26-32.
28. Chrysayi Tokousbalides, M.; Sisler, H. D. Pestic. Biochem. Physiol. 1979, 11, 64-73.
29. Woloshuk, C. P.; Sisler, H. D. J. Pestic. Sci. 1982, 7, 161-6.
30. Georgopoulos, S. G. in "Antifungal Compounds", Vol. 2; Siegel, M. R.; Sisler, H. D., Eds.; Marcel Dekker: New York, 1977; pp. 439-84.
31. Fuchs, A.; Davidse, L. C.; deWaard, M. A.; deWit, P. J. G. M. Pestic. Sci. 1983, 14, 272-93.
32. White, G. A.; Thorn, G. D. Pestic. Biochem. Physiol. 1980, 14, 26-40.
33. Deas, A. H.B.; Clifford, D. R. Pestic. Biochem. Physiol. 1982, 17, 120-33.
34. Gasztonyi, M.; Josepovits, G. Pestic. Sci. 1979, 10, 57-65.
35. Buchenauer, H. in "Abstracts of Papers IX International Congress of Plant Protection", 1979, No. 939.
36. DeWaard, M. A.; van Nistelrooy, J. G. M. Pestic. Biochem. Physiol. 1980, 13, 255-66
37. Siegel, M. R.; Solel, J. Pestic. Biochem. Physiol. 1981, 15, 222-33.

RECEIVED January 15, 1985

Elicitation of Disease Resistance in Plants by the Expression of Latent Genetic Information

STEVEN D. SALT and JOSEPH KUĆ

Department of Plant Pathology, University of Kentucky, Lexington, KY 40546–0091

Cucumber, melon, tobacco, and bean varieties, supposedly lacking significant resistance to particular diseases, can be made highly resistant systemically by limited infection prior to severe exposure to pathogens. Injury and non-specific stress do not elicit persistent systemic resistance. These findings suggest that plants generally possess genetic information for disease resistance mechanisms and that resistance is generally determined by the speed and magnitude of response to pathogens. We shall present aspects of our investigations into immunization of tobacco, cucurbits, and beans (Phaseolus vulgaris L.) and briefly review other workers' investigations into biotically- and chemically-induced disease resistance of plants.

Enhanced resistance to disease in plants after an initial infection has fascinated observers for over 100 years. A review of the subject by Chester in 1933 contains 201 references (1). "Immunization", "acquired systemic resistance", or "induced resistance" of plants have been reviewed in recent years (2-11). We shall not exhaustively review the literature, but shall focus on general principles and phenomena of particular relevance to the use of "plant immunization" for the practical control of disease. This paper will stress examples from our own research program, but will also include literature citations to provide the reader with an appreciation of important research contributions of others previously and presently active in the field. Most examples presented will deal with fungal, bacterial, or viral diseases of crop plants, but similar principles may apply to infestations by nematodes and, possibly, insects.

Protection of Plants Against Pests

Immunization of plants via priming for expression of latent genetic information encoding disease resistance mechanisms may be introduced by contrasting it with other strategies utilized for protecting plants from diseases. The strategies considered can be categorized as: ecological, pesticides, antitoxins, alterations of plant physiology, nonspecific phytoalexin induction, and plant sensitization (immunization).

Ecological. This historically venerable strategy is characterized by tactics designed to minimize exposure of plants to pathogens and make environmental conditions unfavorable for disease development. Measures include crop rotation, sanitation, and quarantine. Such procedures are valuable and worthy of further development, but they are generally of limited effectiveness in providing reliably high yield and quality of crops in areas with severe or persistent disease problems.

Pesticides. The application of exogenous chemicals toxic to plant pathogens and pests has been a major strategy for plant protection in technologically developed areas for several decades. This strategy is well suited to mechanized, high technology agriculture, and has very successfully enhanced crop production. Numerous and serious long-term problems have, however, become evident. Heavy reliance on, and confidence in, pesticides has caused other strategies to become relatively neglected. The expense of pesticides and necessary ancillary equipment, as well as technological expertise required for their effective and safe use, are burdensome and often beyond the reach of farmers in developing nations. Heavy and improper use of pesticides has led to the appearance of resistant pests and consequent loss of pesticide efficacy. Most seriously, the danger of injury to non-target organisms, especially toxicity, teratogenicity, and carcinogenicity to humans, has led to cumbersome and expensive restrictions on the development, sale, and use of pesticides.

Antitoxins. This is a strategy of disarmament in which applied chemicals are not directly toxic to pathogens, but interfere with their mechanisms for pathogenesis, e.g., penetration, maceration of tissues, or induction of wilting or abnormal growth. Few such compounds have been developed for commercial use. Tricyclazoles appear to function by inhibiting the melanization of fungal appressoria and thus reduce penetration into host plants (12-14). Other reported examples include inactivation of picularin, toxin of the rice blast fungus Piricularia oryzae, by ferulic and chlorogenic acids (15), and inhibition of Fusarium wilt symptoms in tomato by catechol, which at effective concentrations appears non-toxic to the fungus and neither prevents nor reduces infection (16). The biochemistry of pathogenesis in plant disease, however, is in an early stage of investigation and the rational design of toxin inhibitors or other antipathogenetic compounds may be even more difficult than that of pesticides. Random screening of compounds for antitoxin activity has inherent difficulties for bioassay, and the application of antitoxins suffers from the difficulties encountered in the environmental release of chemicals as cited for pesticides.

Alterations of Physiology. Susceptible plant species or cultivars may be rendered resistant to disease by so altering their physiology, e.g., hormonal balances, ion fluxes, constitutive secondary metabolism, or respiratory rates, as to render them unfit or hostile environments for pathogen development, i.e., to make them non-hosts. This could be accomplished, for example, by the use of growth regulators or breeding.

This approach may not prove generally practicable, however, because crop plants, over the ages, have been selected precisely for those physiological characteristics which make them more agronomically desirable than frequently hardier wild relatives. Drastic changes in constitutive physiology may reduce agronomic fitness even if disease resistance is enhanced. Breeders are often faced with the problem of reincorporating yield and quality factors back into disease-and insect-resistant plants.

Non-specific or Constitutive Phytoalexin Induction. Many plants are reported to produce a variety of low molecular weight, antimicrobial compounds (phytoalexins) in response to infection. These compounds rapidly accumulate to high concentrations in plant tissues immediately adjacent to sites of infection in resistant cultivars, and they function to restrict development of fungal and bacterial pathogens of plants (17,18). Susceptible cultivars may ultimately accumulate as much or more total phytoalexins during disease development, but accumulation is slower and diffuse, and localized concentrations in advance of the pathogen appear to be insufficient to inhibit pathogen development. Many abiotic agents, including heavy metal salts (19), heat (20), chloroform vapors (21), ultraviolet irradiation (22), fungicides such as maneb and benomyl (23), and even innocuous substances such as sucrose (24) elicit phytoalexin accumulation in some plants. Proposals have been made that general non-specific induction of phytoalexin accumulation in plants may be an effective and "natural" method of plant protection from pests. Unfortunately, phytoalexins, and the effective concentrations of phytoalexin elicitors, are often toxicants which injure or kill plant cells (25). In disease resistance involving phytoalexin accumulation, the plant sacrifices a few of its own cells immediately adjacent to invading pathogens in order to save the entire plant. General or constitutive induction of phytoalexin accumulation may protect plants against infectious disease, but it would likely be disastrously counterproductive to overall plant health because of autotoxicity and the need for continuous diversion of metabolic energy and primary metabolites into secondary metabolism. In addition, phytoalexins require higher concentrations in plant tissues than commercial synthetic pesticides to achieve comparable effectiveness. They also are not translocated within plants, and are rapidly degraded by plants and many pathogens. To maintain effectively high systemic levels of phytoalexins, therefore, would likely require repeated applications of phytotoxic elicitor substances or the breeding of plants with constitutively high levels of phytoalexins. Furthermore, many phytoalexins may be toxic to mammals, and plant tissue containing high total concentrations may be unpalatable or hazardous for consumption (25).

Sensitization of Plants. This strategy for plant protection is conceptually, if not mechanistically, analogous to immunization via vaccination of mammals. An initial stimulus elicits a long-term subtle alteration in a plant such that subsequent exposure to a pathogen or pest results in a vigorous and rapid activation of the plant's endogenous genetically encoded defense mechanisms. However, aside from a relatively brief initial sensitization period, disease resistance mechanisms remain latent until after infection and are generally localized at the sites of infection. There is no gross derangement of plant constitutive metabolism. There is little consumption of energy or metabolites and the expenditure occurs when and where needed. Autotoxicity from non-specific or constitutive defense reactions in the absence of infection is avoided. A "sensitized" plant may be said to possess to some degree a state of "immunity" or "acquired/induced resistance".

The remainder of this paper shall present general principles, examples of applications, and an evaluation of capabilities and limitations of sensitization of plants as a strategy for defense against infectious diseases and damage from other pests.

Statement of Thesis

Virtually all plants, even those considered "susceptible" to particular diseases, possess genetic information for the biochemical pathways responsible for effective disease resistance mechanisms. Susceptibility is due to a failure or delay in recognizing the pathogen, suppression of activation of host defenses by the pathogen, inactivity of gene products, or modification of components of the resistance response. The data presented in this paper support the thesis that, given an appropriate inducing stimulus, susceptible plants can be sensitized to activate defense mechanisms against pathogens so as to become resistant to disease.

However, though enhancement of expression of genetically encoded plant disease resistance mechanisms may be a generally effective means of immunizing plants, we do not claim universal efficacy. In some environments, gene products may be ineffectual in restricting or conditioning the restriction of pathogen development, e.g., activity of gene products for resistance may have specific temperature or light requirements. In plants severely weakened by injury or disease, facultative parasites or normally incompatible pathogens may be able to parasitize normally nonhost or resistant plants due to the plant's inability to develop an effective metabolic response. Some pathogens may activate host defense mechanisms yet develop in host tissues and cause disease by inactivating components of the resistance mechanisms, e.g., metabolism of kievetone, a phytoalexin of beans, by Fusarium (26,27) and the detoxification of pisatin, a phytoalexin of pea, by Nectria hematococca (28).

Methods of Sensitization

Genetic. The differences between many susceptible and resistant cultivars are the abilities of the latter to appropriately respond quantitatively in time and in space to infection, rather than qualitative differences in biochemical pathways. This type of genetic resistance may be considered a form of an endogenous constitutively sensitized state. Genetic sensitization of a cultivar requires both a source of appropriate germplasm and a method of effective transfer and incorporation of the genetic material into the recipient cultivar. Both requirements cannot always be met. Furthermore, undesirable traits often accompany desired resistance traits. This very important area of plant sensitization, broadly construed, lies beyond the scope of this paper.

Physical. Anecdotal accounts abound of enhanced plant resistance to disease achieved by transient exposure to a wide variety of physical stimuli, e.g., heat, light, microwaves, other electromagnetic radiation, electric current, sound waves, and vibration. In our own laboratory, we have made cucumbers resistant to anthracnose by vibration (Stromberg and Kuć, unpublished). However, these phenomena are poorly understood and may include enhanced resistance resulting from non-specific altered (stress) physiology, nonspecific phytoalexin elicitation, modification of the action of gene products, or sensitization. This interesting but little explored area will not be further discussed in this paper.

Biotic. Enhanced resistance to disease which results from prior exposure to avirulent pathogens or nonpathogenic organisms or to virulent pathogens under conditions unfavorable for disease development, is the major focus of this paper and will be discussed at length.

Chemical. There are a number of reports of exogenous chemicals, both synthetic and natural, which enhance endogenous plant defense mechanisms. We shall review some of these reports at a later point in this paper. In some cases, neither the compounds nor their metabolic products are directly toxic to the pathogens under study. In other cases, enhanced plant defense responses appear coexistent with direct pesticidal activity of the compounds or their metabolites. Few of these cases have been thoroughly examined at the biochemical level. However, the possibility that sensitization of plants towards pathogens may be achieved by exogenous chemicals has great potential for practical applications. Of course, all difficulties inherent in releasing alien chemicals into the environment may be encountered in this approach.

   Regardless of the nature of the original sensitizing stimulus, however, induced resistance is mediated and expressed through the endogenous genetically encoded biochemical mechanisms of the plant.

General Occurrence of Resistance Induced by Biotic Agents

Numerous reports of "acquired resistance" or "induced resistance" or "immunization" of plants, especially to diseases caused by viruses, date back as far as a century (1). Only a small sampling of examples is presented in Table I to illustrate the variety of host-disease combinations for which this phenomenon has been reported. Many examples have not been independently verified or extensively investigated. Direct antibiosis or changes in host physiology or pathogen virulence may be confused with, or obscure, sensitization towards subsequent reinfection by the same pathogen. However, enough cases have been closely examined to suggest the generality of biotically induced resistance in plants (2-11). Similar phenomena have also been reported for nematodes (29) and for insects (30).

Non-specificity of Protection

One of the most fascinating and yet bewildering aspects of resistance induced in plants by biotic agents is the frequently observed ability of infection by one organism to induce resistance against disease caused by other organisms distantly related both to the inducing organism and to one another.

Limited foliar infection of cucumber, watermelon, and muskmelon by Colletotrichum lagenarium, a pathogen of leaves and fruits, or by the non-pathogen tobacco necrosis virus (TNV), systemically protected against disease caused by C. lagenarium, Cladosporium cucumerinum, Mycosphaerella melonis, Fusarium oxysporum f. sp. cucumerinum, Pseudoperonospora cubensis, Pseudomonas lachrymans, Erwinia tracheiphila, TNV (local necrosis), or Phytophthora infestans (9,49-55). Thus protection was achieved against necrotrophic and biotrophic pathogens; fungi, viruses, and bacteria; foliar, stem, fruit, and root pathogens; and against various disease types - systemic and local; wilts, rots, blights, leaf spots, mildews, scabs, and local necrosis. The only disease against which we were unable to induce resistance with TNV or C. lagenarium was powdery mildew caused by Sphaerotheca fulginea. However, Bashan and Cohen (56) have more recently reported induction of resistance against powdery mildew of cucumbers by TNV.

Green beans (Phaseolus vulgaris L.) can be systemically protected against cultivar-pathogenic races of the anthracnose fungus, Colletotrichum lindemuthianum, by prior inoculation either with cultivar-incompatible races of the same fungus (57-59) or by cultivar-pathogenic races if the infection was attenuated in situ by heat treatment prior to appearance of symptoms (60,61). Bean cultivars susceptible to all known races of C. lindemuthianum and also those possessing "genetic" resistance to some fungal races were rendered resistant to all races of the pathogen by prior infection with the pathogen of cucurbits, Colletotrichum lagenarium (57,62).

Tobacco was reciprocally protected against TMV, TNV, or Thielaviopsis basicola (34); against Phytophthora parasitica var. nicotianae with TNV (63); and against TMV and Erysiphe cichoracearum (powdery mildew) with Peronospora tabacina (64-66).

Table I. Some Reports of Biotically Induced Resistance In Plants

Plant	Disease	Pathogen	Reference(s)
Tobacco	blue mold	Peronospora tabacina	5,31,32
	tobacco mosaic	virus	1,3
	tobacco ringspot	virus	1,3
	black shank	Phytophthora parasitica	33
	black root rot	Thielaviopsis basicola	34
Cabbage, carnation, tomato, watermelon, flax, silk tree, cotton	wilt	Fusarium spp.	35
Sugar cane	mosaic	virus	1
	corn streak	virus	1
Coffee tree	rust	Hemileia vastatrix	36,37
Euphorbia cyparissias	rust	Uromyces pisi	38
Daisies	gall	Agrobacterium tumefasciens	39
Cedars, apples	rust	Gymnosporangium macropus	40
Carrots	root rot	Botrytis cinerea	41
Cucurbits (cucumbers, melons)	anthracnose	Colletotrichum lagenarium	42,43,54
	mosaic	virus	1
	angular leaf spot	Pseudomonas lachrymans	10
Peaches, plums	canker	Cytospora cincta	44
Green beans	anthracnose	Colletotrichum lindemuthianum	45
Apples, pears	fire blight	Erwinia amylovora	46-48

However, inoculation with E. cichoracearum, TMV, cucumber mosaic
virus, Alternaria solani, Helminthosporium turcicum,or
Pseudoperonospora cubensis failed to protect tobacco against
Peronospora tabacina or E. cichoracearum (65). Thus, while biotic
agents often protect against closely or distantly related
organisms, the effective combinations of plant, biotic inducer, and
challenge pathogen exhibit some specificity. It is puzzling that
reciprocal combinations may be ineffective, and that some pathogens
seem ineffective in eliciting resistance against themselves or
closely related organisms, but highly effective against distantly
related pathogens!
      Some more unusual effective inducer/pathogen combinations
include the use of the nematodes Pratylenchus penetrans or P.
brachyurrus against the black shank fungus P. parasitica in tobacco
(33,67,68), and use of TMV against the aphid Myzus persicae as well
as against TMV, P. parasitica var. nicotianae, Pseudomonas tabaci
(wildfire bacterium), and Peronospora tabacina (69).

Dynamics of Induced Resistance

Initial reaction. In all known cases of effective biotic
sensitization of plants reported to date, a critical factor appears
to be the necrosis of host cells in the zone of initial infection.
However, while non-necrotic infections are ineffective inducers,
necrosis per se is not effective in inducing resistance. Injury by
abiotic agents such as heat, chemicals, dry ice, or various
extracts from plants and microbes does not protect cucumbers
against C. lagenarium (8-10). Infection of tobacco by a wide
variety of Peronosporales fungi other than P. tabacina frequently
causes severe necrosis, but does not induce systemic resistance
against blue mold (Tüzün and Kuć, unpublished).
      The effectiveness of immunization is directly related to the
extent of infection by, or inoculum concentration of, the inducer
organism up to a point of maximal response or saturation
(49,53,70). However, one lesion caused by C. lagenarium or eight
lesions caused by TNV on one inducer leaf can significantly induce
systemic disease resistance in cucumbers.

Spatial and Temporal relationships. A common characteristic of
biotic sensitization of plants is a latent or lag period between
initiation of the inducer infection and manifestation of disease
resistance. In the cucumber/TNV or C. lagenarium system, enhanced
resistance against pathogens is first manifest about 48-72 hr after
initial infection and maximal resistance is achieved by 120-144 hr
(49,53,70). In tobacco, resistance against P. tabacina achieved by
limited stem infection by the same organism is not evident before 9
days and increases until about 21 days after induction (5). An
induction period of 3-6 days is required for expression of
resistance to TMV or TNV induced in tobacco by Thielaviopsis
basicola (34). Removal of the inducer leaf 72-96 hr after
initiation of the inducing infection in cucumbers did not result in
the loss of systemic resistance in the rest of the plant (53).
Likewise, leaves distal from the inducer leaf retained resistance
after detachment from the plant once immunization was established.

Durability of biotically induced resistance varies with the plant/inducer/pathogen combination. In cucumbers, resistance induced by C. lagenarium or TNV gradually declines unless a "booster" inoculation is given 3-6 weeks after induction (70). With a booster inoculation, immunization of cucumbers appears sufficiently durable to protect the plant through flowering and fruiting. However, cucumbers cannot be immunized after initiation of flowering (71). This suggests a hormonal effect upon immunization. In tobacco, immunization against P. tabacina by virulent P. tabacina is hazardous before the plants are at least 20 cm tall due to frequent systemic spread of the fungus in young plants, but a single inoculation of plants over 20 cm in height is effective throughout the plant's life (5,72). Protection of tobacco against Phytophthora parasitica var. nicotianae by TMV seems to require repeated administration of the virus (69).

While biotic sensitization of plants against pathogens may act systemically, not all organs or tissues are necessarily protected equally. Recent work in our laboratory has shown that the extent of resistance against anthracnose or angular leaf spot of leaves in different positions on immunized cucumber plants varies in a complex manner not necessarily directly proportional to proximity of the inducer lesions (73). Removal of epidermal layers from immunized cucumber leaves reduces resistance to anthracnose (74).

Mechanisms of Resistance. The broad spectrum of effectiveness of induced systemic resistance against bacteria, viruses, fungi, and nematodes, makes it seem unlikely that only a single resistance mechanism is activated.

Phytoalexins. Many, but not all, plant families are reported to utilize phytoalexin accumulation as a means of defense against pathogens. A major phytoalexin in green beans, Phaseolus vulgaris, is the isoflavonoid phaseollin (18). Reactions of "genetically resistant" bean varieties against incompatible races of C. lindemuthianum or of immunized "susceptible" bean plants against virulent races are marked by rapid accumulation of high levels of phaseollin in tissues adjacent to the invading fungus, resulting in containment of the fungal hyphae (62,75;Figure 1). On the other hand, while ultimate phytoalexin accumulation on a total plant tissue basis can be greater in nonimmunized infected susceptible plants, the onset of accumulation is delayed (Figure 1); local concentrations are insufficient to contain the fungus, and the plant is successfully colonized by the pathogen. Significantly, as shown in Figure 1, phytoalexins do not accumulate in the immunized "susceptible" bean plant distant from the site of induction prior to exposure to the pathogen. That is, sensitized resistance is not due to high constitutive levels of phytoalexins. Induced resistance through prior biotic infection, therefore, appears to be at least a two-step phenomenon: sensitization followed by a rapid expression of resistance mechanisms after subsequent infection.

Inhibition of penetration and spread within tissues. In cucurbits, classical phytoalexins have not been identified and disease resistance appears due to other mechanisms. Immunization

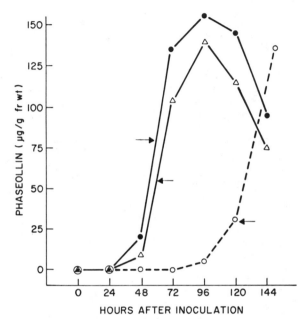

Figure 1.   Time course of phytoalexin (phaseollin)
accumulation in Phaseolus vulgaris upon infection by the
pathogen Colletotrichum lindemuthianum (—△—) "Genetically"
resistant variety; (—●—) "Susceptible" variety immunized
by prior limited infection by an incompatible race of C.
lindemuthianum; (---O---) "Susceptible" variety, nonimmunized.   Arrows indicate onset of necrotization.
Reproduced with permission from Ref. 133.   Copyright 1984,
Ciba Foundation.

of cucumbers by C. lagenarium is reported to inhibit multiplication of the challenge bacterium, Pseudomonas lachrymans (49), though Doss and Hevesi (76) report that immunization of cucumber to P. lachrymans suppresses symptom development, but not bacterial multiplication.

Germination of conidia of C. lagenarium and Cladosporium cucumerinum on the surface of immunized cucumber leaves is not inhibited. Penetration of C. cucumerinum into immunized cucumber leaves is not reduced; however, penetration from C. lagenarium appressoria is reduced by 80-90%. The spread within cucumber leaf mesophyll tissue of immunized plants by either C. cucumerinum or C. lagenarium infiltrated into leaves (thus bypassing surface defensive components) appears restricted (77-79). While the biochemical mechanisms are yet obscure, resistance of tobacco to P. tabacina induced by the same organism appears due to restriction of fungal development within leaves rather than to inhibition of sporangial germination or penetration into the host plant (Tüzün and Kuć, unpublished).

Lignification. Histological and chemical studies reveal rapid lignification around zones of fungal penetration in both immunized cucumber plants and those "genetically" resistant to C. cucumerinum (77-79). Lignification - the generation of free radicals from phenylpropanoid precursors via action of peroxidases and other enzymes, and the nonspecific polymerization of these radical monomers upon matrices such as plant and fungal cell walls - is believed to be a major plant defense mechanism against fungal and bacterial pathogens (80-83).

Lignification in immunized cucumbers is rapid and intense in cells adjacent to penetration of C. lagenarium or C. cucumerinum (Figure 2). Lignification in infected, nonimmunized, susceptible cucumbers also occurs, but is slower and more diffuse. Enhanced lignification is not observed in cucumbers, "genetically" susceptible or resistant, immunized or nonimmunized, in the absence of infection. Thus, constitutive lignification is a defense mechanism in neither "genetically" nor "biotically induced" resistant cucumbers. These results resemble the bean/phytoalexin data cited earlier and further support the concept of "plant immunization" as a sensitization rather than a nonspecific activation of endogenous biochemical defense mechanisms.

Immunization of cucumbers by C. lagenarium, C. cucumerinum, P. lachrymans or TNV generates a systemic increase in peroxidase activities (77,79,84). Like lignification and phytoalexin induction, peroxidase activities also rise more quickly in response to infection in leaves of immunized plants, even though total activity eventually may be highest in infected susceptible leaves (77). Several other stimuli can induce local (mechanical and chemical injury) or systemic (senescence, ethylene) peroxidase increases that are not accompanied by increased disease resistance. Thus, enhanced peroxidase activity per se may not be a defense mechanism, but may be a necessary adjunct with appropriate chemical substrates for processes important in disease resistance, e.g., lignification, suberization, and melanization.

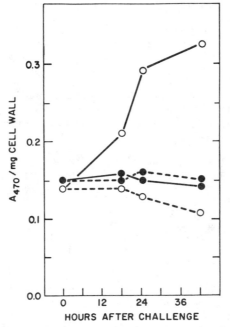

Figure 2. Time course of lignin accumulation in cucumber (Cucumis sativus) leaves in presence or absence of challenge infection by the pathogen Cladosporium cucumerinum. (—O—) immunized by prior limited infection by C. lagenarium, challenged with same pathogen; (--•O---) immunized, not challenged; (—●—) not immunized, challenged; (--●--) not immunized, not challenged. Lignin determined with Gibb's reagent after alkaline hydrolysis (78). Reproduced with permission from Ref. 133. Copyright 1984, Ciba Foundation.

Other mechanisms. Many other biochemical mechanisms are known or postulated to function in disease resistance in plants, e.g., silicization of cell walls, "b" or PR ("pathogenesis-related") proteins, AVF (anti-viral factors) and "phytointerferon", or superoxide anions. These mechanisms have not been investigated as extensively as the above described mechanisms, especially in cases of "induced resistance". Space does not permit their consideration in this article.

Systemic immunization signal. The systemic nature of sensitization against pathogens which is engendered by localized infections makes it evident that plants likely produce or liberate a signal which is transmitted throughout plant tissues.

Girdling of the petiole of the inducer leaf in cucumber prevents immunization of other leaves, and girdling of petioles of leaves other than the inducer leaf at the time of immunization or shortly thereafter also prevents immunization of the girdled leaves (71). However, girdling or even excision of the inducer tissue 80-120 hr after immunization of cucumbers does not prevent systemic immunization. Grafting of susceptible, non-immunized cucumber, muskmelon, or watermelon scions onto immunized cucumber rootstocks results in immunization of the scion tissue (85). Qualitatively similar results for both girdling and grafting are found with P. tabacina-immunized tobacco (Tüzün and Kuć, unpublished). These results strongly suggest that a chemical signal exists. This signal may be transported in the phloem, from the region of the sensitizing infection, throughout the plant.

Presence of active necrosis of plant tissues produced by the inducer organism appears necessary for signal production. This was supported by experiments in which nonimmunized scions were grafted onto immunized cucumber rootstocks which either contained an intact inducer leaf or those which had the inducer leaf excised. Only rootstocks with intact inducer leaves effectively conferred resistance upon susceptible scions (Dean and Kuć, unpublished). Excision of a cucumber inducer leaf infected with C. lagenarium or P. lachrymans before full immunity is attained results in only partial immunization of the remainder of the plant, the extent of immunization being directly proportional to the length of time the inducer leaf is left intact after infection. However, excision of the inducer leaf does not reduce the degree of immunization already attained in the remainder of the plant by the time of excision (Dean and Kuć, unpublished). At least in the cucumber/C. lagenarium system, very little signal substance appears to be present in transit at any given time. The signal also appears to act very rapidly in inducing immunization. Challenging partially protected leaves with virulent pathogens at varying times after excision of inducer tissue reveals no change in level of immunization with time. Such change would be anticipated were a significant amount of signal in transit from the inducer leaf at the time of excision, or were a significant period of time required for signal already arrived in the tissue to elicit a response (Dean and Kuć, unpublished).

Attempts to isolate and characterize the putative signal substances from cucumbers and tobacco have met with mixed results.

However, some success has been achieved with the P. lachrymans/cucumber system (86).

Practical Applications of Biotically Induced Resistance

Reports of field or commercial applications to date of biotically induced disease resistance in plants are few. Attenuated strains of tomato mosaic virus are used to protect field and greenhouse (hothouse) tomatoes from virulent strains of the same pathogen in the U.S.S.R., Japan, China, and the Netherlands (8,87; I. Yamamoto, personal communication at Snowbird Conference 1984). Experimental field studies on protection of cucumbers, watermelons, and muskmelons against anthracnose by prior restricted infection with the same organism or P. lachrymans showed highly effective protection (88). Restricted infection of burley and cigar tobacco plants with P. tabacina in fields in Kentucky and Puerto Rico has proven effective in protecting these plants against blue mold (72).

Chemically Induced Resistance

Natural Products and Biochemicals. In addition to active microbial infections, various preparations of dead microbial cells, subcellular fractions and components, microbial culture filtrates, and diffusates from infected plant tissue have been reported to enhance disease resistance in plants to which they are administered (6,68,86,89-102). Direct antibiosis or nonspecific phytoalexin induction by compounds present in many of the preparations has not been rigorously excluded. However, enhanced endogenous plant defense responses upon pathogen exposure are common features of these reports. Fragments of DNA from Erwinia amylovora were used to protect apple and pear seedlings from fire blight caused by E. amylovora (96,97). Bacterial lipopolysaccharides protected tobacco against bacterial wilt caused by Pseudomonas solanacearum (6,92). Recently, Schönbeck and Dehne found, in screening culture filtrates from over 300 randomly isolated saprophytic soil bacteria and fungi, that more than 10 percent of the isolates enhanced disease resistance in wheat, grapes, rape, lettuce, beets, chrysamthemum, carnations, and cucurbits to downy and powdery mildews without apparent direct antibiosis (98-102). These results suggest that, in addition to direct antibiosis and competition, biotically induced plant resistance to pathogens may be an important and general component of biological control of plant pathogens.

Commercial Synthetic Compounds. The effectiveness of biochemically induced resistance in plants against a broad range of pathogens, including fungi, bacteria, and viruses, has prompted a search for non-toxic synthetic compounds which enhance plant resistance to disease (103-106).

A number of current commercial fungicides have been reported to enhance disease resistance in addition to direct fungitoxicity (103,105,107). Figure 3 presents structures of commercial pesticides reported to enhance endogenous plant disease resistance mechanisms.

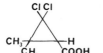

2,2-dichloro-3,3-dimethylcyclopropane-
carboxylic acid (DDCC, WL28325; Shell)

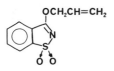

aluminum tris(O-ethylphosphite)
(TEPA, fosetylaluminum, Aliette; Rhone-Poulenc)

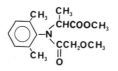

3-(allyloxy)-1,2-benzisothiazole 1,1-dioxide
(probenazole, Oryzemate; Meiji Seika)

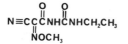

N-(2,6-dimethylphenyl)-N-(2-methoxyacetyl)-
DL-alanine, methyl ester
(metalaxyl, Ridomil; Ciba-Geigy)

2-cyano-N-(ethylcarbomoyl)-2-(methoximino)-
acetamide (cymoxanil, DPX 3217, Curate;
Du Pont)

Figure 3.   Commercial "pesticides" reported to enhance
endogenous plant disease defense mechanisms.

The compound for which the best biochemical evidence has been reported for sensitization of host plant response to pathogens, rather than direct or indirect fungitoxicity or nonspecific phytoalexin induction, is 2,2-dichloro-3,3-dimethylcyclopropane carboxylic acid (108-110). Neither the compound nor any of its metabolites generated after treatment of rice plants are directly inhibitory to the rice blast fungus, Piricularia oryzae. Constitutive phytoalexin production was not induced in rice by the cyclopropane derivative. However, infection of plants treated with the compound results in rapid localized cell death, melanization, and production of the phytoalexins, momilactones A and B.

Probenazole appears to inhibit appressorial formation in vitro by Piricularia oryzae, but is much more effective in vivo than in vitro. In rice plants, probenazole appears to inhibit mycelial elongation, stimulate production of conidial germination inhibitors, promote the accumulation of α-linoleic acid and other fungitoxic compounds, and enhance lignification in response to infection (111-114). Phenylthiourea effectively controls Cladosporium cucumerinum in cucumbers at a concentration of 10-20 µg/ml in hypocotyl sap, whereas, similar in vitro inhibition requires 500 µg/ml (115). No in vivo generated fungitoxic compounds were found in cucumber plants treated with phenylthiourea. Lignification in response to the pathogen was strongly enhanced in treated plants.

Aliette (aluminum tris-O-ethylphosphonate) has been reported to enhance defense reactions and phytoalexin accumulation in grapes and tomatoes in response to infection by Plasmopara viticola and Phytophthora spp., respectively, and to trigger phenolic accumulation and hypersensitive cell death in tomatoes, peppers, and beans in response to infection while possessing little direct fungitoxicity (116,117). However, recent data cast doubt on the earlier reports of the low activity of Aliette as an inhibitor of Phytophthora sporulation in vitro (118), and have attributed the protective properties of the compound to phosphorous acid which is formed in plant tissues or in certain buffer solutions of Aliette (119,120). Toxicity of phosphorous acid to Oomycetes is reversible by phosphate ion, and this may explain Aliette's lack of fungitoxicity in certain growth media.

Metalaxyl is clearly fungicidal in vitro to a variety of phycomycetous fungi. Its activity in vivo may also be enhanced by stimulation of host plant defenses including hypersensitive cell death (104), accumulation of phytoalexins such as glyceollin (121), and callose encasement of hyphae (122) after fungal infection of metalaxyl-treated soybeans. However, the in vivo concentration of the compound in cell sap may be sufficient for fungitoxicity alone to account for its protective activity (123).

Curzate appears to stimulate hypersensitive cell death and phytoalexin accumulation in grapevines infected with Plasmopara viticola (104).

Miscellaneous compounds. A variety of miscellaneous organic compounds have been reported to enhance disease resistance. Especially prominent are the D-isomers of amino acids and nonchiral amino acids. D- and DL-phenylalanine, α-aminoisobutyric acid and D-α-amino-n-butyric acid, though not fungitoxic in vitro, stimulate the accumulation of phenolics, phloretin and phloretic acid, and induce resistance to apple scab, caused by Venturia inaequalis, in several apple varieties (124-126).

Salicylic acid and its acetyl ester (aspirin) (127-129) and polyacrylic acid (130-132) have been reported to induce systemic resistance against TMV and TNV, in tobacco. The resistance appears similar to that in resistant cultivars.

Summary

Biotic and chemical sensitization of plants, i.e., immunization, offers promise as an alternative or complementary method to present strategies for protecting plants from disease and pests. Advantages of biotic sensitization include: a) lack of the need for exogenous chemicals or genetically engineered microbes with possible attendant hazards and regulatory constraints; b) high effectiveness, equivalent to protection by pesticides or of resistant cultivars; c) systemic protection, i.e., chemical spray coverage, or surface inactivation are not problems; d) possibly durable or permanent protection, i.e., protection is not lost after rainfall or decomposed by sunlight, and resistance may persist through the plant's lifetime; e) protection is effective against a broad spectrum of pathogens - a single treatment may protect to varying degrees against fungi, bacteria, viruses, and, possibly, nematodes and insects; f) protection is latent, plant metabolism is not appreciably deranged nor is metabolic energy diverted into mechanisms for disease resistance in the absence of pest or pathogen; and g) supposedly genetically "susceptible" or pesticide-sensitive cultivars with desirable agronomic properties can be successfully protected.

Limitations of plant sensitization include: a) a latent or lag period between delivery of the inducing stimulus and expression of immunization during which period the plant is unprotected; b) requirements for appropriate chemical or biotic inducer/pathogen combinations, not all diseases of all plants have known immunizing agents and known inducers are not necessarily universal; c) efficacy of sensitization may be influenced by the physiological state of plants; and d) delivery of biotic sensitization agents may be technologically difficult. Possible methods for delivery include seed treatment, high pressure sprays of virus suspensions, abrasive dusting or spraying of fungal spores adsorbed to micronized silica particles, tissue culture, and genetic engineering technology. Current reports (5,8-10) demonstrate that inducing agents elicit persistent systemic resistence when applied to young seedlings and parts of older plants.

In many cases, the distinction between alterations in plant physiology or direct antibiosis that results from prior infection, and biotic or chemical sensitization is subtle and probably can only be resolved by rigorous biochemical investigations. However, in practical applications, such subtle distinctions are probably moot - so long as the technique is effective.

Induced resistance has not prevented epiphytotics in plants.

The existence of an immune system in mammals, highly effective in
protecting against disease after recovery from an initial
infection, also has not prevented periodic catastrophic epidemics
and epizootics. However, an astute English country doctor, Edward
Jenner, noted immunological cross protection against deadly
smallpox in English milkmaids who had recovered from mild cowpox.
Despite skepticism, ridicule and even hostility of many of his
peers, Jenner developed vaccination as one of the safest and most
effective mammalian disease preventive measures known. We hope and
believe that the same can be accomplished for plants.

Acknowledgments

The authors' work cited in this paper was supported in part by
grants from the CIBA-Geigy Corp., R. J. Reynolds Corp., Rockefeller
Foundation, Graduate School of the University of Kentucky, and
grant 58-7B30-0-183 of the U.S.D.A./S.E.A.
      This paper is published as Journal paper No. 84-11-209 of the
Kentucky Agricultural Experiment Station, Lexington, Kentucky
40546.

Literature Cited

1.   Chester, K. S. Quarterly Rev. Biol. 1933, 8, 129-154, 275-
      324.
2.   Matta, A. Ann. Rev. Phytopathol. 1971, 9, 387.
3.   Loebenstein, G. Ann. Rev. Phytopathol. 1972, 10, 177.
4.   Johnson, R. Ann. Appl. Biol. 1978, 89, 107.
5.   Kuć, J.; Tüzün, S. Rec. Adv. Tobacco Sci. 1983, 9, 179.
6.   Sequeira, L. In "Recognition and Specificity in Plant Host-
      Parasite Interactions"; Daly, J. M.; Uritani, I., Eds.;
      University Park Press: Baltimore, 1979; p. 231.
7.   Bell, A. A.; Mace, M. E. In "Fungal Wilt Diseases of
      Plants"; Mace, M. E.; Bell, A. A.; Beckman, C. H., Eds.;
      Academic: New York, 1980.
8.   Kuć, J. In "Active Defense Mechanisms in Plants"; Wood, R.
      K. S., Ed.; N.A.T.O. Advanced Study Institutes Series A:
      Life Sciences; Plenum: New York, 1982; Vol. 37; p. 157.
9.   Kuć, J. Bioscience 1982, 32, 854.
10.   Kuć, J. In "Dynamics of Host Defense"; Bailey, J. A.;
      Deverall, B. J., Eds.; Academic: Sydney, 1983; p. 191.
11.   Sequeira, L. Ann. Rev. Microbiol. 1983, 37, 51.
12.   Tokousbalides, M. C.; Sisler, H. D. Pestic. Biochem.
      Physiol. 1978, 8, 26.

13.   Woloshuk, C. P.; Sisler, H. D.; Vigil, E. L. Physiol. Plant
      Pathol. 1983, 22, 245.
14.   Wolkow, P. M.; Sisler, H. D.; Vigil, E. L. Physiol. Plant
      Pathol. 1983, 23, 55.
15.   Tamari, K.; Ogasawara, N.; Kaji, J. In "The Rice Blast
      Disease"; Johns Hopkins: Baltimore; 1965; p. 35.
16.   Chet, I.; Haukin, D.; Katan, J. Phytopathol. Z. 1978, 91,
      60.

17. Bell, A. A. Ann. Rev. Plant Physiol. 1981, 32, 21.
18. Kuć, J. Ann. Rev. Phytopathol. 1972, 10, 207.
19. Hargreaves, J. A. Physiol. Plant Pathol. 1979, 15, 274.
20. Rahe, J. E. Phytopathol. 1973, 63, 572.
21. Bailey, J. A.; Berthier, M. Phytochem. 1981, 20, 187.
22. Langcake, P.; Pryce, R. J. Phytochem. 1977, 16, 1193.
23. Reilly, J. J.; Klarman, W. L. Phytopathol. 1972, 62, 1113.
24. Cooksey, C. J.; Garratt, P. J. Science 1983, 220,1398.
25. Bailey, J. A.; Skipp, R. A. Ann. Appl. Biol. 1978, 89, 354.
26. Cleveland, T. E.; Smith, D. A. Physiol. Plant Pathol. 1983,
    22, 129.
27. Zhang, Y.; Smith, D. A. Physiol. Plant Pathol. 1983, 23,
    89.
28. Van Etten, H. D.; Matthews, P. S.; Tegtmeier, K. J.; Dietert,
    M. F.; Stein, J. I. Physiol. Plant Pathol. 1980, 16, 257.
29. O'Brien, P. C.; Fisher, J. M. Nematologica 1978, 24, 463.
30. Karban, R.; Carey, J. R. Science 1984, 225, 53.
31. Pont, W. Queensland J. Agric. 1958, 16, 299.
32. Cruickshank, I. A. M.; Mandryk, M. J. Austrl. Inst. Agric.
    Sci. 1960, 26, 369.
33. McIntyre, J. L.; Miller, P. M. Phytopathol. 1978, 68, 235.
34. Hecht, E. I.; Bateman, D. F. Phytopathol. 1964, 54, 523.
35. Davis, D. Phytopathol. 1964, 54, 891.
36. Dowson, W. J. Ann. Appl. Biol. 1921, 8, 83.
37. Berotta, M.; Martins, E.; Moraes, W. Summa. Phytopathologica
    1977, 3, 66.
38. Tischler, G. Flora 1911, 4 (n.T.), 1.
39. Smith, E. F. In "Bacteria in Relation to Plant Disease";
    Carnegie Inst.: Washington, D. C., 1911; Vol. 2, 93.
40. Giddings, N. J. W. Va. Agric. Exp. Sta. Bull. 170, 1918, p.
    50.
41. Heale, J. B.; Sharman, S. Physiol. Plant Pathol. 1977, 10,
    51.
42. Andebrhan, T.; Coutts, R. H. A.; Wagih, E. E.; Wood, R. K. S.
    Phytopathol. Z., 1980, 98, 47.
43. Caruso, F.; Kuć, J. Phytopathol. 1977, 67, 1285.
44. Braun, J. W.; Helton, A. W. Phytopathol. 1971, 61, 685.
45. Rahe, J. E.; Kuć, J.; Chuang, C.; Williams, E. B.
    Phytopathol. 1969, 59, 1641.
46. Farcabee, G. J.; Lockwood, J. L. Phytopathol. 1958, 48,
    209.
47. Goodman, R. N. Phytopathol. 1967, 57, 22.
48. McIntyre, J. L.; Kuć, J.; Williams, E. B. Phytopathol.
    1973, 63, 872.
49. Caruso, F.; Kuć, J. Physiol. Plant Pathol. 1979, 14, 191.

50. Gessler, C.; Kuć, J. Phytopathol. 1982, 72, 1439.
51. Hammerschmidt, R.; Acres, S.; Kuć, J. Phytopathol. 1976,
    66, 790.
52. Jenns, A.; Kuć, J. Physiol. Plant Pathol. 1977, 11, 207.
53. Jenns, A.; Kuć, J. Physiol. Plant Pathol. 1980, 17, 81.
54. Kuć, J.; Schockley, G.; Kearney, K. Physiol. Plant Pathol.
    1975, 7, 195.
55. Staub, T.; Kuć, J. Physiol. Plant Pathol. 1980, 17, 389.

56.   Bashan, B.; Cohen, Y. Physiol. Plant Pathol. 1983, 23, 137.
57.   Elliston, J.; Kuć, J.; Williams, E. B. Phytopathol. Z.
      1976, 86, 117.
58.   Skipp, R.; Deverall, B. Physiol. Plant Pathol. 1973, 3,
      299.
59.   Sutton, D. Austral. Plant Pathol. 1979, 8, 4.
60.   Rahe, J. Phytopathol. 1973, 63, 572.
61.   Rahe, J.; Kuć, J. Phytopathol. 1970, 60, 1005.
62.   Elliston, J.; Kuć, J.; Williams, E. B. Phytopathol. Z.
      1976, 88, 43.
63.   McIntyre, J. L.; Dodds, J. A. Physiol. Plant Pathol. 1979,
      15, 321.
64.   Mandryk, M. Austral. J. Agr. Res. 1963, 14, 316.
65.   Cohen, Y. Ann. Appl. Biol. 1978, 89, 317.
66.   Cohen, Y.; Reuveni, M.; Kenneth, R. Phytopathol. 1975, 71,
      783.
67.   Inagaki, H.; Powell, N. Phytopathol. 1969, 59, 1350.
68.   McIntyre, J.; Miller, P. Phytopathol. 1978, 68, 235.
69.   McIntyre, J.; Dodds, J.; Hare, J. Phytopathol. 1981, 71,
      297.
70.   Kuć, J.; Richmond, S. Phytopathol. 1977, 67, 533.
71.   Guedes, M. E.; Richmond, S.; Kuć, J. Physiol. Plant Pathol.
      1980, 17, 229.
72.   Tüzün, S.; Kuć, J. Phytopathol. 1984, 74, 804.
73.   Xuei, X. L.; Kuć, J. Phytopathol. 1984, 74, 849.
74.   Richmond, S.; Kuć, J.; Elliston, J. E. Physiol. Plant
      Pathol. 1979, 14, 329.
75.   Elliston, J. E. Ph.D. Thesis, Purdue University, Lafayette,
      Indiana, 1977.
76.   Doss, M.; Hevesi, M. Acta Phytopathol. Acad. Sci. Hung.
      1981, 16, 269.
77.   Hammerschmidt, R. Ph.D. Thesis, Univ. of Kentucky,
      Lexington, Kentucky, 1980.
78.   Hammerschmidt, R.; Kuć, J. Physiol. Plant Pathol. 1982, 20,
      61.
79.   Hammerschmidt, R.; Nuckles, E.; Kuć, J. Physiol. Plant
      Pathol. 1982, 20, 73.
80.   Hijwegen, T. Neth. J. Plant Pathol. 1963, 69, 314.
81.   Henderson, S. J.; Friend, J. Phytopathol. Z. 1979, 94, 323.
82.   Ride, J. P. Physiol. Plant Pathol. 1980, 16, 187.
83.   Vance, C. P.; Sherwood, R. T.; Kirk, T. K. Ann. Rev.
      Phytopathol. 1980, 18, 259.
84.   Hammerschmidt, R.; Kuć, J. Phytopathol. 1980, 70, 689.
85.   Jenns, A.; Kuć, J. Phytopathol. 1979, 69, 753.
86.   Garas, N. A.; Kuć, J. Phytopathol. 1984, 74, 873.

87.   Rast, A. Th.B. Agric. Research Report No. 834, Inst.
      Phytopathol. Research, Wageningen, Netherlands, 1975.
88.   Caruso, F.; Kuć, J. Phytopathol. 1977, 67, 1290.
89.   Bell, A. A.; Presley, J. T. Phytopathol. 1969, 59, 1147.
90.   Lovrekovitch, L.; Farkas, G. Nature 1965, 205, 823.
91.   Sequeira, L.; Hill, L. Physiol. Plant Pathol. 1974, 4, 447.
92.   Graham, T. L.; Sequeira, L.; Huang, T. S. R. Appl. Environ.
      Microbiol. 1977, 34, 424.

93. Schönbeck, F. In "Active Defense Mechanisms in Plants";
     Wood, R. K. S., Ed.; Plenum: New York, 1982; p. 353.
94. Berard, D.; Kuć, J.; Williams, E. Physiol. Plant Pathol.
     1973, 3, 51.
95. Skipp, R.; Deverall, B. Physiol. Plant Pathol. 1973, 3,
     299.
96. McIntyre, J. L.; Kuć, J.; Williams, E. B. Phytopathol.
     1973, 63, 872.
97. McIntyre, J. L.; Kuć, J.; Williams, E. B. Physiol. Plant
     Pathol. 1975, 7, 153.
98. von Alten, H.; Schönbeck, F. Phytopathol. Z. 1981, 101,
     271.
99. Schönbeck, F.; Dehne, H.-W.; Balder, H. Z. Pflanzenkrank.
     Pflanzenshutz 1982, 89, 177.
100. Balder, H.; Dehne, H.-W.; Schönbeck, F. Med. Fac. Landbouw.
     Rijksuniv. Gent. 1982, 47, 329.
101. Balder, H.; Schönbeck, F. Z. Pflanzenkrank. Pflanzenschutz
     1983, 90, 200.
102. Dehne, H.-W.; Stenzel, K.; Schönbeck, F. Z. Pflanzenkrank.
     Pflanzenschutz 1984, 91, 258.
103. Day, P. R. "Genetics of Host-Parasite Interactions"; W. H.
     Freeman: San Francisco, 1974; p. 169.
104. Langcake, P. Phil. Trans. Royal Soc. London B 1981, 295,
     83.
105. Fuchs, A.; Davidse, L. C.; de Waard, M. A.; de Wit, P. J. G.
     M. Pestic. Sci. 1983, 14, 272.
106. James, J. R. Plant Disease 1984, 68, 651.
107. Guest, D. I. Ph.D. Thesis, University of Sydney, Sydney, N.
     S. W., Australia, 1982.
108. Cartwright, D.; Langcake, P.; Pryce, R. J.; Leworthy, D. P.
     Nature 1977, 267, 511.
109. Langcake, P.; Wickins, S. G. A. J. Gen. Microbiol. 1975,
     88, 295.
110. Cartwright, D. W.; Langcake, P.; Ride, J. P. Physiol. Plant
     Pathol. 1980, 17, 259.
111. Uchiyama, M.; Abe, H.; Sato, R.; Shimura, M.; Watanabe, T.
     Agric. Biol. Chem. 1973, 37, 737.
112. Watanabe, T. J. Pestic. Sci. 1977, 2, 395.
113. Watanabe, T.; Sekizawa, Y.; Shimura, M.; Suzuki, Y.;
     Matsumoto, K.; Iwata, M.; Mase, S. J. Pestic. Sci. 1979, 4,
     53.
114. Sekizawa, Y.; Mase, S. J. Pestic. Sci. 1981, 6, 91.
115. Kaars Sijpesteijn, A. J. Sci. Food Agric. 1969, 20, 403.
116. Bompeix, G.; Ravisé, A.; Raynal, G.; Fettouche, F.; Durrand,
     M. C. Ann. Phytopathol. 1980, 12, 337.
117. Vo-Thi-Hai; Bompeix, G.; Ravisé, A. C. R. Acad. Sci. Ser. D.
     1979, 288, 1171.
118. Farih, H.; Tsao, P. H.; Menge, J. A. Phytopathol. 1981, 71,
     934.
119. Fenn, M. E.; Coffey, M. D. Phytopathol. 1984, 74, 606.
120. Coffey, M. D.; Bower, L. A. Phytopathol. 1984, 74, 738.
121. Ward, E. W. B.; Lazarovits, G.; Stoessel, P.; Barrie, S. D.;
     Unwin, C. H. Phytopathol. 1980, 70, 738.

122.   Hickey, E. L.; Coffey, M. D. Physiol. Plant Pathol. 1980,
        17, 199.
123.   Lazarovits, G.; Ward, E. W. B. Phytopathol. 1982, 72, 1217.
124.   Kuć, J.; Barnes, E.; Daftsios, A.; Williams, E. B.
        Phytopathol. 1959, 49, 313.
125.   Holowczak, J.; Kuć, J.; Williams, E. B. Phytopathol. 1962,
        52, 699.
126.   MacLennan, D. H.; Kuć, J.; Williams, E. B. Phytopathol.
        1963, 53, 1261.
127.   White, R. F. Virology 1979, 99, 410.
128.   van Loon, L. C.; Antoniw, J. F. Neth. J. Plant Pathol.
        1982, 88, 237.
129.   van Loon, L. C. In "Active Defense Mechanisms in Plants"; R.
        K. S. Wood, Ed.; Plenum: New York, 1982; p. 247.
130.   Gianinazzi, S. ibid., p. 275.
131.   Gianinazzi, S.; Kassanis, B. J. Gen. Virol. 1974, 23, 1.
132.   Kassanis, B.; White, R. F. Ann. Appl. Biol. 1975, 79, 215.
133.   Kuć, J. In "Origins and Development of Adaptation"; Evered,
        D.; Collins, G. M., Eds.; Ciba Foundation/Pitman:London,
        1984; p. 100.

RECEIVED November 23, 1984

Use of Subtoxic Herbicide Pretreatments to Improve Crop Tolerance to Herbicides

G. R. STEPHENSON and G. EZRA

Department of Environmental Biology, University of Guelph, Guelph, Ontario, Canada N1G 2W1

Antagonistic combinations of herbicides can lead to the
development of seed applied chemical safeners to
protect crops from herbicide injury. Another approach
has been the development of chloroacetamide compounds
as selctive safeners that can be added to the
formulations of thiocarbamate herbicides to improve
their selectivity in corn. A promising new approach
involves early pretreatments with subtoxic levels of a
particular herbicide to increase crop tolerance to
later, higher rates of that herbicide. In a series of
growth room studies, 0.001 to 0.1X pretreatments with
metribuzin in tomato and pyrazon in red beets increased
the tolerance of these crops to later higher levels of
these particular herbicides. Pretreatments with CDAA
were also highly effective for increasing corn
tolerance to CDAA but similar studies with EPTC,
metribuzin, atrazine, chlorosulfuron, and alachlor in
corn, metribuzin in soybeans, atrazine in sorghum, and
chlorsulfuron in oats were less promising. Elevated
substrate, enzyme activities, and detoxication rates
were involved in some of the cases of improved crop
tolerance via the earlier subtoxic pretreatments with
herbicides.

Chemical herbicides have been available for more than a century
but major impacts on crop production awaited the development of
"truely" selective herbicides or innovations that would permit
use of non-selective herbicides in crop situations. We now have
some form of selective chemical weed control for most of our major
crops. However, continuing problems with herbicide injury to
crops as well as poor control of weeds that are botanically
similar to crops remind us that further improvements in herbicide
selectivity are still needed. Introductions of new selective
herbicides will continue but the high costs of these new chemicals
are stimulating efforts to make wider use of existing herbicide
chemistry. One successful approach has been to genetically
improve the tolerance of new crop cultivars to major herbicides

0097–6156/85/0276–0069$06.00/0

(1). Another approach has been to increase the physiological
tolerance of crop plants to particular herbicides through the
development of chemical antidotes or safeners. Since the
introduction of this latter concept by Otto Hoffman (2) in the
early 1960's, at least five antidotes or safeners have been
developed commercially (Table I).
     Research on chemical antidotes or safeners has been
summarized in several reviews and published symposia (3-9). Most
of the major developments (Table I) have resulted from impirical
screening programs by Industry that may have been stimulated by
observations of herbicide antagonism in plants (3, 10). However,
some of the research on mode of action of antidotes has been
directed at finding new ways to protect crop plants from
herbicides (3). The research to be discussed in this text, namely
the use of subtoxic herbicide pretreatments to improve crop
tolerance to selected herbicides, arises in part from research on
the mode of action of R-25788 as a selective antidote for EPTC or
butylate in corn.

Mode of Action of R-25788

In some of the earliest research, Wilkinson (11) suggested that
opposing actions of EPTC and R-25788 on lipid synthesis in plants
could explain their mechanisms as a herbicide and antidote,
respectively. Ezra et al. (12) have provided strong support for
this idea with their observations that EPTC and R-25788 could have
measurable and opposite effects on lipid synthesis as early as 1
hr after treatment in corn cell suspension cultures. Other
research by Lay and Casida (13) has supported the concept that R-
25788 enhances the metabolic detoxication of EPTC and its
sulfoxide by elevating the glutathione (GSH) levels and GSH-S-
transferase activity involved in the conjugation of EPTC-sulfoxide
with glutathione. Early research in our laboratory on antidote
structure/activity relationships (14, 15) established that
acetamide and carbamate molecules more similar in structure to
EPTC than to R-25788 could have equal or even greater antidote
activity against EPTC in corn with soil free bioassay systems.
The structure/activity studies seemed to support the idea that the
antidote R-25788 (or its analogues) may act as a competitive
inhibitor at site(s) of EPTC action in corn. These latter two
theories of "antidote enhanced herbicide metabolism" versus
"competitive inhibition" were hard to reconcile with each other.
To resolve this apparent disagreement, a series of acetamide
analogues with known antidote activities were tested for their
effects on glutathione levels in corn (Figure 1). In this
comparative study, the dichloro diallyl acetamide (R-25788) was
slightly more effective as an antidote for EPTC in corn than was
the monochloro diallyl acetamide (CDAA) and the trichloro and non-
chlorinated analogues were least effective (Figure 1a).
Surprisingly, this related series of acetamides had virtually the
same spectrum of effects on glutathione levels in corn seedlings
(Figure 1b). The inclusion of CDAA, (N,N-diallyl-2-
chloroacetamide) in this study was very fortunate. This chemical
has been a widely used herbicide for weed control in onions and at
high rates it can also be toxic to corn with symptoms not unlike

Table I. Commercially important herbicide antidotes or safeners.

Chemical name	Application	Crops protected	Herbicides
1,8-naphthalic anhydride (NA)	seed	corn sorghum wheat oats rice	carbamates chloroacetanilides thiocarbamates others
N,N-diallyl-2,2-dichloroacetamide (R-25788)	seed or herbicide formulation	corn sorghum barley wheat rice	thiocarbamates acetanilides carbamates dithiocarbamates others
α-[(cyanomethoxy) imino benzacetonitrile] (cyoxymetrinil)	seed	sorghum wheat rice	chloroacetanilides
α-(1,3-dioxolan-2-yl-methoxy)imino-benzacetonitrile (CGA-92194)	seed	sorghum	chloroacetanilides
2-chloro-2-4-(trifluoromethyl)-5-thiazolecarboxylic acid, benzyl ester (flurazole)	seed	sorghum	chloroacetanilides

From Stephenson and Ezra (3).

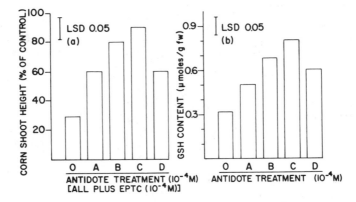

Figure 1. Comparative effects of four diallyl acetamides as
antidotes to EPTC (left) and on GSH levels (right) in corn
seedlings. The various diallyl acetamides [R-CON(CH$_2$CH=CH$_2$]
were R=CH$_3$-, CH$_2$Cl-, CHCl$_2$- and CCl$_3$ for A, B, C, and D,
respectively. Antidote B is better known as the herbicide
CDAA. Antidote C is the antidote R-25788 which has been
developed commercially for EPTC or butylate. Reproduced
with permission from Ref. 17. Copyright 1983 Pergamon Press.

those of EPTC. However, at lower rates it is an effective
antidote or safener for EPTC in corn (Figure 1a, Compound B).
Furthermore, in these studies we learned that like R-25788 it
would also elevate glutathione levels in corn (Figure 1b, Compound
B). Also, like EPTC, it is known to be detoxified in plants by
conjugation with glutathione (16). With this combination of
knowledge, it seemed more likely that CDAA was acting as a less
phytotoxic mimic of EPTC instead of as a competitive inhibitor of
EPTC. It seemed possible that at low rates, both CDAA and EPTC
might be stimulatory on processes such as lipid synthesis, but at
high rates, they might both be inhibitory. If the required rates
for the two compounds were different, it might be possible for a
stimulatory concentration of CDAA (or R-25788) to overcome
inhibitory effects of EPTC. Furthermore, if both CDAA and EPTC-
sulfoxide were conjugated with glutathione, it seemed plausible
that the antidote (CDAA) may act by enhancing the same pathway
needed to detoxify EPTC and its sulfoxide - namely the GSH and
GSH-S-transferase system. One other question also emerged. If
CDAA could elevate GSH levels could other herbicides, known to be
conjugated with GSH, also elevate GSH levels? Some of these
questions have not yet been examined but the latter speculation
proved to be correct. In addition to CDAA, five other herbicides
were shown to elevate glutathione levels in corn seedlings (Table
II) and the metabolism of all these herbicides is known to involve
glutathione conjugation (18).

Table II. Effects of R-25788 and various herbicides on glutathione
(GSH) levels in roots of 5-day-old corn seedlings.

Treatments[3]	Root (GSH) Content[1] (u moles/g fw)			
	0	10^{-6}M	10^{-5}M	10^{-4}M
R-25788	0.19	0.31	0.48	0.71
EPTC	0.19	0.29	0.45	0.41
alachlor	0.30	0.41	0.49	0.71
propachlor	0.20	0.29	0.35	0.42
atrazine	0.18	0.22	0.31	0.34
barban[2]	0.19	0.41	0.48	0.86

From Stephenson et. al. (17).

[1] For all experiments LSD's were 0.05 u moles/g or less.
[2] High rate of barban caused a significant reduction in root fw.
[3] The metabolism of all of these chemicals except R-25788, is
known to involve conjugation with glutathione in plants (18).

Increasing Crop Tolerance to Herbicides with Herbicide Pretreatments

At this point, evidence that similar molecules acted as effective antidotes by inducing needed metabolic pathways for herbicide detoxication was at most very speculative. However, another hypothesis emerged. Could early herbicide pretreatments increase crop tolerance to these herbicides by elevating the substrates and enzymes needed for detoxication? While not a new concept in animal systems, such an idea has received little attention in plant systems and it certainly has not been exploited in any practical way. The whole idea has seemed much more credible with the study by Jacetta and Radosevich (19) of photosynthetic recovery in corn after treatment with atrazine. More specifically, they showed that inhibition of photosynthesis was reduced and the rate of recovery enhanced in corn plants treated for the second or third time with atrazine compared to "first exposed" plants (Figure 2). Furthermore, the faster recovery was related to enhanced rates of atrazine metabolism in the previously treated plants (Table III).

   Stimulated by the antidote research, as well as by the work of Jacetta and Radosevich (19), we decided to examine the influence of pretreatment with subtoxic rates of several herbicides on later crop tolerance to these same herbicides. For these growth room studies (25°C/18°C, 16h photoperiod, light intensity 400 uE/m^2/sec, 75% relative humidity), pretreatments at concentrations of 0.1% to 10% of the final herbicidal rate were given as a root drench or seed treatment to seeds planted in moist vermiculite in styrofoam cups. Herbicidal treatments were later applied to the roots. Plants were harvested 8-14 days after herbicide treatment (Table IV).

   Possibly the most surprising results from these studies were those with atrazine. Even though Jacetta and Radosevich (19) had observed reduced photosynthetic injury and enhanced atrazine metabolism with atrazine pretreatments, in our studies, pretreatments with atrazine did not prevent later dry weight reductions from high rates of atrazine. We were also surprised that, in spite of effects on GSH levels, EPTC pretreatments failed to reduce later EPTC injury to corn. Thus far, antidote research has had little impact on improving the tolerance of broadleaved crops to herbicides. On that basis we were quite encouraged by the effectiveness of metribuzin pretreatments for reducing later metribuzin injury to tomatoes (Table IV, Figure 3) and soybeans (Table IV). Pyrazon pretreatments were slightly more effective for preventing pyrazon injury to red beets (Table IV), but the most promising results were obtained with CDAA (Table IV, Figures 4, 5). Corn shoot dry weights were reduced 40% by CDAA applied at concentrations of 200 uM, 5 days after planting in vermiculite nutrient culture. Optimum prevention of CDAA injury was observed with a CDAA pretreatment of 5 uM, applied 2.5 days earlier. However, significant prevention of injury was also observed with the 1 and 10 uM pretreatments (Figure 4).

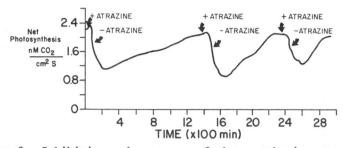

Figure 2.   Inhibition and recovery of photosynthetic rate in corn with 3 successive exposures to atrazine [Illustrated by one representative infrared gas analyzer trace, + and - refer to presence or absence of atrazine].   "Reproduced with permission from Ref. <u>19</u>.   Copyright 1981, 'Weed Science Society of America'".

Figure 3.   Prevention of metribuzin (0.5 mg/L) injury to tomatoes with metribuzin pretreatments (1.0 ug/L).   Left to right, control, 1.0 ug/L pretreatment followed by 0.5 mg/L herbicidal treatment 14 days later.

Table III. Occurrence of ^{14}C in chloroform and aqueous fractions of corn shoots after 1, 2, and 3 successive atrazine treatments.

Treatment						DPM extracted[a]	
Atrazine exposure	Recovery period	Atrazine exposure	Recovery period	Atrazine exposure	Recovery period	Aqueous fraction	Chloroform fraction
------- h -------						------- % -------	
4[1]	0					43.4a	56.6a
4[1]	12					54.6a	45.4a
4	12	4[1]	0			49.3a	50.7a
4	12	4[1]	12			67.9b	32.1b
4	12	4	12	4[1]	0	59.3a	40.7a
4	12	4	12	4[1]	12	72.7b	27.3b

"Reproduced with permission from Ref. 19 Copyright 1981, 'Weed Science Society of America'."

[a] Values adjacent to the same letter in a column are not different at the 5% level of significance.

[1] Supplied as ^{14}C-atrazine.

Table IV. Effects of herbicide pretreatments on subsequent
herbicide injury to crops.

Herbicide	Treatment Concentrations Pre- Herbicidal-		Crop	Prevention of injury by subtoxic pretreatment[1] Dry Wt. Height	
Alachlor	20uM	76uM	corn (cv PAG SX-111)	−*	+
CDAA	5uM	200uM	corn (cv PAG SX-111)	++	+++
Atrazine	5uM	50mM	sorghum (cv De Kalb E59+)	+	−
	5uM	50mM	corn (cv PAG SX-111)	0	0
Chlorsulfuron	50pM	10nM	corn (cv PAG SX-111)	0	+
	8.5nM	50uM	oats (cv centinnel)	0	+
EPTC	15uM	200uM	corn (cv PAG SX-111)	0	0
Metribuzin	4.6nM	2uM	tomatoes (cv H2653)	+	−
	4.6nM	1uM	soybeans (cv EVANS)	+	−
	0.5uM	100uM	corn (cv PAG SX-111)	0	+
Pyrazon	10.0nM	20uM	beets (cv Detroit Dark Red)	++	−

[1]   Results are from at least two experiments. 0 = less than 5%,
+ = 5-10% reduction in injury, ++ = 11-20% reduction in injury,
+++ = greater than 20% reduction in injury.
*   Not recorded in these experiments.

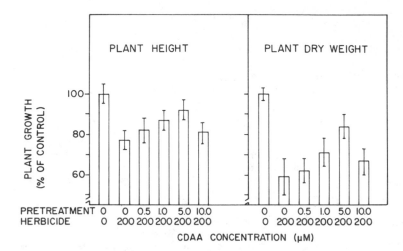

Figure 4. Prevention of CDAA injury to corn with CDAA pretreatments applied 2.5 days earlier. No injury was observed with the pretreatments applied alone except at the highest concentration (50.0 uM).

Figure 5. Prevention of injury to corn from CDAA at 200 uM with 5 uM CDAA pretreatments. Left to right, control, 200 uM CDAA alone, 200 uM CDAA following 5 uM CDAA, and 5 uM CDAA alone.

Effect of CDAA pretreatments on subsequent CDAA metabolism

Corn seeds were germinated between moist paper towels and were then treated with 5 uM CDAA either 1 or 2.5 days prior to assay for GSH at 6 days after planting (Table V). Using the reagent DTNB [2,2-dithiobis (2-nitrobenzoic acid)], root GSH contents were assayed spectrophotometrically as previously described (20).
    Treatment with CDAA (at 5 uM) for either 1 or 2.5 days elevated GSH levels more than 60% (Table V).

Table V. Effect of CDAA on GSH levels in 6 day old corn seedlings.

GSH content in corn roots (ug/g fw)		
Control	5 uM CDAA for 1 day	5 uM CDAA for 2.5 days
65±10	106±6	109±11

Reproduced with permission from Ref. 20. Copyright 1985 Academic Press.

    CDAA also elevated GSH-S-transferase activity in corn roots or shoots (Table VI) when assayed spectrophotometrically with CDNB (1-chloro-2,4-dinitrobenzene) (20). However, in this case, treatment with CDAA for 2.5 days produced the greatest increase in enzyme activity in the roots of 6-day-old corn seedlings.

Table VI. Effect of CDAA pretreatments on GSH-S-transferase activity in roots or shoots of 6-day-old corn seedlings (c.v. PAG SX-111).

5 uM CDAA Pretreatment time (days)	GSH-S-transferase specific activity (nmol/min/mg Protein)	
	Roots	Shoots
0	1094	276
1	1486	315
2.5	1638	258
6	1371	316

    Although CDAA pretreatments elevated both GSH and GSH-S-transferase activity (as assayed by CDNB), it was essential to determine whether these effects would actually result in greater CDAA metabolism. For these studies, we examined [^{14}C]CDAA metabolism in vitro and in excised corn root tips [detailed methods described by Ezra et. al. (20)]. The in vitro assay revealed a significant level of non-enzymatic [^{14}C] CDAA

degradation [as determined by partitioning of water soluble
metabolites from methylene chloride into water and verification by
thin layer chromatography]. However, the enzymatic rate was
double that of the non-enzymatic rate. Furthermore, a 2.5 day
pretreatment with 5uM CDAA increased the GSH-S-transferase
activity for [^{14}C]CDAA metabolism from 62.9+4.6 to 78.9+5.3
nmol/min/mg protein (20). Metabolism of [^{14}C] CDAA was very rapid
in excised root tips from control or pretreated root tips.
However, in all experiments, pretreated roots took up 2-fold more
[^{14}C]CDAA and metabolized it twice as fast as non-pretreated root
tissue.

Effects of R-25788 and EPTC on the GSH/GSH-S-transferase System

In a series of experiments, similar to those with CDAA, R-25788
and EPTC were examined as pretreatments to either enhance their
own metabolism by GSH-S-transferase or prevent later injury (from
toxic doses) in corn seedlings.
   The effects of R-25788 on elevation of GSH and GSH-S-
transferase have been well documented (13). However, unlike CDAA,
R-25788 pretreatments did not prevent or reduce injury to corn
seedlings from later higher doses of R-25788 (Figure 6).
   This may be explained by the fact that even though R-25788
elevates GSH and GSH-S-transferase, it is not actually metabolized
by this system. In fact, R-25788 has been shown to be detoxified
in corn, predominantly to N,N-diallyloxamic acid. However, there
is some N-dealkylatation and hydrolysis to dichloroacetic acid and
some glycoside formation (21). Even though EPTC pretreatments
were ineffective for preventing EPTC injury (Table IV), EPTC did
influence this same system. At 10 ppm (5 x 10^{-4} M), EPTC elevated
both the GSH content (Figure 7) and GSH-S-transferase activity
(Figure 8). However these elevations were not as great and were
not as persistent as those obtained with R-25788. More
specifically, GSH and GSH-S-transferase activity were elevated by
more than 50% for at least 4-days with R-25788 (Figure 8), whereas
this magnitude of elevation occurred only briefly, 2-days after
EPTC pretreatment and then declined. In other studies
(unpublished) we have observed no effect of EPTC pretreatments on
[^{14}C]EPTC metabolism in excised corn root tips. Effects of EPTC
pretreatments on [^{14}C]EPTC sulfoxide metabolism have not yet been
examined.
   One possible explanation for the ineffectiveness of EPTC
pretreatments for enhancing EPTC metabolism and for reducing EPTC
injury is that its conjugation with glutathione may be primarily
non-enzymatic as suggested by other researchers (22). However in
other recent research (unpublished) we have shown that a known
GSH-S-transferase inhibitor is synergistically phytotoxic with
EPTC in corn, indicating that non-enzymatic conjugation may not be
that important in corn. A more likely explanation for the lack of
protection with EPTC pretreatments, is that effects on GSH and
GSH-S-transferase are too transient in nature, and/or that
sulfoxidation is an important and possibly rate-limiting step.

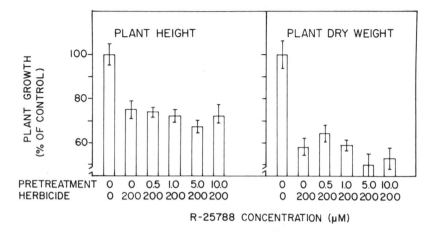

Figure 6.   Effect of R-25788 pretreatments on the toxicity of a later herbicidal amount of R-25788 in corn seedlings.

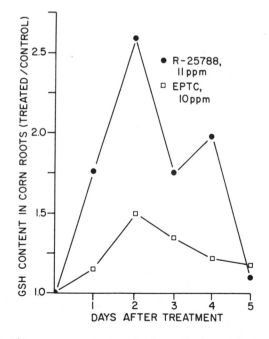

Figure 7.   A time course study of the effects of R-25788 and EPTC on the GSH content of corn seedlings (as assayed with DTNB reagent).

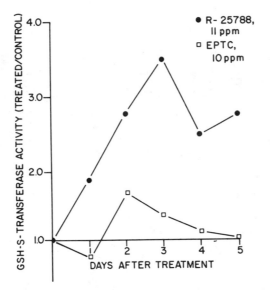

Figure 8. A time course study of the effects of R-25788 and EPTC on the glutathione-$\underline{S}$-transferase activity in corn seedlings (as assayed with CDNB).

Summary

The use of herbicide pretreatments at early seed or seedling
stages is not yet of practical importance for improving herbicide
effectiveness or selectivity in actual field situations. While it
may be possible to document this concept in field use with a
herbicide like CDAA, the use of this herbicide is currently
decreasing even in onions and it certainly is not needed in corn
where our effects were observed. However research on various
innovations with seed dressing or gel seeding systems with the
herbicides discussed here, or with other herbicides, could lead to
important practical applications in the future. In our
experience, we found more success by first determining whether
subtoxic herbicide pretreatments would reduce later growth
inhibitions with higher doses. The studies with EPTC, R-25788,
and atrazine (19) illustrate that it may be easier to observe
effects of pretreatments at the biochemical level, but in many
cases these effects are not subsequently manifested in reduced
herbicide effects on growth. Another obvious possibility is that
analogues of the herbicides may be more effective as pretreatments
to reduce herbicide injury than the herbicides themselves. This
concept brings us full circle, back to the concept that similar
molecules have a good chance of being effective antidotes. Thus
at the very least, this research is another form of evidence in
support of that theory.

Acknowledgments

The authors are indebted to Dr. A. Ali for his consultations and
participation in the earlier stages of this research. The
technical assistance of E. Pelly and C. Kubanek is also gratefully
acknowledged.

Literature Cited

1. LeBaron, H.; Gressel, J. "Herbicide Resistance in Plants"
    Wiley: New York, 1982; 386pp.
2. Hoffman, O.L. Weed Sci. 1962, 10, 322-3.
3. Stephenson, G.R.; Ezra, G. In "Plant Growth Regulating
    Chemicals"; Nickell, L.G., Ed., CRC Press: Boca Raton,
    Florida, 1983; Vol. II, pp. 193-212.
4. Blair, A.M.; Parker, C.; Kasasian, L. P.A.N.S. 1976, 22, 65-
    74.
5. Pallos, F.M.; Casida, J.E. "Chemistry and Action of
    Herbicide Antidotes"; Academic Press: New York, 1978; pp.
    171.
6. Gressel, J.; Ezra, G.; Jain, S.M. In "Chemical Manipulation
    of Crop Growth and Development"; McLaren, J.D., Ed.;
    Butterworth: London, 1982; pp. 79-91.
7. Anonymous. British Crop Protection Conference - Weeds.,
    1982; 2, p. 429.
8. Parker, C. Pestic. Sci. 1983, 14, 40-8.
9. Hatzios, K.K. Adv. Agron., 1983, 36, 265-316.
10. Gray, R.A.; Green, L.L.; Hoch, P.E.; Pallos, F.M. Proc.
    Brit. Crop. Prot. Conf. - Weeds, 1982, II, 431-7.

11. Wilkinson, R.E. In "Chemistry and Action of Herbicide
     Antidotes", Pallos, F.M.; Casida, J.E., Eds.; Academic Press:
     New York, 1978; pp. 85-108.
12. Ezra, G.; Gressel, J.; Flowers, H.M. Pestic. Biochem.
     Physiol. 1983, 18, 225-234.
13. Lay, M.M.; Casida, J.E. In "Chemistry and Action of
     Herbicide Antidotes"; Academic Press: New York, 1978; pp.
     151-160.
14. Stephenson, G.R.; Bunce, N.J.; Makowski, R.I.; Bergsma, M.D.;
     Curry, J.C. J. Agric. Food Chem. 1979, 27, 543-7.
15. Stephenson, G.R.; Bunce, N.J.; Makowski, R.I.; J.C. Curry.
     J. Agric. Food Chem. 1978, 26, 137-140.
16. Lamoureux, G.L.; Stafford, L.E.; Tanaka, F.S.
     J. Agric. Food Chem. 1971, 19, 346-
17. Stephenson, G.R.; Ali, A.; Ashton, F.M. In "IUPAC Pesticide
     Chemistry - Human Welfare and the Environment"; Miyamoto,
     J., Kearney, P.C., Eds.; Pergammon Press: Oxford, 1983; pp.
     219-224.
18. Shimabukro, R.H.; Lamoureux, G.L.; Frear, D.S. In "Chemistry
     and Action of Herbicide Antidotes"; Pallos, F.M.; Casida,
     J.E., Eds.; Academic Press: New York, 1978; pp. 133-149.
19. Jacetta, J.J.; S.R. Radosevich. Weed Sci. 1981, 29, 37-44.
20. Ezra, G.; Rusness, D.G.; Lamoureux, G.L.; Stephenson, G.R.
     Pestic. Biochem. Physiol. 1985, 23, 108-15.
21. Miaullis, J.B.; Thomas, V.M.; Gray, R.A.; Murphy, J.J.;
     Hollingworth, R.M. In "Chemistry and Action of Herbicide
     Antidotes"; Pallos, F.M. and Casida, J.E. Eds.; Academic
     Press, New York, 1978; pp. 109-131.
22. Leavitt, J.R.C.; Penner, D. J. Agric. Food Chem. 1979, 27,
     533-6. 536. (1979).

RECEIVED November 23, 1984

Regulation of Plant Growth and Development by Endogenous Hormones

THOMAS C. MOORE

Department of Botany and Plant Pathology, Oregon State University, Corvallis, OR 97331

The roles and mechanisms of action of all the major
kinds of hormones in normal regulation of growth and
development of seed plants are discussed. Auxins
promote growth by inducing cell-wall loosening and
nucleic acid (mRNA) and protein synthesis. Like
auxins, gibberellins also have a dual cell-wall plas-
ticizing effect and gene-activation function which is
manifested as synthesis of specific mRNAs and proteins. The mode of action of free cytokinins is
unknown; however, the cytokinin moieties which occur
in certain molecular species of tRNA may modulate
protein synthesis by influencing the binding of
aminoacyl-tRNAs to the mRNA-ribosomes. In the barley
aleurone system and dormant tissue, ABA seems to
function by opposing the action of GA and inhibiting
the synthesis of RNA, whereas evocation of rapid
stomatal closure is due to an effect on membrane
permeability. It also affects protein synthesis
qualitatively in responsive tissues. Ethylene par-
ticipates in growth regulation of plants throughout
ontogeny. The nature of the hormonal regulation of
flowering in angiosperms has never been fully elucidated on the basis of the known hormones. There
may be unique florigens or anthesins for the regulation of the flowering process.

The subject of this paper is natural plant growth regulators, i.e.,
endogenous hormones, and the mechanisms by which they regulate plant
growth and development. The commonly recognized plant hormones are
the auxins, gibberellins, cytokinins, abscisic acid, ethylene and
the hypothetical florigens or anthesins.

As Paleg (1) stated, there is merit in applying different conno-
tations to two terms, "mechanism" and "mode" of action, which often
are used synonymously. When a hormone acts upon a responsive plant
system, it of course enters into some direct and specific molecular
interaction which results, eventually, in the manifestation of a

0097–6156/85/0276–0085$06.00/0

measurable effect (biochemical or physiological response, e.g., cell
elongation). But there really are two aspects of the hormone action
involved: (1) the direct and specific molecular interaction; and
(2) the succeeding series of events which result in the measurable
biochemical or physiological response. The former is the "mechanism"
of action; the latter is the "mode" of action. It is the former
which will be discussed here.

Auxins

The natural auxins are represented by a single compound, indole-3-
acetic acid (IAA), which occurs as the free acid and in various
"bound" forms, including a thioglucoside, glycosyl esters, and IAA
peptides. For a recent discussion see the book chapter by Bandurski
and Nonhebel (2) in Wilkins (3).
      When auxin is supplied to a responsive system (e.g., excised
coleoptile or stem segment), there are two basic responses: first,
cell wall and media acidification and cell wall loosening; and
second, stimulation of RNA and protein synthesis. The mechanism of
the cell wall and media acidification is unknown. However, it has
been postulated that IAA activates a plasma membrane "pump" or ATPase
thereby stimulating active proton efflux from the plasma membrane
(4,5). The lowered pH then either causes breakage of hydrogen bonds
between certain cell-wall polymers (e.g., xyloglucans) and cellulose
microfibrils, or activates enzymes capable of hydrolyzing certain
wall polysaccharides to soften the wall and allow cell enlargement.
In auxin-depleted, excised stem sections (e.g., soybean hypocotyl)
this initial burst of elongation is transient, lasting only some 30
to 90 min, depending upon the species and conditions. The rate of
enlargement first increases, after a lag period of 15 min or less,
then begins to decrease, with kinetics resembling (perhaps identical
to) acid-induced growth.
      The observation that IAA stimulation of growth may be preceded
by an increase in RNA synthesis (6) led to the early idea that IAA
might act by derepressing certain genes, thus causing altered RNA and
protein synthesis. Key and associates (see 7) and others reported
extensively on an IAA-stimulation of incorporation of radiolabeled
nucleotides into RNA. Recently Zurfluh and Guilfoyle (8) and
Theologis and Ray (9) reported that certain mRNA sequences appeared
following application of IAA and 2,4-D to soybean and pea seedling
stem segments. A few mRNA sequences increased in amount or translation activity within 15 to 20 min of exposure of pea and soybean
tissues to IAA or 2,4-D; that is, within about the same time as the
auxin stimulation of cell enlargement (9). In Zurfluh's and
Guilfoyle's (8) work, the levels of translatable mRNA for at least
ten in vitro translation products were increased by 2,4-D in sections
of soybean hypocotyl. The induction by auxin occurred rapidly (within 15 min), and the amounts of the induced in vitro translation products increased with time of auxin treatment. Theologis and Ray (9)
stated that, "Although for several reasons, it seems unlikely that
those mRNAs are actually causative in the auxin induction of cell
enlargement, their increase seems to be relatively close to primary
auxin action and might well serve in maintaining a steady rate of
cell enlargement over the longer term ('second phase' of auxin action
in cell enlargement). Other mRNAs increase substantially subsequent

to 0.5, 1 and 2 hr of auxin treatment, and beyond about 2 hr certain
other mRNA sequences become repressed by auxins." Obviously, it will
be very important ultimately to elucidate the physiological roles of
auxin-regulated mRNAs and the mechanism of their regulation. The
model presented by Vanderhoef and Dute (10) summarizes some earlier
evidence reconciling the effects of wall acidification and RNA and
protein synthesis as a two-phase action of auxin following addition
of the hormone to auxin-depleted, responsive tissues.

Gibberellins

The more than 50 gibberellins (GAs) occur as free 19-carbon or 20-
carbon diterpenoid mono-, di-, or tri-carboxylic acids and as the
glucosides and glucosyl esters. See Jones and MacMillan (11) for a
recent discussion.
      Much of our knowledge about the mechanism of action of GAs con-
cerns the induction by GA of de novo synthesis of certain hydrolytic
enzymes in the aleurone layer of germinating barley grains. In the
course of natural germination of a barley grain, the embryo is the
source of some endogenous GA. Radley (12) showed in 1967 that GA-
like substances are produced by the scutellum during the first 2
days of germination and thereafter by the embryo axis. Either the
GA itself, liberated by the embryo, or some unknown factor capable
of stimulating GA synthesis, diffuses across the endosperm to the
aleurone cells, in which the several hydrolytic enzymes are produced
de novo under the action of GA.
      When isolated aleurone layers of barley are incubated in a solu-
tion containing GA, they produce and secrete several hydrolytic
enzymes. GA-dependent de novo synthesis has been demonstrated for
α-amylase (13), protease (14), and β-1,3-glucanase and ribonuclease
(15). In addition, a GA-dependent release of ribonuclease and β-1,3-
glucanase has been demonstrated. The increase in activity of at
least one hydrolase, β-amylase, in the presence of GA is due to release of preformed enzyme and not to de novo synthesis.
      Evidence that the increase in activity of at least four hydro-
lases induced by GA actually occurs as a result of de novo synthesis,
rather than by activation of preformed enzyme, has been obtained in
various ways. However, the most unequivocal proof comes from density
labeling experiments, first performed by Filner and Varner in 1967
(13). Later Jacobsen and Varner (14) proved by the same procedures
that protease also is synthesized de novo in response to GA. And
Bennett and Chrispeels (15) proved, using D_2O, GA-induced de novo
synthesis of ribonuclease and β-1,3-glucanase in barley aleurone
cells.
      Induction of de novo synthesis of α-amylase by GA in isolated
aleurone layers is evident after a lag period of approximately 8 hr
following administration of the hormone. In keeping with hormone
responses generally, GA must be present continuously if the de novo
synthesis of hydrolases is to be sustained. Synthesis of new RNA is
essential to the GA-induction of de novo synthesis of hydrolases.
Actinomycin D, an inhibitor of RNA synthesis, inhibits the synthesis
and release of α-amylase if the inhibitor is presented during the
first 7 to 8 hr after treatment. Inhibitors of protein synthesis,
such as cycloheximide, also inhibit GA-induction of hydrolases. And,
interestingly, abscisic acid, a growth-inhibiting hormone, inhibits
GA-induced α-amylase synthesis as well.

A close look at the events which occur during the lag period in GA induction of de novo synthesis of new enzymes has provided some important clues as to whether GA acts at the transcriptional or translational level (16 and papers cited therein). An increase in polyribosome formation and an increased synthesis of ribosomes and endoplasmic reticulum membranes were found. All of these effects begin within 2 to 4 hr after application of GA. Their observations led Evins and Varner (16) to conclude that the GA-stimulated increases in the number of monoribosomes and the percentage of polyribosomes probably are prerequisite for the hormone induction of protein synthesis.

Higgins et al. (17) provided a quite direct link between GA-stimulated de novo enzyme synthesis and appearance of the complementary mRNA. They demonstrated convincingly that the level of translatable α-amylase mRNA increased in GA-treated tissue in parallel with the increased rate of enzyme synthesis. These results provide still more evidence that GA acts to induce selective mRNA and de novo enzyme synthesis in aleurone cells.

On the basis of collective data now available, it may confidently be concluded that in the barley aleurone system, GA evokes selectively the synthesis of particular molecular species of mRNA which in turn leads to de novo synthesis of certain hydrolytic enzymes. Or, as Jones and MacMillan (11) put it, "There is compelling evidence that GA3 affects the synthesis of hydrolytic enzymes in cereal aleurone and that it does so by controlling the transcription and possibly translation of new mRNAs." Left unanswered unequivocally at this stage of our knowledge are two important questions. One is whether the action of GA just described alone accounts exclusively for the fundamental mechanism of action of GA in the aleurone system. The other question is whether the action of GA in stimulating de novo synthesis of particular enzymes as described for the barley aleurone system is universal. That is, is the mechanism of GA action in the barley aleurone system directly indicative of the mechanism by which GA hormones participate generally in the regulation of the growth and development of higher plants? Obviously, the answers to these questions must await additional research.

Cytokinins

Natural cytokinins occur as free bases and as the ribonucleosides and ribonucleotides of N^6-substituted adenine derivatives and also as constituents of particular molecular species of tRNA (18).

Cytokinins are unique among plant hormones in that adenine compounds of identical structure occur in nucleic acids. More specifically, particular cytokinin-active ribonucleosides occur as components of certain molecular species of tRNA. Zachau et al. (19), during the determination of the base sequences of serine tRNA in yeast, first reported an "odd" base immediately adjacent to the 3' end of the anticodon. In collaboration with Biemann et al. (20), this "odd" base was identified as the natural cytokinin isopentenyladenosine, which is one of the most highly active cytokinins known. In fact, in all cases, where the specific location of a cytokinin moiety in tRNA has been determined, it invariably is immediately adjacent to the 3' end of the anticodon. And, interestingly, all tRNA species which are known to contain a cytokinin recognize codons with the initial base as uracil (U), although not all tRNA species recognizing codons beginning with U contain a cytokinin. The cytokinin-containing tRNA species are tRNA^{Ser}, tRNA^{Phe}, tRNA^{Cys}, tRNA^{Trp}, tRNA^{Leu}, and tRNA^{Tyr}. In tRNA from microbial sources, cytokinin-active ribonucleosides appear to be present in most, if not all, of the tRNA species that recognize codons beginning with U. But in higher organisms, the distribution appears to be more restricted. In the cases of wheat germ and etiolated bean seedlings, for examples, cytokinin-active ribonucleosides are limited to only tRNA^{Ser} and tRNA^{Leu}.

An important question is: Are the cytokinins present in tRNA incorporated intact into tRNA, in a manner like all other nucleoside triphosphates that are polymerized in the synthesis of tRNA, or does the presence of the cytokinins result from alkylation (transfer of isoprenoid side chain) of existing adenosine moieties already present in preformed tRNA? In fact, experimental evidence has been reported for both processes (21,22). But the amounts of actual incorporation were very small, and the observed incorporation may be merely the result of transcriptional errors. Meanwhile, the evidence for cytokinin moieties in tRNA arising by alkylation of specific adenosine residues in preformed tRNA is conclusive (23).

There is good evidence that the cytokinin moieties in tRNA are functionally significant and that they do affect the behavior of those tRNA molecules in the process of protein synthesis. One of the earliest and most conclusive investigations from which this important fact emerged was conducted by Gefter and Russell (24) with Escherichia coli.

Gefter and Russell (24) studied three forms of tRNA^{Tyr} which had the same nucleotide sequence and differed only in the degree of modification of the adenosine residue adjacent to the 3' end of the anticodon: (1) unmodified adenosine (A), (2) N^6-(Δ^2-isopentenyl)adenosine (i^6A), and (3) the methyl-thio derivative of i^6A (ms^2i^6A). All three forms of the tRNA were then tested for tyrosine-acceptor activity and for binding of the tyrosyl-tRNA to an mRNA-ribosome complex. No significant differences were found among the three forms of tRNA^{Tyr} as regards amino acid acceptor capacity. However, the tRNA^{Tyr} containing an unmodified adenosine residue adjacent to the anticodon was markedly less effective, in the in vitro experiments, in the binding of the aminoacyl-tRNA^{Tyr} to the mRNA-ribosome complex than tRNA^{Tyr} containing i^6A or ms^2i^6A. Thus, in this important case, a cytokinin moiety evidently must be present adjacent to the anticodon of tRNA^{Tyr}, if this tRNA species is to function effectively in protein synthesis. By whatever mechanism the regulation is achieved, it thus appears that cytokinin-active bases in certain types of tRNA may have a modulating effect on protein synthesis at the translational level.

There are good reasons presently for believing also that free cytokinins have important biological activity independently of any direct association with tRNA: (1) there are results showing direct effects of exogenous cytokinin independently of any apparent incorporation into tRNA; (2) ethanolic extracts of corn kernels contain the trans-isomer of zeatin, while the tRNA hydrolysates of corn kernels contain the cis-isomer, suggesting that zeatin is not a precursor in the synthesis of tRNA in that material; (3) dihydrozeatin

is a major free cytokinin in beans, but it apparently does not occur
in bean tRNA; (4) even tissues which contain potent cytokinins in
their RNA still require exogenous cytokinin for growth in vitro; and
(5) the data on direct incorporation of exogenous cytokinins into
tRNA, as previously mentioned, do not show conclusively that this
process is anything more than transcriptional error.

A prevalent idea about mechanisms of action of plant hormones,
cytokinin free bases and others, is that the hormone first binds--by
weak hydrogen or ionic bonds--to some receptor. The receptor commonly is envisaged to be an allosteric protein which, as a consequence
of binding the hormone, can, as the altered receptor or as a recep-
tor-hormone complex, evoke hormonal action (see 25).

Since about 1970, progress along these lines with respect to
cytokinins has been made by Fox and associates. A basic observation
that Fox and Erion (26) made was that cytokinins would bind, with
rather high specificity, to ribosomes isolated from higher plants.
Later, Fox and Erion (27) described actually three cytokinin-binding
proteins isolated from ribosomal preparations from wheat germ. Of
particular interest was a fraction of medium molecular weight of
93,000 (CBF-1) with high binding affinity for cytokinins. A large
number of cytokinins, cytokinin nucleotides, analogs, related purines,
and other plant growth substances were tested for their ability to
compete with 6-benzylaminopurine (or N^6-benzyladenine) for binding
sites on CBF-1. Apparently, wheat germ ribosomes contain one copy of
CBF-1 per ribosome. Competition studies showed a high degree of
specificity for cytokinin-active moieties. More recently, a very
high-affinity binding factor has been found associated with a mitochondrial fraction from mung bean (Phaseolus aureus) seedlings (28).
Unfortunately, it has not been possible yet to assign any biological
function to the cytokinin binding described above, and these binding
factors cannot yet be considered to be true cytokinin receptors.

Abscisic Acid

The mechanism of action of abscisic acid (ABA) has been studied to
the greatest extent in the barley aleurone system (29), in which ABA
counteracts the effect of GA in the induction of hydrolases. This
action of ABA has largely been the basis for speculating that ABA
may act specifically to inhibit, by some unknown mechanism, DNA-
dependent RNA synthesis. Much evidence indicates that ABA acts at
the transcriptional level, but it also has been proposed that the
inhibition of induction of α-amylase synthesis is caused, at least in
part, by an effect on translation because ABA still inhibited the
formation of α-amylase at 12 hr when cordycepin (an inhibitor of RNA
synthesis) no longer had an effect (30).

Some other recent work indicates an apparent qualitative effect
on protein synthesis. Ho (31) and Jacobsen et al. (32) detected new
peptides when aleurone layers were treated with ABA. These peptides
turned over rapidly, and the quantities formed were reduced by inhibitors of transcription and translation. Little else is known
currently about the peptides.

Some physiological effects of ABA are correlated with alteration
of plasma membranes. This effect is manifested in a change in bio-
electric potential across the membranes and a leakiness and efflux
of K^+, which is involved in some actions of ABA, e.g., stomatal

regulation. At this time the effects on nucleic acid and protein
synthesis cannot be reconciled with these effects involving changes
in membrane permeability.

Ethylene

Ethylene, C_2H_4, has many diverse effects on plant growth and development. Yet the mechanism of action of this hormone is not understood.
A major obstacle to discovering its mechanism of action has been the
lack of an isolated subcellular system that responds to ethylene in
a way that clearly reflects its action in vivo (33).
     Fruit tissues respond to ethylene by exhibiting increases in the
activities of enzymes that catalyze ripening reactions, and in some
cases, the increases in enzyme activity probably are the result of
de novo synthesis, rather than activation of preexisting enzymes.
Other target tissues respond similarly to ethylene. But it is not
known whether ethylene acts directly to evoke new enzyme production.
Interpretation of results with inhibitors of RNA and protein synthesis is inconclusive, because it could be merely that RNA and protein
synthesis are essential to maintain the cells in a state competent
to respond to ethylene. Moreover, there are some responses to ethylene, besides fruit ripening, which occur under conditions which
apparently do not directly involve RNA and protein synthesis (e.g.,
membrane permeability changes). It has been proposed that the in
vivo ethylene receptor site contains a metal such as copper (34,35).

Anthesins

Within less than two decades after the discovery of photoperiodism in
1920, the hypothesis developed that one or more specific flowering
hormones are responsible for floral initiation (23,36-39).
Chailakhyan (36) coined the word "florigen" for the hypothetical
flowering hormone. A short time later, Melchers (40) suggested the
term "vernalin" for the hypothetical stimulus thought to develop during vernalization of cold-requiring plants. Many years later,
Chailakhyan proposed the term "anthesins" for hypothetical flowering
hormones (38) and gave emphasis to the role GAs play in flowering of
many plants.
     Early experiments, conducted in the 1930s and 1940s, with
classical obligate short-day plants such as cocklebur (Xanthium
strumarium) yielded results which provided a very logical basis for
the florigen (anthesin) concept. Experiments involving a large
number of photoperiodically sensitive species suggested strongly that
the leaf is the organ which perceives the photoperiodic stimulus,
that phytochrome is the photoreceptor, and that a substance or substances is formed in the leaves of short-day plants which is translocated to vegetative meristems and causes their morphogenetic transformation into floral meristems. From numerous grafting experiments
between two species (e.g., Kalanchoë blossfeldiana, SDP, and Sedum
spectabile, LDP), it appeared that there was only one florigen, or
if two or more substances, they evidently were physiologically equivalent among many species. The various photoperiodic response types
would differ not in their requirement for florigen, but in the environmental requirements for florigen production.
     With such evidence compatible with a florigen concept, a logical

next step would be to attempt to isolate and chemically characterize
the flowering stimulus or stimuli. In fact, there have been many
such efforts. Generally the attempts have ended in failure; occasionally there have been modest successes (41,42). Thus, there is
much circumstantial evidence that flower initiation is controlled,
at least in part, by anthesins. Meanwhile, certain of the known
hormones definitely are involved in the regulation of flowering, as
indicated by numerous investigations conducted with GAs since circa
1957 and with ABA since circa 1967.

Lang (43) first reported in 1957 that exogenous GA could cause
flowering in numerous species of long-day and vernalization-requiring
plants under noninductive environmental conditions. However, GA is
not the long sought after florigen. One reason is that GAs do not
cause flowering of short-day plants under noninductive conditions, or
even in all long-day plants. Moreover, there is now good evidence
that flowering and flower-bearing stem elongation (bolting) are separate processes in plants such as Silene armeria (44), with GA directly promoting stem elongation only. While GA is not florigen, it con-
ceivably might be vernalin.

During the early investigations of the effects of exogenous ABA
on plants, it was reported that this growth-inhibiting hormone could
cause flowering of certain short-day plants (e.g., Chenopodium
rubrum, Pharbitis nil, and some others) under long-day conditions
(23). But ABA is not a florigen because there is no effect of exogenous ABA on flowering of some short-day plants or on long-day
plants. It may be one of the postulated flowering inhibitors, which
have long been thought to be produced in the leaves of long-day
plants subjected to noninductive short-day conditions.

Literature Cited

1. Paleg, L. G. Ann. Rev. Plant Physiol. 1965, 16, 291-322.
2. Bandurski, R. S.; Nonhebel, H. M. In "Advanced Plant Physiol-
    ogy"; Wilkins, M. B., Ed.; Pitman: Marshfield, MA, 1984; pp.
    1-20.
3. Wilkins, M. B. (Ed.) "Advanced Plant Physiology"; Pitman:
    Marshfield, MA, 1984; 514 pp.
4. Rayle, D. L.; Cleland, R. Plant Physiol. 1970, 46, 250-3.
5. Hager, A.; Menzel, H.; Krauss, A. Planta 1971, 100, 47-75.
6. Silberger, J.; Skoog, F. Science 1953, 118, 443-4.
7. Key, J. L. Ann. Rev. Plant Physiol. 1969, 20, 449-74.
8. Zurfluh, L. L.; Guilfoyle, T. J. Plant Physiol. 1982, 69,
    332-7.
9. Theologis, A.; Ray, P. M. In "Plant Growth Substances 1982";
    Wareing, P. F., Ed.; Academic: London, 1982; pp. 43-57.
10. Vanderhoef, L. N.; Dute, R. R. Plant Physiol. 1981, 67, 146-9.
11. Jones, R. L.; MacMillan, J. In "Advanced Plant Physiology";
    Wilkins, M. B., Ed.; Pitman: Marshfield, MA, 1984; pp. 21-52.
12. Radley, M. Planta 1967, 75, 164-71.
13. Filner, P.; Varner, J. E. Proc. Natl. Acad. Sci. U.S.A.
    1967, 58, 1520-6.
14. Jacobsen, J. V.; Varner, J. E. Plant Physiol. 1967, 42, 1596-
    1600.
15. Bennett, P. A.; Chrispeels, M. J. Plant Physiol. 1972, 49,
    445-7.

16. Evins, W. H.; Varner, J. E. Plant Physiol. 1972, 49, 348-52.
17. Higgins, T. J. V.; Zwar, J. A.; Jacobsen, J. V. Nature 1976, 260, 166-9.
18. Horgan, R. In "Advanced Plant Physiology"; Wilkins, M. B., Ed.; Pitman: Marshfield, MA, 1984; pp. 53-75.
19. Zachau, H. G.; Dütting, D.; Feldmann, H. Hoppe-Seylers Zeit. Physiol. Chem. 1966, 347, 212-35.
20. Beimann, K.; Tsunakawa, S.; Sonnenbichler, J.; Feldmann, H.; Dütting, D.; Zachau, H. G. Angew. Chem. Int. Edit. 1966, 5, 590-1.
21. Fox, J. E. Plant Physiol. 1966, 41, 75-82.
22. Armstrong, D. J.; Murai, N.; Taller, B. J.; Skoog, F. Plant Physiol. 1976, 57, 15-22.
23. Moore, T. C. "Biochemistry and Physiology of Plant Hormones"; Springer-Verlag: New York, Heidelberg, Berlin, 1979; 274 pp.
24. Gefter, M. L.; Russell, R. L. J. Mol. Biol. 1969, 39, 145-57.
25. Kende, H.; Gardner, G. Ann. Rev. Plant Physiol. 1976, 27, 267-90.
26. Fox, J. E.; Erion, J. L. Biochem. Biophys. Res. Commun. 1975, 64, 694-700.
27. Fox, J. E.; Erion, J. L. In "Plant Growth Regulation"; Pilet, P. E., Ed.; Springer-Verlag: New York, 1977; pp. 139-46.
28. Keim, P.; Erion, J.; Fox, J. E. In "Metabolism and Molecular Activities of Cytokinins"; Guern, J.; Péaud-Lenoël, C., Eds.; Springer-Verlag: Berlin, 1981; pp. 179-90.
29. Milborrow, B. V. In "Advanced Plant Physiology"; Wilkins, M. B., Ed.; Pitman: Marshfield, MA, 1984; pp. 76-110.
30. Varner, J. E.; Ho, D. T. H. In "The Molecular Biology of Hormone Action"; Papaconstantinou, J., Ed.; Academic: New York, 1976; pp. 173-94.
31. Ho, D. T. H. Plant Physiol. (Supplement) 1979, 63, 79.
32. Jacobsen, J. V.; Higgins, T. J. V.; Zwar, J. A. In "The Plant Seed--Development, Preservation and Germination"; Rubenstein, J., Ed.; Academic: New York, 1979; pp. 241-62.
33. Beyer, E. M., Jr.; Morgan, P. W.; Yang, S. F. In "Advanced Plant Physiology"; Wilkins, M. B., Ed.; Pitman: Marshfield, MA, 1984; pp. 111-26.
34. Burg, S. P.; Burg, E. A. Science 1965, 148, 1190-6.
35. Burg, S. P.; Burg, E. A. Plant Physiol. 1967, 42, 144-52.
36. Chailakhyan, M. Kh. Comptes Rendus (Doklady) Acad. Sci. U.S.S.R. 1936, 4, 79-83.
37. Chailakhyan, M. Kh. Ann. Rev. Plant Physiol. 1968, 19, 1-36.
38. Chailakhyan, M. Kh. Plant Sci. Bull. 1970, 3, 1-7.
39. Chailakhyan, M. Kh. Bot. Rev. 1975, 41, 1-29.
40. Melchers, G. Ber. Dtsch. Bot. Ges. 1939, 57, 29-48.
41. Lincoln, R. G.; Mayfield, D. L.; Cunningham, A. Science 1961, 133, 756.
42. Hodson, H. K.; Hamner, K. C. Science 1970, 167, 384-5.
43. Lang, A. Proc. Natl. Acad. Sci. U.S.A. 1957, 43, 709-17.
44. Cleland, C. F.; Zeevaart, J. A. D. Plant Physiol. 1970, 46, 392-400.

RECEIVED November 15, 1984

Plant Bioregulators: Overview, Use, and Development

JOHANNES JUNG

BASF Agricultural Research Centre, D-6703 Limburghof, Federal Republic of Germany

The action of most exogenous plant bioregulators (PBRs) consists of interference with the plant's
hormone system. Accordingly, the action of these
substances may be related to the five phytohormone
groups, known up to now, as homologs, synergists,
antagonists, or inhibitors of hormone biosynthesis.
Beside this group of PBRs there are, however, other
important compounds which have not so far been shown
to have a definite relationship to a phytohormone.
    On this basis, an overview of substances with
bioregulatory effects is presented.
    The specific influence of exogenous PBRs on
crop plants has in some cases already been integrated into the crop production system. Examples
of this are the use of chlormequat chloride in cereal growing, ethephon for influencing the development and maturity of various crops, and mepiquat
chloride in cotton.
    The development of PBRs is discussed under
the aspect of relevant concepts for the synthesis
and screening of new compounds and other factors
involved.

This state-of-the-art contribution on plant bioregulators (PBRs) is
directed to the following four questions:

1. What are plant bioregulators and what is plant bioregulation?
2. What substances and what principle of action are available?
3. How have plant bioregulators been used so far in crop production
   and where are they employed in particular?
4. What factors will have a major influence on determining the development of new plant bioregulators and on opening up further
   possibilities for use in crop production?

0097–6156/85/0276–0095$06.00/0
© 1985 American Chemical Society

What are plant bioregulators and what is plant bioregulation?

It is necessary to define the term "plant bioregulators" in order
to provide a general characterization of their properties and mode of
action, and to distinguish them from other agrochemicals. An endogenous substance may be considered to be a bioregulator if, at a low
concentration (for example below 1 mM) and without having a biocidal
effect, it exercises an influence on the growth, development, and
composition of plants, without being a nutrient.

This term covers a broader spectrum of effects on plants than
the term "plant growth regulators" that has been commonly used in
English-speaking countries. This new definition should do greater
justice to the variety of effects that are expected from this class
of substances. These include not only an influence on the growth and
development processes of crop plants or their specific organs, but
also the modification of metabolic processes or the formation of certain constituents, as well as a modified stress behavior.

Another characteristic of a bioregulator is that the modifications that it produces must not affect the genome, i.e., its action
must be of a temporary nature.

Naturally, phytohormones have a special position among plant
bioregulators, because they are lead compounds for regulatory functions in the plant system. Since, however, these endogenous substances have already been dealt with by Thomas C. Moore in the previous chapter, they are only touched on as far as interactions with
exogenous bioregulators and the characterization of their effects
are concerned. Namely, an important guideline, which continues to
substantially govern the search for new synthetic bioregulators, is
influencing the plant's hormone status. This can be achieved not only
by an effect of the applied exogenous compound analogous to that of
the particular phytohormone, but also by synergistic and antagonistic
effects, or by promoting or inhibiting hormonal biosynthesis by means
of exogenous substances.

A classification related to the "phytohormonal interaction principle" is also appropriate in a systematic overview of various groups
of exogenous PBRs.

What substances and what principles of action are available?

Table 1 shows a list and classification of the range of exogenous bioregulators presently available, drawn up on the basis of the
aspects already outlined. Even if there is a substantial limitation
to groups of active compounds rather than to individual substances,
the overview given cannot be considered complete. It does, however,
contain the majority of synthetic bioregulator groups that are already in use or are in the developmental stage.

Of the analogous compounds related to a particular hormone, attention must be drawn to the large group of synthetic auxins, to the
synthetic cytokinins (- recently not only adenine, but also urea derivatives -) and to the ethylene generators (1, 2). An extension of
the range of analogous compounds is indicated in the case of abscisic
acid, as well (3).

The groups of substances that exhibit antagonistic behavior toward a phytohormone or inhibit its biosynthesis are experiencing

intense and expanding diversification. Here, particularly the group
of gibberellin antagonists or inhibitors (anti-gibberellins) should
be emphasized. This group represents the so-called growth retardants,
some of which have attained considerable importance in practice. It
is surprising, but characteristic of the plant bioregulators presently used in crop production, that most of the beneficial effects
result from growth retardation rather than from growth stimulation
(4). For this reason, this group of PBRs is discussed in more detail.

Of the substances that interfere with gibberellin biosynthesis,
mention should first be made of the so-called onium compounds, which
are substances with a charged central atom (5, 6). They include the
already extensively employed bioregulators chlormequat chloride (CCC)
and mepiquat chloride (DPC) (Figure 1).

Other substances with an anti-gibberellin action are found in
the groups of pyrimidines (ancymidol), norbornenodiazetines (tetcyclacis) and triazoles (paclobutrazol), etc. (Figure 2). It is an interesting fact that some of these substances originate from groups
that have produced potential fungicides.

The retardation effect of all these substances can be reversed
by gibberellins. Furthermore, it has been possible, in the case of
some of these substances, to detect precisely the point of attack or
site of action in the gibberellin biosynthesis sequence. As can be
seen from Figure 3, it is assumed that the onium compounds inhibit
the cyclization of geranylgeranyl pyrophosphate to copalyl pyrophosphate (7), whereas it has been demonstrated in cell-free systems that
pyrimidines, norbornenodiazetines, and triazoles inhibit the sequential oxidation of ent-kaurene to ent-kaurenoic acid (8, 9).

Morphologically this biochemical process manifests itself in the
"anti-gibberellin habitus" of the treated plants that corresponds to
compact growth with shortened internodes and to a more intensive
color of leaves. However, it is also worth mentioning the changes in
the shoot-root ratio that lead to a pronounced shift in favor of root
growth, especially after treatment with norbornenodiazetines (10).

The induced inhibition of longitudinal growth by tetcyclacis
results both from inhibited cell elongation and from reduced cell
division. The relative proportions to which cell elongation and cell
division contribute to the shortening effect varied in trials with
maize, sunflowers, and soybeans, i.e., the influence effected via
cell division increased with an increase in concentration of the PBR.
The inhibitory effect on cell division detected in intact plants has
been confirmed in cell suspension cultures of the same plant species
(11).

After these detailed remarks about anti-gibberellins and their
mode of action, it should be emphasized that the action of not all
bioregulatory active substances can be assigned so precisely to an
interaction with a specific phytohormone. This partially applies to
daminozide (succinic acid 2,2-dimethylhydrazide), which is frequently
classified as a retardant (7).

An overview of some representatives of this heterogeneous group
is given in Table II. The first group again includes substances with
an inhibitory action, which may also be used in some cases for sucker
control and for pinching (12). Attention should also be drawn to the
substances that increase the sugar content of sugarcane (13), especially those of the glycine type.

Table I. Overview of Exogenous Plant Bioregulators Classified
according to Their Phytohormonal Interaction (Abstract)

Reference phytohormones	Compounds with homologous or synergistic activity	Compounds with an antagonistic activity or inhibitors of biosynthesis
Auxins	Synthetic Auxins	Triiodobenzoic acid Hydroxyfluorene carboxylates
Gibberellins	Phthalimides Steroids Steviol	Onium compounds (N, S, P) Pyrimidines Norbornenodiazetines Triazoles
Cytokinins	Benzyl- and Furfuryl-aminopurine Phenylurea derivatives	Pyrrolo- and Pyrazolepyrimidines
Abscisins	Terpenoic analogues of ABA Farnesol	
Ethylene	Chloroethyl phosphonic acid Aminocyclopropane carboxylic acid	Aminoethoxivinyl glycine

Table II. Survey on Synthetic PBR's with Different Modes of
Action (Examples)

effect	compound
Retardation of growth	
Dwarfing	Succinic acid 2,2-dimethyl hydrazide (daminozide) Trifluormethanesulfonanilides 6-Azauracil Maleic hydrazide
Sucker control	Maleic hydrazide higher alcohols and fatty acids (C_8-C_{14})
Pinching	Dikegulac-Na
Stimulation of growth	4-Hydroxybenzoic acid (phthalimides, steroids)
Enhancement of sucrose content (sugarcane)	Phosphonomethylglycine (glyphosine, glyphosate)
Enhancement of isoprenoids (Guayule)	2-Diethylamino-ethyl-3,4-dichlorophenylether
Defoliation (cotton)	N-phenyl-N'-1,2,3-thiadiazol-5-yl urea S,S,S-tributylphosphorotrithioate
Pollen suppression	1-(p-chlorophenyl)1,4-dihydro-6-methyl-4-oxopyridazine-3-carboxylate

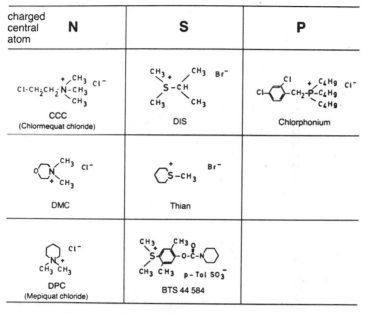

Figure 1.   Growth retardants: Onium compounds.
CCC   =   N-2-chloroethyl-trimethylammonium chloride
DMC   =   N-dimethylmorpholinium chloride
DPC   =   N-dimethylpiperidinium chloride
DIS   =   S-dimethyl-isopropylsulfonium bromide
Thian   =   S-methylthianium bromide
BTS 44 584   =   S-2,5-dimethyl-4-pentamethylenecarbamoyl-
                 oxyphenyl-S,S-dimethylsulfonium p-toluene
                 sulphonate
Chlorphonium =   P-tributyl-2,4-dichlorbenzylphosphonium
                 chloride

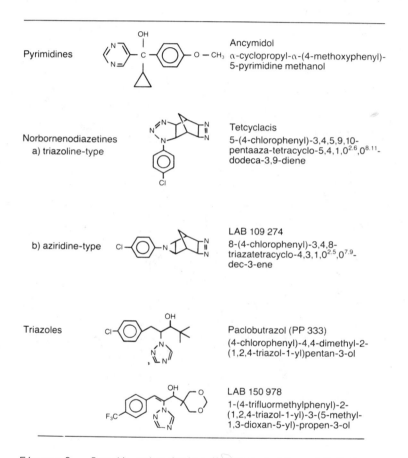

Pyrimidines

Ancymidol
α-cyclopropyl-α-(4-methoxyphenyl)-
5-pyrimidine methanol

Norbornenodiazetines
a) triazoline-type

Tetcyclacis
5-(4-chlorophenyl)-3,4,5,9,10-
pentaaza-tetracyclo-5,4,1,0$^{2.6}$,0$^{8.11}$-
dodeca-3,9-diene

b) aziridine-type

LAB 109 274
8-(4-chlorophenyl)-3,4,8-
triazatetracyclo-4,3,1,0$^{2.5}$,0$^{7.9}$-
dec-3-ene

Triazoles

Paclobutrazol (PP 333)
(4-chlorophenyl)-4,4-dimethyl-2-
(1,2,4-triazol-1-yl)pentan-3-ol

LAB 150 978
1-(4-trifluormethylphenyl)-2-
(1,2,4-triazol-1-yl)-3-(5-methyl-
1,3-dioxan-5-yl)-propen-3-ol

Figure 2.   Growth retardants: Kaurene oxidase inhibitors.

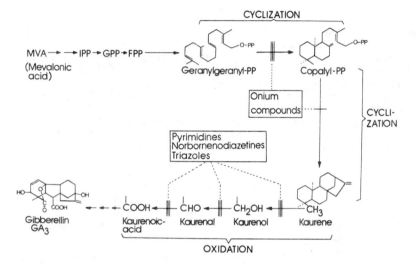

Figure 3.   Interference of growth retardants with gibberellin biosynthesis.

In addition to promoting the production of rubber (latex) by ethephon, which is already being practised (14), increasing the isoprene content of other plants by different active substances should also be mentioned (15), as well as the trials for enhancing the quality of vegetable oils by treatment with pyridazinones (16).

In the transition area to a biocidal effect, substances that have a defoliating action and those that suppress pollen formation should also be mentioned (17).

How have plant bioregulators been used so far in crop production and where are they employed in particular?

Among the various groups of agrochemicals, bioregulators are the youngest class of products following fertilizers, fungicides, insecticides, and herbicides. For this reason, they obviously also come last in volumes of sales. With worldwide sales of crop protection chemicals at approximately US $ 14.3 billion, the proportion of bioregulators is between 3 and 4%, i.e., an average of 3.5%, which is roughly equal to $ 500 million. This consumption, contrary to originally higher expectations, has been increasing at an annual rate that is on a level with the real average growth in crop protection chemicals namely 4 - 6% per annum. In Western Europe, the market proportion of bioregulators to crop protection chemical sales of US $ 3 billion is estimated to be 5.5%, i.e., US $ 165 million.

This relatively high percentage of bioregulators in West European agriculture compared with the world average is probably primarily due to the fact that this group of products has gained greater access to main agricultural crops, particularly in the cultivation of cereals, than elsewhere (18, 19). In West Germany, for example, 60 - 70% of the wheat is treated with chlormequat chloride (CCC) as an antilodging agent. In barley, too, a significant proportion of the acreage is treated with Terpal, a combination of mepiquat chloride and ethephon, or with ethephon on its own. The bioregulators mentioned have been fully integrated into the crop production concept and are coordinated with the use of other agrochemicals.

This report on the situation in European agriculture stands in obvious contrast to the present level of use of PBRs in the United States, which of course has a somewhat different main crop spectrum. In this connection, the following comment (20) can be quoted: "Scientists have long recognized the potential market that exists for a plant growth regulator on large acreage economic crops. Despite years of research by both industry and government scientists there are only a few PBRs registered for use on the major crops today. Of the many current uses of PBRs, effect on yield are often indirect ... Studies with economic crops have frequently identified compounds capable of altering inidvidual characteristics such as lodging, plant height, seed number, maturity etc.; however, these changes have not always resulted in improved yield or quality." Nevertheless, the optimistic conclusion is drawn that, "without a doubt, innovative growers are anxious to utilize PBR technology in increasing profitability of agronomic crops".

With reference to crop production, the following objectives for the development and use of PBRs can be listed (19):

Aims for Plant Bioregulators

Objectives	Possible results
Sequence and duration of development steps during vegetative and reproductive growth	Enhanced establishment and biomass production
Germination Tillering Shooting Flowering Fruit development	Increased or improved formation of yield-determining organs
Anatomical and morphological structure	Enhanced "harvest index" Improved standing ability Improved light interception Improved shoot/root ratio
Behavior towards drought, low temperature and diseases	Enhanced resistance to environmental stress and infections
Photosynthesis and metabolism	Increased photoproductivity
	Enhanced content of yield- and quality-determining ingredients
Senescence, ripening, abscission, and dormancy	Control of ripening processes Promotion of mechanical harvesting
	Influence on post-harvest behavior and shelf life

To deal with the question of what are the optimum R + D concepts and procedures for developing new PBRs, the literature section of this paper lists contributions from seven associates of companies engaged in PBR research, namely Beyer/Dupont (28); Geissbühler et al./Ciba Geigy (29); Lürssen/Bayer (30); Lever/ICI (31); Nickell/Velsicol (32); Raven/Duphar (33) and Sacher/Monsanto (34).

I would like to quote the last of these authors concerning the strategies to discover plant bioregulators for agronomic crops: "When attempting to evaluate the overall difficulty in discovering a plant growth regulator it becomes readily apparent that the task is substantially more demanding than the one involved in search for a new herbicide. In fact, the overall complexity is estimated to be 10 to 100 times larger. The key factor involved in the discovery process of PBRs is the reorientation of thinking away from classical screens for herbicides, fungicides and insecticides, towards the non-conventional approach."

This prospective statement can be supported by the actual use of mepiquat chloride (Pix) in cotton - as an example of the introduction of a bioregulator in a main agricultural crop. Certain possibilities, at least in some fields, are emerging in the form of seed soaking in the important tropical cereal crop rice (21, 22).

It is more difficult at present to make a prediction for maize and
soybeans as main crops, although the latter crop is being intensively
investigated for the use of PBRs (23).

Of the synthetic PBRs favored up to now, reference should also
be made, after chlormequat and mepiquat chloride, to ethephon from
the group of ethylene generators, a substance with a particularly
wide range of uses, including the following: induction of flowering,
hastening of maturity, stimulation of latex flow, promotion of leaf
curing (tobacco), fruit degreening (citrus), and prevention of lodging (small grains) (4, 17).

The progress in research on ethylene biosynthesis and the identification of 1-aminocyclopropane-1-carboxylic acid (ACC) as an ethylene precursor have stimulated the search for new active substances
in the group of both ethylene generators and inhibitors of ethylene
biosynthesis (24). ACC itself is basically an interesting synthetical-
ly accessible bioregulator. However, there seems to be an unreliabil-
ity factor in use due to its ready conjugation and metabolization as
an endogenous compound. Nevertheless, it may be assumed that this
substance will provide further incentives for extending the ethylene-
related spectrum of active compounds.

What factors will have a major influence on determining the
development of new plant bioregulators and on opening up further
possibilities for use in crop production?

It will first have to be established, in more general terms, that
contributions are necessary in reasonable proportions both from the
academic and from the industrial side, for the development of individual bioregulators. Chlormequat chloride may be cited as an example
worth following (25, 26, 27).

When developing projects in this field, one is naturally con-
fronted with the necessity of a concrete:

research and development (R + D) concept

If the quantitative aspect, i.e., the personnel and financial
resources, is initially left out of account, it is the basic approach
that must be given primary attention, i.e., the objectives and the
ways in which they can be realized.

The goal that is easiest to define, but hardest to achieve, is
an increase in yield. After several years of intensive R + D work it
is difficult - if not even impossible - to find bioregulators that
increase the yield directly, and that any such increase will have to
be achieved by influencing partial processes in a plant's growth and
development (4).

Innovation in the search for synthesis of active substances. In this
connection, "empirical approach" and "biorational design" are two
relevant concepts for bioregulators. Up to now, the empirical approach or mass screenings of chemicals have been the dominant and
most successful concept for discovering candidates with adequate activity, when a specific goal was recognized. This does not mean,
however, that the input of crop physiology, and thus the "biorational
design", could not lead to a substantial increase in efficiency in

the search for bioregulators. Basically, the need for a biorational
design arises logically from the diversity of structural variation.
However, bioregulator research is still in the initial phase as far
as biorational design is concerned. As Geissbühler (29) pointed out,
there is not sufficient basic information on the mode of action of
agrochemicals at the biochemical and molecular levels to be system-
atically employed for biorational improvements and modifications.

At present, there are the following possibilities and alternatives as guidelines for discovering a new PBR:

Guidelines for discovering a new PBR
a) Empirical approach
              Random chemistry
              Discovery of activity via mass screening
              Optimization and modification of a lead structure
              Discovery and evaluation of side-effects
b) Biorational design
              Determination and utilization of structure-
              activity relationships
              Modification of endogenous substances with phytohor-
              monal and metabolic activity

Efficiency of the screening systems. The qualitative and quantitative
efficiency of the screening system is of great importance for the
identification of new PBRs among a large number of individual chemicals.

The following characterization by C.W. Raven (33) is probably
typical of the present status of PBR screening techniques: "Selection
from and a mixture of a number of company philosophies, theories,
strategies, and experience forms the basis of the modern industrial
screening operation and make it a profession of its own!"
Briefly, the methods and techniques used may be grouped as follows:

Methods and techniques of PBR screening systems
- laboratory bioassays including biochemical enzyme assays, cell
   and tissue culture, detached plant organs, etc.;
- indoor biological evaluations on major crop species;
- advanced pot and model experiments (e.g. the Mitscherlich system);
- field trials.

The test systems that have been outlined here are being constantly modified and adapted to the particular tasks. In general,
however, they have already been developed to the extent that the
PBR effects that are of interest can be reliably realized with them.

The following additional determining factors in the development
and performance of PBRs should also be mentioned:
- the efficient projection of a PBR into crop management
   and
- the coordinated adaption to the state of development in plant
   breeding and to the problems with the particular plant species
   that have not yet been solved by it.

Particular emphasis should be placed on the last point, not
least because the view is often expressed that the genetic approach
to modifying crop characteristics may be superior to the use of PBRs

(30). This no doubt applies to many cases; on the other hand, there should be an interesting chance for chemical bioregulation precisely where the same modifications are possible and desirable over a broad spectrum of different varieties of a plant species or where genetically fixed negative correlations between certain properties can be compensated for by bioregulators.

Finally, the question of the expenditure that must be expected in developing a PBR nowadays should not be neglected. The overall costs for research, development, and registration are estimated to be in the order of $ 30 million. A figure of this order of magnitude naturally limits the number of companies interested in PBR research and development, and selects not only in terms of innovative capability but also in terms of available resources.

Acknowledgments

My thanks are given to Dr. C. Rentzea from BASF's Central Research Laboratory for the synthesis of the compound LAB 150 978 and his helpful co-operation and to Mr. B.E. Byrt for the translation of the manuscript.

Literature Cited

1. Moore, T.C. "Biochemistry and Physiology of Plant Hormones"; Springer-Verlag: New York, Heidelberg, Berlin, 1979.
2. Luckwill, L.G. "Growth Regulators in Crop Production"; Studies in Biology 129, Edward Arnold Publishers Ltd.: London, 1981.
3. Grossmann, K.; Jung, J. J. Agronomy & Crop Science 1984, 153, 14-22.
4. Morgan, P.W. Bot. Gaz. 1980, 141 (4), 337-46.
5. Zeeh, B.; König, K.-H.; Jung, J. Kemia (Helsinki), 1974, 1 (9), 621-3.
6. Sauter, H. In "Bioregulators: Chemistry and Use"; Ory, R.L.; Rittig, F.R., Eds.; ACS SYMPOSIUM SERIES No. 257, American Chemical Society: Washington, D.C., 1984; pp. 9-21.
7. Dicks, J.W. In "Recent Developments in the Use of Plant Growth Retardants"; Clifford, D.R.; Lenton, J.R., Eds.; British Plant Growth Regulator Working Group, MONOGRAPH 4, Wessex Press: Wantage, 1980; pp. 1-14.
8. Graebe, J.E. In "Plant Growth Substances"; Wareing, P.F., Ed.; Academic: London-New York, 1982; pp. 71-80.
9. Rademacher, W.; Jung, J.; Hildebrandt, E.; Graebe, J.E. In "Regulation des Phytohormongehaltes und seine Beeinflussung durch synthetische Wachstumsregulatoren"; Bangerth, F., Ed.; Ulmer-Verlag: Stuttgart, 1983; pp. 132-44.
10. Jung, J. In "Bioregulators: Chemistry and Use"; Ory, R.L.; Rittig, F.R., Eds.; ACS SYMPOSIUM SERIES No. 257, American Chemical Society: Washington, D.C., 1984, pp. 29-43.
11. Nitsche, K. Diploma dissertation, University of Heidelberg, West Germany, 1983.
12. Steffens, G.L. In "Plant Growth Regulating Chemicals"; Nickell, L.G., Ed.; CRC Press: Boca Raton/FL, 1983; Vol. I, pp. 71-88.
13. Nickell, L.G. In "Plant Growth Regulating Chemicals"; Nickell, L.G., Ed.; CRC Press: Boca Raton/FL, 1983; Vol. I, pp. 185-205.

14. Bridge, K. In "Plant Growth Regulating Chemicals"; Nickell, L.G., Ed.; CRC Press: Boca Raton/FL, 1983, Vol. I pp. 41-58.
15. Yokoyama, H.; Hsu, W.J.; Hayman, E.; Poling, S.M. In "Plant Growth Regulating Chemicals"; Nickell, L.G., Ed.; CRC Press: Boca Raton/FL, 1983; Vol. I, pp. 59-70.
16. St. John, J.B.; Christiansen, M.N.; Terlizzi, D.E. In "Bioregulators: Chemistry and Use"; Ory, R.L.; Rittig, F.R., Eds.; ACS SYMPOSIUM SERIES No. 257, American Chemical Society: Washington, D.C., 1984; pp. 65-73.
17. Nickell, L.G. "Plant Growth Regulators - Agricultural Uses"; Springer-Verlag: Berlin, Heidelberg, New York, 1982; pp. 15-18.
18. Jung, J. In "Plant Regulation and World Agriculture"; Scott. R.K., Ed.; Plenum Press: New York, London, 1979; pp. 279-307.
19. Jung, J.; Rademacher, W. In "Plant Growth Regulating Chemicals"; Nickell, L.G., Ed.; CRC Press: Boca Raton/FL, 1983; Vol. I, pp. 253-71.
20. Oplinger, E.S. Beltsville Symposium in Agricultural Research VIII, 1983 (in press).
21. Jung, J.; Koch, H.; Rieber, N.; Würzer, B. J. Agronomy & Crop Science 1980, 149 128-36.
22. Schott, P.; Knittel, H.; Klapproth, H. In "Bioregulators: Chemistry and Use"; Ory, R.L.; Rittig, F.R., Eds.; ACS SYMPOSIUM SERIES No. 257, American Chemical Society: Washington, D.C., 1984; pp. 45-63.
23. Stutte, C.A.; Davis, M.D. In "Plant Growth Regulating Chemicals"; Nickell, L.G., Ed.; CRC Press: Boca Raton/FL, 1983; Vol.II, pp. 99-112.
24. Lürssen, K. In "Chemical Manipulation of Crop Growth and Development"; McLaren, J.S., Ed; Butterworth Scientific: London, 1982; pp. 67-78.
25. Tolbert, E. Plant Physiol. 1960, 35, 380-5.
26. Linser, H.; Kühn, H. Z. Pflanzenern., Düng., Bodenk. 1962, 96, 231-47.
27. Jung, J.; Sturm, H. Landw. Forsch. 1964, 17, pp. 1-9.
28. Beyer, E.M. Proc. 8th Ann. Mtg. Plant Growth Soc. Amer., 1981, pp. 161-2.
29. Geissbühler, H.; Müller, U.; Pachlatter, J.P.; Waespe, H.R. In "Chemistry and World Food Supplies"; Shemilt, L.W., Ed.; Pergamon Press: Oxford, 1983; pp. 643-56.
30. Lürssen, K. In "Aspects and Prospects of Plant Growth Regulators"; Jeffcoat, B., Ed.; British Plant Growth Regulator Working Group, MONOGRAPH 6, Wessex Press: Wantage, 1981; pp. 241-9.
31. Lever, B.G. In "Opportunities for Chemical Plant Growth Regulation", Brit. Crop Prot. Council MONOGRAPH 21, 1978; pp. 17-24.
32. Nickell, L.G. In "Chemistry and World Food Supplies"; Shemilt, L.W., Ed.; Pergamon Press: Oxford, 1983; 601-6.
33. Raven, C.W. In "Aspects and Prospects of Plant Growth Regulators"; Jeffcoat, B., Ed.; British Plant Growth Regulator Working Group, MONOGRAPH 6, Wessex Press: Wantage, 1981; pp. 229-40.
34. Sacher, R.M. In "Chemical Manipulation of Crop Growth and Development"; McLaren, J.S., Ed.; Butterworth Scientific: London, 1982; pp. 13-15.

RECEIVED November 23, 1984

Effects of Allelopathic Chemicals on Crop Productivity

F. A. EINHELLIG

Department of Biology, University of South Dakota, Vermillion, SD 57069

Biochemical interactions among plants (allelopathy) result
from the activity of a diverse group of compounds synthesized by higher plants and microorganisms. Commonly
accepted representatives include scopoletin, ferulic
acid, p-hydroxybenzoic acid, catechin, amygdalin, patulin,
and juglone. Allelopathic regulation of plant growth and
development depends on the concentration, combination of
substances, edaphic and climatic factors, interaction with
other stresses, and species sensitivity. The source of
allelochemicals in agricultural fields may be the weeds,
crops, or microorganisms. Yields may be affected by (a)
the inhibitory or stimulatory effect of a crop on the
subsequent crop, (b) the capacity of crop plants to
inhibit weeds, and (c) production losses due to allelopathic weeds. Allelochemical interference with germination or growth of a crop can occur from direct effects
on metabolism, or indirectly through effects on nitrogen
fixation and other microorganism activity. Many physiological processes are altered by allelopathic chemicals,
but it has been difficult to determine the primary
mechanism involved for a specific compound. Both avoidance and application strategies may be employed to
utilize allelochemicals for improving crop production.
These include management of crop sequences, utilization
of allelopathic crop residues, breeding crops for weed
control, and development of allelochemicals as
herbicides.

Investigations over the last three decades have provided abundant
evidence that plants and animals often produce products that affect
the growth, development, distribution, and behavior of other organisms (1,2). Collectively, these natural substances are termed
allelochemicals, or allelochemics. They often impart plant resistance to insects, nematodes, and pathogens. Likewise, following their
release into the environment through volatilization, leaching, root
exudation, or tissue decomposition, some allelochemicals regulate the

distribution of plants and the vigor of plant growth. These interactions are the phenomenon of allelopathy, which includes all biochemical interrelationships among plants, both higher plants and
microorganisms. Allelopathic interference occurs in agroecosystems,
and it is one of the many factors that influence crop productivity
(3). The effects of such interactions may be either stimulatory or
inhibitory, but the major documentation of allelochemical effects in
agronomy has been that of growth inhibition. The purpose of this
paper is to discuss the major generalizations that can be made about
allelopathic interactions, provide examples of the role in agriculture, and focus on some considerations for the future. Only a few of
the salient investigations will be cited in illustrating major
principles.

Implicit in the concept of allelopathy is the recognition that
there are significant differences in (a) the capacity of species and
varieties to produce allelochemicals, (b) the sensitivities of
various plants to allelopathic compounds, and (c) plant responses
during the various stages of the life cycle. The producing and
receiving plants may be the same or different species. The receiving
plants may be growing concurrently with the producers, or they may be
found sequent to the producers. Crop plants most often contact
allelochemicals by the presence of these substances in the soil compartment. The schematic of Figure 1 illustrates some of the potential field interrelationships and complexities. In addition to these
aspects, as allelochemicals move through the physical environment
their quantity, residence time, and biological activity fluctuate
widely.

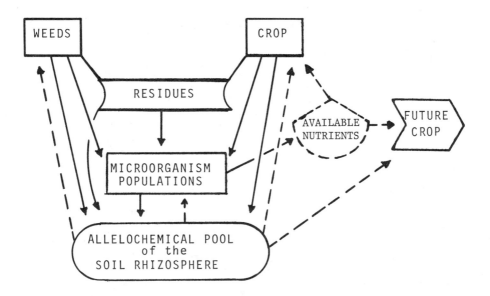

Figure 1. Sources (boxes), transfers (⟶), and direct and indirect
impacts (⎯ ⎯ ⟶) of allelopathic chemicals in agriculture.

Some investigators have postulated that allelopathic substances
in higher plants are immediately detoxified after release. While
such transformations occur, at times the resultant products can be of
higher toxicity. For example, hydrojuglone is oxidized to juglone, a
very potent quinone that is inhibitory to some species at 10^{-6} M
levels (4). A cyanogenic glucoside of peach roots, amygdalin, yields
hydrogen cyanide, benzaldehyde, and subsequently other inhibitors
associated with the peach replant problem (1). Microbial activity in
transformations, plus the metabolic production of diverse allelochemicals by many microorganisms, add to the complications in determining the role of allelopathic interference in crop production.
During residence in the soil compartment, impacts of such compounds
on crop plants may also be modified by moisture, temperature, and
other soil factors (5-10).

The number and diversity of compounds implicated in allelopathy
are rapidly growing. Acetic acid and a few other major intermediates
of metabolism have been reported as allelopathic agents, but most of
those that have been identified are secondary compounds that arise
from the shikimic acid and acetate pathways, or result as hybrids
from these synthesis pathways. Rice (1) classified the compounds
into fourteen chemical categories, plus a miscellaneous group.
Several major groups of his scheme are the terpenoids and steroids,
alkaloids and cyanohydrins, long-chain fatty acids and polyacetylenes, unsaturated lactones, tannins, cinnamic acid derivatives,
benzoic acid derivatives and other simple phenols, coumarins, and
flavonoids. Each category is not equally important, and certainly
only a fraction of the naturally occurring compounds that could
be named under each category are allelopathic in nature. Some,
such as many of the terpenoids and polyacetylenes, may function in a
volatile state, but most of the current documentation in agroecosystems involves water-soluble compounds. Numerous phenolic compounds have been implicated in allelopathy, and derivatives of
cinnamic acid, benzoic acid, and coumarin have been those most often
identified from higher plants (11). Common ones include scopoletin,
esculetin, and the phenolic acids; ferulic, p-coumaric, caffeic,
vanillic, p-hydroxybenzoic, and chlorogenic.

The known list of chemicals involved in allelopathy continues
to expand and examples given should not be automatically assumed to
be the most important agents. Better isolation techniques are
expediting the identification of additional substances, some with
higher biological activity than those noted. Over 10,000 secondary
plant compounds are known (12), most have not been tested as allelopathic agents, and many thousands more are probably present in plants.
The infancy of this area of research is illustrated by the fact that
presently more than 2,000 different alkaloids have been isolated from
over 3,000 species of plants (13), yet only a few have been evaluated for their activity in growth regulation (1). Similarly,
flavonoids are probably the largest class of phenolic compounds in
flowering plants, yet only a few, such as myricitin, quercetin, and
kaempferol, have been tested for potential allelopathic effects.
Assessment of allelochemicals for specific involvement in growth
inhibition or stimulation and analyses of the mechanisms of such
actions are urgently needed to determine their roles in crop
production.

Actions and Interactions of Allelochemicals

Interactions Involving Allelochemicals. Production of many secondary substances by higher plants is modified by a number of environmental factors. Mineral deficiencies, cold treatment, UV light,
herbicides, and other stress conditions typically cause an increase
in the quantity of some of the common allelopathic chemicals in
plants (14-21). Limited work in the U.S.S.R. even suggests that
allelochemicals which are received by a plant may regulate gene
activity that controls the quantity of chemical production by that
plant (22). These interactions between the plant and its environment do not preclude the fact that the basic genotypic capacity for
synthesis of allelochemicals varies extensively even among cultivars
of a crop (23,24).
    It should be emphasized that a complex of substances is
generally involved when allelopathic interferences occur, often with
each below a threshold level for impact. This is illustrated by the
combinations of phenolic acids found in decomposing crop residues
(25-27) and from soils (28-34). In allelopathic situations which
implicate phenolic acids, soil quantities of ferulic, p-coumaric,
and caffeic acids have ranged from below 10 to above 1,000 ppm for
each compound (11,35). The lower end of this spectrum is below a
concentration required for an effect in current bioassays. However,
additive and synergistic effects have been documented for combinations of cinnamic acids (35), benzoic acids (36), benzoic and
cinnamic acids (37), and p-hydroxybenzaldehyde with coumarin (38).
Each of the allelochemicals in these tests was not equally toxic,
but they contributed incrementally to inhibition of germination and
growth. Whereas combinations of many allelochemicals have not been
determined, it appears that both additive and synergistic interactions are extremely important under field conditions.
    Allelopathic chemicals may also act in concert with residual
quantities of herbicides. Our tests demonstrated that a combination
of triflurin and ferulic acid inhibited sorghum [Sorghum bicolor
(L.) Moench] germination and seedling growth more than either alone
(39). Likewise, atrazine stress acted cooperatively with ferulic
acid in stunting oat (Avena sativa L.) seedlings. Obviously,
allelopathy is only one of the several stress factors of the crop
environment. Stress conditions from herbicides, allelochemicals,
temperature extremes, and moisture deficits may work in conjunction
as they impact on crop production (10,39,40).

Indirect Modes of Action. Allelochemicals may either affect crop
plants directly by interference with metabolic functions, or the
effects may be indirect through actions on associated organisms.
Examples of the latter result from effects on organisms of the
nitrogen cycle, on mycorrhizal fungi, and on disease susceptibility
and resistance. Plants subjected to allelopathic stress have less
vigorous growth and are often more susceptible to disease (6). It
also has long been apparent that plants in natural systems have
considerable defense against disease, insect damage, and herbivore
grazing, with a good part of this being due to the quantity and
quality of secondary compounds they contain. Crop breeding programs
have resulted in more than 75% of the agricultural land in the U.S.

being planted to varieties resistant to some bacteria, fungi, or
virus (41,42). Information on these relationships has been compiled
by Hedin (43), and the role of secondary compounds and mechanisms in
protective functions have been recently reviewed (12,44). Thus, in
this summary I will only address allelochemical effects other than
those involving disease relationships.

Rice and co-workers clearly demonstrated that when cropped-out
fields in Oklahoma were abandoned, the succession of plant colonization was partially controlled by allelopathic substances produced by
early invading weeds which reduced the activity of nitrogen-fixing
organisms, including free-living fixers, *Rhizobium* spp., and blue-
green algae (1). These chemicals were also responsible for reducing
nitrification. Evidence from studies of several natural ecosystems
suggests that nitrification may be increasingly inhibited during
succession, so that climax communities retain more nitrogen in the
reduced form (1). Similar factors can alter nitrogen levels in
agricultural fields. When rice (*Oryza sativa* L.) residue is left in
the field in Taiwan and not burned, yield of the subsequent crop of
soybeans [*Glycine max* (L.) Merr.] is depressed. This is due to the
release of phenolic acids, and perhaps other substances which
inhibit nodulation and heme production, suppressing activity of the
nitrogen-fixing bacteria (45). Although the exact cause of autotoxicity of alfalfa (*Medicago sativa* L.) and several other field
legumes has not been determined (46,47), to some degree it may relate
to allelopathic inhibition of their symbiont associates.

Several indirect allelopathic effects have been reported in
forestry. Walnut (*Juglan nigra* L.) plantations in the Central U.S.
often have European black alder [*Alnus glutinosa* (L.) Gaertn.], a
host for nitrogen fixation, interplanted with walnut as a nurse crop.
However, in 8-13 years the alders die out on poorly drained sites due
to juglone toxicity (4,48). Some of this effect may be on the
nitrogen-fixing organisms. Also, there is growing evidence that an
important allelopathic impact may occur on mycorrhizal fungi. Brown
and Mikola (49) reported that reindeer lichen, particularly *Cladonia
alpestris*, inhibited mycorrhizal symbionts, resulting in less
phosphorus uptake and suppression of pine and spruce seedling growth
in Finland forests. As pointed out by Fisher (50) and Rose et al.
(51), allelopathic suppression of fungal growth and root colonization
may explain failures of reforestation by conifer species in disturbed
sites. Unfortunately, similar influences on mineral nutrition of
agronomic crops have not been assessed.

Direct Modes of Action. Evaluation of the effects of allelochemicals
on crop plants has generally been in terms of alterations in germination or some aspect of seedling growth. Often seedling growth is
diminished by lower levels of an inhibitor than germination, and
long-term growth may reflect effects not evident in short-term
bioassays (35,4). Neither germination nor seedling growth bioassays
provides evidence for the mechanisms of growth regulation, and these
mechanisms are currently not well understood (11,52). It appears
evident that with the many different categories of compounds that
have been identified, a variety of mechanisms of action must exist.
Another difficulty in defining mechanisms of action is that a
specific compound may affect several metabolic functions, and as

a result, it has been seldom possible to sort primary from secondary effects.

Reports on physiological effects are most numerous for phenolic compounds that are derivatives of cinnamic and benzoic acids, or closely related compounds with the coumarin skeleton. Two primary mechanisms of action that have been suggested for phenolic acids are their interactions with phytohormones (53,54) and their effects on membranes (55). Early work indicated that diphenolic compounds synergize IAA-induced growth by counteracting IAA (indole-3-acetic acid) destruction, and monophenols stimulate decarboxylation, with both actions reducing growth (56). Numerous subsequent investigations have demonstrated that phenols may interact with the oxidase enzyme system and control the level of IAA (57). A recent study reported that oxidation of IAA was coupled to the cooxidation of phenols through the formation of H_2O_2 as an intermediate (58). Polyphenols may block the action of gibberellic acid, antagonizing growth (59). Indirect hormonal effects on crops, such as elevated ethylene levels in the soil as the result of microbial metabolism of plant residue, have also been implicated in allelopathy (60). However, the evidence that phenolic acids and polyphenols cause their effect through action on phytohormones is not the complete picture.

Phenolic acids are known to alter photosynthetic and respiration rates, cause stomatal closure, reduce chlorophyll content, modify the flow of carbon into various metabolic pools, and alter nutrient uptake in affected tissue (61-73). A common denominator for these multiple effects appears to be the action of phenolic compounds on membranes. They are soluble in membranes, and cause a reduction in ion accumulation in cells (71-73). Several phenolic acids cause membrane depolarization, especially at low pH, increasing membrane permeability to ions (72,73). This action undoubtedly impairs the proton gradient and ATP-driven ion transport. Logically, the effects phenolic acids have on membranes could disturb the water balance and mineral nutrition of seedlings, and research in my laboratory has established such a relationship.

Ferulic and p-coumaric acids altered the water balance of grain sorghum and soybean seedlings (55,74). Even at treatment levels that were below the threshold for short-term growth reduction, water stress was evident by the depression of seedling water potential. Sorghum treated with 0.25 mM ferulic or p-coumaric acids had midday leaf water potentials of approximately -10 bars, compared to -5 bars for the controls. Treatments with phenolic compounds that reduced growth also were correlated with partial stomatal closure (55,63,64).

Visible symptoms of allelopathic effects are often quite similar to those found with nutrient deficiencies, and in a few cases interference with nutrient uptake has been documented. In laboratory experiments designed to test the effects of ferulic acid on the mineral content of sorghum, we found lower tissue levels of phosphorus, potassium, and magnesium accompanying reduced growth of seedlings (75). Whether such growth effects can be modified by increased fertility is not clear. Our preliminary work indicated that inhibition by ferulic acid was not overcome by increased nutrient levels in the growth medium, but Stowe and Osborn (76) reported nutrient augmentation could alleviate growth reductions from phenolic acids. The evidence cited shows that phenolic acids can reduce the uptake of major nutrients, yet in certain situations these same substances may be exuded to benefit plant nutrition. In times of stress, some plants release orthodihydroxyphenols, such as caffeic acid, into the root medium and thus mediate the movement of iron into the root by chelation (77,78). Thus, it is difficult to make generalizations about the effects of phenolic allelochemicals on mineral nutrition, even though we know more about the mode of action of these compounds than any other allelopathic agents.

In subsequent discussions, other examples will be given which support the view that allelopathic interference with crops may operate through effects on water balance and mineral nutrition. A number of allelopathic chemicals other than phenolic acids may also have their initial effects on cellular and organelle membranes. A disruption of membranes not only affects nutrient transport, but respiratory coupling, photosynthesis reactions, and other membrane associated processes.

Allelochemical Interference in Agricultural Fields

Allelopathic Weeds. Weed infestations are a major factor reducing crop yields, and it is not surprising that a recent estimate of costs of weed control and weed-incurred losses was $14 billion annually in the U.S. (79). Interference from weedy species is often due both to competition and allelopathy, and field studies have not generally separated these factors. However, allelopathic potential has been reported for more than 70 weedy species, and the list of weed-crop interactions is growing. These include some of the toughest weeds to control, and some causing the greatest economic damage in major crops. Examples of the allelopathic effects of a few weeds will illustrate the breadth of their impact.

One of the most complete studies of allelopathic interference has been of *Parthenium hysterophorus* L., a tropical American weed which has taken over many cultivated areas of India (80-83). Allelochemicals identified from *Parthenium* include caffeic, vanillic, ferulic, chlorogenic, and anisic acids, plus the sesquiterpene lactone, parthenia. Both aerial plant parts and the roots contribute to the problem. Field studies were conducted to ascertain allelopathic effects on yield. Dried *Parthenium* leaves mixed into the soil of field plots reduced the yield of cowpea (*Vigna sinensis* L.), tomato (*Lycopersicon esculentum* L.), ragi (*Eleusine coracana* Gaertn.), and beans (*Phaselous vulgaris* L.), with nodulation also reduced in the legumes. However, growth of bajra (*Pennisetum typhoideum* Rich.) was stimulated, indicating the subtle differences among species.

Corn (*Zea mays* L.) production systems are an example where a number of allelopathic weeds have been identified that can potentially reduce yield. Corn growth is inhibited by root residues and whole plant leachates of giant foxtail (*Setaria faberii* Herrm.); root residues of giant foxtail, yellow foxtail [*S. glauca* (L.) Beauv.] and crabgrass [*Digitaria sanguinalis* (L.) Scop.]; rhizomes and residues of quackgrass [*Agropyron repens* (L.) Beauv]; and residues of velvetleaf (*Abutilon theophrasti* Medic.) and yellow nutsedge (*Cyperus esculentus* L.) (84-88). Velvetleaf residue caused the greatest suppression of corn height and weight in sandy and light textured soil, with reductions up to 50%. Yellow nutsedge resulted

in greater inhibition as the percentage of sand in the soil mixture
increased, and when the corn seeds were in close proximity to the
residue. These findings confirm the importance of soil conditions
and reinforce an earlier conclusion that the extent of crop damage
may b e related to the chance encounter of roots with localized areas
of high toxicity (6,89). Interestingly, a recent opinion survey
ranked foxtails, velvetleaf, and quackgrass among the five weeds
causing the most economic damage in corn (90). Johnsongrass [Sorghum
halpense (L.) Pers.] was another one of this group, and its allelopathic effects against several plants are well known (1).

My own investigations have focused on aggressive weeds common to
the North Central Plains. The weeds investigated are common in grain
sorghum and soybean fields, thus seedlings of these crops have been
utilized in bioassays. Allelopathic effects have been documented
from curly dock (Rumex crispus L.), giant ragweed (Ambrosia trifida
L.), velvetleaf, cocklebur (Xanthium strumarium L.), Jerusalem artichoke (Helianthus tuberosus L.), and Kochia [Kochia scoparia(L.)
Schrad.] (55,91-94). Aqueous extracts from each of the weeds
inhibited the growth of sorghum seedlings in nutrient solution when
the extract from 1 g fresh leaf weight was incorporated in 60 ml of
nutrient medium. Progressively greater growth reductions and other
visible toxicity symptoms occurred with higher levels of extracts.
Species differences were apparent, with some weeds reducing sorghum
growth at less than 1 g in 120 ml of medium. When dried-shoot residue from Kochia, Jerusalem artichoke, or cocklebur were added to
soil, germination was not modified. However, after two weeks, both
shoot and root dry weights of sorghum were less in soil containing
0.63% (w/w) weed residue, or higher. Based on field samplings from
weed-infested areas, these residue levels could easily occur in the
surface decimeter of the soil.

In several instances, soybeans had a different sensitivity to
these allelopathic weeds than sorghum. For example, soybean growth
was significantly reduced in extract treatments containing 1 g fresh
weight of Kochia in 240 ml of the nutrient medium, indicating that
soybeans were more sensitive than sorghum to allelochemicals in the
Kochia extract. In contrast, inhibition of soybeans grown in soil
containing Kochia required a higher residue level (2.5%) than the
inhibition threshold for sorghum. The lowest Kochia-residue amend-
ment to soil (0.63%) stimulated the growth of soybeans.

Subsequent tests with velvetleaf, Kochia, Jerusalem artichoke,
and cocklebur showed that their allelopathic action altered water
balance (55,94,95). Growth reductions in sorghum and soybean seedlings in nutrient solution amended with extracts from these weeds
correlated with high diffusive resistances and low leaf water potentials. Stomatal closure occurred in plants treated with the more
concentrated extracts. Depressions in water potential were due to a
reduction in both turgor pressure and osmotic potential. A lower
relative water content was also found in velvetleaf-treated plants.
These impacts on water balance were not from osmotic factors.
Allelochemicals from these weeds have not been thoroughly ascertained,
but the present evidence shows that some contain phenolic inhibitors.
Lodhi (96) reported that Kochia contains ferulic acid, chlorogenic
acid, caffeic acid, myricetin, and quercetin. As noted earlier, an
effect on plant-water relationships is one mechanism associated with
the action of ferulic acid.

Residue amendments to soil changed the water status of seedlings
in some treatments. However, the growth-inhibition threshold for a
particular weed residue in soil was typically lower than the level
required to reduce water potential and increase diffusive resistance.
This probably resulted from the combined action of a number of
allelopathic chemicals, with certain ones disrupting other aspects
of metabolism. It is too early to suggest how often allelopathic
weed-crop interactions affect water uptake, but it is interesting to
note that a recent report indicated that production losses from
quackgrass could be overcome by irrigation (97).

Several allelopathic weeds alter mineral nutrition (98,99),
and such an action on a crop would likely result in lower yields.
Chambers and Holm (100) reported beans took up less phosphorus when
they were grown in association with pigweed (*Amaranthus retroflexus*
L.). Corn stunted by quackgrass was low in nitrogen and potassium,
and yet heavy fertilization did not improve yield (86,101).
Similarly, Bhowmik and Doll (88) reported above-ground residues of
common lambsquarter (*Chenopodium album* L.), pigweed, velvetleaf, and
yellow foxtail reduced corn and soybean growth independently of
nitrogen and phosphorus augmentation. Their data did not indicate
that growth inhibition was related to nutrient uptake. Undoubtedly,
a better understanding of mechanism of weed interference with
metabolism could lead to avoidance management strategies.

Crop-crop Interactions. Crop relationships involve both autotoxicity
(self-inhibition) and the effects that one crop may have on a different crop. The latter can be stimulatory as well as inhibitory. While
the knowledge base for most of these interactions is embryonic, a few
examples illustrate their potential.

Farm operators have recognized for quite some time that a drop
in production can occur under continuous cropping in some fields.
This has been referred to as soil sickness, or a sod-bound condition
for grasses. Often, the cause is unknown, but the problem for
several legumes may be due to allelopathy. In soil sickness of red
clover (*Trifolium pratense* L.), isoflavonoid and phenolic acid
toxins have been implicated (102). Katznelson (46) tenuously concluded that the problem in Berseem clover (*T. alexandrinin* L.) was
from disturbances in phosphorus uptake, but nematodes were the major
cause of Persian clover (*T. resupinatum* L.) soil sickness. Autotoxicity in alfalfa appears to occur (47), although the specific
allelochemicals have not been determined and the extent of the
problem varies with soil and climatic conditions. Another example of
autotoxicity has been observed for the growth of pigeon pea ⌊*Cajanus
cajan* (L.) Millsp.⌋ , a tropical bush legume cultivated in Puerto
Rico. Hepperly and Diaz (103) reported instances where yield
dropped drastically under continuous planting and there was little
response to supplemental fertility. They suggested phytotoxicity
from terpenoids, polyphenols, and other allelochemicals was responsible for the decline in yield of pigeon pea.

Major cereal crops may suffer similar problems. Planting of
wheat (*Triticum aestivum* L.) each year in the production belts of
the U.S., Australia, and the U.S.S.R. sometimes results in declining
yields, especially when the straw is left and a cool, wet season
occurs. Research in Nebraska indicated that patulin, an unsaturated

lactone produced by several fungi, was involved (8,104,105). A major source of patulin was *Penicillium urticae* Bainier which flourishes during decomposition of wheat straw. About 40% of all soil microorganisms which they isolated produced substances that reduced plant growth. Thus, other microbial toxins may also be involved. Kimber (106) found that both immobilization of nitrogen and phytotoxic effects were important in suppression of wheat germination by wheat straw. Statistics on corn following corn the previous year show that yield is about 10 bu/acre less than corn following soybeans, and this is not due to fertility (107). The difference in yield may be due to inhibition from corn residue (108), stimulation from soybean residue, or a combination of both. Certainly the yield differential indicates that the historical practice of crop rotation was beneficial.

Crop impact on a subsequent planting of a different species was alluded to previously in the discussion of yield reductions in soybeans following rice. Early work of Patrick and colleagues demonstrated extensive phytotoxicity from a variety of decomposing crop residues (6,89). Guenzi and McCalla (25) reported that oats, wheat, sorghum, and corn residues contained water-soluble allelochemicals, including ferulic, p-coumaric, syringic, vanillic, and p-hydroxybenzoic acids, which affected the germination and growth of wheat, corn, and sorghum. They also found much higher levels of these phenolic acids under farm operations that left significant residues on the surface (29,109). Other examples demonstrate that crop allelopathic problems encompass a range of situations, from vegetables to forestry. Asparagus (*Asparagus officinalis* L.) is autotoxic as well as allelopathic to other vegetables, and part of the toxic effects may result from interactions with pathogenic *Fusarium* spp. (110). Walters and Gilmore (111) found that fescue (*Festuca arundinacea* Shreb.) interfered with establishment and growth of sweetgum (*Liquidambar styraciflua* L.). Studies where comparative effects were eliminated through use of a stairstep apparatus indicated an allelopathic mechanism caused the growth reductions. Chemical analysis of sweetgum seedlings treated with fescue leachates showed that growth inhibition was associated with an impaired absorption of phosphorus an nitrogen. Obviously, conditions of decomposition, allelochemical enhancement of disease, the nature of the secondary products from microbial activity, and interactions among allelochemicals are all significant variables in intercrop allelopathy. The uniqueness of the chemical environment for each crop sequence and situation will continue to confound precise analyses of effects on yield.

Reinforcement needs to be given to the fact that certain crop residues may be stimulatory. In 1975, it was reported that small amendments of chopped alfalfa to the soil stimulated the growth of tomato, cucumber (*Cucumis sativum* L.), lettuce (*Lactuca sativa* L.), and several others. Triacontanol, a long-chain alcohol which is a component of the waxy coat of some leaves, was isolated as the effective compound (112). Unfortunately, numerous field tests with triacontanol have not consistently given growth stimulation. Agrostemmin was isolated from corn cockle (*Agrostemma githago* L.), and strong claims have been made about its capability for enhancing wheat yield (113). Brassinolide, a steroid that has growth stimulatory capabilities in exceedingly small amounts, was isolated from

rape pollen. Commercial analogues, called brassinosteroids, have
been synthesized and some suggest these are a prototype of a new
group of plant growth regulators (114). Whatever the future holds,
certain crop sequences may improve yields even when the details of
allelochemical involvement remain a mystery.

Crop Inhibition of Weeds. The allelopathic capacity of crops to
suppress weed growth has immediate utility for management strategies
and, if heritable factors for allelochemicals can be identified,
these could be incorporated into commercial cultivars. The several
thrusts of research on weed control have been to (a) identify cultivars and accessions with high allelopathic potential, (b) isolate
primary allelochemics, and (c) develop field protocol for capitalizing on allelochemicals from crop plants.

The capacity for Putnam and Duke (115) screened for phytotoxicity in seed sources
from 526 accessions of cucumber and found 3% of the accessions
inhibited the indicator species, proso millet (*Panicum miliaceum* L.)
and white mustard (*Brassica hirta* Moench), by more than 75%. In the
field, some accessions reduced weed populations more than 50%, but
less weed suppression occurred under periods of increased rainfall
(116). Fay and Duke (23) approached the question of allelopathic
expression from oats in a different way. They screened 3,000
accessions of *Avena* spp. germ plasm for output of scopoletin, a known
inhibitor from oats. Some exuded up to three times as much
scopoletin as a standard cultivar, and these were the most active in
suppressing wild mustard [*Brassica kaber* (D.C.) L.C. Wheeler] in
sand culture. However, in loamy-sand soil this activity was lost.

The capacity for weed suppression also has been shown for
soybeans and sunflowers (*Helianthus annuus* L.). Greenhouse evaluation of 141 accessions of soybeans on two weeds, [*Helminthia
echioides* (L.) Gaertner] and *Alopecurus myosuroides* Huds., illustrated
that some lines stimulated these weeds, whereas others were inhibitory, and the sensitivity of the two test species was quite different
(117). Leather (24,118) reported allelopathic-mediated weed suppression in a five year field study using a sunflower-oat-sunflower
rotation. Weed density increased in all plots during the study, but
in the rotation with sunflowers the increase in weed density was
significantly less than in plots without sunflowers. Work is now in
progress on isolation and characterization of the chemicals involved.
Unfortunately, sunflowers may also be deleterious to other crops.
Lower yields than expected have been reported for crops following
sunflowers in South Dakota, and our laboratory tests showed that
sorghum grown in soil amended with sunflower residue was stunted
and exhibited water stress (119).

The potential for using allelopathic crop residues in weed
control has been evaluated in Michigan over the last decade. Putnam
and DeFrank (120) reported that residues of barley (*Hordeum vulgare*
L.), oats, wheat, rye (*Secale cereale* L.) and sorghums were very
effective in reducing weed populations in several vegetable crops.
These grass cover crops were desiccated by freezing, glyphosate,
or paraquat. Sorghum residues were the most effective, suppressing
some weeds more than 90%. The larger seeded vegetables, particularly
legumes, grew normally or were stimulated by the residues, whereas
some of the smaller seeded vegetables were injured. Parallel

approaches in orchards and vineyards have shown that interspace planting with rye or wheat in the fall does not interfere with tree growth, yet provides excellent weed control (121). There is some indication that the long-term studies now in progress may demonstrate an actual increase in orchard production from this practice, compared to conventional weed control. Cover crop residues left on the soil surface leach allelochemicals, are a source for production of microbial phytotoxic products, and provide a number of physical modifications that help suppress weed seed germination.

Exploiting Allelopathy and Allelochemicals

Knowledge of allelopathy and allelochemical functions obviously offers several attractive possibilities for agricultural practices. I anticipate that the immediate future will involve a transition from research evidence to management recommendations. The hope is that strategies can be devised to increase crop yields, reduce production expenses, or circumvent a decline in environmental quality that sometimes occurs in conjunction with current production methods. Efforts to capitalize on allelopathy include ways to (a) avoid negative impacts; (b) benefit from stimulatory effects; (c) utilize allelopathic crops for weed control, including enhancement of their genetic capacity to this end; and (d) identification and production of allelochemics or their derivatives as herbicides and growth regulators. Several scientists have speculated on such opportunities to utilize allelopathy (2,3,122-125), and at the present time some of these goals are feasible. Fortunately, not all allelopathy-based management strategies require a precise identification of allelochemicals and their mechanisms of action.

Tillage Scenarios. A definite complementary relationship is apparent between recent tillage trends and capitalization on allelopathy. No-till and low-tillage practices are becoming more widespread, with nearly 25% of the U.S. crop acreage under some form of conservation tillage (126). Minimum tillage operations reduce cost, are energy efficient, and have conservation advantages that include decreased erosion, increased water infiltration and reduced runoff, surface protection from the impact and compaction of rain, reduced evaporation, and improvement of binding of soil particles. These conservation advantages occur from leaving crop residues on the surface, a factor that is also often associated with an increase in allelochemical activity.

Previously cited examples illustrated that desiccation of an allelopathic cover crop or nurse crop prior to planting provides considerable weed control (120,121,127). Similarly, experiments in North Carolina have shown weed control from planting corn, tobacco (*Nicotiana tabacum* L.), sunflower, and full-season and double-crop soybeans into the previous crop residue or in a herbicide-killed cover (128-129). For example, corn planted into a desiccated green wheat cover crop had 79% less *Ipomoea* spp. compared to non-mulched, tilled treatment. Biomass of common lambsquarter, common ragweed (*Ambrosia artemisiifolia* L.), and redroot pigweed were reduced over 90% by planting into desiccated green rye. Allelochemical suppression was implicated as a causative factor reducing weeds in these

studies. Two compounds from rye not previously implicated in allelopathy were isolated, β-phenyllactic and β-hydroxybutyric acid. Wheat
extracts yielded ferulic acid, and this was found to be decarboxy-
lated by microorganisms to a more phytotoxic styrene, 2-methyl-4-
ethenylphenol. Physical factors associated with minimal disturbance
and surface residues are also important, since seeds are not reposi-
tioned in a germination horizon and the light stimulus required by
some may be absent.

Allelopathic effects differ significantly according to edaphic
factors, climatic conditions, species sensitivity, etc., and reduced
tillage and allelopathic mulches certainly cannot be considered a
panacea for either weed control or conservation measures. In double-
crop systems of Southern U. S., leaving plant residues from the first
crop on the soil surface has been reported to reduce soybean yield,
just as has been documented from rice straw in Taiwan (45). Likewise, stubble-mulch farming sometimes has resulted in suppression
of crop yields (8). In wet years in South Dakota, current reports
indicate wheat fields planted no-till with straw on the surface have
had lower yields than with conventional plowing, whereas in dry years
the yields under these two systems have been comparable. Thus, in
some situations rice and wheat straw removal may be necessary to
avoid allelopathic effects on a crop, and the weed control value of
residues maintained on the soil surface cannot be the only concern.
These differences suggest the necessity for matching tillage
practices with local observations and field tests. When this is
done, strategies can be developed that consider a spectrum of
objectives.

Allelochemic and Herbicide Integration. I expect that future
planning will involve the coordinated use of allelopathic residues
and herbicides. One necessity for such an integration is that
allelopathic mulches suppress certain weeds while having little
action on others, indicating additional chemical weed control may
be necessary. For example, weed control is a problem without the
use of tillage between crops in continuous winter wheat production
systems in Oklahoma. Herbicides have been used to mitigate this
weed problem in no-till wheat, but adequate herbicide efficacy has
not been obtained in all cases (130).

Another example of the coordinated use of herbicides and
allelopathic residues is the situation where a herbicide is used to
desiccate a cover crop. Lehle and Putnam (131) recognized that the
allelochemical content of a residue, such as sorghum, was dependent
on the stage of growth at the time of desiccation. However, no
studies have been undertaken on other factors that may control the
allelochemical content of the residue. It is possible that the
quantity and type of allelochemicals in a cover crop can be manipulated by adjusting the formulation and application rate of the
herbicides used for desiccation. Evidence in the literature demonstrates that certain herbicide treatments and other stress factors
can result in elevations of several coumarins and phenolic compounds
(14-19,21). Thus, it may be reasonable to assume that a herbicide
used to kill a cover crop can also be used as a stimulus for the
synthesis of allelochemicals prior to senescence.

   Alternatively, if a residue interferes with the next crop it
may be possible to minimize this allelopathic effect by suppressing
the content of inhibitors. Duke (132) summarized evidence showing
that glyphosate blocks the shikimic acid pathway and inhibits
synthesis of chlorogenic acid, rutin, procyanidin, anthocyanin, and
other phenolics that may be active in allelopathy. This indicates
that the production of certain allelochemicals can be suppressed.
Maintenance of a reduced rate of bacterial decomposition of a residue, and thus a lower concentration of substances released, may also
be a method for circumventing allelopathic inhibition of the next
crop. Allelochemical effects are concentration dependent and generally interference with crop growth is less after the seedling stage.
Therefore, keeping the released toxins and microbial origin toxins
below the threshold for inhibition may only require a short-term
delay. Presently, chemicals are available to suppress nitrifying
bacteria, suggesting that it is reasonable to attempt to modify the
activity of microorganisms involved with the release and transformation of allelopathic chemicals from residue. Newman (124) pointed
out that where organic acids arise from residue under anaerobic
conditions, even an alteration in microbial metabolism toward production of acetic acid, compared to the more toxic butyric acid,
could reduce toxicity.

Crop Sequence Strategies. A basic axiom of allelopathy is that
plants are not equally sensitive to allelochemicals. Thus, it is
probable that crop rotations will be useful in avoiding negative
impacts from allelochemicals and capitalizing on stimulatory aspects.
Considerations involving soil conservation, soil fertility, and pest
and disease control, interwoven with economic factors, are also providing pressures that will force a trend toward more extensive use
of rotations. Unfortunately, adding allelopathy into the decision-
making process will often require more information than is
presently available.
   It was noted in this manuscript that allelopathy has been
implicated in yield depressions that may occur in crops following
sunflower (119), when corn follows corn (107), soybeans following
rice (45), continuous crop alfalfa (47), and several others.
Possibly picking the right crop sequence can result in yield
stimulation, as suggested for corn following soybeans. While the
list of observations on inhibitory and stimulatory effects can be
extended, little documented evidence on causative factors or the
actual impact on productivity is available.
   Dzyubenko et al. (133) found that monoculture growth for a
number of crop plants resulted in an accumulation of toxins in both
the roots and soil, and this correlated with a decrease in crop
production. Work of Soviet scientists on mixed cultures illustrated
that interactions in these situations are hard to predict. In mixed
culture, root releases from barley and oats reduced the uptake of
nutrients by pea (*Pisum sativum* L.) and hairy vetch (*Vicia villosa*
Roth), while these two legumes released compounds that stimulated
the uptake of several nutrients in the cereals (134). Petrova (135)
found that coplanting of several legumes with corn stimulated yield
of corn, and the evidence suggested that volatile compounds from
the legume shoots and roots were involved. Rice (1) summarized more

extensive evidence from the Soviet literature that demonstrates certain advantages to rotations and mixed plantings, but also illustrates that an understanding of each of the interactions between any two crops merits separate consideration.

Breeding for Weed Control. Breeding crops capable of producing their own herbicides is a longer-term consideration than cover crop and crop rotation strategies. However, the proven successes of breeding programs for pest and disease resistance suggest a similar potential as one way to capitalize on allelopathy. Work by Fay and Duke (23) showing decided differences in scopoletin production among oat germ plasm illustrates that there is genetic diversity that could be utilized. Obvious major stumbling blocks are that (a) autotoxicity must be avoided, (b) allelopathic effects are generally due to a complex of compounds, (c) classical breeding programs involve multiple generations, and (d) crop breeders may have numerous other goals of higher priority. The future for breeding crops that have enhanced herbicide activity as an objective will very likely be linked to advancements in genetic engineering. According to a recent summary, attempts by biotechnology to produce glyphosate resistance and other herbicide resistance in crops are now in progress, suggesting that genetic manipulation of crop metabolism may be close (136).

Allelochemicals in the Marketplace. The array of allelochemicals in plants is large, and some should be adaptable as herbicides and plant growth regulators. Shettel and Balke (137) have demonstrated that even the common phenolic acids and coumarins can be applied as herbicides. Salicylic acid, p-hydroxybenzoic acid, hydroquinone, and umbelliferone were evaluated on several crop and weed species in greenhouse experiments. These chemicals effectively suppressed the growth of several weeds when applied preplant incorporated, preemergence, or postemergence. Unfortunately, the application rate required to reduce weed growth was quite high and some reduction in the growth of crop seedlings occurred. The hope is that some allelochemicals will have herbicidal activity at concentrations that make them amenable for economic consideration.

Allelochemics have previously served as structural models for herbicide development, and natural compounds will undoubtedly be used as prototypes for new herbicides. Banvel T (Tricamba), trichlorobenzoic acid (2,3,6-TBA), 2,3,5-triiodobenzoic acid (TIBA), and other derivatives of benzoic acid have been successfully marketed. The chemical skeleton of picloram (Tordon) is the microbial alkaloid α-picolinic acid. In a recent symposium, van Aller et al. (138) presented data on oxygenated fatty acids extracted from aquatic plants that they suggested had potential use for control of blue-green algae growth. Martin and Dooris (139) proposed that natural products isolated from cypress [*Taxodium distichum* (L.) Rich.] may be feasible for control of *Hydrilla verticillata* (Royal), a serious problem in waterways and lakes of Florida. However, the actual testing of these compounds has not been done.

Clearly, a good bit of chance is involved in finding a compound that may be useful. One approach is to thoroughly extract and

analyze fractions from microorganisms and higher plants that are known
to cause allelopathic inhibition. Several of the most toxic compounds previously isolated, such as patulin, are from microbial
metabolism. Cutler et al. (140) have isolated several fungal products, including cladosporin and pergillin, and pursued making more
active derivatives of these (e.g., cladosporin diacetate, dihydro-
pergillin). Also note Cutler's presentation at this conference.
Certainly, many soil organisms appear to produce substances that
have herbicidal activity (141). Allelopathic weeds previously noted
might also be good targets for success in isolating chemicals that
can be utilized as herbicides. However, there is no way to predict
the best plant prospects. Isolation and identification procedures
are time consuming, typically very small quantities of a compound are
obtained, and the evaluation of biological activity is dependent on
the bioassay utilized. Using duckweed (*Lemna* spp.), we have developed a bioassay that is quite sensitive and requires minimal
quantities of an unknown (142). Current high resolution separation
techniques and improved bioassays should make emphasis on identification and evaluation of allelochemics for growth regulator activity
rewarding.

Finding allelochemicals that can stimulate crop yield is
probably even more difficult than looking for herbicidal activity.
Modification of metabolism in this way is complicated by the
difficulties of hitting just the right physiological state, yet not
overdosing. Whereas triacontanol, brassinolide, and agrostemmin
have all shown yield enhancement potential on major crops, incon-
sistency between locations, genotypes, and spraying dates, plus
difficulties in formulation, have hindered their commercial use.
Applications of synthetic products, such as the use of 2-sec-butyl-
4, 6-dinitrophenol (dinoseb) on corn, have suffered similar problems
(143). However, the future portends more extensive evaluation of
allelochemicals as yield stimulants, and it is exciting to think that
breakthroughs may occur.

Conclusions

It is my position that we are on the threshold of effective efforts
to manipulate and use allelochemicals, and it is going to be done.
Some strategies in agriculture may profit from the regulatory
functions of allelopathic chemicals without exhaustive identification
of the compounds involved or their mechanisms of action. The first
practices to avoid negative impacts on crops, provide yield stimulation, or utilize allelopathy for weed control, involve management
of crop sequences and utilization of residues. Capitalizing on
allelopathy may be approached by removing or providing residues and
mulches, planting an allelopathic crop in a rotation, or using companion plantings where one crop either stimulates the other or yields
products that are active in weed control. Certainly, consideration
for the activity of allelochemicals will be only one of several goals
recognized in making these decisions. The potential for using
herbicides to complement management strategies that attempt to
utilize allelopathy should not be overlooked. Efforts to chemically
control the biochemical pathways of a crop for an increase or decrease in production of specific allelopathic components, or to

control microorganism activity to avoid the build up of allelochemicals in the rhizosphere seem feasible, but research in this area has hardly begun. Likewise, enhancement of the genetic capacity of crop plants to produce their own herbicides, either through traditional breeding programs or biotechnology, is a long-term goal.

The search is on for herbicide activity and yield stimulants from plant compounds. Marketing allelochemicals in the interest of benefiting crop production will require continued attention toward screening and testing of the many biologically active secondary compounds extant. The continued study of allelochemicals should result in some of them serving as the basis of new, unique, and more biodegradable herbicides. Undoubtedly, the timetable for success is integrally tied to the intensity of the search among natural products for all types of plant growth regulators. Utilization of allelochemicals is a new and challenging frontier that has positive implications for aiding crop productivity.

Literature Cited

1. Rice, E. L. "Allelopathy"; Academic Press, Inc.: Orlando, Florida, 1984; 2nd ed.; p. 422.
2. Rice, E. L. "Pest Control with Nature's Chemicals"; University of Oklahoma Press: Norman, Oklahoma, 1983; p. 224.
3. Einhellig, F. A. Proc. Plant Growth Reg. Soc. Am. 1981, 8, 40-51.
4. Rietveld, W. J. J. Chem. Ecol. 1983, 9, 295-308.
5. Patrick, Z. A.; Koch, L. W. Can. J. Bot. 1958, 36, 621-47.
6. Patrick, Z. A.; Toussoun, T. A.; Koch, L. W. Annu. Rev. Phytopathol. 1964, 2, 267-92.
7. Wang, T. S. C.; Cheng, S. Y.; Tung, H. Soil Sci. 1968, 104, 138-44.
8. McCalla, T. M.; Norstadt, F. A. Agric. Environm. 1974, 1, 153-74.
9. Bhowmik, P. C.; Doll, J. D. J. Chem. Ecol. 1983, 9, 1263-80.
10. Einhellig, F. A.; Eckrich, P. C. J. Chem. Ecol. 1984, 10, 161-70.
11. Einhellig, F. A. In "Handbook of Natural Pesticides: Methods, Vol. 1"; Mandava, H. B., Ed.; CRC Press, Inc.: Boca Raton, Florida, 1984; pp. 161-200.
12. Swain, T. Annu. Rev. Plant Physiol. 1977, 28, 479-501.
13. Williams, C. M. Weeds Today. 1983, 14(2), 6-7.
14. Dieterman, L. J.; Lin, C. Y.; Rohrbaugh, L. M.; Thiesfeld, V.; Wender, S. H. Anal. Biochem. 1964, 9, 139-45.
15. Dieterman, L. J.; Lin, C. Y.; Rohrbaugh, L. M.; Wender, S. H. Arch. Biochem. Biophys. 1964, 106, 275-9.
16. Koeppe, D. E.; Rohrbaugh, L. M.; Wender, S. H. Phytochemistry 1969, 8, 889-96.
17. Koeppe, D. E.; Rohrbaugh, L. M.; Rice, E. L. Wender, S. H. Physiol. Plant. 1970, 23, 258-66.
18. Armstrong, G. M.; Rohrbaugh, L. M.; Rice, E. L.; Wender, S. H. Phytochemistry 1970, 9, 945-8.
19. del Moral, R. Oecologia 1971, 9, 289-300.
20. Gilmore, A. R. J. Chem. Ecol. 1977, 3, 667-76.
21. Hall, A. B.; Blum, U.; Fites, R. C. Am. J. Bot. 1982, 69, 776-83.

22. Baranetsky, G. G.; Moroz, P. A. In "Rol allelopatii v rastenievodstve"; Grodzinsky, A. M., Ed.; Sci. Coll. Kiev: Naukova Dumka, 1982, pp. 15-21. (In Russian, English Summary).

23. Fay, P. K.; Duke, W. B. Weed Sci. 1977, 25, 224-8.

24. Leather, G. R. Weed Sci. 1983, 31, 37-42.

25. Guenzi, W. D.; McCalla, T. M. Agron. J. 1966, 58, 303-4.

26. Chou, C. H.; Lin, H. J. J. Chem. Ecol. 1976, 2, 353-67.

27. Chou, C. H.; Patrick, Z. A. J. Chem. Ecol. 1976, 2, 369-87.

28. Whitehead, D. C. Nature 1964, 202, 417-8.

29. Guenzi, W. D.; McCalla, T. M. Soil Sci. Soc. Am. Proc. 1966, 30, 214-6.

30. Hennequin, J. R.; Juste, C. Ann. Agron. 1967, 18, 545-69.

31. Wang, T. S. C.; Yang, T. K.; Chuang, T. T. Soil Sci. 1967, 103, 239-46.

32. Wang, T. S. C.; Yeh, K. L.; Cheng, S. Y.; Yang, T. K. In "Biochemical Interactions Among Plants"; U. S. Nat. Comm. for IBP, Eds.; National Academy Science, Washington, D.C., 1971; pp. 113-9.

33. Carballeira, A. J. Chem. Ecol. 1980, 6, 593-6.

34. Whitehead, D. C.; Dibb, H.; Hartley, R. D. J. Appl. Ecol. 1982, 19, 579-88.

35. Einhellig, F. A.; Schon, M. K.; Rasmussen, J. A. J. Plant Regul. 1982, 1, 251-8.

36. Einhellig, F. A.; Rasmussen, J. A. J. Chem. Ecol. 1978, 4, 425-36.

37. Rasmussen, J. A.; Einhellig, F. A. Plant Sci. Lett. 1979, 14, 69-74.

38. Williams, R. D.; Hoagland, R. E. Weed Sci. 1982, 30, 206-12.

39. Hamm, J. H., M. A. Thesis, University of South Dakota, Vermillion, 1984.

40. Eckrich, P. C., M. A. Thesis, University of South Dakota, Vermillion, 1983.

41. Nelson, R. R., Ed. "Breeding Plants for Disease Resistance: Concepts and Applications"; Pennsylvania State University Press: Univeristy Park, Pennsylvania, 1973; p. 401.

42. DeBach, P. "Biological Control by Natural Enemies"; Cambridge University Press, London, 1974; p. 323.

43. Hedin, P. A., Ed. "Host Plant Resistance to Pests"; ACS SYMPOSIUM SERIES No. 62, American Chemical Society: Washington, D. C., 1977; p. 286.

44. Bell, A. A. Annu. Rev. Plant Physiol. 1981, 32, 21-81.

45. Rice, E. L.; Lin, C. Y.; Huang, C. M. J. Chem. Ecol. 1981, 6, 333-4.

46. Katznelson, J. Plant Soil 1972, 36, 379-93.

47. Miller, D. A. J. Chem. Ecol. 1983, 9, 1059-72.

48. Rietveld, W. J.; Schlesinger, R. C.; Kessler, K. J. J. Chem. Ecol. 1983, 9, 1119-34.

49. Brown, R. T.; Mikola, P. Acta Forestalia Fennica. 1974, 141, p. 23.

50. Fisher, R. F. J. For. 1980, 78, 346-50.

51. Rose, S. L.; Perry, D. A.; Pilz, D.; Schoeneberger, M. M. J. Chem. Ecol. 1983, 9, 1153-62.

52.   Horsley, S. B.   In "Proc. 4th N. Am. Forest Biol. Workshop";
      Wilcox, H. E.;   Hamer, A. F., Eds.;   School of Continuing
      Education, College of Environmental Science and Forestry,
      Syracuse, N. Y., 1977;   pp. 93-136.
53.   McClure, J. W.   Rec. Adv. Phytochem.   1979, 12, 525-56.
54.   Swain, T.   Proc. Plant Growth Reg. Soc. Am.   1981, 8, 275-90.
55.   Einhellig, F. A.;   Muth, M. Stille;   Schon, M. K.   In "The
      Chemistry of Allelopathy";   Thompson, A. C., Ed.;   ACS
      SYMPOSIUM SERIES No. 268, American Chemical Society: Washington,
      D.C., 1985;   Ch. 12.
56.   Tomaszewski, M.;   Thimann, K. V.   Plant Physiol.   1966, 41,
      1443-54.
57.   Miidla, Kh.;   Khaldre, Y.;   Padu, E.;   Yaakma, Yu.
      Fiziologiya Rastenii 1982, 29, 649-54.
58.   Grambow, H. J.;   Langenbeck-Schwich, B.   Planta 1983, 157,
      131-7.
59.   Jacobson, A.;   Corcoran, M. R.   Plant Physiol.   1977, 59,
      129-33.
60.   Harvey, R. G.;   Linscott, J. J.   Soil Sci. Soc. Am. 1978,
      42, 721-4.
61.   Einhellig, F. A.;   Rice, E. L.;   Risser, P. G.;   Wender, S. H.
      Bull. Torrey Bot. Club 1970, 97, 22-33.
62.   Kadlec, K. D.   M. A. Thesis, University of South Dakota,
      Vermillion, 1973.
63.   Patterson, D. T.   Weed Sci.   1981, 29, 53-9.
64.   Einhellig, F. A.;   Kaun, L. Y.   Bull. Torrey Bot. Club.
      1971, 98, 155-62.
65.   Van Sumere, C. F.;   Cottenie, J.;   DeGreef, J.;   Kint, J.
      Rec. Adv. Phytochem.   1971, 4, 165-221.
66.   Demos, E. K.;   Woolwine, M.;   Wilson, R. H.;   McMillan, C.
      Am. J. Bot.   1975, 62, 97-102.
67.   Einhellig, F. A.;   Rasmussen, J. A.   J. Chem. Ecol. 1979,
      5, 815-24.
68.   Danks, M. L.;   Fletcher, J. S.;   Rice, E. L.   Am. J. Bot.
      1975, 62, 311-7.
69.   Danks, M. L.;   Fletcher, J. S.;   Rice, E. L.   Am. J. Bot.
      1975, 62, 749-55.
70.   Glass, A. D. M.;   Dunlop, J.   Plant Physiol. 1974, 54, 855-8.
71.   Glass, A. D. M.;   J. Exp. Bot.   1974, 25, 1104-13.
72.   Glass, A. D. M.;   Plant Physiol.   1973, 51, 1037-41.
73.   Harper, J. R.;   Balke, N. E.   Plant Physiol. 1981, 68, 1349-53.
74.   Stille, M.   M. A. Thesis, University of South Dakota,
      Vermillion, 1979.
75.   Kobza, J.   M. A. Thesis, University of South Dakota, Vermillion,
      1980.
76.   Stowe, L. G.;   Osborn, A.   Can. J. Bot. 1980, 58, 1149-53.
77.   Olsen, R. A.;   Clark, R. B.;   Bennett, J. H.   Am. Scientist
      1981, 69, 378-84.
78.   Julian, G.;   Cameron, H. J.;   Olsen, R. A.   J. Plant Nutrition
      1983, 6, 163-75.
79.   Chandler, J. M.   In "The Chemistry of Allelopathy";   Thompson,
      A. C., Ed.;   ACS SYMPOSIUM SERIES No. 268, American Chemical
      Society: Washington, D.C., 1985;   Ch. 2.

80. Kanchan, S. D.; Jayachandra. Plant Soil 1979, 53, 37-47.
81. Kanchan, S. D.; Jayachandra. Plant Soil 1979, 53, 27-35.
82. Kanchan, S. D.; Jayachandra. Plant Soil 1980, 55, 61-6.
83. Kanchan, S. D.; Jayachandra. Plant Soil 1980, 55, 67-75.
84. Bell, D. T.; Koeppe, D. E. Agron. J. 1972, 64, 321-5.
85. Schreiber, M. M.; Williams, J. L., Jr. Weeds 1967, 15, 80-1.
86. Buchholtz, K. P. In "Biochemical Interactions Among Plants";
     U. S. Natl. Comm. for IBP, Ed.; National Academy Science:
     Washington, D. C., 1971; pp. 86-9.
87. Drost, D. C.; Doll, J. D. Weed Sci. 1980. 28, 229-33.
88. Bhowmik, P. C.; Doll, J. D. Agron. J. 1982, 74, 601-6.
89. Patrick, Z. A. Soil Sci. 1971, 111, 13-18.
90. "National Survey Pinpoints the Most Destructive Weeds,"
     Agrichem. Age 1984, May; pp. 10-11.
91. Einhellig, F. A.; Rasmussen, J. A. Am. Midl. Nat. 1973, 90,
     79-86.
92. Rasmussen, J. A.; Einhellig, F. A. Am. Midl. Nat. 1975, 94,
     478-83.
93. Rasmussen, J. A.; Einhellig, F. A. Southwest. Nat. 1979, 24,
     637-44.
94. Colton, C. E.; Einhellig, F. A. Am. J. Bot. 1980, 67, 1407-
     13.
95. Einhellig, F. A.; Schon, M. K. Can. J. Bot. 1982, 60, 2923-
     30.
96. Lodhi, M. A. K. Can. J. Bot. 1979, 57, 1083-8.
97. Young, F. L.; Wyse, D. L.; Jones, R. J. Weed Sci. 1984, 32,
     226-34.
98. Newman, E. I.; Miller, M. H. J. Ecol. 1977, 65, 399-411.
99. Fisher, R. F.; Woods, R. A.; Glavicic, M. R. Can. J. For. Res,
     1978, 8, 1-9.
100. Chambers, E. E.; Holm, L. G. Weeds 1965, 13, 312-4.
101. Kommedahl, T.; Old, K. M.; Ohman, J. H.; Ryan, E. W.
     Weed Sci. 1975, 23, 29-32.
102. Chang, C. F.; Suzuki, A.; Kumai, S.; Tamura, S. Agric. Biol.
     Chem. 1969, 33, 398-408.
103. Hepperly, P. R.; Diaz, M. J. Agric. Univ. Puerto Rico. 1983,
     67, 453-63.
104. Norstadt, F. A.; McCalla, T. M. Science 1963, 140, 410-11.
105. Ellis, J. R.; Norstadt, F. A.; McCalla, T. M. Plant Soil
     1977, 47, 679-86.
106. Kimber, R. W. L. Plant Soil 1973, 38, 543-55.
107. Anderson, I. C. Soybean News. 1984. 35(1), 3.
108. Yakle, G. A.; Cruse, R. M. Can. J. Plant Sci. 1983, 63,
     871-7.
109. McCalla, T. M. In "Biochemical Interactions Among Plants";
     U. S. Natl. Comm. for IBP, Ed.; National Academy Science:
     Washington, D. C., 1971, pp. 39-43.
110. Hartung, A. C.; Stephens; C. T. J. Chem. Ecol. 1983, 9,
     1163-74.
111. Walters, D. T.; Gilmore, A. R. J. Chem. Ecol. 1976, 2, 469-79.
112. Ries, S. K.; Went, V.; Sweeley, C. C.; Leavitt, R. A.
     Science 1977, 195, 1339-41.
113. Gajić, D.; Malencic, S.; Vrbaski, M.; Vrbaski, S. Fragm.
     Herb. Jugoslavica. 1976, 63, 121-41.

114. Maugh, T. H., II. Science 1981, 212, 33-4.
115. Putnam, A. R.; Duke, W. B. Science 1974, 185, 370-2.
116. Lockerman, R. H.; Putnam, A. R. Weed Sci. 1979, 27, 54-7.
117. Massantini, F.; Caporali, F.; Zellini, G. Proc. EWRA Symposium: Methods of Weed Control and Their Integration, 1977, pp. 23-28.
118. Leather, G. R. J. Chem. Ecol. 1983, 9, 983-89.
119. Schon, M. K.; Einhellig, F. A. Bot. Gaz. 1982, 143, 505-10.
120. Putnam, A. R.; DeFrank, J. Crop Protection 1983, 2, 173-81.
121. Putnam, A. R.; DeFrank, J.; Barnes, J. P. J. Chem. Ecol. 1983, 9, 1001-10.
122. Putnam, A. R.; Duke, W. B. Annu. Rev. Phytopathol. 1978, 16, 431-51.
123. Klein, R. R.; Miller, D. A. Commun. Soil Sci. Plant Anal. 1980, 11, 43-56.
124. Newman, E. I. Pesticide Sci. 1982, 13, 575-82.
125. Putnam, A. R. Chem. Eng. News, 1983, 61, 34-45.
126. "No-till Gets Mixed Reviews But Keeps Growing," Successful Farming 1983, Nov., p. 46.
127. Barnes, J. P.; Putnam, A. R. J. Chem. Ecol. 1983, 9, 1045-57.
128. Liebl, R. A.; Worsham, A. D. J. Chem. Ecol. 1983, 1027-43.
129. Shilling, D. G.; Liebl, R. A.; Worsham, A. D. In "Chemistry of Allelopathy"; Thompson, A. C., Ed.; ACS SYMPOSIUM SERIES No. 268, American Chemical Society: Washington, D. C., 1985; Ch. 17.
130. Cleary, C. L.; Peeper, T. F. Weed Sci. 1983, 31, 813-8.
131. Lehle, F. R.; Putnam, A. R. Plant Physiol. 1982, 69, 1212-6.
132. Duke, S. B. In "Chemistry of Allelopathy"; Thompson, A. C., Ed.; ACS SYMPOSIUM SERIES No. 268, American Chemical Society: Washington, D. C., 1985; Ch. 8.
133. Dzyubenko, N. N.; Krupa, L. I.; Boiko, P. I. In "Interactions of Plants and Microorganisms in Phytocenoses", Grodzinsky, A. M., Ed.; Naukova, Dumka, Kiev; 1977, pp. 70-77. (In Russian, English Summary).
134. Rakhteenko, I. N.; Kaurov, I. A.; Minko, I. F. In "Physiological-Biochemical Basis of Plant Interactions in Phytocenoses, Vol. 4" Grodzinsky, A. M., Ed.; Naukova, Dumka, Kiev.; 1973, pp. 16-19. (In Russian, English Summary).
135. Petrova, A. G. In "Interactions of Plants and Microorganisms in Phytocenoses", Grodzinsky, A. M., Ed.; Naukova, Dumka, Kiev.; 1977, pp. 91-7. (In Russian, English Summary).
136. Valiulis, D. Agrichem. Age 1984, 28 (8), 53-60.
137. Shettel, N.; Balke, N. E. Weed Sci. 1983, 31, 293-8.
138. van Aller, R. T.; Pessoney, G. F.; Rogers, V. A.; Watkins, E. J.; Leggett, H. G. In "The Chemistry of Allelopathy"; Thompson, A. C., Ed.; ACS SYMPOSIUM SERIES No. 268, American Chemical Society: Washington, D. C., 1985; Ch. 26.
139. Martin, D. F.; Dooris, P. M. In "The Chemistry of Allelopathy", Thompson, A. C., Ed.; ACS SYMPOSIUM SERIES No. 268, American Chemical Society: Washington, D. C., 1985; Ch. 25.
140. Cutler, H. G.; Springer, J. P.; Cox, R. H. Proc. Plant Growth Reg. Soc. Am. 1981, 8, 132-6.

141. Heisey, R. M.; DeFrank, J.; Putnam, A. R. In "Chemistry of
      Allelopathy"; Thompson, A. C., Ed.; ACS SYMPOSIUM SERIES
      No. 268, American Chemical Society: Washington, D. C., 1985;
      Ch. 22.
142. Einhellig, F. A.; Leather, G. R.; Hobbs, L. I. J. Chem. Ecol.
      1984, 10, (in press).
143. Basabe, P. J.; Oplinger, E. S. Proc. Plant Growth Reg. Soc.
      Am. 1981, 8, 231-9.

RECEIVED November 15, 1984

CONTROL OF INSECT GROWTH
AND DEVELOPMENT

INTRODUCTION TO CONTROL
OF INSECT GROWTH
AND DEVELOPMENT

THE THEME OF THE CONFERENCE, "New Concepts in Pesticide Chemistry," is most timely in terms of increasing regulatory and social demands for safer, biodegradable, specific, and highly effective crop protection chemicals. Most insecticides that were introduced commercially since the 1940s are nerve poisons that intrinsically affect both target and nontarget living organisms, including humans. Selective toxicity has usually been conferred on a number of these chemicals by virtue of differential uptake, metabolism, and excretion. This applies to such diverse classes of insecticides as the chlorinated hydrocarbons, organophosphorus esters, carbamates, and synthetic pyrethroids.

The focus of this session was to examine the scientific merits of alternate strategies having their genesis in biorational approaches to exploit physiological and biochemical processes unique to insects or arthropods with the aim of providing biochemical leads for the discovery of unique new chemical control agents.

The contributors to this session are well-known scientists whose research over the years has added markedly to our understanding and development of biorational models. Some models have been further developed via synthetic optimization to yield uniquely selective control chemicals. Notable examples in this respect, particularly hydroprene, are described here by Staal, as novel reproduction control agents in the German cockroach. In this vein, Bowers discusses a variety of plant sources from which he has extracted unique compounds with juvenoid and antijuvenile hormone (AJH) activity (most notably the chromenes, precocenes I and II). These AJH compounds have stimulated intense synthetic effort in several industrial and academic laboratories with the aim of optimizing and broadening the bioactivity of the prototype compounds.

Camps reviews elegant synthetic work on precocene-type chemistry directed toward optimization of AJH activity. Hammock applies transition-state theory to discover inhibitors of juvenile hormone esterases that hydrolyze juvenile hormone in the cabbage looper. He further elaborates on one such inhibitor, 3-(octylthio)-1,1,1-trifluoro-2-propanone (OTFP), a slow, tightly binding inhibitor. When applied topically, OTFP inhibits juvenile hormone esterase in vivo and induces morphogenetic effects similar to those observed following treatment with juvenoids. It is hoped that this lead will be followed in search of more potent juvenile hormone esterase inhibitors.

Inhibition of reproductive capacity in insects has long been a goal of research scientists of the U.S. Agricultural Research Service. Borkovec reviews here research with juvenoids and chitin synthesis inhibitors affecting both immature and adult insect stages. Indeed, viable examples of this research involve the practical use of diflubenzuron, a chitin synthesis inhibitor, in sterilizing the boll weevil and hydroprene in sterilizing cockroaches.

Advances in insect neurophysiology and analytical chemistry facilitating rapid identification in minute amounts of bioactive substances have given considerable impetus to the discovery of behavior-modifying chemicals from plants, such as antifeedants described here by Kubo and sex pheromones described by Yamamoto.

Kubo discusses the isolation, structure elucidation, and antifeedant activity in the cotton bollworm of abyssinin derived from an African medicinal plant, *Bersama abyssinica*. Unquestionably, this example, and many others described in the recent literature, suggests that natural products offer an increasingly rich source of novel chemical compounds with selective action in insects.

A most novel approach to insect control exploiting induced deviant sexual behavior in the Azuki bean weevil is described by Yamamoto. The female of the species produces a sex pheromone that induces erection of the male organ and copulation. Yamamoto named the substance Erectin, which is a synergistic mixture of C_{26} to C_{35} hydrocarbons or octadecane and (*E*)-3,7-dimethyl-2-octene-1,3-dioic acid, named callosobruchusic acid. Yamamoto devised a "confusion strategy" by applying Erectin to various surfaces, causing the males to forgo their natural mates in favor of the "sham females."

The bioactivity of the various chemical control prototypes described in this section is greatly influenced by their uptake, transport, and eventual degradation in the insect. The latter event is highly influenced by the action of the mixed-function oxidase enzymes (MFOs) in the midgut and fat body of the insect. Wilkinson discusses the role the MFO enzymes in insects, their comparative aspects to mammalian enzymes, and their role in detoxification and activation of xenobiotics in living organisms. A thorough understanding of these enzymes is most useful in designing more active compounds containing blocking groups substituted as metabolic blockers at critical junctions in the target molecule.

Blomquist discusses propionate and methyl malonate metabolism, Feyereisen reports on the suicidal destruction of cytochrome P-450, Mullin describes detoxification enzyme relationships, and Roe contributes to this book with a chapter on the effects of δ-endotoxin in mice and insects. Also contributing to this section are Sparks, describing a bioassay of anti juvenile hormone compounds, and Van Emon, discussing immunoassay applications to pesticides.

It is inteded that the ideas and experimental models described in the following pages will stimulate further intelligent research useful in developing rational leads for selective insect control chemicals.

JULIUS J. MENN
Zoecon Corporation
Palo Alto, CA 94304

RECEIVED November 15, 1984

Use of Transition-State Theory in the Development of Bioactive Molecules

YEHIA A. I. ABDEL-AAL[1] and BRUCE D. HAMMOCK

Departments of Entomology and Environmental Toxicology, University of California, Davis, CA 95616

The application of transition state theory to enzyme catalysis is presented in terms of mechanistic and energetic approaches. These treatments emphasize the rationale of using transition state analogs as bioactive molecules. Included in this chapter are some criteria for determining if a compound is a transition state analog. A hypothesis is then presented that the initial binding of some insecticidal carbamates and organophosphates to acetylcholinesterase is dependent in part upon their mimicking the transition state configuration of acetylcholine. As an indication of how transition state theory could be applied to problems in agricultural chemistry, we present data from this laboratory that some trifluoromethylketone inhibitors of insect juvenile hormone esterase(s) are transition state mimics and extend this argument by applying a quantitative structure-activity relationship approach to these analogs.

There has been an obvious decline in the number of commercial agricultural and pharmaceutical chemicals during the last decade. There are two major reasons for this decline. First, there are increasingly tighter requirements for market development. To use agricultural chemicals as an example, we see an increasingly narrow margin of profit for farmers and increasing concern over environmental health effects. The cost of registration, production and marketing also continues to increase. Secondly, the cost of discovery has increased dramatically. There are numerous components to this observation as well, but simplistically one can see that to meet the tight requirements for market development, more complex and expensive syntheses and bioassays are required. Random screening with the aid of some serendipity served effectively in the past for the discovery of new biological activities, yet we have reached the

[1]On leave from the Department of Plant Protection, College of Agriculture, Assiut University, Assiut, A.R. Egypt.

0097–6156/85/0276–0135$07.50/0

point of diminishing returns with regard to its application to
future problems. Therefore, a theme of several of the manuscripts
from this meeting relates to the development of new paradigms for
the discovery of biological activity.

Molecules can react with a biological matrix in many ways. At
least some biological activity is common among chemically reactive
species. For instance, Michael acceptors and acetylating agents are
commonly irritating agents or general cytotoxins. Molecules may
reversibly react with biological systems based upon only solubility
or weak molecular interactions with biological depressants as prime
examples. However, these molecules lack the specificity desired for
field application and the potency needed for economic feasibility.
However, if an alkylating agent has a high affinity for nucleic
acids, if an acetylating agent binds well to an enzyme active site,
or if an inert molecule can form several weak molecular interactions
with a receptor site, these molecules may display very high biological activity indeed.

During the last several years we have become impressed with the
utility of transition state theory in the optimization as well as
the discovery of biologically active structures. Empirical
approaches indicate that a mimic of an enzyme-substrate transition
state could have a K_I below 10^{-15} M (1,2). Although one is not
likely to precisely mimic a transition state, even a vague mimic
could still be a very potent inhibitor. Therefore, we feel that
transition state theory is likely to be one of several approaches
useful in agricultural chemistry. Thus in the following pages we
present a brief review of the application of transition state theory
in enzyme catalysis in terms of both mechanistic and energetic
treatments. These treatments introduce the rationale of using
transition state analogs as bioactive molecules. Data from this
laboratory were presented as an example of the use of transition
state theory in the development of potent inhibitors of insect
juvenile hormone esterase(s).

Transition State Theory

It is well known from structural and kinetic studies that enzymes
have well-defined binding sites for their substrates (3), sometimes
form covalent intermediates, and generally involve acidic, basic and
nucleophilic groups. Many of the concepts in catalysis are based on
transition state (TS) theory. The first quantitative formulation of
that theory was extensively used in the work of H. Eyring (4,5).
Noteworthy contributions to the basic theory were made by others
(see (6) for review). As an elementary introduction, we will apply
the fundamental assumptions of the TS theory in simple enzyme
catalysis as follows.
(a) In every chemical reaction the reactants are in equilibrium
with an unstable activated complex, the transition state complex,
which decomposes to give products. In this complex chemical bonds
are in the process of being formed or broken. Therefore, it occurs
at the peak of the reaction coordinate diagram, i.e. at the saddle
points of potential energy surfaces (7). In contrast, intermediates, whose bonds are fully established, occupy the troughs in
the diagram (Figure 1). These intermediates can either be transient

or actual isolatable intermediates such as an acyl enzyme.
(b) It is postulated that the starting materials (in case of enzyme
catalysis the enzyme (E) and its substrate (S)) are in equilibrium
with all complexes which occur before the activated complex (ES*)
and also with the activated complex itself "Equation 1".

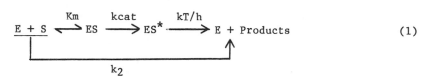

$$E + S \underset{Km}{\overset{}{\rightleftharpoons}} ES \xrightarrow{kcat} ES^* \xrightarrow{kT/h} E + Products \tag{1}$$

$$k_2$$

(c) The important postulate is made that in this theory all the
activated complexes decompose to products at exactly the same rate
for a given temperature. This means that the rate is proportional
to the concentration of ES* with a universal proportionality constant (kT/h) where k is the Boltzmann's constant, T is the absolute
temperature and h is the Plank's constant. At 25° (kT/h) equals
6.212×10^{12} sec^{-1}.

$$\text{rate} = \frac{kT}{h} [ES^*] \tag{2}$$

However ES* is in equilibrium with E and S and is governed by the
equilibrium constant K_t^* as follows:

$$\frac{[ES^*]}{[E][S]} = K_t^* \tag{3}$$

Substitution for the value of [ES*] in "Equation 2" from
"Equation 3" results in the following equation:

$$\text{rate} = [E][S]K_t^*(\frac{kT}{h}) \tag{4}$$

However, from "Equation 1" the rate also should be proportional to
[E] and [S] and a second order rate constant (k_2):

$$\text{rate} = k_2[E][S] \tag{5}$$

Comparing "Equation 5" with "Equation 4" indicates that:

$$k_2 = \frac{kT}{h} K_t^* \tag{6}$$

(d) A special property of the activated TS complex is that it has a
unique, very loose internal mode of vibration, which is unstable
with respect to dissociation into products. This "vibration" occurs
along the reaction coordinate (Figure 1). Therefore we will consider the TS complex as the end point on the energy profile for
simplicity.

Free Energy Change and TS Theory. We will now express the mechanism
of enzyme catalysis "Equation 1" in terms of the change in the free
energy as follows:

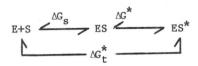

$$\text{E+S} \underset{\Delta G_s}{\longleftrightarrow} \text{ES} \underset{\Delta G^*}{\longleftrightarrow} \text{ES}^* \qquad (7)$$

By applying the well-known relationship between the Gibbs-energy change and the equilibrium constant (8) to the above scheme, one finds:

$$\Delta G_t^* = -RT \ \ln K_t^* \qquad (8)$$

where K_t^* is the equilibrium constant between the reactants and the TS. Analogous equations relate the equilibrium constant of any step and the corresponding free energy change. Upon rearrangement of "Equation 8" and introducing the value of K_t^* in terms of ΔG_t^* into "Equation 6", the following equation results:

$$k_2 = \frac{kT}{h} \cdot e^{-\Delta G_t^*/RT} \qquad (9)$$

Using the familiar relationship between Gibbs free energy change (ΔG_t^*) and the change in the enthalpy (ΔH_t^*) and entropy (ΔS_t^*),

$$\Delta G_t^* = \Delta H_t^* - T\Delta S_t^* \qquad (10)$$

"Equation 9" can be expressed in terms of ΔH_t^* and ΔS_t^* as follows:

$$k_2 = \frac{kT}{h} \cdot e^{-\Delta H_t^*/RT} \ e^{\Delta S_t^*/R} \qquad (11)$$

"Equation 11" indicates, from the theoretical point of view, that k_2 for a particular enzyme can be accelerated by either a decrease in the enthalpy and/or an increase in the entropy of the overall reaction E+S $\rightarrow$ ES* so that ΔH_t^* will be negative and ΔS_t^* will be positive in the above equation. The role of entropy change in terms of different entropy vectors, translational, rotational and internal entropies in enzyme catalysis is not an easy task and the reader should refer to special textbooks in physical organic chemistry (see (8) and references therein). Theoretically forming a TS complex (ES*) from an enzyme and substrate (2 $\rightarrow$ 1 reaction) would lead to a loss of the entropy of three degrees of translational freedom. However, a loose transition state with a high potential energy may, perhaps, be considered as two molecules in close juxtaposition but retaining considerable entropy freedom. Furthermore, enzymic reactions are different from pure organic chemical reactions since the former reaction takes place in the confines of the enzyme substrate complex. Therefore, the remarkably high catalytic activity of enzymes could be attributed to localization of S or S* within the active site with susceptible bonds of the substrate or its TS configuration optimally oriented to the appropriate catalytic moieties of the enzyme (3). Due to this localization and orientation, the substrate would be roughly considered as part of the same molecule as the catalytic group so there is no extensive loss in entropy as the reaction would be assumed to be intramolecular. If there is a great loss of entropy, it would be on forming the ES rather than ES* complex which results in increasing K_m (8). The entropy change in terms of translational, rotational and internal rotational entropy has been discussed in more detail (9,10) in favor of contribution to the binding of E and S* rather than S. However, the Van der Waals attraction (hydrophobic bonding) in enzyme-substrate interactions might increase the entropy of either ES or ES* as compared to that of the reactants by ejecting water molecules formerly bound to the catalytic site(s) (11). As illustrated in the above discussion, S* binds more tightly to the enzyme than S. Thus more molecules of solvating water would be released upon the formation of ES* resulting in an additional entropic advantage for its formation. Such hydrophobic bonding and entropy change in the reaction of enzyme with substrate or inhibitors will be fully explained in the next section.

Transition State and Binding Energy. There are at least two major factors which account for enzyme catalysis. The first one is a combination of entropic, acid base catalysis and electrostatic effects. The second one depends mostly on the enzyme substrate complementarity which results in a large amount of binding energy which may be used to distort the substrate to the structure of the products (12). The latter factor seems to be of high importance in lowering the activation energy of the overall reaction and subsequently increasing the magnitude of k_2 which is defined as k_{cat}/K_m in the simple Michaelis-Menten mechanism where $K_s = K_m$. Let us now put the simple scheme for enzyme catalysis in terms of both mechanistic and energetic entities.

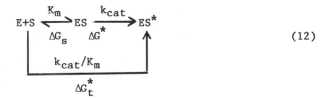

$$(12)$$

The above scheme can be seen clearly from Figure 2 in which ΔG_s is algebraically negative, i.e. favorable reaction due to the realization of binding energy. However, ΔG^* is positive (unfavorable reaction) due to the activation energy of bond rearrangement in the activated TS complex. That is the activation energy ΔG_t^* for the whole process (E+S $\rightleftharpoons$ ES*) would be

$$\Delta G_t^* = \Delta G^* + \Delta G_s \tag{13}$$

Substitution for the value of ΔG_t^* in "Equation 9" by $\Delta G^* + \Delta G_s$ "Equation 13" results in the following equation:

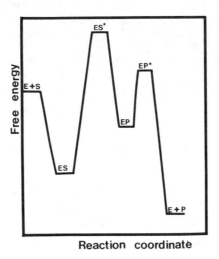

Figure 1. Schematic representation of the free energy changes in an enzyme-catalyzed reaction.

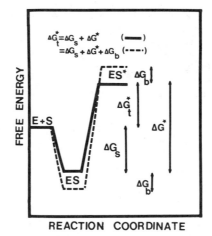

Figure 2. Schematic representation of the free energy changes in an enzyme-catalyzed reaction where the enzyme is complementary to either the substrate (broken lines) or to its transition state configuration (solid lines).

$$k_2 = k_{cat}/K_m = \frac{kT}{h} \; e^{-\Delta G^*/RT} \; e^{-\Delta G_s/RT} \tag{14}$$

From the above equation it can be shown that the binding energy of ES complex and/or of ES* complex can play a crucial role in lowering ΔG_s or ΔG^* respectively and eventually increasing the catalytic activity (k_{cat}/K_m). However, since the structure of the substrate (based on TS theory) changes throughout the reaction, it is likely that the undistorted enzyme can have maximum complementariness to only one species (S or S*) of the substrate (8). At this point an important question arises regarding which form of substrate would have the highest complementarity and the highest binding energy in its reaction with the enzyme. The importance of the above question comes from the fact that its answer is considered to be the main rationale for the development of TS-analogues (TSA) as enzyme inhibitors. The qualitative answer to this question was introduced by Haldane (12) and Pauling (1,2) as the favored complementarity would be between the enzyme and the transition state portion (S*) of the substrate rather than between the enzyme and the substrate itself. In supporting the great insight of Haldane and Pauling, Fersht (8) proved collectively that the intrinsic binding energy in driving k_{cat}/K_m would be in favor of the transition state. In the following paragraphs we will slightly modify Fersht's approach to show that it is energetically expensive for enzyme catalyzed reactions to have maximum binding interactions with the substrate rather than its TS configuration.

Assume that the extra binding energy resulting from complementarity to the substrate or to its TS equals ΔG_b. If this extra binding energy is in the enzyme substrate complex, it will decrease the value of K_m and reduce its free energy change to become $\Delta G_s - \Delta G_b$. However, since the formation of TS complex will lead to a reduction in binding energy as the substrate geometry changes to give poorer fit and eventually this will increase ΔG^* for k_{cat} to be $\Delta G^* + 2\Delta G_b \cdot \Delta G_t^*$ for the overall reaction (k_{cat}/K_m) can be calculated (see Figure 2) to be the sum of ΔG values for the preceding steps as follows:

$$\Delta G_t^* = \Delta G^* + 2\Delta G_b + \Delta G_s - \Delta G_b =$$
$$\Delta G^* + \Delta G_s + \Delta G_b \tag{15}$$

Substitution for the components of ΔG_t^* ("Equation 15") into "Equation 9":

$$k_2 = k_{cat}/K_m = \frac{kT}{h} \cdot e^{-\Delta G^*/RT} \; e^{-\Delta G_s/RT} \; e^{-\Delta G_b/RT} \tag{16}$$

Comparing "Equation 16" with "Equation 14" where in the latter the maximum intrinsic binding energy was assumed to be in the activated TS complex and taking into consideration ΔG_b to be originally

positive in "Equation 16" one finds that the favorable maximum fit
is likely to be with the transition state configuration rather than
the substrate itself. In fact the enzyme can have an extra
catalytic activity by a factor of $e^{\Delta G_b/RT}$ (by dividing "Equation 14"
over "Equation 16") just by using the same amount of binding energy
in interaction with S^* rather than with S. It is not hard to
realize the above arguments since k_{cat}/K_m is independent of the
interactions in the initial enzyme-substrate complex (Figure 2).

Although the overall catalytic activity of an enzyme catalyzed
reaction can be accounted for both by the affinity of the substrate
to the enzyme (K_m or more accurately $1/K_m$) and the substrate reactivity (k_{cat}), the latter value seems to be more important in
reflecting the extra binding energy in the ES^* complex. This
binding energy, as mentioned before, will decrease the energy of
activation for the reactivity process (k_{cat}).

This hypothesis has been supported with some serine proteases
where increasing the length of the leaving group increased k_{cat} for
chymotrypsin (13,14) or increasing the length of the polypeptide
chain of the substrate increased k_{cat} for elastase (15). The
catalytic rate constant (k_{cat}) for juvenile hormone esterase from
the larval hemolymph of Trichoplusia ni using JH_I, JH_{III} (16) and
JH_{II} (17) as substrates was kinetically measured to be respectively
37.1, 19.4 and 31.8 min^{-1}. Mumby and Hammock (18) reported the
partition coefficient (log P values) for 3 series of geranyl derivatives and JH_I. The log P value for the latter compound was
3.71. Log P values for the 2,3-unsaturated-6,7-epoxides, as the
closest series to the structure of JH homologs, were subjected to
Hansch's approach (19,20) to measure the substituent hydrophobicity
(π). The π value for CH_3 was calculated from different comparisons to be 0.47, 0.49, 0.54 with an average value of 0.5 in an
excellent agreement with the reported values for CH_3 (0.49-0.56)
calculated from the octanol/water partition coefficients of four
different systems (21). Therefore it is reasonably accepted to
calculate the expected log P values for JH_{II} and JH_{III} from that
of JH_I using the π value for CH_3 group. Interestingly a good
correlation between log P and k_{cat} for the three homologs was
obtained (Figure 3). This correlation might indicate that the
binding energy through hydrophobic interactions is likely to be
involved in the TS complex (ES^*). Since values of k_{cat} were
obtained from different hemolymph pools, the activity of the enzyme
from one single hemolymph pool was measured towards the three
homologs at a final molar concentration of $5x10^{-6}$ and the data were
plotted against log P. The same relation was obtained as with k_{cat}
which is expected since at the substrate concentration used
(~two orders of magnitude greater than the values of K_m for the
three homologs), the velocity would approximate V_{max} and the
latter equals $k_{cat}[E_t]$. An excellent linear function relationship
($r^2=0.999$) between the velocity and k_{cat} (Figure 3, inset) was
obtained and supports the above discussion. One of the advantages
of the above approach is that one can estimate the molar equivalency
of a specific enzyme in crude preparations. Since at enzyme-
substrate saturation conditions $v \simeq V_{max}=k_{cat}[E_t]$, the above
relation (Figure 3, inset) enables the calculation of $[E_t]$ from the

slope of the inset (1.22 nmoles/ml plasma) which is equivalent to $1.22 \times 10^{-6}M$ of juvenile hormone esterase in the hemolymph of T. ni. In fact this average number is in close agreement with that calculated from using each substrate separately using different hemolymph pools (16). The ability to determine the molarity of a catalytic site in situ is of tremendous benefit in the elucidation of physiological or pharmacokinetic parameters.

Transition State Analogs (TSA)

The qualitative description of enzyme catalysis in terms of the TS theory (Pauling, 1,2) that the enzyme is complementary to an unstable molecule with only transient existence; namely, the activated complex (ES*) for which the power of attraction by the enzyme is much greater than that of the substrate itself has been discussed energetically (8) and mechanistically (10,22-25). Pauling's assertion has opened a new era in enzymology, and relevant to our discussion is the stabilization of the activated complex and TSA as powerful enzyme inhibitors. As the transition state is a mathematical construction (with a typical half-life of 10^{-10} msec., (22)), its structure cannot be defined in common chemical sign language. In fact difficulty in isolating and defining transition state complexes was the major reason behind the recreation of Pauling's concept to design stable analogs approaching the structure of the altered substrate in the transition state without undergoing catalytic conversion. Eventually these TSA are expected to be stabilized upon the reaction with the enzyme to such a degree that they approach ground state energy minima (Figure 4). This stabilization would in fact enable indirect observation of the structure of the enzyme-TS by using the available techniques. Furthermore, if a TSA takes advantage of the additional, favorable binding interactions that are inherent in ES* interactions, it could be an extremely powerful inhibitor to the limit of stoichiometric reaction. Theoretically a perfect TSA would have the same affinity for the enzyme as the transition state of the substrate (Figure 4). The question at issue now is: what is the magnitude of enzyme-TS affinity as compared to enzyme substrate affinity? This question has been fully declared by several workers (10,22,23,25). In the following paragraphs we will summarize their mathematical approach in a simple way. Note that in this section all equilibria are defined as association constants.

The two reactions to be compared are:

$$E+S \xrightleftharpoons{K_{ES}} ES \qquad (17)$$

where K_{ES} is the equilibrium association constant ($= 1/K_m$) between the enzyme and substrate.

$$E+S^* \xrightleftharpoons{K_{TX}} ES^* \qquad (18)$$

However, there should be a hypothetical step that precedes "Equation 18" as follows:

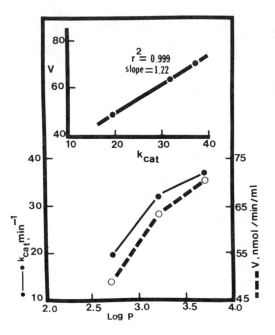

Figure 3. Plot of JH-hydrophobicity and JH-esterase hydrolytic
activity of JH I, JH II and JH III.

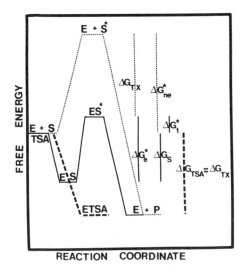

Figure 4. Schematic representation of the free energy changes
in non-enzymatic and enzymatic reactions and in the reaction
of a hypothetical transition state analog (TSA) with the enzyme.

$$E+S \xrightleftharpoons{K^*_{ne}} E+S^* \qquad (19)$$

In order to quantitate the relative affinity of the enzyme to S^* and S, a reasonable assumption is made that "Equation 19" stems from non-enzymatic reaction of the substrate, i.e. the mechanism of nonenzymatic reaction is the same as enzyme catalyzed reaction. It is useful to combine the above three reactions in a thermodynamic box as follows:

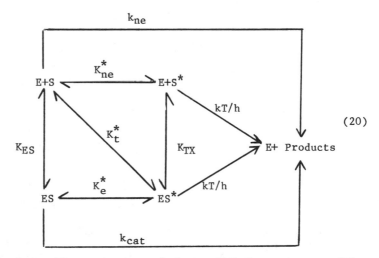

$$(20)$$

In the following table, a summary of the equilibrium constants (K) in relation to the changes of free energy of activation ($-RT \ln K$) and the corresponding rate constants is displayed.

Table I.   Equilibrium Constants and Free Energy Changes for Non-
Enzymatic and Enzymatic Catalyzed Reactions

Equilibrium constant (K)	ΔG	Rate constant (k)
$K^*_{ne} = [S^*]/[S]$	ΔG^*_{ne}	k_{ne} (t^{-1})
$K_{ES} = [ES]/[E][S]$	ΔG_S	$--$
$K_{TX} = [ES^*]/[E][S^*]$	ΔG_{TX}	$--$
$K^*_e = [ES^*]/[ES]$	ΔG^*_e	k_{cat} (t^{-1})
$K^*_t = [ES^*]/[E][S]$	ΔG^*_t	$k_{cat} \cdot K_{ES} = k_{cat}/K_m$ $(M^{-1} t^{-1})$

From the above table an energy profile can be drawn for enzymatic vs. non-enzymatic reaction (Figure 4). From this figure, one easily can see how important it is that the enzyme should have un-

usual affinity for the altered substrate to make the transition
state easier to reach, i.e. by decreasing the activation energy or
the energy barrier of the overall reaction by a fraction equivalent
to ΔG_{TX}. This amount of energy change must be due to the additional, favorable binding interactions that are inherited in the
theoretical interaction between E and S^* ("Equation 18") and used
to decrease the activation energy of the reaction as the collision
complex or ground state complex (ES) approaches its transition state
conformation (ES^*). Since K_{TX} is rather hypothetical, in order
to make a prediction of its magnitude as compared with K_{ES}, the
rate constants of the corresponding reactions (Table I) can be used
according to the general relationship which is generated from the TS
theory:

$$K_{ne}^* \cdot kT/h = k_{ne} \tag{21}$$

$$K_e^* \cdot kT/h = k_{cat} \tag{22}$$

$$K_t^* \cdot kT/h = k_{cat} \cdot K_{ES} \tag{23}$$

However, since ES^* is in equilibrium with the whole thermodynamic
box, that is the overall K_{eq} of the reaction between E, S and ES^*
must be the same regardless of the path, i.e.

$$K_{ne}^* \cdot K_{TX} = K_{ES} \cdot K_e^* = K_t^* \tag{24}$$

These relations can also be reached from the first column of
Table I.

   "Equation 24" is considered to be a cornerstone to the understanding of enzyme catalysis in terms of the magnitude of enzyme-
substrate affinity in the transition state and how this affinity
affects the energy of activation in favor of enzymatic catalysis
rather than non-enzymatic reaction.

   The relative affinity of E to S^* and to S can be calculated
from "Equation 24" to be

$$K_{TX}/K_{ES} = K_e^*/K_{ne}^* \tag{25}$$

Substituting for K_e^* and K_{ne}^* by their corresponding rate constants ("Equations 21 and 22") results in the following equation:

$$K_{TX}/K_{ES} = k_{cat}/k_{ne} \tag{26}$$

Thus "Equation 26" shows that the tighter binding of S^* than S to
the enzyme as expressed by K_{TX}/K_{ES} must be equivalent to the rate
constant ratio (k_{cat}/k_{ne}). Since the latter ratio is typically
10^{10} or more (10, 23,26), the binding of the enzyme to the transition state configuration of the substrate should be at least 10
orders of magnitude tighter than the binding of the enzyme to the
substrate itself. On the other hand by applying the free energy
relationship ($\Delta G = -RT \ln K$) to "Equation 24", the changes in the
free energy for the thermodynamic box, "Equation 20", would be calculated as follows:

$$\Delta G^*_{ne} + \Delta G_{TX} = \Delta G_S + \Delta G^*_e = \Delta G^*_t \tag{27}$$

where ΔG^*_{ne}, ΔG^*_e and ΔG^*_t are algebraically positive and ΔG_{TX} and ΔG_S are negative (Figure 4). From the above equation one can see that the activation energy for the reaction is less in the presence of enzyme (ΔG^*_t) than in its absence (ΔG^*_{ne}) by an amount of energy equivalent of ΔG_{TX} due to the greatest power of attraction between the enzyme and the TS configuration of the substrate. Accordingly the enzyme would speed the reaction. This picture explains the main concept of TSA as extraordinarily powerful enzyme inhibitors. The rationale for this approach is to design molecules which are structured in such a way that they resemble the transition state configuration of the substrate to the extent that they can take advantage of the binding interactions similar to ΔG_{TX} without undergoing catalytic conversion, and therefore would exhibit tighter binding to the enzyme than the substrate itself (Figure 4). However, it would be unrealistic to suppose than an ideal TSA, perfectly resembling the substrate in its transition state can ever be synthesized so that K_I/K_M would be in the same order as k_{ne}/k_{cat} (10^{-10} or lower). This is an important point from the practical point of view since it raises the question of which criteria one can use to differentiate between substrate analogs and TSA as inhibitors of a particular enzyme. In general TSA should exhibit abnormally low dissociation constants with the enzymes when compared with their K_m values (3). Exploitation of this approach has been reviewed (10,22-25,27). The simple numerical comparison of K_I with K_M fails to distinguish between either a TSA and a substrate analog for at least two reasons. First, no inhibitor will be a perfect mimic of either the substrate or its TS configuration (28). Second, most enzymes show overlapping specificity to different substrates and one enzyme can have a range of K_m values over several orders of magnitude. For example K_m values for the hydrolysis of N-acyl-L-amino acid esters by α-chymotrypsin (29) varied from 0.018 mM for benzoyl tyrosyl methyl ester to 862 mM for acetyl glycyl methyl ester indicating a difference of about five orders of magnitude. Fortunately, enzymes with overlapping substrate specificity can be of great help in distinguishing TSA from substrate analogs if a series of related inhibitors can be compared with the corresponding members of a series of related substrates. By taking the log of both sides of "Equation 26" and rearranging the log values,

$$-\log K_{TX} = -\log K_{ES} - \log k_{cat} + \log k_{ne} \tag{28}$$

However, $-\log K_{ES} = \log K_M$ and $-\log K_{TX}$ can be substituted by $\log K_I$ in the case of TSA, where K_I is the enzyme-inhibitor dissociation constant.

$$\log K_I = \log(K_M/k_{cat}) + \log k_{ne} \tag{29}$$

If one assumes that the rate of the nonenzymatic reaction does not vary among the substrates used, application of "Equation 29" can be a rigorous criterion for justifying the TSA since a linear function relationship would be obtained between $\log K_I$ for these inhibitors and $\log(K_M/k_{cat})$ for the corresponding parallel substrates. In case of using the same enzyme preparation the substrates' $\log (K_m/V_{max})$ can be used in the above relationship. Correlations of this kind have been reported for TSA of different enzymes (28,30-33). On the other hand, if a better correlation between $\log K_I$ and $\log K_M$ were obtained, the inhibitors would likely be acting as substrate analogues.

Application of TS Theory to Organophosphate and Carbamate Insecticides

It is well known that organophosphates, carbamates and sulfonates are acid-transfer-inhibitors of serine hydrolases because they transfer the acid moiety of the inhibitor to the serine hydroxyl of the enzyme active site (34). Extensive evidence indicates that the reaction of these inhibitors with acetylcholinesterases (AChE) appears to involve the same reaction pathway as that for the esters of carboxylic acids, i.e. acetylcholine (see (35) for review), and in fact these inhibitors are considered to be poor substrates of AChE (36), especially the carbamic acid esters ("Equation 30").

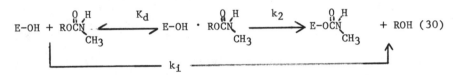

Therefore, it is not surprising that the basic kinetics ("Equation 31") for the inhibition by these compounds (37,38) has precisely the same form as the Michaelis-Menten rectangular

$$\rho = k_2 I/(K_d + I) \tag{31}$$

hyperbolic equation $(v = \dfrac{V_{max}S}{K_m + S})$ by replacing ρ, the first order inhibition rate constant, with v, k_2 with V_{max}, K_d with K_m and I with S (39). In this respect it came to us that these compounds might in fact resemble the transition state configuration of the substrate and accordingly part of their effect might be due to acting as TSA somewhere along the reaction coordinate of the catalytic reaction. Recently it has been found that the N-hydroxy (40), and N-methoxy (40,41) in contrast to N-methyl, and N,N-dimethylcarbamates inhibit AChE reversibly in a complete competitive manner. These findings stimulated our hypothesis that N-methyl and N,N-dimethylcarbamates, as carbamylating agents of AChE (36), can be considered poor substrates and the corresponding N-hydroxy N-methyl or N-methoxy N-methylcarbamates may be truly TSA. Supportive to our hypothesis is that m-trimethylammonium phenyl N-methyl N-methoxy-carbamate, the most active reversible competitive inhibitor among the whole series (41) has a dissociation constant (K_I) of about three orders of magnitudes lower than the K_m values of the three tested enzymes. To test the above hypothesis "Equation 29" was applied to the inhibition of housefly head AChE by substituted phenyl N-methyl-carbamates using the I_{50} values reported by Metcalf

(42) and the K_I values of the corresponding N-methyl-N-methoxy-carbamates measured by Wustner et al. (41). The equivalent values of substrate (K_m/k_{cat}) would be the reciprocal of the bimolecular reaction constant ($1/k_i$) since k_i equals k_2/K_d (37). k_i values were calculated from the I_{50} values and 15 minute preincubation time (t) from the following relationship (43):

$$k_i = \frac{0.693}{I_{50}t} \qquad (32)$$

A plot of log K_I against log $1/k_i$ (Figure 5) clearly reveals an excellent linear relationship (r=0.97) indicating that these N-methoxy N-methylcarbamates may in fact act as TSA, and the corresponding N-methylcarbamates are poor substrates and part of their inhibitory potency is likely to be due to their resemblance to the TS complex in the reaction of the enzyme with its carboxy ester substrate. Additional supportive evidence is that when Wustner et al. (41) tried to prove that the ring substituents in both series of compounds interact similarly with the bovine erythrocyte AChE active site they found a disappointing correlation between K_I and K_d and a better correlation between K_I and k_i respectively for N-methoxy N-methyl and N-methylcarbamates.

A similar but intuitive argument can be made that the binding of organophosphorous compounds to the acetylcholinesterase may involve some aspects of TS theory. In the development of organophosphorous toxins, we classically assume that one is attempting to synthesize a molecule which mimics the substrate acetylcholine. With VX gas as an example one can assume that back bonding between oxygen and phosphorous leads to an electrophilic center mimicking the carbonyl carbon and that the tertiary amine mimics the quaternary ammonium of acetylcholine. However, one could argue that the tetrahedral phosphorous mimics the tetrahedral intermediate formed during hydrolysis of acetylcholine. It should be pointed out that such tetrahedral "transition states" often mentioned in the hydrolysis of esters are really transient intermediates occupying a slight dip near the peak of the reaction coordinates. The true transition states occur during the formation and breaking of the bonds of the tetrahedral intermediate. Thus, it is not surprising that some tetrahedral organophosphorous compounds can display an apparent affinity for an esterase which is even higher than that shown by the trigonal substrate. Perhaps an appreciation of TS theory may lead to better optimization of such organophosphorous and carbamate structures.

TSA as Inhibitors of Juvenile Hormone Esterase(s)

Background. The juvenile hormones (JHs) regulate a myriad of developmental and reproductive events in insects, and metamorphosis certainly is among the most striking of these events (44). The reduction in JH titer to initiate metamorphosis in Lepidoptera examined appears to be caused by degradative metabolism as well as reduction in the rate of biosynthesis (44,45). Ester cleavage of JHs is apparently the major route of metabolism. In the cabbage looper, Trichoplusia ni (T. ni) hydrolysis of JH is due largely to a

single enzyme (JH esterase) mainly present in the hemolymph and fat
body (46-49). Although the specification of the precise binding
mode through crystallographic or other means is eagerly awaited for
JHEs, inhibition kinetics indicate indirectly that JHEs from T. ni
(50) and Manduca sexta (51) are likely to be serine hydrolases.
Typically, drug metabolizing carboxylesterases are serine hydrolases
(EC 3.1.1.1.) and as such associate with the electrophilic moieties
of esters or amides, organophosphates (52) and halomethyl ketones
(53). Based on these indirect lines of evidence hydrolysis of the
JHs by JHEs is thought to proceed with a change in the bond order of
the substrate (54) to form tetrahedral adduct with the enzyme from
the trigonal structure of the ester. The mode of operation of
serine proteases is now fairly well understood in terms of their
three dimensional geometries down to the level of individual atomic
position (55). There appears to be a general agreement on the role
of the charge relay system (proton relay system) in the catalytic
activity of serine proteases with some little confusion in the
detailed mechanism or the relative importance of each step in proton
transfer (see (55) for review). Since JHs are considered to be less
polarized at the carbonyl of the methyl esters due to the lack of
electron withdrawing properties of the JH acids, one expects with
caution a similar proton relay system is likely to be evolved in
increasing the nucleophilicity of the serine oxygen to be reactive
toward the less polarized carbonyl of the JH ester. This, of
course, does not exclude the importance of the bulk of the long
chain acid moiety and the methyl group of the alcohol moiety in the
binding sites for the stabilization of the tetrahedral adduct.
Based on the above discussion it was thought that the trifluoromethyl ketones would be more polarized and thus create a great
electrophilicity on the carbonyl carbon which facilitates -OH
attack by the serine residue. Yet there is no carbon-oxygen bond
to be cleaved in the ketone moiety, and therefore the enzyme-
trifluoromethyl ketone transition state complex does not undergo
catalytic conversion. The above rationale seems reasonable as
trifluoromethyl ketones were found to be extraordinary selective and
potent inhibitors of cholinesterases (56) of JHE from T. ni (57) and
of meperidine carboxylesterases from mouse and human livers (58).
Since JH homologs are alpha-beta unsaturated esters, a sulfide bond
was placed beta to the carbonyl in hopes that it would mimic the
2,3-olefin of JHs and yield more powerful inhibitors (54). This
empirical approach was extremely successful since it resulted in
compounds that were extremely potent inhibitors of JHEs from
different species (51,54,59).

Quantitative Structure-Activity Relationships (QSAR) of Trifluoro-
ketones as JHE Inhibitors. The great structural variety of trifluoromethyl ketones, and the reported biological response against
JHE from T. ni (54,57) makes these results particularly well suited
for a QSAR investigation. Except for the values of molar refractivity (MR), all the physicochemical parameters used in the present
work were from the recent compilation of Hansch and Leo (21). To
the best of our knowledge this will be the second QSAR study for the
inhibition of JHE. The first one was done by Magee (60) in
analyzing the data of Hammock et al. (61) for the inhibition of JHEs

from cockroach (Blaberus giganteus) by some phosphoramidothioates.

In the first approach the activity of 3-alkyl and 3-alkylthio 1,1,1-trifluoro-propan-2-ones was considered for their structure-activity relationship (SAR) with MR. MR was used in the present study to model the enzyme-inhibitor attraction forces since MR is related to London dispersion forces (21,62,63) and has been also proposed to be really a corrected form of the molar volume (21). Figure 6 shows a clear parabolic relation between the molar I_{50} value and MR values for R groups in the general structure $R\overset{O}{\overset{\|}{C}}CF_3$. It is worth noting that both series show an optimum MR value for strong inhibitory activity and that activity decreases when MR values are either larger or smaller than this value. This parabolic relation indicates that the compounds with an MR value smaller than the optimum value might have a positive coefficient for significant MR term which would indicate productive binding for JHE inhibition. However the rest of the compounds appeared to have a negative coefficient which might be due to steric inhibition of binding to the enzyme active site. Although the above discussion seems reasonable since a considerable collinearity between MR and Hancock's steric parameter (64) was found by Dunn (63), the attraction by dispersion forces for another region beyond the catalytic site of the enzyme could be involved. In general, it can be concluded that the affinity of these compounds for the enzyme is a linearly decreasing function of the sum of non-overlappable volumes of the inhibitor molecule and a receptor cavity on the enzyme active site. Since MR is an additive and constitutive property of the molecules (21), it was calculated from the fragment values of the atoms (C = 2.418, H = 1.1; ethereal oxygen = 1.643 and an increment for C = C double bond of 1.733) for both series except that the experimental MR value for C_4H_9S-(31.5) as quoted by Balaban et al. (65) was used as a basic group for the sulfide series. In an attempt to relate the MR value for these inhibitors with that of the corresponding straight chain of JHs, values of 47.1 and 52, respectively, for JH III and JH I were calculated. Considering the optical exaltation due to the conjugated double bond to be 2.85 (ΔMR between the enol and keto forms of ethyl acetoacetate) (66), the calculated value for the above JH homologs would be 50 and 54.9 respectively. When the above values scaled by 0.1 as in Figure 6, the optimum MR value for the most active compound in each series is identical with that of JH I and slightly larger than that of JH III. If one assumes that the octyl sulfide compound, as the most active compound in both series, is practically and roughly a perfect TSA it would be expected to have most of the available binding energy to the enzyme and eventually would behave as a tight binding inhibitor and can be used to titrate the active site of JHE. In fact this compound was found kinetically to act as a slow tight binding (16) and was used to evaluate the molar equivalency of JHE from T. ni to be 1.6×10^{-6}M in the hemolymph. Since in our laboratory a 1:500 diluted hemolymph is always used and if the above molarity is correct, this means that a molar equivalency of 3.2×10^{-9} is used in the inhibition assay. The molar I_{50} value of this compound was found to be consistent in at least five measurements (2.3×10^{-9}M) and almost identical with half the enzyme concentration, a behavior

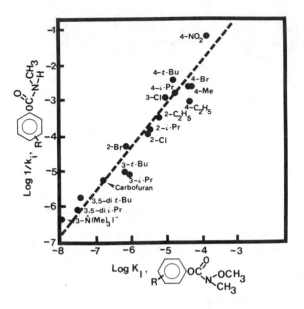

Figure 5. Relationship between the bimolecular inhibition
reaction constant (k_i) of substituted phenyl N-methylcarbamates
and the inhibition dissociation constant (K_I) of the corresponding N-methyl N-methoxycarbamates in their reaction with
house fly acetylcholinesterase. k_i Values were recalculated
from (42) and K_I values were from (41).

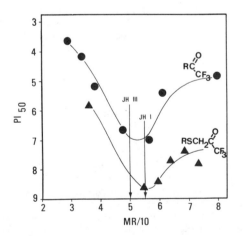

Figure 6. Relationship between the inhibitory activity of
trifluoromethylketones against JH-esterase from <u>Trichoplusia
ni</u> and the molar equivalents of these inhibitors.

of titrimetric and stoichiometric enzyme inhibitor. Additional
supportive evidence for considering these compounds or more accurately the most active ones of the series to be TSA is that JH I
which has the same MR as the octyl sulfide compound showed
faster turnover number than JH III, with smaller MR than optimum, in con-
sistence with the discussion mentioned before for the binding energy
of the TS complex. The above relation is not consistent with K_m
values for both homologs which excludes the possibility that these
compounds act as substrate analogs. Therefore the unusually high
activity of these compounds can be explained by forming hemiketal
structures which mimic a transition state near the tetrahedral
intermediate complex in ester hydrolysis (Figure 7). In moving
from SAR to QSAR, log MR was used instead of 0.1 MR where the latter
was used to generate Figure 6. This was done for at least two
reasons. First, better correlation was obtained when log MR was
used. Second, almost all the physicochemical parameters used for
QSAR are originally from log terms in order to be linear free energy
parameters. The alkyl and alkyl sulfide derivatives were analyzed
separately at first and the following equations were obtained
respectively for the following structures:

$$\overset{\overset{O}{\|}}{RCCX_3} \qquad\qquad \text{(series no. 1)}$$

where R = alkyl, X = H or F

$$\text{and } R\overset{\overset{O}{\|}}{SCH_2CCF_3} \qquad\qquad \text{(series no. 2)}$$

$$PI_{50} = -140.5 + 167.7 \log MR + 2.5\ \Sigma F_{x3} - 48.9\ (\log MR)^2 \qquad (33)$$

n	r	s
9	0.95	0.6

$$PI_{50} = -226.6 + 269.9 \log MR - 77.5\ (\log MR)^2 \qquad (34)$$

n	r	s
6	0.94	0.44

where n represents the number of data points, r is the correlation
coefficient and s is the standard deviation from the regression. It
is rather interesting that the F term (polar effect) has a positive
coefficient in "Equation 33" in agreement with our belief that the
activity of these compounds might be due to the polarized ketone
induced by the fluorine atoms on the α-carbon which facilitates
the electrophilicity towards lone pair electrons on the oxygen atom
of serine hydroxyl in the enzyme catalytic site. On the other hand,
using the continuous variables MR, F in the above equations leaves a
relatively large amount of variance in the data unaccounted for
which might indicate that the basic group in the two series
(trifluoromethyl) has the key structural requirement for their
potency. At this point, structural modification of this group using
different congeners might yield insights into whether the fluorine
atoms act mainly through electronic transmission and/or steric

interaction with the catalytic site. In comparing members of the
sulfide series with those lacking sulfide of the same MR, the former
seems to have an average bestowed potency of 67.5. Therefore an
indicator variable (X) was used to analyze both series in one
regression, "Equation 35". X = 1.83 for the sulfide series and = 0
for the series lacking the sulfide bond is simply a device for
merging the above two equations. To this point it is not certain
how the sulfur increases the inhibitory potency of these compounds.
The crystallographic study of a pure enzyme might offer clear opportunity to know whether the sulfur mimics the olefin structure of JH
homologs or it contributes to the binding forces for example through
hydrogen bonding since the extra bestowed activity is equivalent of
about 2.54 Kcal/mole.

$$PI_{50} = -143.4 + 170.0 \log MR + 2.39 \ \Sigma F_{x3} - 49.1 \ (\log MR)^2 + 0.89X$$

$$
\begin{array}{ccc}
n & r & s \\
15 & .962 & 0.61
\end{array}
\qquad (35)
$$

If the above equations represent a real model of quantitation to the
interactions of these compounds with JHE, which seems to be the case
since "Equation 35" holds for two series from two separate studies,
one can assume that MR which dominates the above equations can explain more than 90% of the variation in the activity of members of
both series. Interestingly, the hydrophobicity (π term) did not
give better correlation when used instead of log MR in the above e-
quations. Since the first use of MR for correlation of substituent
effects (62), greater effort has been made to discuss the theory
behind the use of this parameter in correlating ligand interactions
(67) and to discuss the nature of interaction in which either MR
and/or π model such interaction (68,69). Generally, MR reflects
apolar as well as polar interactions where desolvation is not the
main driving force. However, when π and not MR models the interaction, desolvation appears to play the dominant role (69). In
relating the regression equations with the above working hypothesis,
desolvation in the interaction of the above compounds with JHE
appears not to be the driving force in the interaction process.
Whether the interaction is apolar or polar needs further study in
which π and MR should be orthogonal so that one can see exactly the
role of each parameter in the interaction without being so
collinear. Figure 8 shows the relation between the observed PI_{50}
values vs. the expected values according to "Equation 35".
        The activity of 1,1,1-trifluoro 3-mercapto substituted phenyl
propan-2-ones was not simply correlated with MR alone, therefore, a
combination of Hammett σ constant, Taft steric parameter (E_s) and
Hansch hydrophobic constant (π) was included in the regression
analysis. "Equation 36" was found to be of the best fit for compounds substituted in the meta and para positions.

$$PI_{50} = 5.15 + 1.07 \ \Sigma\pi + 1.23 \ \Sigma \log MR - 1.08 \ \Sigma\sigma + 0.59 \ (\Sigma\sigma)^2 + 0.35 \ \Sigma E_s$$

$$
\begin{array}{ccc}
n & r & s \\
13 & 0.951 & 0.39
\end{array}
\qquad (36)
$$

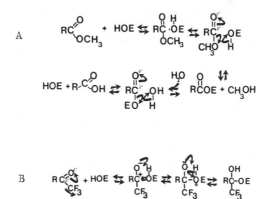

Figure 7.   Proposed scheme for the reactions of JH-esterase with JH homologs (A) and with trifluoromethylketones (B).

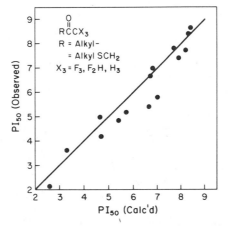

Figure 8.   Relationship between the observed PI_{50} values and the PI_{50} values calculated from "Equation 35".

From the slope of the physicochemical parameters in "Equation 36" it
appeared that π, MR and σ have about the same weight in modeling the
activity of the 13 meta and para substituted compounds. The importance of E_s and σ^2 seems to be less although high covariance
between E_s and both π and MR is a serious obstacle preventing the
true delineation of E_s effects. Meanwhile the equation reveals
that hydrophobic substituents (positive π term) with large MR and e-
lectron donating properties (negative σ term) would enhance the
inhibitory potency of the trifluoromercaptophenylacetone as a parent
compound. In fact this was the rationale behind synthesizing the
para t-butyl analog, and it came out to be the most active congener
among the tested compounds. Since covariance between MR and π is
low ($r^2 = 0.46$), they are considered to be orthogonal and have
different substituent effects. As discussed earlier for the alkyl
compounds, these substituent effects might establish different type
of interactions, i.e. polar and apolar interactions. Since these
compounds are substituted in the same positions, it is not clear
whether those types of interactions are in two different types of
enzymic spaces or one space which is open to surrounding solution.
In any case this leads to further research in which crystallographic
structure of JHE might shed some light on the binding space(s) and
the role of desolvation in their interaction with these novel inhibitors. At this moment, the coefficient of near 1 with π in
"Equation 36" should not be ignored and as has been discussed before
(70) for papain ligand interactions, brings out the close parallel
between binding to JHE and partitioning into octanol. This indicates that there might be a true hydrophobic pocket on the enzyme
active surface where interaction is driven by desolvation and the
interaction is likely to be entropic in nature. This conclusion
should not be oversimplified since extensive thermodynamic studies
certainly are required to shed some light on the role of both
entropy and enthalpy changes for the inhibition process by these
compounds.

The selective inhibition of JHE and α-naphthyl acetate
esterase(s) by ortho, meta and para substituted compounds showed
that meta and para substitution offered selectivity towards JHE,
however, the ortho substituted compounds favored inhibition of
α-naphthyl acetate esterases (54). It was thought that substitution
in the ortho position might be detrimental for inhibition of JHE.
Therefore we decided to add E_s value for the ortho substituents in
the regression analysis to merge the ortho compounds with their
meta and para analogs in the same regression, "Equation 37".

$$PI_{50}=5.05 + 0.74 \ \Sigma\pi + 1.14 \ \Sigma \log MR - 0.61 \ \Sigma\sigma - 0.11 \ (\Sigma\sigma)^2 + 0.61 \ E_s$$

(ortho)

n	r	s	
18	0.92	0.45	(37)

By comparing "Equations 36 and 37" one can see that the coefficients
of π and MR are more stable to the addition of the ortho compounds
to the regression analysis. However, the coefficients of σ terms
became small upon the addition of E_s ortho. The stability of the

coefficients of π and MR in addition to the fact that the intercepts in both equations are so similar to the observed PI_{50} value (5.08) for the parent unsubstituted compound might suggest the validity of these equations and support the above discussion in modeling the interaction of these compounds with JHE in terms of their hydrophobicity and molar refractivity. Figure 9 shows the relation between the observed PI_{50} values for the members of the aromatic series and the PI_{50} values calculated from "Equation 37".

In conclusion, although our QSAR for the inhibition of JHE with trifluoromethyl ketones does not offer an excellent and sharp fit of the data to the regression ("Equations 33–37"), it might provide the chemist with a model and rough base line for testing new compounds. In general the interactions of these compounds with JHE are likely to be hydrophobic and nonhydrophobic. Whether the latter type is separable from the former or in fact it is hydrophobic without being dependent on desolvation is not clear. Nevertheless, as these compounds are considered to be stable TSA, they offer a valuable tool in studying the x-ray crystallography of JHE and its TS complex from which valuable information can be coupled with the QSAR study and would greatly increase our understanding of JH–JHE interaction. It is worth reporting that one of these analogs was attached to insoluble support and proved to be an excellent ligand for the purification of JHE by affinity chromatography (51). This is the first step in approaching the three dimensional structure of JHE and JHE–inhibitor complex.

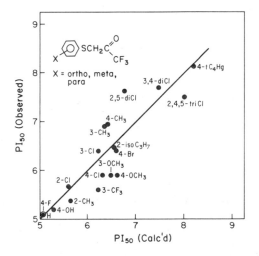

Figure 9. Relationship between the observed PI_{50} values and the PI_{50} values calculated from "Equation 37".

Acknowledgments

This work was supported, in part, by NIEHS Grant R01 ES02710-03 and and award in Agricultural Chemistry from the Herman Frasch Foundation. B. D. Hammock by NIEHS Research Career Development Award 5K04 ES00107-05. The illuminating discussion and technical assistance of Terry Hanzlik and Jim Ottea, and the patience of Ms. Peggy Kaplan and Kathleen Dooley in typing and revising the manuscript are gratefully appreciated.

Literature Cited

1. Pauling, L. Chem. Eng. News 1946, 24, 1375.
2. Pauling, L. Am. Scient. 1948, 36, 51.
3. Kirsch, J. F. Ann. Rev. Biochem. 1973, 42, 205.
4. Eyring, H. J. Chem. Phys. 1935, 3, 107.
5. Eyring, H. Chem. Rev. 1935, 17, 65.
6. Laidler, K.J. "Theories of Chemical Reaction Rates"; McGraw-Hill: New York, 1969.
7. Moss, S.J.; Coady, C.J. J. Chem. Educ. 1983, 60, 455.
8. Fersht, A. "Enzyme Structure and Mechanism"; W.H. Freeman: San Francisco, 1977; Chap. 2, 9, 10.
9. Page, M.I.; Jencks, W.P. Proc. Nat. Acad. Sci. U.S.A. 1971, 68, 1678.
10. Lienhard, G.E. Science 1973, 180, 149.
11. Laidler, K.J. "Some Kinetic and Mechanistic Aspects of Hydrolytic Enzyme Action"; Discussion Farady Soc. 1955, 20, pp. 83-95.
12. Haldane, J.B.S. "Enzymes"; Longmans, Green and Co: London; 1930.
13. Baumann, W.K.; Bizzozero, S.A.; Dutler, H. FEBS Letts. 1970, 8, 257.
14. Baumann, W.K.; Bizzozero, S.A.; Dutler, H. Eur. J. Biochem. 1973, 39, 381.
15. Thompson, R.C.; Blout, E.R. Biochemistry 1973, 12, 57.
16. Abdel-Aal, Y.A.I.; Hammock, B.D. Insect Biochem. (in press).
17. Abdel-Aal, Y.A.I.; Hammock, B.D., unpublished data.
18. Mumby, S.M.; Hammock, B.D. J. Agric. Fd. Chem. 1979, 27, 1223.
19. Fujita, T.; Iwasa, J.; Hansch, C. J. Am. Chem. Soc. 1964, 86, 5175.
20. Leo, A.; Hansch, C.; Elkins, D. Chem. Rev. 1971, 71, 525.
21. Hansch, C.; Leo, A. "Substituent Constant for Correlation Analysis in Chemistry and Biology"; Wiley-Interscience: New York, 1979.
22. Wolfenden, R. Nature 1969, 223, 704.
23. Wolfenden, R. Acc. Chem. Res. 1972, 5, 10.
24. Lienhard, G.E. Ann. Rep. Med. Chem. 1972, 7, 249.
25. Lienhard, G.E.; Secemski, I.I.; Koehler, K.A. In "Cold Spring Harbor Symposia on Quantitative Biology"; 1972, vol. 36, 45-51.
26. Jenks, W.P. "Catalysis in Chemistry and Enzymology"; McGraw-Hill: New York, 1969.
27. Wolfenden, R. Ann. Rev. Biophys. Bioeng. 1976, 5, 271.
28. Bartlett, P.A.; Marlowe, C.K. Biochemistry 1983, 22, 4618.
29. Berezin, I.V.; Kazanskaya, N.F.; Klyosov, A.A. FEBS Letts. 1971, 15, 121.

30. Thompson, R.C. Biochemistry 1973, 12, 47.
31. Westerik, J.O.; Wolfenden, R. J. Biol. Chem. 1972, 247, 8195.
32. Wolfenden, R.; Wentworth, D.F.; Mitchell, G.N. Biochemistry 1977, 16, 5071.
33. Thompson, R.C.; Bauer, C.-A. Biochemistry 1979, 18, 1552.
34. Smissaert, H.R. In "Toxicology, Biodegradation and Efficacy of Livestock Pesticides"; Khan, M.A.; Haufe, W.O., Eds.; Swets & Zeitlinger: Amsterdam, 1970; pp. 43-72.
35. Rosenberry, T.L. In "Adv. in Eznymol."; Meister, A. Ed.; Wiley: New York, 1975, Vol. 43, pp. 103-218.
36. Aldridge, W.N.; Reiner, E. "Enzyme Inhibitors as Substrates: Interaction of Esterases with Esters of Organophosphorus and Carbamic Acids"; North Holland-Amsterdam, 1975.
37. Main, A.R. Science 1964, 144, 992.
38. Main, A.R.; Iverson, F. Biochem. J. 1966, 100, 525.
39. Main, A.R. In "Introduction to Biochemical Toxicology"; Hodgson, E.; Guthrie, F.E., Eds.; Elsevier: New York, 1980, Chap. 11.
40. Chiu, Y.C.; Fahmy, M.A.H.; Fukuto, T.R. Pestic. Biochem. Physiol. 1973, 3, 1.
41. Wustner, D.A.; Smith, C.; Fukuto, T.R. Pestic. Biochem. Physiol. 1978, 9, 281.
42. Metcalf, R.L. Bull. WHO 1971, 44, 43.
43. O'Brien, R.D. "Toxic Phosphorus Esters"; Academic: New York, 1960.
44. Hammock, B.D.; Quistad, G.B. In "Progress in Pesticide Biochemistry"; Huston, D.H.; Roberts, T.R., Eds.; Wiley: New York, 1981, Vol. 1, p. 1.
45. de Kort, C.A.D.; Granger, N.A. Ann. Rev. Entomol. 1981, 26, 1.
46. Sparks, T.C.; Hammock, B.D. Insect Biochem. 1979, 9, 411.
47. Wing, K.D.; Sparks, T.C.; Lovell, V.M.; Levinson, S.O.; Hammock, B.D. Insect Biochem. 1981, 11, 473.
48. Wing, K.D.; Rudnicka, M.; Jones, G.; Jones, D.; Hammock, B.D. J. Comp. Physiol. B 1984, 154, 213.
49. Rudnicka, M.; Hammock, B.D. Insect Biochem. 1981, 11, 437.
50. Abdel-Aal, Y.A.I.; Roe, R.M.; Hammock, B.D. Pestic. Biochem. Physiol. 1984, 21, 232.
51. Abdel-Aal, Y.A.I.; Hammock, B.D., unpublished data.
52. Junge, W.; Krisch, K. CRC Crit. Rev. Toxic. 1975, 3, 371.
53. Powers, J.C. In "Methods in Enzymology"; Jakoby, W.B.; Wilchek, M., Eds.; Academic: New York, 1977, Vol. XLVI, pp. 197-208.
54. Hammock, B.D.; Abdel-Aal, Y.A.I.; Mullin, C.A.; Hanzlik, T.N.; Roe, R.M. Pestic. Biochem. Physiol. (in press).
55. Kraut, J. Ann. Rev. Biochem. 1977, 46, 331.
56. Brodbeck, U.; Schweikert, K.; Gentinetta, R.; Rottenberg, M. Biochim. Biophys. Acta 1979, 567, 357.
57. Hammock, B.D.; Wing, K.D.; McLaughlin, J.; Lovell, V.M.; Sparks, T.C. Pestic. Biochem. Physiol. 1982, 17, 76.
58. Lotti, M.; Ketterman, A.; Waskell, L.; Talcott, R.E. Biochem. Pharmacol. 1983, 32, 3735.
59. Roe, R.M., personal communication.
60. Magee, P.S. In "Insecticide Mode of Action"; Coates, J.R. Ed.; Academic: New York, 1982, Chap. 5, pp. 101-161.

61. Hammock, B.D.; Sparks, T.C.; Mumby, S.M. Pestic. Biochem.
    Physiol. 1977, 7, 517.
62. Pauling, L.; Pressman, D. J. Am. Chem. Soc. 1945, 67, 1003.
63. Dunn, III, W.J. Eur. J. Med. Chem. 1977, 12, 109.
64. Hancock, C.K.; Meyers, E.A.; Yager, B.J. J.Am. Chem. Soc.
    1961, 83, 4211.
65. Balaban, A.T.; Chiriac, A.; Motoc, I.; Simon, Z. In "Lecture
    Notes in Chemistry"; Berthier, G. et al., Eds.; Springer-
    Verlag: New York, 1980, Vol. 15, pp. 1-178.
66. Glasstone, S. "Textbook of Physical Chemistry"; 2nd Ed.; D.
    Van Nostrand: New York; 1946, p. 531.
67. Agin, D.; Hersh, L.; Holtzman, D. Proc. Nat. Acad. Sci. U.S.A.
    1965, 53, 952.
68. Franks, F. In "Water"; Franks, F. Ed.; Plenum Press: New York,
    1975, Vol. IV, Chap. 1.
69. Yoshimoto, M.; Hansch, C. J. Org. Chem. 1976, 41, 2269.
70. Hansch, C.; Smith, R.N.; Rockoff, A.; Calef, D.F.; Jow, P.Y.C.;
    Fukunaga, J.Y. Arch. Biochem. Biophys. 1977, 183, 383.

RECEIVED February 11, 1985

Role of Mixed-Function Oxidases in Insect Growth and Development

C. F. WILKINSON

Department of Entomology, Cornell University, Ithaca, NY 14853

In addition to their well established role in catalyzing the metabolism of a wide variety of naturally
occurring and synthetic xenobiotics, cytochrome P-450-
mediated mixed-function oxidases are of critical
importance in the biosynthesis and regulation of the
major hormones (ecdysteroids and juvenile hormone) that
control insect growth and development. The characteristics of the mixed-function oxidases involved in the
synthesis of insect hormones are described and the
possibility that the enzymes might represent potential
targets for insect control is discussed.

Mixed-function oxidases constitute a group of evolutionarily
ancient metalloproteins capable of catalyzing a wide variety of
oxidative reactions in which one atom of molecular oxygen is
inserted into an appropriate substrate and the other reduced to
water (1). The Overall reaction typically involves the transfer of
two electrons from a suitable reducing agent, often a reduced
pyridine nucleotide (e.g., NADPH, NADH), and the formation of a
highly reactive, electrophilic species of oxygen that is subsequently transferred to the substrate. Consequently, the main
catalytic function of the enzymes is to activate molecular oxygen;
the substrate to be oxidized plays a relatively passive role in the
process (2).
    Mixed-function oxidases are ubiquitously distributed through-
out the plant and animal kingdoms as well as in many aerobic
procaryotes (3). They play a number of important biological roles
and, indeed, are biochemically unique in their ability to catalyze
the activation of non-activated carbon-hydrogen bonds.
    The following discussion will center on mixed-function oxidases
involving the hemoprotein cytochrome P-450, the active center of
which consists of protoporphyrin IX. Mixed-function oxidases based
on cytochrome P-450 are perhaps best known for their role in the
primary metabolism of lipophilic xenobiotics in mammals, birds,
fish and many invertebrates including insects. While this function
is often critical in determining the survival of an organism

0097–6156/85/0276–0161$06.00/0

following exposure to toxic chemicals and is frequently involved in insect resistance to insecticide chemicals (4), it will be discussed only briefly in this presentation. Most of the discussion will focus on the physiological role of cytochrome P-450-mediated mixed-function oxidases in the biosynthesis and regulation of the hormones controlling insect growth and development.

Xenobiotic Metabolism

The oxidative enzymes involved in xenobiotic metabolism in insects are associated mainly with the microsomal fractions of the midgut and fat body, that are derived from the endoplasmic reticulum of the intact tissues (4-8). They catalyze the oxidative metabolism of a remarkable range of lipophilic compounds through reactions involving numerous functional groups; these reactions include aromatic, alicyclic and aliphatic hydroxylation reactions, dealkylation of ethers and substituted amines, epoxidation of double bonds and desulfuration of phosphorothionates. While the majority of these reactions constitute important detoxication pathways several can lead to enhanced biological activity through formation of reactive intermediates or products (9). All require NADPH and oxygen.

The system depends on an electron transport pathway that transfers electrons from NADPH through a flavoprotein (NADPH cytochrome P-450 reductase) to cytochrome P-450 that is the terminal oxidase of the chain (10). The xenobiotic first forms a complex with the oxidized form of cytochrome P-450 which is reduced by an electron passing down the chain from NADPH. The reduced cytochrome P-450/substrate complex then reacts with and activates molecular oxygen to an electrophilic oxene species (an electron deficient species similar to singlet oxygen) that is transferred to the substrate with the concommitant formation of water. Cytochrome P-450 thus acts primarily as an oxene transferase (2). Substrate binding is a relatively nonspecific, passive process that serves to bring the xenobiotic into close association with the active center and provide the opportunity for the oxene transfer to occur.

The unusual degree of nonspecificity of the xenobiotic-metabolizing cytochrome P-450 system results in part from the generally low level of specificity of the substrate binding sites at the active center of the cytochrome and in part from the existence of a number of isozymes of cytochrome P-450 each catalyzing a limited, perhaps overlapping spectrum of oxidative reactions with a variety of substrates (11). This has led to the development of the "cytochrome P-450 family" concept where the blend of isozymes may differ not only between different species but also between the different tissues of a single species (3,4).

It is now generally believed that the cytochrome P-450 system that metabolizes xenobiotics has evolved specifically in response to selection pressure from the many naturally-occurring materials to which organisms are exposed in their diets or otherwise encounter in the environment (12). The ability of the enzymes to metabolize modern synthetic organic chemicals such as pesticides is simply a reflection of the evolutionary success that has been achieved in the development of a highly versatile system that is fully prepared for any eventuality (3,4,12).

The mixed-function oxidases of mammals, insects and other organisms are remarkably well adapted for their role in biochemical defense. In addition to their metabolic versatility resulting from nonspecificity and the presence of numerous isozymes, the enzymes are located in those tissues that represent the major routes of entry of xenobiotics into the body. Furthermore, the ability of the enzymes to be induced enables organisms to make temporary qualitative and quantitative adjustments in the isozyme composition of their tissues to meet the ever changing challenges of their external chemical environment; this is clearly energetically advantageous since it precludes the necessity of maintaining high titers of enzyme protein at all times. It is particularly important to polyphytophagous insects since it allows them to adjust the composition of their protective enzymes depending on the host plant on which they are feeding (12).

Role of Mixed-function Oxidases in Steroidogenic Reactions

In addition to their obvious importance as a primary biochemical defense against a large variety of naturally occurring and synthetic lipophilic xenobiotics, mammalian mixed-function oxidases play a critical physiological role in the production and regulation of steroid hormones; it is becoming increasingly apparent that this is also true in insects. Since in comparative biochemistry, it is often the similarities rather than the differences that exist between organisms that are the most striking, a brief review of mammalian steroid hormone biosynthesis might be of both interest and relevance in interpreting the rather sparse information available with insects.

The major sites of steroid hormone production in mammals are the adrenal cortex and the gonads, (i.e., the testis and the ovary). The major hormones synthesized and secreted by the adrenal cortex are the corticosteroids. These are of two types, the mineralocorticoids (e.g., aldosterone) that regulate fluid and ion balance, and the glucocorticoids (e.g, cortisol) that control carbohydrate, fat and protein metabolism (Figure 1). The major hormones produced by the gonads are, of course, the sex hormones, androgens (primarily testosterone) in the male testis, and estrogens (e.g., estradiol) in the female ovary (Figure 1) (13).

Since the major precursor of all of these hormones is cholesterol it becomes clear that, while the total biosynthetic routes are quite complex, they all involve a number of specific hydroxylation reactions involving both the steroid nucleus and the side chain at C-17. Most of these are catalyzed by mixed-function oxidases involving cytochrome P-450 (Figure 1) (13-15).

The cytochome P-450 systems of the steroid producing tissues have many characteristics in common with those of the liver and other xenobiotic-metabolizing tissues. They require NADPH and molecular oxygen, and contain a flavoprotein, NADPH cytochrome P-450 reductase. Unlike the systems in the liver, however, the steroid hydroxylases are associated with both the endoplasmic reticulum and the mitochondria and are quite specific for the different reactions that they catalyze. Thus, adrenal microsomes

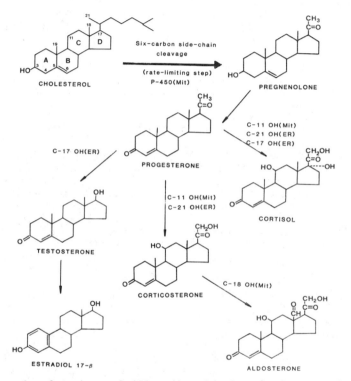

Figure 1.   Cytochrome P-450-mediated hydroxylation reactions in biosynthesis of major mammalian steroid hormones.   (MIT) = mitochondria; (ER) = endoplasmic reticulum.

contain 17α- and 21-hydroxylases while adrenal mitochondria catalyze 11β- and 18-steroid hydroxylations as well as cholesterol side
chain cleavage that is considered to be the primary rate-limiting
step in mammalian steroidogenesis (14,16). The mitochondrial
system differs markedly from that in adrenal microsomes in that the
former contains an iron-sulfur protein (adrenodoxin) between the
reductase and cytochrome P-450 and in that the adrenal flavoprotein
is immunochemically distinct from that in the microsomal system;
the latter appears identical with that in liver microsomes. In
passing, it is interesting to note that the mitochondrial steroid
hydroxylases of the mammalian endocrine system are remarkably
similar to many of the procaryotic hydroxylases such as that in
Pseudomonas putida that also contain iron-sulfur proteins (e.g.,
putidaredoxin) similar to adrenodoxin (13).

As with the mixed-function oxidases involved in xenobiotic
metabolism, the substrate specificity of the steroid hydroxylases
is dictated, in part, by the existence of multiple forms of both
microsomal and mitochondrial cytochrome P-450s and further opportunities for specificity are provided by the distinct localization
of the various enzymes in either the mitochondria or the endoplasmic reticulum.

While the hydroxylases of the adrenals and other steroid
producing tissues show some (variable) catalytic activity towards
xenobiotics (e.g., benzo[a]pyrene) (17) this is probably fortuitous
in nature and it seems that, in general, the enzymes are quite
specific for their respective steroid substrates. Furthermore, the
steroid hydroxylases of the endocrine system are not susceptible to
the inductive effects of xenobiotics as are the cytochrome P-450-
mediated oxidases of the liver and other tissues (17,18); indeed,
this is not unexpected, since if they responded to the external
environment in this way their critical homeostatic role would
rapidly be compromized.

In keeping with this role, the activity of the mixed-function
oxidases in the mammalian producing steroid tissues is regulated
primarily by a series of hormones that are released from the
anterior pituitary in response to hormone releasing factors originating in the hypothalamus of the brain (Figure 2). The hormones
controlling enzyme activities in the adrenal cortex, testis and
ovary are the adrenal corticotropic hormone (ACTH), the follicle
stimulating hormone (FSH) and the luteinizing hormone (LH), respectively. The arrival of these hormones at their appropriate endocrine tissues serves to trigger the enzymes involved in steroid
hydroxylation and in each case the metabolic process is initiated
by the side chain cleavage of cholesterol that is stored in the
tissues in the form of cholesterol esters (16).

Steroidogenesis in Insects

The steroids known to play major regulatory roles in insect development and metamorphosis all fall into the class of polyhydroxylated
ketosteroids called ecdysones (19-22). With the exception of
Makisterone A (a C_{28} ecdysteroid identified from the milkweed bug
Oncopeltus fasciatus) the known insect ecdysteroids constitute a
group of eight or nine C_{27} steroids that differ from one another

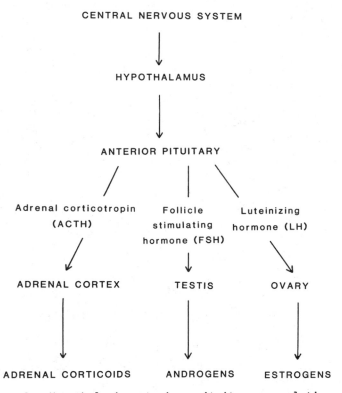

Figure 2. Hypothalamic-anterior pituitary regulation of mammalian steroidogenesis (see text for explanation).

primarily in the number and position of the hydroxyl groups they contain; several others have been isolated but not yet fully identified.

To date, interest in the ecdysteroids has focused primarily on their role in insect molting and metamorphosis and it is perhaps somewhat unfortunate that bioassays for ecdysteroid activity have been restricted almost exclusively to evaluating their effect on the molting process. It now appears probable that molting is just one of several important regulatory functions (e.g., embryogenesis, reproduction) performed by the ecdysteroids and that we are only just beginning to scratch the surface of an insect steroid system that is equally as complex as that in vertebrate species.

The major ecdysteroids known to play a role in insect molting and/or metamorphosis are ecdysone, 20-hydroxyecdysone, and 20,26-dihydroxyecdysone (Figure 3); 20-hydroxyecdysone (β-ecdysone) is the most active in terms of molting hormone activity. Ecdysones have been extracted in varying amounts from late larval and pupal stages of several lepidopteran species [e.g., Bombyx mori (23) and Manduca sexta (19-22)] as well as from the penultimate nymphal stages of several hemimetabola such as Locusta migratoria (24), Schistocerca gregaria (25) and other species. 26-Hydroxyecdysone (Figure 3) has been identified as the major ecdysteroid (about 80% of total) present in embryonated eggs of Manduca, although it is devoid of molting hormone activity (20). This suggests that it might have an important embryogenic function or that it represents a metabolic degradation product of ecdysone.

It is now well established that the predominant, perhaps only, source of ecdysone in larval insects is the prothoracic glands (19,20). Since, unlike mammals, insects are incapable of de novo synthesis of the steroid ring, the precursors of ecdysone must be one or more of several compounds, such as phytosteroids ingested in the diet (19-21). The precise pathway of ecdysone synthesis from precursors such as cholesterol remains unknown but clearly involves a number of steps including C_7 desaturation (e.g., to dehydrocholesterol), C_5 saturation, keto formation at C_6, cis fusion of the A/B rings, and a variety of hydroxylations at C_2, C_{14} C_{22} and C_{25}. While dehydrocholesterol has not been detected in extracts from whole Manduca larvae, it has been extracted from the protharacic gland and may constitute 50% of the total steroids present in this organ (26). Although the exact sequence in which these ring hydroxylations take place is not clear and the enzymes responsible have not yet been characterized, it is generally thought that the last two reactions are at the 25- and 22- positions, respectively (Figure 4); another suggested pathway indicates the hydroxylation of 2-deoxyecdysone as the final step. The enzymes catalyzing these reactions have not yet been identified. It is probable, however, that they are cytochrome P-450 mediated and that they will be found to exhibit a good deal of regiospecificity with respect to the site of attack on the sterol molecule. One of these may be rate limiting as is the side-chain cleavage of cholesterol in mammalian steroidogenesis.

The synthesis of ecdysone, the only ecdysteroid formed by the prothoracic glands, is under the control of the prothoracicotropic

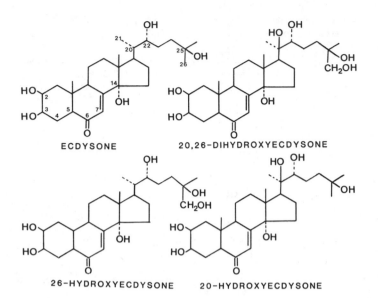

Figure 3.   Major ecdysteroids.

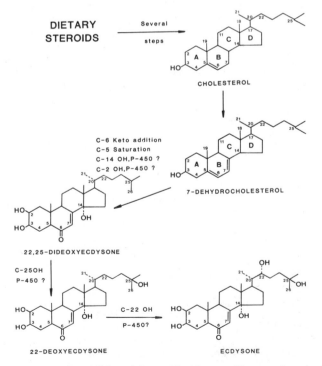

Figure 4.   Possible biosynthetic pathway from dietary steroids to α-ecdysone.

hormone (PTTH) a peptide hormone that is produced by the neurosecretory cells of the brain and released by the corpora cardiaca (<u>21</u>). The fact that ecdysone synthesis can be initiated rapidly in response to PTTH suggests that the prothoracic gland may have the ability to sequester an appropriate ecdysone precursor (possibly 7-dehydrocholesterol or a conjugate) in a manner similar to the storage of cholesterol esters in the tissues of the mammalian endocrine system. It is not established whether ecdysone itself exerts any feedback control on the release of PTTH by the corpora cardiaca (<u>21</u>).

While ecdysone, itself, may elicit a general hormonal response affecting many tissues, it is now well established that ecdysone is further metabolized to other, perhaps more active, ecdysteroids such as 20-hydroxyecdysone, 20,26-dihydroxyecdysone and 26-hydroxyecdysone in a number of specific tissues peripheral to the prothoracic gland. In this sense, ecdysone may be serving as a prohormone that can be metabolically activated to a variety of derivatives each regulating the specific endocrine functions of one or more target organs. Indeed, ecdysone may serve primarily as a general precursor for the synthesis of other ecdysteroids and may be sequestered in specific target tissues in inactive form until required.

Balance between the active and inactive forms of ecdysone and/or other ecdysteroids may be accomplished by the formation of conjugates such as sulfate esters or glucosides (<u>20,27</u>). The sulfation of phenols and a variety of sterols has been demonstrated in insect tissues and this, in close association with an appropriate sulfatase, would constitute a readily reversible mechanism whereby the required balance between active and inactive forms of ecdysteroids could be regulated (<u>27</u>).

The further metabolism of ecdysone to other more active or less active ecdysteroids constitutes another potential mechanism for hormone regulation and it is likely that many of the hydroxylation reactions involved are catalyzed by cytochrome P-450-mediated mixed function oxidases (Figure 5). Since the enzyme ecdysone 20-hydroxylase catalyzes the conversion of ecdysone to the more potent molting hormone 20-hydroxyecdysone (Figure 5), it has received considerable attention in recent years. Ecdysone 20-hydroxylase activity has been measured in the tissues of a number of insect species; these include the fat body and several other tissues of lepidopterous larvae such as <u>Manduca</u> <u>sexta</u> (19,20,28,29) and the midgut, fat body and Malpighian tubules of penultimate nymphal instars of <u>Locusta</u> <u>migratoria</u> (24,30) and <u>Schistocerca</u> <u>gregaria</u> (<u>25</u>).

Most of these studies have established that the enzyme is located in the mitochondrial fraction of tissue homogenates, although the enzyme in the fat body and Malpighian tubules of <u>Locusta</u> is reportedly associated with the microsomal fraction (<u>30</u>). All studies concur that the enzyme is a cytochrome P-450-mediated mixed-function oxidase and its requirements for NADPH and O_2 and sensitivity to inhibitors such as carbon monoxide, metyrapone, etc., support this conclusion. As yet, there are no reports as to whether the enzyme is associated with an iron sulfur protein similar to the adrenodoxin of the mammalian mitochondrial steroid synthesis system although it is entirely probable that such a component will eventually be identified.

It is generally assumed that the cytochrome P-450s involved in steroidogenesis exhibit a considerably greater degree of substrate specificity than their counterparts involved in xenobiotic metabolism and certainly this is true for the mammalian enzymes. From the results of limited studies based on the ability of various ecdysteroids to competitively inhibit the conversion of ecdysone to 20-hydroxyecdysone in vitro by the Manduca fat body mitochondrial enzyme, it has been concluded that the enzyme is not highly specific for ecdysone (19). However, this is not a conclusive test for substrate specificity and may have little or no relevance to the situation occurring under physiological conditions in the intact cell.

At the present time, ecydsone 20-hydroxylase is the only major ecdysteroid synthesizing enzyme that has been isolated and characterized to any extent from insect tissues and its regulatory role, if any, has not been established.

While the prothoracic gland appears to be the only tissue capable of synthesizing ecdysone in immature insects, ecdysone and several other ecdysteroids are also produced by the ovary in adult female insects (31). In most species this occurs at the end of oocyte maturation and seems to be controlled by a folliculotropic factor released by the corpara cardiaca. Ecdysone synthesis appears to occur de novo from cholesterol and ecdysone and possibly other ecdysteroids are stored in the oocyte tissue as polar conjugates rather than being released into the blood (31). It seems probable that ovarian synthesis of ecdysone is the major source of the various ecdysteroids (20-hydroxyecdysone, 26-hydroxyecdysone, 20,26-dihydroxyecdysone, 2-deoxyecdysone and 3-dehydroecdysone) that have been identified in insect eggs and developing embryos. These hormones derived from the adult female may well control critical events in the early stages of embryonic development prior to the development of the embryonic prothoracic gland. At least, one of the functions of embryonic ecdysteroids may be the control of cuticulogenesis (31).

Insect ecdysteroids may also play an important role in the hormonal control of reproduction (31,34). Thus, in the case of the mosquito, Aedes aegypti, ecdysone appears to be responsible for stimulating the synthesis of vitellogenin (egg protein) by the fat body (32,33). The source of the ecdysone in this case is again the ovary and ecdysone biosynthesis is initiated in response to a neurosecretory hormone (EDNH, egg development neurosecretory hormone) released following ingestion of the blood meal (34). As in the case of the ecdysone sequestered in the eggs, the mosquito ovary appears to synthesize ecdysone de novo from cholesterol. In the latter case, however, the ecdysone is released into the blood, and passed to the fat body where it is converted into 20-hydroxyecdysone and possibly other ecdysteroids before stimulating vitellogenin synthesis. While not yet studied at the enzymatic level, it is almost certain that one or more cytochrome P-450-mediated monooxygenases are involved in the ovarian synthesis of ecdysone and may thus play an important regulatory role in reproductive maturation in adult female insects. It is possible that ecdysone also plays a role in the reproductive maturation in male insects since the testes of last-instar larvae of the tobacco budworm effect the synthesis and release of several ecdysteroids (35).

The similarities between ecdysone synthesis in the insect prothoracic gland and the ovary are obvious and in each case synthesis is initiated in response to a hormone originating in the brain. Both bear a striking resemblance to the mammalian system where steroid hormone synthesis in the various endocrine tissues is initiated in response to the release of appropriate hormones from the anterior pituitary.

Juvenile Hormone

The molting and other hormonal activities of the ecdysteroids are, of course, modulated by the titer of the insect juvenile hormone and the two materials typically function in close concert with one another in dictating insect molting and metamorphosis as well as in reproductive maturation (21).

The three major, known insect juvenile hormones (JH I, JH II and JH III) (Figure 6) are all methyl esters of terminally epoxidized homologs of farnesoic acid. They are present in varying amounts in different insects at different stages of development and it has been suggested, though not determined, that they may play different hormonal roles.

The juvenile hormones are synthesized and rapidly released into the hemolymph from the neuroendocrine glands known as the corpora allata; synthesis appears to be under control of a hormone originating in the brain.

The farnesoic acid precursors of the juvenile hormones are synthesized by appropriate isoprenoid routes from acetyl CoA and/or propionyl CoA and subsequently undergo 10,11-epoxidation and methylation of the carboxylic acid group by O-methyl transferase (36) (Figure 7). It is not yet fully ascertained whether in vivo farnesoic acid is epoxidized prior to methylation or whether the order of these two terminal reactions is reversed.

The epoxidase catalyzing the 10,11-epoxidation of methyl farnesoate in homogenates of corpora allata from Locusta migratoria has been studied in detail and has been shown to be a cytochrome P-450-mediated monooxygenase associated with the microsomal fraction. The enzyme is strictly dependent on NADPH and requires oxygen (36). It is sensitive to inhibition by carbon monoxide (36) and to other compounds such as methylenedioxyphenyl compounds and imidazoles that are well established inhibitors of the cytochrome P-450-mediated monooxygenases involved in xenobiotic metabolism (37-39).

In recent years, considerable attention has been given to the possibility of developing synthetic inhibitors of juvenile hormone synthesis (anti-juvenile hormones) that would result in a lethal block of JH synthesis. Although a large number of compounds have been screened as potential anti-juvenile hormone agents and several have been found to be effective inhibitors in vitro, few have exhibited significant in vivo morphogenetic activity. The possibility of causing premature metamorphosis and lethality through chemical allatectomy has been given further impetus, however, by

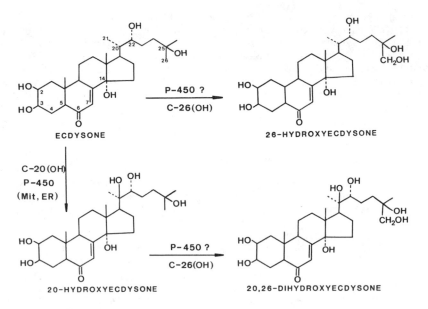

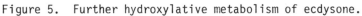

Figure 5. Further hydroxylative metabolism of ecdysone.

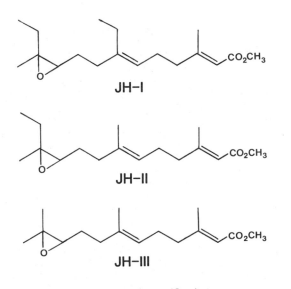

Figure 6. Insect juvenile hormones.

the finding that, in sensitive species, the anti-allatal precocenes (40) are bioactivated by the cytochrome P-450 system of the corpora allata to potent cytotoxins that result in irreversible necrosis of the corpora allata. The mechanism of bioactivation appears to be the formation of a highly reactive electrophilic 3,4-epoxide that immediately alkylates critical macromolecular site(s) in the corpora allata (42) (Figure 8).

The suicidal bioactivation of the precocenes clearly suggests the possibility that, while the cytochrome P-450 system involved in JH synthesis may be quite specific with respect to the substrates it will accept, it represents a potentially important and viable target for future pest control agents.

Summary

In conclusion, mixed-function oxidases based on cytochrome P-450 are extremely important to insects, not only in providing a measure of protection against a wide variety of naturally occurring and synthetic xenobiotics, but also in the biosynthesis of the hormones that determine the complex patterns of insect growth, development and sexual maturation.

While the mixed-function oxidases involved in hormone biosynthesis have not yet been characterized in any detail, it is probable that like the enzymes involved in mammalian steroidogenesis, they will be found to differ from those involved in xenobiotic metabolism in both substrate specificity and in the regulatory mechanisms through which they are controlled. It seems inconceivable, for example, that the enzymes involved in hormone synthesis are inducible by xenobiotics since this would destroy homeostatic control and in effect would open up critical physiological development to the whims of the external environment; clearly, this does not occur. Instead, the enzymes are undoubtedly regulated by a variety of endogenous factors (hormones) released from the central neuroendocrine system in response to a variety of endogenous and exogenous stimuli.

In view of their critical importance in regulating the growth and development of insects, the enzymes should continue to be viewed as potentially valuable targets around which to develop new pest control agents. This possibility is likely to become more realistic as the enzymes are further characterized and the full extent of their many roles in insect development are more fully understood.

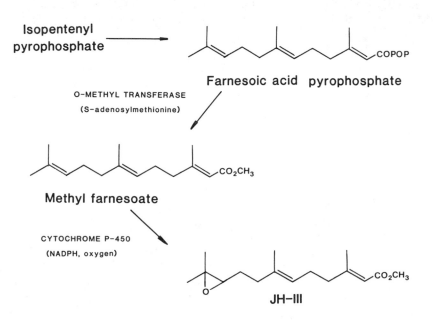

Figure 7.   Biosynthesis of insect juvenile hormone III.

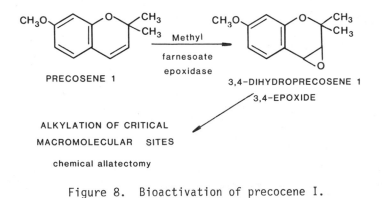

Figure 8.   Bioactivation of precocene I.

Literature Cited

1. Mason, H. S. In "Homologous Enzymes and Biochemical Evolution"; Thoai, N.V. and Roche, J., Eds.; Gordon and Breach: New York, 1968; pp. 69-91.
2. Trager, W. F. In "Concepts in Drug Metabolsim"; Jenner, P. and Testa, B., Eds.; Marcell Dekker: New York, 1981; Part A, Chap. 3.
3. Wilkinson, C. F., In "The Scientific Basis of Selective Toxicity"; Witschi, H. R., Ed.; Elsevier/North Holland: Amsterdam, 1980; pp. 251-268.
4. Wilkinson, D. F., In "Pest Resistance to Pesticides"; Georghiou, G. P. and Saito, T., Eds; Plenum: New York, 1983; pp. 175-205.
5. Agosin, M. and Perry, A. S. In "The Physiology of Insecta"; Rockstein, M., Ed.; Academic: New York, 1974; Vol. V., pp. 537-596.
6. Hodgson, E. and Plapp, F. W. J. Agr. Food Chem. 1970, 1048-1052.
7. Wilkinson, C. F. and Brattsten, L. B. Drug Metab. Revs. 1972, 1, 153-228.
8. Wilkinson, C. F. In "Xenobiotic Metabolism: In Vitro Methods"; Paulson, G. D., Frear, D. S. and Marks, E. P., Eds.; ACS SYMPOSIUM SERIES No. 97, American Chemical Society: Washington, D.C. 1979; pp. 249-284.
9. Cummings, S. W. and Prough, R. A. In "Biological Basis of Detoxication"; Caldwell, J. and Jakoby, W. B., Eds.; Academic: London, 1983; Chap. 1.
10. Estabrook, R. W., Werringloer, J. and Peterson, J. A. In "Xenobiotic Metabolism: In Vitro Methods"; Paulson, G. D., Frear, D. S. and Marks, E. P., Eds.; ACS SYMPOSIUM SERIES No. 97, American Chemical Society: Washington, D. C. 1979; pp. 149-179.
11. Lu, A.Y.U. and West, S. B. Pharmacol. Revs. 1980, 31, 277-295.
12. Brattsten, L. B. In "Herbivores: Their Interaction with Secondary Plant Metabolites"; Rosenthal, G. A. and Janzen, D. H., Eds.; Academic: New York; 1979; pp. 199-270.
13. Armstrong, F. B. "Biochemistry"; Oxford University Press: Oxford, 1979; p. 386.
14. Gunsalus, I. C., Pederson, T. C. and Sligar, S. G. Ann, Rev. Biochem. 1975, 44, 377-407.
15. Dempsey, M. E. Ann. Rev. Biochem. 1974, 43, 967-990.
16. Boyd, G. S. and Trzeciak, W. H. In "Multienzyme Systems in Endocrinology: Progress in Purification and Methods of Investigation"; Cooper, D. Y. and Salhanick, H. A., Eds.; Annals N.Y. Academy Sciences: New York, 1973; Vol. 212, pp. 361-377.
17. Guenther, T. M. and Nebert, D. W. Mol. Pharmacol. 1979, 15, 719-728.
18. Feuer, G., Sosa-Lucero, J. C., Lumb, G. and Moddel, G. Toxicol. Appl. Pharmacol. 1971, 19, 579-585.
19. Smith, S. L., Bollenbacher, W. E. and Gilbert, L. K. In "Progress in Ecdysone Research"; Hoffmann, J. A., Ed.; Elsevier/North Holland: Amsterdam, 1980; pp. 139-162.

20. Kaplanis, J. N., Weirich, G. F., Svoboda, J. A., Thompson, M.
    J. and Robbins, W. E. In "Progress in Ecdysone Research";
    Hoffmann, J. A., Ed.; Elsevier/North Holland: Amsterdam, 1980;
    163-186.
21. Riddiford, L. M. and Truman, J. W. In "Biochemistry of
    Insects"; Rockstein, M., Ed.; Academic: New York; 1978, Chap.
    7.
22. Thompson, M. J., Kaplanis, J. N., Weirich, G. f., Svoboda, J.
    A. and Robbins, W. E. Proc. International Conference on
    Regulation of Insect Development and Behavior; Wroclaw Tech-
    nical Univ.: Poland, 1981; Conf. No. 7, pp. 107-124.
23. Chino, H., Sakurai, S., Ohtaki, T., Ikekawa, N., Miyazaki, H.,
    Ishibashi, M. and Abuki, H. Science 1974, 183, 529-530.
24. Feyereisen, R. and Durst, F. Mol. Cell. Endocrinol. 1980, 20,
    157-169.
25. Johnson, P. and Rees, H. H. Biochem. J. 1977, 168, 513-520.
26. Thompson, M. J., Kaplanis, J. N., Robbins, W. E. and Svoboda,
    J. A. Adv. Lipid Res. 1973, 11, 219-265.
27. Yang, R.S.H. and Wilkinson, C. F. Biochem. J. 1972, 130,
    487-493.
28. Mayer, R. T., Svoboda, J. A. and Weirich, G. F. Hoppe-Seyler's
    Z. Physiol. Chem. 1978, 359, 1247-1257.
29. Bollenbacher, W. E., Smith, S. L., Wielgus, J. J. and Gilbert,
    L. I. Science 1977, 268, 660-663.
30. Feyereisen, R. and Durst, F. Eur. J. Biochem. 1978, 88, 37-47.
31. Hoffmann, J. A., Lageux, M., Hetru, C., Charlet, M. and
    Goltzene, F. In "Progress in Ecdysone Research"; Hoffman, J.
    A., Ed.; Elsevier/North Holland: Amsterdam, 1980; pp. 431-465.
32. Hagedorn, H. H. and Fallon, A. M. Nature 1973, 244, 103-105.
33. Hagedorn, H. H., O'connor, J. D., Fuchs, M. S., Sage, B.,
    Schlaeger, D. A. and Bohm, M. K. Proc. Nat. Acad. Sci. U.S.A.
    1975, 72, 3255-3259.
34. Hanaoka, K. and Hagedorn, H. H. In "Progress in Ecdysone
    Research"; Hoffmann, J. A., Ed.; Elsevier/North Holland:
    Amsterdam, 1980; pp. 467-480.
35. Loeb, M. J., Woods, C. W., Brandt, E. P. and Borkovec, A. B.
    Science 1982, 281, 896-898.
36. Feyereisen, R., Pratt, G. E. and Hamnett, A. F. Eur. J.
    Biochem. 1981, 118, 231-238.
37. Hammock, B. D. and Mumby, S. M. Pestic. Biochem. Physiol.
    1978, 9, 39-47.
38. Brooks, G. T., Pratt, G. E. Ottridge, A. P. and Mace, D. W.
    Proc. Brit. Crop Protection Conf. 1984, in press.
39. Feyereisen, R., Langry, K. C. and Ortiz de Montellano, P. R.
    Insect Biochem. 1984, 14, 19-26.
40. Bowers, W. S., Ohta, T., Cleere, J. S. and Marsella, P. A.
    Science 1976, 193, 542-547.
41. Masner, P., Bowers, W. S., Kalin, M. and Muhle, T. Gen Comp
    Endocrinol. 1979, 37, 156-166.
42. Hamnett, A. F., Ottridge, A. P. Pratt, G. E., Jennings, R. C.
    and Stott, K. M. Pestic. Sci. 1981, 12, 245-254.

RECEIVED November 15, 1984

Inhibition of Reproduction in Insect Control

ALEXEJ B. BOŘKOVEC

Insect Reproduction Laboratory, Agricultural Research Service, U.S. Department of Agriculture, Beltsville, MD 20705

Sexual sterilization is now one of the accepted
insect control strategies but its limitations invite
a fresh look at other chemical treatment methods
that reduce reproduction. The physiological effects
of insect growth regulators (IGRs) sometimes include
the disruption of reproduction but the significance
of this effect and its contribution to the overall
population reduction is variable. Juvenoids, chitin
synthesis inhibitors, and other IGRs with as yet
unknown mode of action are examples of compounds with
different activities in immature stages and in adults.
However, current research on neurochemical regulation
of growth and reproduction offers challenging
opportunities for the development of entirely new
types of insect control agents.

Some of the developments outlined briefly here have been already
mentioned in more detail by others at this Conference but my
purpose is to discuss the historical background that led to changes
in our concept of utilizing chemically induced sexual sterility in
the control and management of insect populations. The conclusion
reached by Knipling (1) in his comprehensive treatise on this
subject remain valid. The sterile insect release technique is now
a generally accepted procedure even if its practical application
has been so far limited to the screwworm, Cochliomyia hominivorax
(Coquerel), Mediterranean fruit fly, Ceratitis capitata
(Wiedemann), and boll weevil, Anthonomus grandis grandis Boheman.
On the other hand, the direct sterilization of insects in their
natural environment (2) has found only marginal utility because the
chemosterilants suitable for this application are effective only in
females. From the standpoint of a population, a sterile female is
equivalent to a dead female even if she remains capable of mating.
In contrast, a sterile male is much more important because his
mating may incapacitate several females (the so called bonus
effect). Therefore, a female chemosterilant is at its best no more
effective than an insecticide, whereas a male chemosterilant can be
considerably more effective than an insecticide of comparable
activity.

It should be noted at this point that the first highly active
chemosterilants were developed as alternatives to the radiation-
induced mutagenesis employed in the sterile male release method.
For this application, an acceptable compound had to be highly
specific in its action on the gametes or gonads; any side-effects
were undesirable because of their possible interference with mating,
locomotion, pheromone production, or longevity. Mutagens, and
particularly alkylating agents derived from aziridine (3), were
almost ideal in this regard and they sterilized males or both sexes
of numerous insect species. Unfortunately, these successful
developments created a false impression that all chemosterilants
were mutagenic and that a safe application of sterilizing chemicals
in the field was impossible. That this impression was indeed false
will become clear later in this review but let us turn to another
point that was responsible for reexamining the scope of the field
sterilization techniques.

The requirement for a high action specificity of a
chemosterilant is fully justified when the compound is to be used
for sterilizing insects that are to be released. However, a
sterilant that is to be applied in the field where it may contact
not only adults but also all the immature stages of the insect does
not have to be specific; in fact, its effectiveness would
drastically increase if it would sterilize the adults and kill or
otherwise incapacitate the immature. To turn around this argument,
one can also say that a sterilizing activity in adults is a most
desirable property of any ovicide or larvicide that is used in
situations where all stages of the pest are present simultaneously
or in short succession.

Strictly speaking, compounds that possess this type of
multifarious activity cannot be any longer classified as just
chemosterilants or toxicants. They are population controlling
agents that interfere with several critical physiological processes
at different stages of the insect's development. In some instances,
for example when only certain developmental stages of the insect are
available for contact with the compound or when a certain kind of
effects require much smaller doses of the agent than others, the
action may be primarily sterilizing or toxic; nevertheless, the
potential for the multifarious activity remains the decisive
criterion. From the standpoint of reproduction, two classes of such
agents are known today: (1) compounds that interfere with the
development and/or function of the gonads and (2) compounds that
prevent the development of offspring.

Compounds that Interfere with the Development and/or Function of the Gonads

In this class the prominent mode of action is the induced variation
of the normal juvenile hormone (JH) or molting hormone (MH) titres.
Although the detailed processes by which these hormones regulate
ovarian development, oogenesis, and early embryogenesis are still
unclear, there are critical periods during which any major deviation
from the normal JH or MH titer results in pathological developments
and eventually in decreased fertility. JHs themselves (4-7) as well
as their analogs (JHAs) such as methoprene (7-12) function as

sterilants in many species. Similarly, the ecdysteroids and their
analogs (9,13,18) may induce sterility by interfering with
vitellogenesis and early embryogenesis in a manner seemingly not
very different from that of JHs and JHAs. The toxicity of these
agents is a consequence of morphogenetic effects that are expressed
in immature stages and frequently at doses that are much lower than
those required for the induction of sterility. As a result, the
still rather restricted practical use of compounds in this class is
limited to their larvicidal and ovicidal activity. However, a
notable exception to this generalization will be described later in
this Conference (19).

The titer of JHs can be manipulated by other means than by the
action of their structural analogs. The biosynthesis as well as the
deactivating metabolism of JH is now sufficiently well understood
so that specific inhibitors of either process can be designed (20-
22). Although in theory this type of manipulation of the natural JH
titer could lead to sterility, specific examples are lacking in the
literature. In contrast, there is ample evidence for inducing
sterility in females of several species after treatments with
precocene (23-35). This natural product derived from Ageratum
houstonianum (36) selectively destroys the tissues of corpora
allata, the organ that produces JH.

Compounds that Prevent the Development of Progeny

It may be argued that sterility, i.e., the inability to reproduce,
does not properly describe situations in which the development
proceeds beyond the egg stage. Consequently, compounds that are
applied to adult females and reduce the hatchability of their eggs
or the survival of their larvae could be called ovicides or
maternally applied larvicides rather than sterilants. Because this
distinction appears primarily cosmetic rather than substantive, it
will be disregarded here. The distinctive feature of these
compounds is their accumulation in the eggs prior to their
deposition. Perhaps the best understood examples are the
derivatives of phenyl benzoyl urea (37). These materials,
exemplified by diflubenzuron (38), inhibit the biosynthesis of
chitin and thus disrupt the formation and reformation of insect's
exoskeleton (39,40). In the female house fly, Musca domestica L.,
unmetabolized diflubenzuron is translocated to the eggs and when its
accumulation is sufficiently high, malformed embryos that do not
hatch are produced (41). Similar sterilizing effects of
diflubenzuron or its analogs were confirmed in over 50 species of
Diptera, Coleoptera, Hemiptera, and Lepidoptera.

Another class of compounds that prevent reproduction by
interfering with the survival of offspring are the derivatives of
s-triazine (42). 2,4-Diamino-6-(2-furyl)-s-triazine, one of the
most active chemosterilants of female house flies, was investigated
in considerable detail (43,44) but its mode of action remains
unknown. Like diflubenzuron, this triazine is transported through
the hemolymph to the eggs where it disrupts embryogenesis and
reduces the survival of larvae that do hatch. Unlike diflubenzuron,
however, it has no demonstrable effects on chitin synthesis (45,46).
In house fly larvae, this and related triazines exhibit strong
toxicity (47) that is remarkable by its delayed onset (44).

Closely related larvicides CGA 19255 (6-azido-N-cyclopropyl-N'-ethyl-
s-triazine-2,4-diamine) and CGA 72662 (N-cyclopropyl-s-triazine-
2,4,6-triamine) do not function as sterilants of adult house flies
but they were reported active chitin synthesis inhibitors in the
cockroach leg regenerate system (48). However, since the latter
activity could not be confirmed in another chitin synthesis bioassay
(46) it may be possible that all the larvicidal s-triazines have a
common mode of action and that the differences in their sterilizing
properties depend on the ease of their metabolism and deactivation.

Sterilization by Neuroregulators

Recent advances in insect neurochemistry (49) are opening new
approaches to disrupting reproduction. The existence of a
neuropeptide that affects oogenesis has already been confirmed and
intensive efforts are underway for its isolation and structural
identification. The egg development neurosecretory hormone (EDNH,
50,51) regulates vitellogenesis and its removal or deactivation
results in infecundity. Effects of this neuroregulator on other
physiological processes, particularly those occurring in immature
stages remain to be determined but indications of its
crossreactivity with the prothoracicotropic hormone (52) suggest the
possibility of a multifarious activity that would include
sterilizing effects in adults and toxic effects in immature stages.
      Neuroregulation may indeed be the last frontier of chemical
insect control since the multitude of suspected neurohormones,
estimates range from tens to hundreds, cover all important aspects
of insect physiology and behavior. It is tempting to speculate
whether the hardest problem of insect sterilization, the
interference with male reproduction, could also yield to the
neurochemical approach. Thus far, the regulation of insect
spermatogenesis remains one of the darkest mysteries of reproductive
physiology. However, recent work (53,55) indicates that various
factors, some originating in the brain, may be the sought for
regulators. Once these materials are isolated and characterized,
the possibility of developing a nonmutagenic and highly effective
male chemosterilant may become reality. Neurochemistry offers new
opportunities for developing insect control agents that are specific
and narrow in their effects but it also provides a way for finding
highly active materials that are from a developmental standpoint
nonspecific. The historical example of choline esterase inhibitors
is too well known to require elaboration but it is to be hoped that
the growing understanding of neuroendocrinology will facilitate an a
priori selection of processes specific to insects with a similar
control potential. I have little doubt that reproduction will be
one of these processes.

Literature Cited

1. Knipling, E. P. "Basic Principles of Insect Population
    Suppression and Management"; USDA Agric. Handbook 512,
    Washington, DC, 1979; pp. 281-314.
2. Borkovec, A. B. Environ. Letters 1975, 8, 61-9.

3. Bořkovec, A. B.; Woods, C. W. <u>Israel J. Entomol</u>. 1976, 11, 53-9.
4. Morgan, P. B.; LaBrecque, G. C. <u>J. Econ. Entomol</u>. 1971, 64, 1479-81.
5. Abdallah, M. D.; Zaazou, M. A.; El-Tantawi, M. <u>Z. Angew. Entomol</u>. 1975, 78, 176-81.
6. Schooneveld, H.; Abdallah, M. D. <u>J. Econ. Entomol</u>. 1975, 68, 529-33.
7. Judson, C. L.; De Lumen, H. Z. <u>J. Med. Entomol</u>. 1976, 13, 197-201.
8. Rohdendorf, E. P. <u>Acta Entomol. Bohemoslov</u>. 1975, 72, 209-20.
9. Denlinger, D. L. <u>Nature</u> 1975, 253, 347-8.
10. Marzke, F. O.; Coffelt, J. A.; Silhacek, D. L. <u>Entomol. Exp. Appl</u>. 1977, 22, 294-300.
11. Martinez-Pardo, R.; Ribo, J.; Primo-Yufera E. <u>J. Econ. Entomol</u>. 1979, 72, 537-540.
12. van Sambeek, J. W.; Bridges, J. R. <u>J. Georgia Entomol. Soc</u>. 1981, 16, 83-90.
13. Robbins, W. E.; Kaplanis, J. N.; Thompson, M. J.; Shortino, T. J. ; Cohen, C. F.; Joyner, S. C. <u>Science</u> 1968, 161, 1158-60.
14. Robbins, W. E.; Kaplanis, J. N.; Thompson, M. J.; Shortino, T. J.; Joyner, S. C. <u>Steroids</u> 1970, 16, 105-25.
15. Řežábová, B.; Hora, J.; Landa, V.; Černý, V.; Šorm,, F. <u>Steroids</u> 1968, 11, 475-96.
16. Wright, J. E.; Kaplanis, J. N. <u>Ann. Entomol. Soc. Amer</u>. 1970, 63, 622-3.
17. Wright, J. E.; Chamberlain, W. F.; Barrett, C. C. <u>Science</u> 1971, 172, 1247-8.
18. Dorn, A.; Buhlman, K. J. <u>J. Econ. Entomol</u>. 1982, 75, 935-8.
19. Staal, G. B., this Conference.
20. Monger, D. J.; Lim, W. A.; Kerzdy, F. J.; Law, J. H. <u>Biochem. Biophys. Res. Commun</u>. 1982, 4, 1374-80.
21. Quistad, G. B.; Cerf, D. C.; Schooley, D. A.; Staal, G. B. <u>Nature</u> 1981, 289, 176-7.
22. de Kort, C. A. D.; Granger, N. A. <u>Ann. Rev. Entomol</u>. 1981, 26, 1-28.
23. Bowers, W. S.; Martinez-Pardo, R. <u>Science</u> 1977, 197, 1369-71.
24. Pener, M. P.; Orshan, L.; De Wilde, J. <u>Nature</u> 1978, 272, 351-3.
25. Judson, P.; Rao, B. K.; Thakur, S. S.; Revathy, D. <u>Indian J. Exp. Biol</u>. 1979, 17, 947-9.
26. Pound, J. M.; Oliver, Jr., J. H. <u>Science</u> 1979, 206, 355-7.
27. Rankin, M. A. <u>J. Insect Physiol</u>. 1980, 26, 67-73.
28. Belles, X.; Messeguer, A. <u>Dev. Endocrinol. (Amsterdam) Juv. Horm. Biochem</u>. 1981, 15, 421-4.
29. Fagoone, I.; Umrit, G. <u>Insect Sci. Applic</u>. 1981, 1, 373-6.
30. Feldlaufer, M. F.; Eberle, M. W.; McCelland, G. A. H. <u>Insect Sci. Applic</u>. 1981, 1, 389-92.
31. Matolcsy, G.; Farag, A. I.; Varjas, L.; Belai, I.; Darwish, Y. M. <u>Dev. Endocrinol. (Amsterdam) Juv. Horm. Biochem</u>. 1981, 15, 393-402.
32. Samaranayaka-Ramasamy, M.; Chaudhury, M. F. B. <u>Experientia</u> 1981, 37, 1027-9.

33. Samaranayaka-Ramasamy, M.; Chaudhury, M. F. B. J. Insect Physiol. 1982, 28, 559-69.
34. Bowers, W. S.; Evans, P. H.; Marsella, P. A.; Soderlund, D. M. Science 1982, 217, 647-8.
35. Tarrant, C.; Cupp, E. W.; Bowers, W. S. Amer. J. Trop. Med. Hyg. 1982, 31, 416-20.
36. Bowers, W. S.; Ohta, T.; Cleere, J. S.; Marsella, P. A. Science 1976, 193, 542-7.
37. Wellinga, K. R.; Mulder, R.; van Daalen, J. J. J. Agric. Food Chem. 1973, 21, 993-8.
38. Anonymous. Technical information, TH6040 insect growth regulator. Thompson-Hayward Chem. Co., Kansas City, KS, 1974, 25 pp.
39. Post, L. C.; Vincent, W. R. Naturwissenschaften 1973, 9, 431-2.
40. Cohen, E.; Casida, J. E. In "Pesticide Chemistry; Human Welfare and the Environment"; Miyamoto, J.; Kearney, P. C.; Matsunaka, D. H.; Murphy, S. D.; Eds.; Pergamon: Oxford, 1983; Vol. III, pp. 25-32.
41. Chang, S. C.; Bořkovec, A. B. J. Econ. Entomol. 1980, 73, 285-7.
42. Bořkovec, A. B.; DeMilo, A. B.; Fye, R. L. J. Econ. Entomol. 1972, 65, 69-73.
43. Matolín, S.; Landa, V. Acta Entomol. Bohemoslov. 1971, 68, 1-5.
44. Pessah, I. N. Ph.D. Thesis, University of Maryland, College Park, MD, 1983.
45. Marks, E. P., personal communication.
46. Gelman, D. B., unpublished data.
47. DeMilo, A. B.; Bořkovec, A. B.; Cohen, C. F.; Robbins, W. E. J. Agric. Food Chem. 1981, 29, 82-4.
48. Miller, R. W.; Corley, C.; Cohen, C. F.; Robbins, W. E.; Marks, E. P. Southwest. Entomol. 1981, 6, 272-8.
49. Bořkovec, A. B.; Kelly, T. J., Eds. "Insect Neurochemistry and Neurophysiology"; Plenum Press: New York, 1984; 523 pp.
50. Lea, A. Gen. Comp. Endocrinol. 1972, Suppl. 3, 602-8.
51. Masler, E. P.; Hagedorn, H. H.; Petzel, D.; Bořkovec, A. B. Life Sci. 1983, 33, 1925-31.
52. Hagedorn, H. H.; Shapiro, J. P.; Hanaoka, K. Nature 1979, 282, 92-4.
53. Dumser, J. B. Ann. Rev. Entomol. 1980, 25, 341-69.
54. Friedlander, M. J. Insect Physiol. 1982, 28, 1009-12.
55. Loeb, M. J.; Woods, C. W.; Brandt, E. P.; Bořkovec, A. B. Science 1982, 218, 896-8.

RECEIVED November 8, 1984

Potent Insect Antifeedants from the African Medicinal Plant *Bersama abyssinica*

ISAO KUBO and TAKESHI MATSUMOTO

Division of Entomology and Parasitology, College of Natural Resources, University of California, Berkeley, CA 94720

The chemical investigation of pest insect control agents from the East African medicinal plant Bersama abyssinica (Melianthaceae) has led us to the isolation and characterization of four new bufadienolide steroids; abyssinin, abyssinol-A, -B, and -C. One of these, abyssinin, exhibits very strong antifeedant activity (PC_{95}=5µg/disk) against the cotton pest insect, Heliothis zea. This is one of the most potent antifeedants isolated from botanical sources so far. 2D-COSY NMR was very useful in the structural assignment of these insect antifeedants.

In the course of our search for pest insect control agents based on natural products our interests have focused upon plants which are not attacked by insects as these plants may produce defensive chemicals for deterring herbivory. An example is the East African medicinal plant Bersama abyssinica (Melianthaceae), which is widely used for various diseases such as dysentery, epilepsy and hemorrhoid. The extract from young twigs is also drunk for the treatment of roundworm (1). In a preliminary experiment, using the leaf disk bioassay with a glandless cotton cultivar (2), the root bark extract of B. abyssinica exhibited potent insect antifeedant activity against the important North American cotton insect pest Heliothis zea. The isolation of the active principles, guided by a leaf disk bioassay, produced the four bufadienolides abyssinin, abyssinol-A, -B, and -C as insect antifeedants. Their structures were established as (I), (II), (III) and (IV), respectively. The chemical components of B. abyssinica were previously examined by Kupchan (3). By monitoring the cytotoxic activity against KB tissue cultures four bufadienolides, (bersaldegenin 3-acetate (V), bersaldegenin 1,3,5-orthoacetate (VI), hellebrigenin 3-acetate (VII) and hellebrigenin 3,5-diacetate (VIII)) were isolated as active principles. However, none of these were detected by the bioassay methods used in the present study.

0097–6156/85/0276–0183$06.00/0
© 1985 American Chemical Society

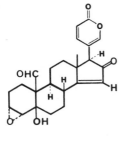

II

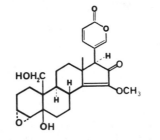

III

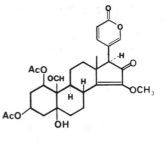

IV

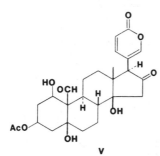

V

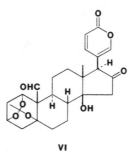

VI

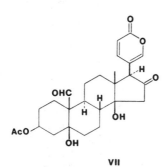

VII

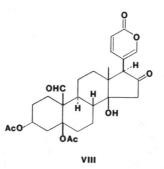

VIII

Isolation

The fresh root bark (2 kg) of B. abyssinica was collected near Kakamega, Kenya, in 1981, and extracted with aqueous methanol. The isolation scheme is depicted in Figure 1. The crude extract was concentrated, and the residue was extracted with hexane, methylene chloride, and ethyl acetate. The bioactive methylene chloride extract was chromatographed on silica gel. A $CHCl_3-CH_3OH$ (20 : 1) elution gave crude abyssinin (200 mg). Further elution with $CHCl_3-CH_3OH$ (10 : 1) gave a mixture of crude abyssinol A, B, and C (20 mg). This mixture was further separated by TLC (SiO_2, $CHCl_3-CH_3OH$; 10 : 1) into the three fractions composed of crude abyssinol A, B, and C, respectively. Each of the crude fractions of abyssinin, abyssinol A, B, and C still showed two spots with a ca. 1 : 1 ratio on TLC. The ^1H NMR spectra of these crude fractions indicated that each fraction contained two structurally similar compounds epimeric at C-17. The separation of these two components was troublesome because these were convertible to each other gradually under both normal- and reversed-phase chromatographic conditions with a protic solvent such as methanol. Final purification of the crude of abyssinin was performed by flash chromatography on silica gel using ether to give pure crystalline abyssinin and 17-epiabyssinin. Similarly, pure abyssinol A, B, and C were separated from their C-17 epimers by silica gel TLC without the use of protic solvents. The ethyl acetate fraction, which exhibited antimicrobial activity against Bacilus subtilis other than antifeedant activity, was subjected to droplet counter current chromatography (DCCC) to give, as the major compounds, gallic acid (72 mg) and methylgallate (45 mg) as shown in Figure 2.

Structure of abyssinin

Abyssinin (I), M.P. 278°C (from EtOH), $C_{27}H_{30}O_8$ (elemental analysis), possess the following physical constants: EI-MS, m/z 482(M^+), 454(M-CO), 439(M-COCH$_3$) and 422 (M-CH$_3$COOH); UV (EtOH) 228(ϵ 7700, sh.), 256 (ϵ 13400) and 296 nm (ϵ 7100, sh.); CD (EtOH) 330($\Delta\epsilon$+3.4) and 335 nm($\Delta\epsilon$+3.2, sh.); IR (CHCl$_3$) 2850, 1735, 1724, 1710, 1643, 1630, 1545 and 1250 cm^{-1}. The ^{13}C NMR data summarized in Figure 3a shows the presence of three CH$_3$, six CH$_2$, five CH, three quaternary, six olefinic, and four carbonyl carbons. These results were based on a combination of; proton-noise decoupling (PND), continuous wave decoupling (CWD), and partially relaxed Fourier transform (PRFT) techniques (4). The 400 MHz ^1H NMR data shown in Figure 3b were obtained mainly by 2D COSY techniques and by LIS experiments using Eu(fod)$_3$. These spectral data showed abyssinin to be a bufadienolide type steroid (5-8) containing the following groups: an acetate ester (g), an aldehyde (b), and epoxide (c), a quaternary methyl (d), a fully substituted α-methoxy enone moiety (e), and an α-pyrone group (f, λ_{max} 296 nm, ν_{max} 1710, 1630, 1545 cm^{-1}) (see Figure 4).

The 2D NMR spectrum shown in Figure 5 indicates the three low field proton signals (7.30, 7.04 and 6.33 ppm) are coupled to each

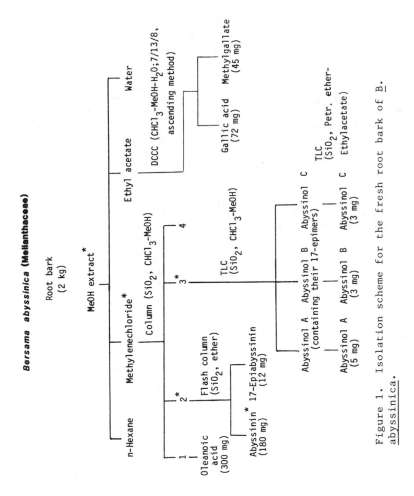

Figure 1. Isolation scheme for the fresh root bark of B. abyssinica.

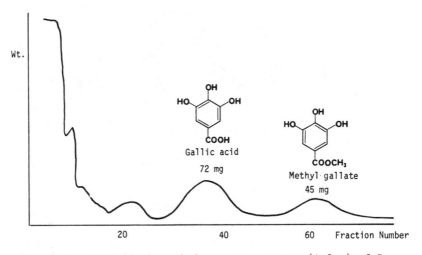

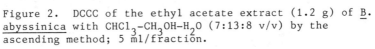

Figure 2.   DCCC of the ethyl acetate extract (1.2 g) of B.
abyssinica with $CHCl_3$–CH_3OH–H_2O (7:13:8 v/v) by the
ascending method; 5 ml/fraction.

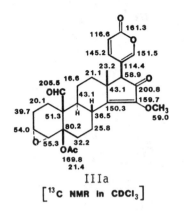

IIIa

$[^{13}C$ NMR in $CDCl_3]$

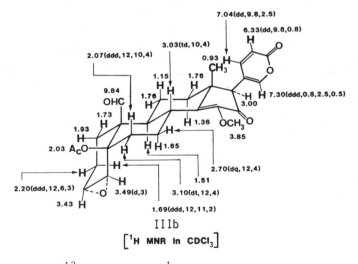

IIIb

$[^1H$ MNR in $CDCl_3]$

Figure 3. ^{13}C NMR (a) and ^1H NMR (b) data for abyssinin.

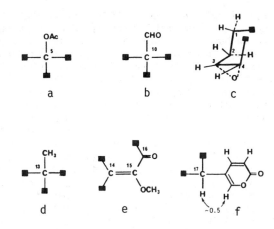

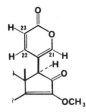

Figure 4.   Groups contained by abyssinin as shown by the spectral
data.

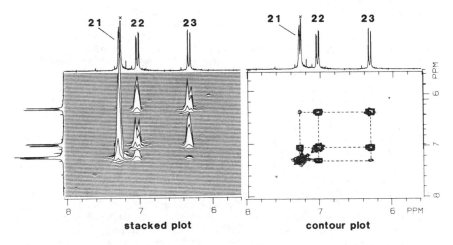

Figure 5.   Two-diminsional NMR spectrum.

other, and these signals were assigned to the protons on the
α-pyrone moiety. The high field shift of 22-H in comparison with
those of all known bufadienolides (6) is probably due to a
diamagnetic anisotropic effect caused by the C-16 carbonyl group.
The broad singlet proton signal at 3.00 ppm was identified as 17-H
by the presence of an allylic coupling (0.5 Hz) with 21-H (7.30
ppm). The contour plot shown in Figure 6 also showed the presence
of this allylic coupling. Similarly, a long-range coupling
between the methyl shift at 0.93 ppm and 17-H was observed in the
2D COSY spectrum shown in Figure 7, indicating their trans
relationship, and locating this CH_3 group at C-13. This
relationship also confirmed the configuration of the α-pyrone ring
to be β, and cis to the 13-CH_3. Upon catalytic hydrogenation
(over 5 Pd/C in ethanol), abyssinin absorbed two moles of
hydrogen to give tetrahydroabyssinin (IX). Tetrahydroabyssinin
(IX), $C_{27}H_{34}O_8$, possessed the following physical properties:
UV(EtOH), 248 nm(ε 9700); CD(10 dioxane/EtOH) 353(Δε+3.4), 349 nm
(Δε+3.2, sh.); EI-MS m/z 486(M^+); ^1H NMR(CDC1$_3$) 9.84(1H, s,
19-H), 4.62 (1H, t, J=13 Hz, 21-H_a), 4.20(1H, dd, J=13, 4.5 Hz,
21-H_e), 3.78(3H, s, OCH_3), 3.49(1H, d, J=3.5 Hz, 4-H),
3.41(1H, br.s, 3-H), 3.09(1H,dt,J=12, 4 Hz,6-H_e), 3.02(1H, td,
J=10, 4Hz, 8-H), 2.74(1H, ddd, J=15, 9, 6 Hz, 23-H_a), 2.64(1H,
dq, J=12, 4 Hz, 7-H_e), 2.48(1H, dt, J=15, 6 Hz, 23-H_e), 2.32
(1H, m, 20-H), 2.17(1H, ddd, J=12, 6, 3 Hz, 2-H_e), 2.08 (1H, m,
6-H_a), 2.03(3H, s, $OCOCH_3$), 1.87(1H, d, 3 Hz, 17-H), 1.15(3H,
s, 13-CH_3).
The low field shift (Δ+0.22 ppm) of the 13-CH_3 signal in going
from I to IX, without any other group shifting, supports the
β-configuration of the pyrone group. A molecular model of I shows
that the 13-CH_3 group is situated over the pyrone ring, and the
shift is probably due to an anisotropic effect induced by a ring
current of the pyrone moiety. The stereochemistry at position 17
in I is consistent with those of all bufadienolides known to date.
     The α-methoxy enone moiety (λ_{max} 256 nm, ν_{max} 2850, 1770,
1643, and 1250 cm^{-1}), isolated from the contiguous proton
systems, is positioned on the D ring on the basis of biogenetic
considerations of this class of steroids. The unusually low
chemical shift of the equatorial proton signal (2.70 ppm) at C-7
is due to deshielding by a through space effect of the methoxy
group. Abyssinin (I) is unstable in a protic solvent such as
ethanol, and epimerizes at C-17 to give a mixture of I and
17-epiabyssinin (X) (3:2). A significant low field shift (Δ+0.41
ppm) of the C-13 CH_3 signal in X vs I clearly shows that the
configuration of the α-pyrone ring is α. 17-Epiabyssinin (X) was
also isolated, and it also epimerized to give an equilibrium
mixture of X and I (2:3) in ethanol. It is presumable that X is
an artifact. The aldehyde group and the acetate group must be
located on the remaining quaternary carbons, C-10(51.3 ppm) and
C-4(80.2 ppm) respectively. The addition of Eu(fod)$_3$ to the
CDC1$_3$ solution spread out the congested spectrum, and induced
large changes of in the absorption due to proton 4-H. Since these
shifts showed complexation of the lanthanide reagent with the
acetate oxygen, the configuration of the epoxide was established
as α.

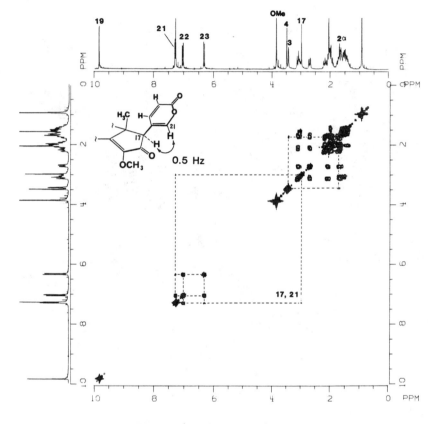

Figure 6.   Contour plot.

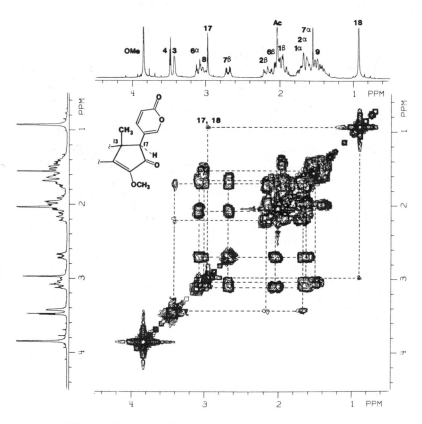

Figure 7.   Two-dimensional COSY spectrum.

The absolute configuration of abyssinin (I) was determined from CD studies of the tetrahydroabyssinin (IX) and its 3,4-bis (p-N,N-dimethylaminobenzoate) (XIII). Acid catalyzed epoxy-hydrolysis of abyssinin (I) (7N H_2SO_4/THF, 50°C), followed by hydrogenation (5 Pd/C, EtOH) afforded the tetrahydroglycol (XI) and its C-17 epimer (XII). In general, p-N,N-dimethylamino-benzoates are prepared from by p-N,N-dimethylaminobenzoyl chloride in pyridine. Because the preparation and storage of this reagent is troublesome due to its reactivity with water, we prepared the more stable reagent p-N,N-dimethylaminobenzoyl nitrile, which reacts easily with most alcohols but not water (9). A solution of p-N,N-dimethylaminobenzoyl nitrile in dry CH_3CN was added to a CH_3CN solution of tetrahydroglycol (XI), containing a catalytic amount of quinuclidine and stirred at room temperature for 17 hr. 3,4-Bis(p-N,N-dimethylaminobenzoate) (XIII) was purified directly from the reaction mixture by preparative TLC. The configuration of (XIII) was assigned by the ^{1}H NMR coupling constant (J=10 Hz) of the protons on carbon 3 and 4 (see Figure 8). The negative first Cotton effect at 324.8 nm ($\Delta\epsilon$-38.4) in Figure 8 showed the absolute stereochemistry of abyssinin to be that as indicated in structure (I). This also establishes the absolute configuration of the bufadienolides that were isolated from the same source as antitumor agents: bersaldegenin 3-acetate (V), bersaldegenin 1,3,5-orthoacetate (VI), hellebrigenin 3-acetate (VII), and hellebrigenin 3,5-diacetate (VIII).

Structure of Abyssinol A, B, and C

The structures of these antifeedants were established mainly through the comparison of their spectroscopic data with those of abyssinin (I). ^{1}H NMR data (400 MHz) of these bufadienolides, summarized in Table I, showed the characteristic signals of an α-pyrone ring and of a proton adjacent to an unsaturated carbonyl like abyssinin. The signals (0.86-0.99 ppm) of their C-13 methyl group suggest a β pyrone ring in all three compounds, in contrast to the chemical shift of the C-13 methyl group in 17-epiabyssinin (1.34 ppm).

Abyssinol A (II), m.p. 146°C (amorphous) has the following physical constants: IBEI-MS m/z 410(M^+), 392(M^+-H_2O), 364, 346 and 336, DCI-MS(NH_3) 428(M^++NH_4) and 411(M^++H), IR($CHCl_3$) 3530, 2860, 1745, 1720, 1700, 1638, 1610, 1535, 1450, 1375 and 1305 cm^{-1}, UV(EtOH) 227(ϵ 12000) and 292(ϵ 7700) nm. The ^{1}H NMR of II is quite similar to that of abyssinin, except for the following: (i) the lack of the acetoxy methyl signal, (ii) the presence of a hydroxyl proton signal, (iii) the lack of the methoxy methyl signal and (iv) the presence of an olefinic proton signal at 5.92 ppm. Together with the EI-MS data, (i) and (ii) indicate that 5β-acetoxyl group in abyssinin should be replaced by a 5β-hydroxyl group, which is also supported by the change in the shift (Δ-.41 ppm) of the proton attached to C-4. A molecular model of abyssinin shows that the C-4 proton is close enough to the acetoxy-carbonyl group as to be subject to an anisotoropic deshielding, although this isn't possible in the case of II. Since the 5.92 ppm olefinic proton of II is coupled to the

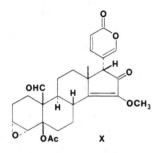

IX

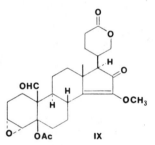

X

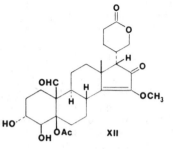

XVI

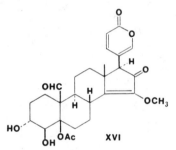

XII

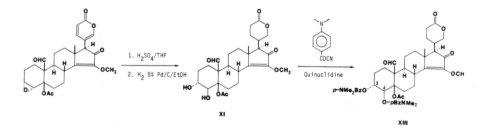

XI

XIII

Table 1.   1H NMR spectral data of abyssinol A. B and C and abyssinin in CDC13.

Assignment	Abyssinol A	Abyssinol B	Abyssinol C	Abyssinin
H-1			5.76(1H,m,$w_{1/2}$=5)	1.73(1H,m),1.93(1H,m)
H-2		4.23(1H,d,J=11.5,OH) 3.25(1H,brs.OH)	2.31(1H,m), 1.80(1H,m)	2.20(1H,ddd,12,6,3), 1.69(1H,ddd,12,11,2)
H-3	3.53(1H,m)	3.56(1H,m)	5.31(1H,m)	3.43(1H,m)
H-4	3.08(1H,d,3.7)	3.09(1H,d,3.6)	2.36(1H,dd,4,16.1), 1.92(1H,dd,2,16.1)	3.49(1H,d,3)
H-8	3.05(1H,dddd, 11.7,11.7,2.9,1.8			3.03(1H,ddd,10,10,4)
H-15	5.92(1H,d,1.8)			
H-17	3.14(1H,s)	2.98(1H,s)	2.96(1H,s)	3.00(1H,s)
H-19	9.83(1H,d,1.8)	3.59(1H,d,12), 3.33(1H,d,12)	10.18(1H,s)	9.84(1H,s)
H-21	7.30(1H,brd,2.5)	7.29(1H,d,2.7)	7.29(1H,d,2.6)	7.30(1H,d,2.5)
H-22	7.02(1H,dd,9.5, 2.5)	7.04(1H,dd,9.6, 2.7)	7.02(1H,dd,9.5, 2.6)	7.04(1H,dd,9.8, 2.5)
H-23	6.32(1H,dd,9.6, 1.1)	6.32(1H,d,9.6)	6.32(1H,d,9.5)	6.33(1H,d,9.8)
OMe OAc		3.83(3H,s)	3.84(3H,s), 2.06(3H,s), 2.05(3H,s)	3.85(3H,s) 2.03(3H,s)
Me	0.99(3H,s)	0.91(3H,s)	0.86(3H,s)	0.93(3H,s)

isolated allylic 8-H at 3.05 ppm (1.8 Hz), this olefinic proton
was placed at C-15 to replace the methoxy group found in abyssinin.
The UV spectral data are consistent with this.

Abyssinol B (III), EI-MS m/z 442(M⁺), 424(M⁺-H₂O),
406(M⁺-2H₂O), 394 and 287, UV(EtOH) 220(ε 9700), 253 (ε 14300)
and 290(ε 7200, sh.) nm, contains two hydroxyl groups. One was
placed at C-5 for the same reason as in the case of abyssinol A.
The other was assigned to C-19 since the ¹H NMR signal due to
the aldehyde group in abyssinin was not observed in abyssinol B,
but was replaced by a new isolated AB system corresponding to
-CH₂OH appearing at 3.59 and 3.33 ppm (J=12 Hz). Thus, abyssinol
B was determined to be structure III.

Abyssinol C (IV), m.p. 154°C (amorphous), EI-MS m/z 542(M⁺),
524(M⁺-H₂O), 482(M⁺-CH₃COOH), 464, 422(M⁺-2CH₃COOH),
404, 376 and 286, UV(EtOH) 225(ε 7900, sh.), 253(ε 13400) and
290(ε 6600, sh.) nm, differed from abyssinin in the following
respects: (i) the C-5 acetoxyl group is replaced by hydroxyl group
as in the case of abyssinol A and B, (ii) two secondary acetoxyl
groups (2.06 and 2.05 ppm) are present in ring-A instead of the
epoxy group found in Abyssinin. Since the coupling values of the
acetate bearing methine protons (5.76 and 5.31 ppm) are consistent
with the expected values for equatorial protons, the stereochemistry of the two acetoxyl groups were assigned as axial.
Further, a 1,3-diaxial relationship was disclosed by double
resonance experiments. Of the two possible structures (1,3-diaxial
structure XIV or XV in ring A), XIV was chosen because the aldehyde
proton signal underwent a low field shift (Δδ+0.34 ppm) compared
with that of abyssinin. The structure of abyssinol C was therefore
established as shown in IV.

Antifeedant Activity

Abyssinin (I), 17-epiabyssinin (X), abyssinol A (II), B (III),
C (IV) and abyssinin derivatives (IX and XVI) were tested for
antifeedant activity in a 'choice' situation (10). Leaf disks (1
cm²) were punched out from a glandless cotton cultivar (Pima S-4
of Gossypium barbadanse, a favored host of the cotton budworm
Heliothis zea), randomized, and arranged on moistened filter paper
in polyethylene form grids inside glass petri dishes as shown in
Figure 9. Alternating disks were treated with either 25 μl acetone
or with from 1 μg to 100 μg of test compounds dissolved in 25 μl
acetone. Three newly-molted third instar larvae of the cotton
budworm were then placed in the disks at 22°C and 80 RH in a dark
incubator. After 48 hrs, the larvae were removed and the disks
were visually examined. Table II indicates a PC₉₅ of (I)-(IV),
(IX), (X) and (XVI) for H. zea larvae. PC₉₅ values are
concentrations of samples resulting in 95 "protection" of treated
disks when compared to untreated disks. Cotton leaf disks which
had been treated with 5 μg of abyssinin (I) or 17-epiabyssinin (X)
were not eaten by H. zea.

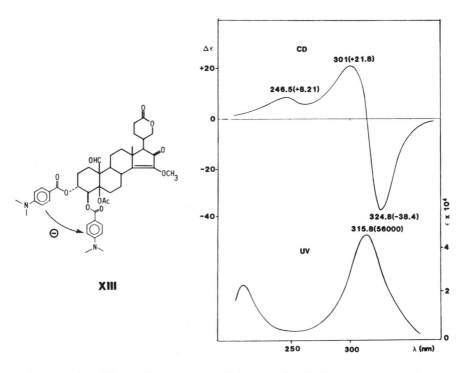

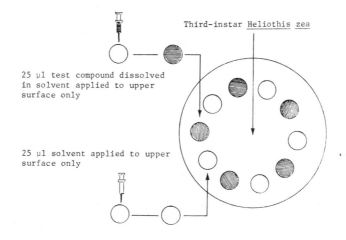

Figure 8.   CD and UV spectra of 3,4-bis(p-(N,N-dimethylamino)-benzoate) in EtOH.

Figure 9.   Cotton leaf disk "choice" bioassay.

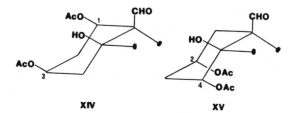

XIV   XV

Table II.   Antifeedant activities (PC_{95}) of bufadienolides
against H. zea.

Compounds	PC_{95}μg/disk)
(I)	5
(X)	5
(II)	25
(III)	20
(IV)	20
(IX)	>100
(XVI)	>100

Abyssinol A (II), B (III) and C (IV), without the acetoxyl moiety,
were less active than abyssinin (I) and 17-epiabyssinin (X) with
the acetoxyl moiety. We had observed a similar decrease of
antifeedant activity due to the removal of the acetoxyl group when
we investigated antifeedant activities of the ajugarins (11).
Tetrahydroabyssinin (IX) and its epoxy-hydrolyzed derivative (XVI)
were found to be inactive even at 100 μg/disk. This suggests both
the α-pyrone ring and the epoxide group are required for the
strong antifeedant activity of abyssinin (I). We also conducted a
'no-choice' artificial diet feeding assay (12), since compounds
not showing antifeedant activity but having growth inhibitory
activity would be overlooked in the previous 'choice' situation.
As a result, we isolated only abyssinin as an insect growth
inhibitor while monitoring by an artificial diet feeding assay.
In fact, abyssinol A (II), B (III) and C (IV), isolated monitoring
by the leaf disk assay, did not show any insect growth inhibitory
activity, as is shown in Table III.

Table III.   Growth inhibitory activity (ED_{50} in ppm) of
bufadienolides against Pectinophora gossypiella.

Compounds	ED_{50}(ppm)
(I)	10
(II)	>50
(III)	>100
(IV)	>100
(IX)	>150
(XVI)	>150
Gallic acid	>500
Methyl gallate	>500

Thus, the growth inhibitory activity of abyssinin (I) compared to abyssinols (II-IV) could be attributed to a 'potent' antifeedant effect which can cause starvation. Accordingly, the antifeedant activities of the abyssinols (II-IV) appear not to be strong enough as to force starvation.

Gallic acid and methylgallate, isolated monitoring by antimicrobial activity against B. subtilis, were inactive against P. gossypiella even at a concentration of 500 ppm. On the other hand, abyssinin (I) did not show microbial activity against B. subtilis even at a high concentration. This shows another example where different type of secondary compounds act in a variety of ways as defense adaptations of a plant.

Acknowledgments

Insects for the bioassay were kindly supplied by the agencies of the USDA in Phoenix, AZ; Tifton, GA; and Brownsville, TX. The authors are grateful to Dr. H. Naoki and Professor A. S. Kende for NMR measurements, Professor N. Harada for CD measurement, and Dr. J. A. Klocke for antifeedant bioassay.

Literature Cited

1. J. O. Kokwaro, "Medicinal Plants of East Africa"; East African Literature Bureau, Nairobi, 1976; p. 159.
2. I. Kubo and K. Nakanishi, "Host Plant Resistance to Pests"; ed. by P. A. Hedin, ACS Symposium Series 62, American Chemical Society: Washington, D.C., 1977; pp. 165-178.
3. S. M. Kupchan and I. Ognyanov, Tetrahedron Lett., 1709 (1969), S. M. Kupchan, R. J. Hemingway and J. C. Hemingway, J. Org. Chem., 34, 3894 (1969).
4. P. Zanno, I. Miura, K. Nakanishi and D. Elder, J. Am. Chem. Soc., 97, 1975 (1975).
5. P. J. May, Terpenoidas and Steroids, 1, 527 (1971).
6. L. Gsell and Ch. Tamm, Helv. Chim. Acta., 52, 551 (1969).
7. K. Nakanishi, T. Goto, S. Ito, S. Natori and S. Nozoe, "Natural Products Chemistry"; Academic Press: New York, 1974; p. 469-476.
8. J. Meinwald, D. F. Wiemer and T. Eisner, J. Am. Chem. Soc., 101, 3055 (1979).
9. J. Goto, N. Goto, F. Shamsa, M. Sato, S. Komatsu, K. Suzaki and T. Nambara, Anal. Chim. Acta., 147, 397 (1983).
10. I. Kubo and J. A. Klocke, l'Colloques de l'I.N.R.A., 7, 117 (1981).
11. B. G. Chan, A. C. Waiss, Jr., W. L. Stanley and A. E. Goodban, J. Econ. Ent., 71, 366 (1978).

RECEIVED November 8, 1984

Cockroach Control with Juvenoids

GERARDUS B. STAAL, CLIVE A. HENRICK, DAVID L. GRANT, DAVID W. MOSS, MICHAEL C. JOHNSTON, ROBIN R. RUDOLPH, and WILLIAM A. DONAHUE

Zoecon Corporation, Palo Alto, CA 94304

The potential of juvenoids to control cockroach populations depends on their ability to stop reproduction. The assessment of potency should therefore relate to reproduction or correlate this property with short term responses such as morphogenetic abnormalities. In this study, several dienoates with high activity on various other insect targets and a few other prominent juvenoids were compared in a feeding assay for activity on the German cockroach. Of the compounds tested, hydroprene was the most potent, whereas methoprene was considerably less active against this target species. Further studies on the persistence of residues on various substrates of technical hydroprene confirmed the volatility of hydroprene. Through formulation, the volatility could be greatly reduced with retention of full activity. However, volatility may well play a contributory role in the efficacy of practical household applications. The coapplication of conventional insecticides contributes to the short term reduction in population and thus enhances consumer acceptance of juvenoids.

Studies in the early and mid seventies on the effect of several juvenoids on cockroaches have indicated their activity in terms of inhibition of metamorphosis, increased melanization, development of ovaries, mortality of adults and progeny, effects on regeneration, deficiencies in mating behavior, delay of the final molt and, last but not least, the inhibition of reproduction (1-11). The potential of certain juvenoids, particularly that of hydroprene, to control cockroach populations by inhibiting the reproduction through exposure to the juvenoids in a critical period before metamorphosis was clearly suggested by this earlier work. However, this was not followed up at that time because measures requiring several months to achieve control were not considered to be commercially attractive. With juvenoids alone it would indeed take several months for the total disappearance by attrition of the last normal and sterile adults, since the average longevity of female adults is about 5 months.

0097–6156/85/0276–0201$06.00/0
© 1985 American Chemical Society

At least two events were responsible for a change in this attitude and provided the incentive to select and develop an effective juvenoid for cockroach control: a) the success achieved with the juvenoid methoprene in the control of breeding flea populations in household carpets and upholstery (17, 18) b) subjective observations by tenants of apartments treated with methoprene foggers during "field" experiments conducted in 1981 designed to establish the efficacy on flea populations indicating that cockroach populations also appeared to be affected.

Over the past decade, the outlook for cockroach control had undergone changes due to the realization that the most effective pesticide treatments, even if applied by experienced pest control operators (P.C.O.'s) although seemingly effective for a short while, never appeared to yield long lasting control. This is probably related in part to cockroach behavioral adaptations such as hiding in cracks and crevices when insecticide presence is sensed or to temporary relocation to untreated adjacent habitats until the residues have disappeared. In addition to this, several of the most efficacious pesticides have been banned in recent years and resistance against the remaining ones have become widespread (19). Hence, an increasing interest developed for the introduction of less toxic products with a novel mode of action.

In a further development of the juvenoid control approach, there appeared to be a very good promise also for combinations of conventional insecticides (whether really effective or only perceived as such) with effective long term juvenoid control agents.

Therefore, we undertook to confirm whether hydroprene (as yet unregistered for any insect control use) was indeed the most efficacious of the juvenoids proprietary to Zoecon and whether the use of methoprene, another dienoate which was already registered for household insect control targets (i.e. fleas), could be equally efficacious if applied at higher rates. We also wanted to compare our most effective dienoates with juvenoids under consideration for commercial development by other parties. The activity of various juvenoids on a spectrum of insect targets has been described in several review papers (20-22). Furthermore, if the study would suggest that a dienoate juvenoid could indeed be successfully applied against cockroaches, further studies on the most effective mode of application, longevity of residues, optimized formulations, combinations with adulticides, etc. would be undertaken.

Our principal target was the German cockroach, (Blattella germanica) as the most widespread domestic cockroach in the USA, but a limited number of confirmatory studies were done on the American cockroach (Periplaneta americana) and the oriental cockroach (Blatta orientalis).

Assay for Comparative Activity

To study the relative intrinsic activity of a series of dienoate juvenoids, a forced feeding test was selected in which the materials were incorporated in the only food available to Blattella germanica populations throughout the duration of the experiment. This choice was made because of the difficulty in synchronizing the peak of sensitivity in the last nymphal instar due to the long duration of that (and other) stages in cockroaches. The absence of definite developmental markers other than the molting process and variations in the duration of the last nymphal instar make determination of the sensitive period unreliable. Food administration would at least secure the presence of the compound throughout the feeding period of the last nymphal stage, covering the window of sensitivity.

Since many years of juvenoid testing have confirmed that the results of feeding tests parallel those of topical applications in several insect species (primarily phytophagous Lepidoptera) with a better synchronisable sensitive period, we felt reasonably secure with a food test. However, it was known already at that time that a food bait is an ineffective way of administering juvenoids when the bait is in competition with a choice of other household food sources (unpublished studies). Other test possiblities such as continuous substrate contact, seemed more relevant but were not pursued because their accuracy and reproducibility was unknown at the time, at least for juvenoids. The starch experiments from Cruickshank & Palmere (23) and Riddiford et al (14) (whether called "bait" or "bedding") did not appeal to us because, in our experience, starch could act as a sorptive dust contributing to the cockroaches' demise by causing dehydration. As will become apparent later, substrate residue exposure has many parameters that would indeed have interfered with a comparative evaluation. We considered that the use of a food substrate to administer the juvenoids would probably minimize the uncertainties in the persistence of residues in the bedding material or on other surfaces in the habitat. However, the chosen method of food incorporation could have one potential flaw: if any of the incorporated materials were repellant at the applied concentration, death by starvation could result. Careful observation could probably rule this out. Having no real estimate of the effective concentrations under our conditions, the initial phase of this experiment was carried out at 1000 ppm for all materials investigated. In following test cycles, the most active materials were applied at lower doses for further differentiation. The results of these successive cycles are here described together as one experiment.

Materials & Methods. Batches of fifty 4th instar mixed sex B. germanica were placed in standard 28x17x12 cm disposable mouse cages. The inside walls of the cages were treated with Fluon (Fluon AD-1 liquid teflon suspension. Northeast Chemical Co., Inc. 153 Hamlet Avenue, P.O. Box 1175, Woonsocket, RI 02895) to prevent the cockroaches from escaping. Water and harborage was provided, and cages were fitted with fine mesh screen tops (white polyester sheer curtain fabric. Chandler Enterprises, P.O. Box 4934 Foster City, CA 94404) to prevent ingress or egress. Cages were maintained at 27°C in a 16 hr photoperiod and approximately 50% RH in an air conditioned room (no air recycled.) These tests were all performed in triplicate.

The food was prepared by pulverizing Gaines burger dog food in a Waring blender, then mixing this with the active ingredient dissolved in acetone. After mixing, the food was left in an uncovered storage jar for 24 hours to allow for evaporation of the solvent. Approximately 10 g of food was supplied to each cage, and the rest (40 grams) was stored at 5°C in sealed jars for use as needed to resupply consumed food in the test cages.

The dienoates #1, #2, #4-15, #17, #18, including methoprene (#9) and hydroprene (#1; 25) have been described by Henrick et al including their synthesis and biological activity against various insect targets, and were synthesized in Zoecon Corporation (24, 25). Compound #3 (R-20458) was supplied by Stauffer Chemical Company (26); compound #16 was supplied by M. Schwartz (USDA) (27).

All populations were regularly scored for either presence or absence of unambiguous morphogenetic abnormalities (ranging from slightly deformed or crumpled wings to completely nymphal supernumerary instars) as well as for the number of F_1 progeny. It was deemed essential to observe the population for a sufficiently long period after the molt to adult, since it is a characteristic of most juvenoids to delay the metamorphic molt. In all cases, the final reproductive scoring was done after the first ovarian cycle in all treatments was completed and the first round of egg capsules (viable or unviable) were dropped. We have observed that non reproductive females would remain sterile in subsequent ovarian cycles under continuous exposure to the juvenoid. The possibility of a reversal after removal from exposure is the subject of ongoing studies. However, earlier studies (29) have indicated that this would not occur.

Results. In nearly all cases, the triplicate cages for each dose yielded identical results. The order of activity on reproduction obtained in these food experiments is indicated by their ranking order in Table I. At 10 ppm, hydroprene (#1) appears to be somewhat more effective than the other dienoates and the Stauffer cpd (#3). Methoprene (#9) failed even at 1000 ppm, indicating a lesser efficacy as compared to hydroprene by a factor between 10 and hundred. Several other dienoates and Schwartz's compound (#16), although exhibiting at least fair to good activity on other insect targets (22, 24), failed in this test even at 1000 ppm. For the 2,4-dienoate esters tested, the presence of an 11-methoxy or an 11-hydroxy group produces a sharp drop in activity (#1 vs #13 and #7 vs #9). A similar result is obtained in Table I for the S-isopropyl esters (#5 vs #11). The ketone #2, which has a shape similar to that of the ethyl ester #1 has an activity very similar to #1. For the dienamides, the N-ethyl analogs are more active than the N,N-diethyl compounds (#4 vs #8 and #6 vs #10). The sec-butyl ester #12 is much less active than the corresponding isopropyl ester #7. It should be noted that these results differ from those obtained by Radwan & Sehnal (12), who found methoprene to be more active than hydroprene in topical applications to Nauphoeta cinerea. They found that methoprene, hydroprene and the isopropyl ester #7 were all very active at inhibiting metamorphosis of N. cinerea, with methoprene being the most active. The aromatic ether #3, and several related analogs, showed much lower activity

Table I. Percent Abnormal and Number of First Cycle Progeny (in Parentheses) Resulting from "No Choice" Feeding of Blattella germanica Nymphs on Various Juvenoids Incorporated in Food

Compound No.	1000 ppm	500 ppm	250 ppm	100 ppm	10 ppm	5 ppm
1	100 (0)	100 (0)	100 (0)	100 (0)	11 (98)	10 (313)
2	100 (0)	100 (0)	100 (0)	100 (0)	12 (253)	6 (469)
3	100 (0)	100 (0)	100 (0)	100 (0)	13 (260)	11 (322)
4	100 (0)	100 (0)	100 (0)	100 (0)	11 (391)	0 (402)
5	100 (0)	100 (0)	98 (0)	54 (0)	4 (397)	2 (493)
6	100 (0)	100 (0)	100 (0)	100 (0)	13 (421)	8 (494)
7	100 (0)	100 (0)	100 (0)	87 (0)	0 (448)	0 (438)
8	100 (0)	100 (0)	99 (0)	31 (188)	8 (350)	4 (315)
9	24 (266)	11 (400+)	9 (400+)	-	-	-
10	98 (13)	94 (400+)	76 (400+)	-	-	-
11	100 (0)	99 (400+)	92 (400+)	-	-	-
12	100 (10)	98 (400+)	47 (400+)	-	-	-
13	0 (116)	-	-	-	-	-
14	0 (194)	-	-	-	-	-
15	0 (215)	-	-	-	-	-
16	0 (148)	-	-	-	-	-
17	0 (400+)	-	-	-	-	-
18	0 (150)	-	-	-	-	-
19 (untreated)	0 (400+)	-	-	-	-	-

Note: - = not tested at this level.

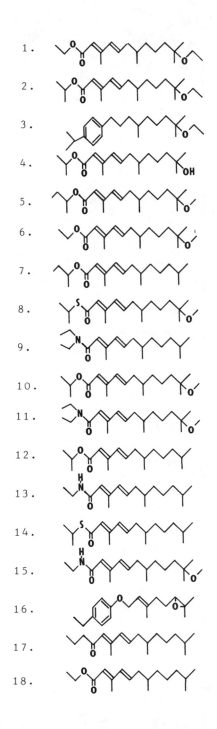

against this species (12). Riddiford <u>et al</u> had reported previously that under their testing conditions (constant contact with the compound deposited on cornstarch bedding material) hydroprene (#1) was substantially more active than methoprene (#9), kinoprene (22) and #3 against <u>Blattella germanica</u> (14). Triprene (22) was found to be reasonable active under their conditions. The low activity they observed for #3 (cf. Table I) could be due to its relatively low volatility and/or increased binding to the cornstarch.

This experiment also allowed us to draw an important conclusion that could be substantiated in all further work on hydroprene: morphogenetic abnormality and full reproductive inhibition were correlated with each other at all dose levels in the sense that morphogenetically affected adults always failed to reproduce (compare Table I). The reverse was not always true; at marginal doses of juvenoids, morphogenetically normal adults were found occasionally that were unable to reproduce. However, the observed correlation may not hold true for still other types of juvenoids. We found no evidence of repellancy for any of the juvenoids at the concentrations used in this experiment, although methoprene had shown this in preliminary experiments at 10,000 ppm.

<u>Residue test on latex painted surface</u>

In this test series technical hydroprene and methoprene were compared with a residual dust formulation (RF 10%) on a latex paint substrate.

<u>Materials & Methods</u>. The bottoms of standard disposable mouse cages were first painted with a flat latex interior paint. After letting the paint dry for one to three days, the test materials were applied by spreading either acetone dilutions dribbled on from a syringe or by spreading the dust formulation with a spatula over the painted surfaces. One day later, these cages were infested with fifty 4th instar nymphs of <u>Blattella germanica</u> and further treated and scored as described in Experiment 1. All test doses were set up in duplicate. Final scoring was performed 100 days after initiation of the experiment.

<u>Results</u>. Technical hydroprene as well as the RF formulation were both fully effective in inhibiting reproduction and producing a high or complete incidence of morphogenetic abnormalities at a rate as low as 1 $\mu g/cm^2$ (Table II). Technical methoprene required 1000 $\mu g/cm^2$, and thus was 100 to 1000 times less effective than hydroprene technical. Methoprene RF was at least 10 times more effective than technical methoprene and therefore approximately 100 times less effective than hydroprene in a similar formulation. High concentrations of the inerts of RF formulations appeared to be toxic by themselves, at least when formulated with dienoates. A formulation blank was not included because the blank RF formulation was found to be physically not comparable (only a formulation with an inert look alike dienoate would have made a relevant comparison).

Table II. The influence of formulation on the efficacy of
hydroprene and methoprene on Blattella germanica nymphs
when applied to a latex paint substrate.

	% abnormal morphogenesis				ug/cm^2
	0.1	1	10	100	1000
Hydroprene Tech	3 *r*	87 *o*	100 *o*	100 *o*	100 *o*
Hydroprene RF 10%	6 *r*	100 *o*	100 *o*	$-^a$	−
Methoprene Tech.	−	7 *r*	4 *r*	3 *r*	75 *o*
Methoprene RF 10%	−	6 *r*	64 *r*	100 *o*	[100 *o*]

r = *reproduction observed;* *o* = *no reproduction;* − = not done.
$-^a$ no survival; [] few survivors only.

Discussion. The increase in activity for the RF formulations of
methoprene and the retention of activity for hydroprene was a
surprise since formulations with a long residual generally tend to
decrease acute activity. However, the increased stability, the
decrease in volatility and the intense contact with a dust
formulation by grooming cockroaches may have contributed to this
result. It also appeared that methoprene (RF 10%) at 100 ug/cm^2
was an effective control dose.

Persistence of hydroprene residues on various household surface
substrates

The effect of three different formulations on the persistence of
hydroprene for aging periods of up to 32 weeks was studied on four
different substrates under average household exposure conditions.

Preliminary observations had indicated that residues of
technical hydroprene of 0.5 ug/cm^2 on glass (after overnight
exposure to ambient room conditions) yielded only marginal results
probably due to the volatility of this dienoate. This led us to
investigate two existing commercial formulations and one
experimental formulation at higher rates for their persistence
throughout extended periods of exposure to ambient conditions.

Materials & Methods. Rectangles (14x23 cm) of vinyl floor tile,
unfinished plywood, glass, and vinyl tiles painted with enamel
paint (U.S. Navy) were treated with the following formulations:
a) emulsifiable concentrate (5E) diluted in water sprayed on
with a De Vilbiss sprayer at 5 p.s.i. at rates of 1 and 2 ug/cm^2
b) A 10% microencapsulated suspension (M 10%) equivalent in
manufacture to ALTOSID SR10 was applied in the same way diluted in
water, yielding the same rates of A.I. c) The RF dust
formulation (9-10% A.I.) was applied as dry powder and spread with
a spatula. This formulation was only applied to plywood and
glass. All plates were then kept for stated intervals at ambient
room conditions similar to those in experiment 1, with ambient

light from fluorescent fixtures (equipped with infrared absorbing shields).   Plates treated with a) and b)   were stored upright, plates with c) were kept horizontally to prevent the loss of powdery residues.

After the exposure period (actually 24 hrs for the 0 weeks of exposure), the substrate plates were then inserted in standard mouse cages, covering most of the bottom.   In each cage twenty-five 4th instar <u>Blattella germanica</u> nymphs were introduced, after which the rearing conditions and scoring techniques described under experiment 1 were applied.

Results.   Table III indicates that the residual persistence varied with the formulation and the substrate.   Vinyl tile and painted surfaces retained residues better than glass.   Bare plywood was better for M than for the EC.   The RF formulation was a clear

Table III.   Residual persistence of hydroprene formulations on various household surfaces applied at equal rates of active ingredient

Substrate	Formulation	1 $\mu g/cm^2$ Weeks of aging						2 $\mu g/cm^2$ Weeks of aging					
		0	2	4	8	16	32	0	2	4	8	16	32
Vinyl tile	EC 60%	0	0	0	r	r	r	0	0	0	0	r	r
	M 10 %	0	0	r	r	r	r	0	0	0	r	r	r
	Untreated	r	r	r	r	r	r	r	r	r	r	r	r
Enamel Paint	EC 60 %	0	r	r	r	r	–	0	0	0	0	r	–
	M 10%	0	0	r	r	r	–	0	0	0	r	r	–
	Untreated	r	r	r	r	r	r	r	r	r	r	r	r
Plywood	EC 60 %	0	r	r	r	r	–	0	r	0	r	r	–
	M 10 %	0	0	r	r	r	–	0	0	0	r	r	–
	R F 10 %	0	0	0	0	0	r	0	0	0	0	0	0
	Untreated	r	r	r	r	r	r	r	r	r	r	r	r
Glass	EC 60 %	0	r	r	r	r	–	0	r	r	r	r	–
	M 10 %	0	r	r	r	r	–	0	r	r	r	r	–
	RF 10%	0	0	0	0	0	0	0	0	0	0	0	0
	Untreated	r	r	r	r	r	r	r	r	r	r	r	r

0 = no reproduction; r = reproduction; – = not tested.
EC = emulsifiable concentrate, M = microencapsulated formulation (suspension),   R F = residual formulation (dry powder).

winner. Only at 32 weeks and at the 1 µg dose is there a beginning of breakdown of reproductive control. We concluded that the RF formulation afforded a superior protection against evaporation and possibly also against photo-instability without decreasing biological availability.

In this table only the complete inhibition of reproduction is represented since we concluded that partial failure of reproduction could not be assessed quantitatively (it was not known, for instance, from how many surviving adults a low number of offspring could have originated).

Vapor test

In order to test the volatility and the containment of such vapors in the usual mouse cages with screen tops, 50 mg of technical hydroprene or 50 mg A.I. of hydroprene RF 10% were placed in aluminum boats with 1/2 cm upturned edges in the top 3 cm of the cages in such a way that no direct contact with the population of cockroaches could take place. All other conditions were similar to those described in the other experiments. The test was scored for morphological as well as reproductive effects well after completion of the first reproduction cycle in the controls (Table IV).

Results. The results convincingly demonstrate the vapor activity of technical hydroprene in both types of scoring and the absence of it in the RF formulation. Although one of the RF replicates produced a sizable number of morphogenetically affected individuals, the reproduction was not noticeably affected. Even though the doses of hydroprene (equivalent to 100 $\mu g/cm^2$) were on

Table IV. Morphogenetic inhibition and effect on reproduction as a result of exposure to hydroprene as vapor only at 50 mg/cage (equivalent to 100 $\mu g/cm^2$).

Cage	Compound	Abnormal	Normal	Reproduction
1	untreated	2	27	R
2	"	3	18	R
3	"	2	18	R
4	hydroprene	23	0	0
5	"	18	0	0
6	"	21	0	0
7	hydroprene RF 10%	18[a]	4	R
8	"	2	33	R
9	"	0	30	R

a) Possible false positive, see text.

the high side, the RF formulation failed to produce any vapor effect. The false positive was probably caused by a very marginal spill of powder formulation or a vapor contamination from an adjacent cage.

Aquarium and chamber experiments

3 different and independent long term tests with hydroprene were run either in experimental chambers simulating household cockroach habitats or in containers (aquariums) that had been exposed to methoprene or hydroprene under simulated field conditions.

Dallas Aquarium Tests. In October of 1981, tests were initiated at Zoecon Industries (Dallas, Texas), using three 10 gallon aquariums, each infested with 5 adult male and 5 adult female B. germanica. One aquarium, including a harborage, was placed in a 3000 cu ft room which was then fogged with 3 oz of 0.15% hydroprene or approximately 0.23 µg/cm^2. The second aquarium was fogged in similar fashion with 3 oz of 0.15% methoprene, and the third aquarium was left untreated. After fogging, the aquariums with the fogging residues were removed to a clean room and maintained at ambient temperature (18-24°C) and 30-60% RH. Food and water were supplied routinely to all three chambers.

Since reproduction of adult cockroaches was already known to be unaffected by juvenoid type compounds applied in that stage, the original 10 adults were removed from each aquarium on day 65, leaving an established F_1 target population. Total population counts were made periodically over a six month period.

Figure 1 shows the changes in the total population size with time. By the end of six months the hydroprene population was reduced to near zero; the few remaining cockroaches were non reproductive adults. The control population increased from 10 to approximately 1000 during the test period.

Dallas Chamber Test. In February of 1982, two 1000 ft^3 chambers were set up as simulated kitchens in Zoecon Industries (Dallas, Texas). These chambers were supplied with ready made kitchen cabinets with unfinished plywood tops. The temperature was kept near 27°C using a small space heater and a built in air conditioner as necessary and humidity was maintained at 65%. The diurnal cycle was set at 16 hrs light, 8 hrs dark. Food (Waynes Pro Mix dry dog food), water and partitioned harborage boxes were placed in 6 locations throughout the chambers.

The chambers were infested with approximately 500 Blattella germanica (mixed population) each. These populations were allowed to acclimatize and build up for about a month. Chamber A was treated with a 1.2% hydroprene fogger at the rate of 2 oz per 1000 ft.^{3} Chamber B was treated with a blank fogger (no A.I.) at the rate of 2 oz per 1000 ft.^{3} or approximately 0.62 µg/cm^2. Both chambers were monitored monthly by means of counts of cockroaches present in the six harborages. These cockroaches were returned to the chamber after counting.

Figure 2 shows that the hydroprene treated population remained nearly constant at about 1000 over the entire one year duration of the test whereas the untreated population increased to over 60,000 cockroaches, which constitutes 98.5% control by comparison. The stable population in the treated chamber suggests that progeny of the original hydroprene population were indeed unable to reproduce. In completely isolated chambers the population could have been expected to disappear by attrition in approximately 6 months. The population remaining in the hydroprene chamber probably reflected a continuing low level of reinfestation with untreated cockroaches because the chambers appeared to be not fully cockroachproof.

<u>Purdue University Kitchen Test.</u>   In the fall of 1982, E.S. Runstrom, a research associate working with Dr. G.W. Bennett of Purdue University (West Lafayette, Indiana), initiated tests of hydroprene against <u>Blattella germanica</u> in four imitation kitchens set up in a vacant campus building. These structures, designed to simulate the household environment of the German cockroach, were built out of wood, and measured 10'x8'x6'. Each chamber was made escape proof and contained a base cabinet with two doors under the sink, two side units, each with one drawer and one door, and two wall mounted cabinets above the base units. Lights were installed in each chamber and were programmed for a 12:12 photophase. Two chambers served as untreated controls, while two were treated with hydroprene at the rate of 5.95 ml of hydroprene 5E (65% A.I.) per gallon of water applied at 1 gallon per 1000 ft.^{2} amounting to approximately 4.4 µg/cm^2. Food and water were supplied continuously.

Each chamber was infested before treatment with a population of 750 <u>Blattella germanica</u> consisting of 250 nymphs of mixed age, 250 adult males, and 250 adult females. Populations were sampled monthly with monitoring traps made from baby food jars baited with bread soaked in beer. These escape proof jars were placed in the chambers for a period of 24 hours once per month. Trapped cockroaches were counted as male, female, gravid female, adults showing juvenoid effects, large and small nymphs.

Since the ultimate goal was to reduce the population to zero through the sterilizing effect of hydroprene, only the monthly total trap counts are shown. The data points in figure 3 represent the combined total trap counts of two hydroprene treated chambers and of two untreated ones. The striking divergence in population levels began at the sixth month, which is the time when unaffected cockroaches, which were adults at the time of treatment, started to die out. By month 10 the treated populations were near zero and those remaining were only sterile adults.

The results obtained in these chamber and aquarium tests clearly confirmed the potential of hydroprene as a cockroach control agent with consumer oriented application methods.

<u>Field Experiments</u>

During 1981-83 field experiments were conducted in single family residences and in apartment complexes in several locations in the USA in compliance with Experimental Use Permits (E.U.P.'s) from the

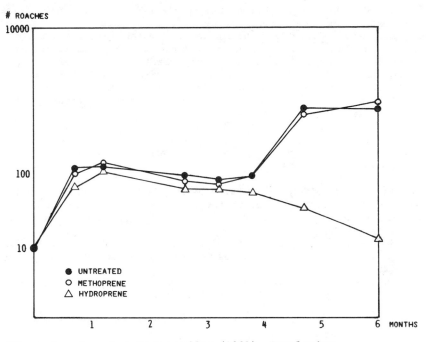

Figure 1. Aquarium tests Dallas (1981) Population
development over 6 months in aquaria exposed to fogging with
hydroprene and methoprene.

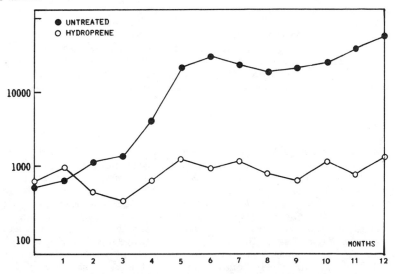

Figure 2. Chamber tests Dallas (1982). Population
development over 12 months after hydroprene application.

US Environmental Protection Agency (E.P.A.). In all cases, apartment complexes were selected that harbored a sizable population of <u>Blattella germanica</u> and in which full cooperation of owner and tenants for a lengthy treatment and observation schedule could be secured. Uniform treatments were applied in all units of a complex (4-6 units). The treatments were generally 0.6 % or 1.2% hydroprene total release aerosol foggers, often combined with simultaneous spray treatment of cracks and crevices with a conventional insecticide.

The great variety of treatment schedules and variations between locations etc make a comprehensive discussion of the results within the scope of this paper impossible. However, the following general conclusions could be drawn:

1. Full eradication of cockroach populations in single family residences with hydroprene foggers alone or in combination with conventional insecticides has been consistently achieved. (Figure 4).
2. Simultaneous insecticide applications do contribute substantially to the early demise of cockroach populations (29).
3. Full success in apartment complexes is equally feasible but is often delayed by compliance failures, tenants moving in and out, etc.
4. Retreatment schedules at 4-6 months intervals usually yield the best results.

As an illustration, population curves obtained in single residences in various locations (Dallas, Denver, Chicago) were depicted in figure 4. The treatments in this group of field experiments were a) hydroprene 0.6 % foggers b) hydroprene 1.2% foggers plus dichlorvos/propoxur foggers c) dichlorvos/propoxur foggers only. The dichlorvos/propoxur treatments were applied at twice the usual treatment rate in both b) and c). The results clearly indicated that hydroprene either alone or in combination with insecticides provided a more effective and longer lasting control than the conventional insecticides by themselves. The combination treatment appears to yield superior control at least initially.

Discussion

Hydroprene has proven to be the juvenoid of choice for the control of the major domestic cockroach species. Its success appears to be due to its very high intrinsic activity as compared with most other juvenoids and probably also to its volatility which may allow for a penetration of vapor in inaccessible cockroach harborages. It is likely that hydroprene vapors readily translocate between many household surface materials and thus remain accessible to resident cockroach populations. This volatility of hydroprene could also be a liability for persistence in places with high air displacement. In this case, hydroprene formulated for slow release in a dust (RF 10%) could provide a very persistent residue, albeit without the vapor benefits.

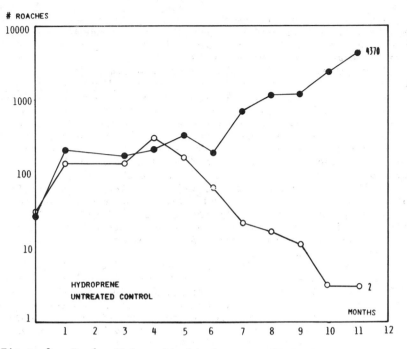

Figure 3. Purdue University kitchen test (E.S. Runstrom and
G.W. Bennett 1982). Population development over 11 months
after hydroprene application.

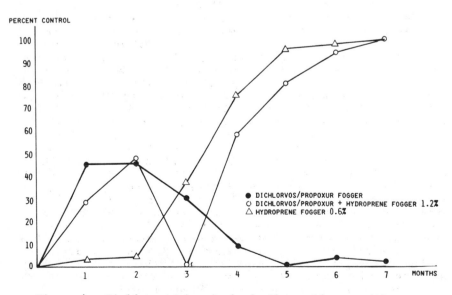
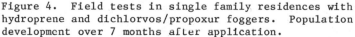

Figure 4. Field tests in single family residences with
hydroprene and dichlorvos/propoxur foggers. Population
development over 7 months after application.

Methoprene, a compound already registered for many applications, can be used for cockroach control but at higher rates than for hydroprene and only if formulated for optimal effectiveness, such as in the RF 10% formulation.

Many different effects have been described as a result of juvenoid treatment on cockroaches, primarily Blattella germanica. In none of the tests we undertook have we seen a significant mortality attributable to either a direct effect of the compounds tested, (13, 14, 23), an inability to molt properly, or a failure to molt (5. 8, 11). On the contrary, many of the morphogenetically affected adults or adultoids appeared to suffer no ill effect from the indignities inflicted by the juvenoids, other than failure to reproduce. Although the treatment methods may be partly responsible for differences between the results of different authors, it is more likely that qualitative differences between types of juvenoids are partly responsible. Delays in the final molt were frequently observed in batches affected by juvenoids. Most, if not all, of these nymphs made it eventually to viable, but non reproducing, adultoids.

The sensitive period for morphogenetic and sterility induction was clearly limited to the last nymphal instar, excluding the last 10 days before the molt. (If batches of last instar nymphs of mixed age are exposed to hydroprene, adults molting in the first ten days after the treatment appear to be morphogenetically as well as reproductively unaffected). Treatment of adults themselves was also without consequences, although a temporary reduction in reproduction rate was observed on several occasions. However, it remains possible that such treatment does not affect total reproductive capacity of the females. The literature on effects on the males is very scant. However, there are indications that males may also be affected and reproductively incapacitated. Whether this would take place at the same dose level as that for the females, remains to be established. The true nature of the reduced fertility effect has never been established with certainty. However, since only exposure during the sensitive period in the last nymphal instar induces permanent sterility, there must be a link with metamorphosis. In our trials we established a link between external metamorphic inhibition and sterility for hydroprene in the sense that adultoids (characterized by deformed or crumpled wings) were always sterile. However, females could sometimes be non reproductive while not visibly affected. Non fertile females did produce oothecae containing no eggs and having a shriveled appearance. Yet, on dissection, hypertrophied ovaria were often found. The opposite effect, atrophied ovaria, was described by Das & Gupta (3). In one of our preliminary trials, females kept separated from males as adults, did produce equally affected and sterile oothecae. It is therefore not possible to state that the observed infertility could not result from a failure to mate, either because of morphological or behavioral inadequacy or the absence of chemical signals. Cruickshank & Palmere (23). observed that exposure to certain amide juvenoids caused the premature dropping of oothecae. This is likely to be an irritation response, since it is also observed as a result from exposure to many other conventional insecticides.

We have been unable to interpret Edwards' finding that exposure to earlier larval instars intensifies the response to methoprene (28). Other authors, working with more active juvenoids usually reported increasing sensitivity during the last nymphal instars (10). Since methoprene is only marginally active at 1000 ppm and mortality was prevalent in Edwards' treatments, a confirmation is much needed. It is very obvious that more research into the nature of the infertility induced by juvenoids in general is much overdue.

The volatility of hydroprene was easily established, but it was a surprise that activity could be easily contained for long periods in cages with a screen top. Our results suggest that hydroprene is migrating between the various substrates in these cages without significant loss. Cockroaches appear to be easily affected by such residues either by contact or perhaps also by vapor. Volatile residues may also increase the efficacy by penetrating into inaccessible harborages, thus increasing the cockroaches' exposure during the inactive (daytime) periods. The volatility was almost completely repressed in a stable and residual dust formulation, which proved to be also very effective in cage trials. The relative efficacy of various formulations in household habitats has not yet been established, but it appears that compounds such as hydroprene can be tailored to the requirements of practically any cockroach habitat.

Literature cited

1. Bell, W.J.; Barth, R.H. J. Insect Physiol. 1970, 16, 2303–13.
2. Das, Y.T. PH.D. Thesis, Rutgers University, 1975.
3. Das, Y.T.; Gupta, A.P. Experientia 1974, 30, 1093–5.
4. Das, Y.T.: Gupta, A.P. Experientia 1977, 33, 968–70.
5. Hangartner, W.; Masner P. Experientia 1973, 19, 1358–9.
6. Kunkel, J.P. J. Insect Physiol. 1973, 19, 1285–97.
7. Masner, P.; Hangartner, W.; Suchy, M. J. Insect Physiol. 1975, 21, 1755–62.
8. Masner, P.; Dorn, S.; Vogel, W., Kahn, M.; Graf, O.; Gunthart, E. In: Scientific Papers of the Institute of Organic and Physical Chemistry of Wroclaw Technical University, 1981, 22, 809–818.
9. Pincus, D.S. J. Insect Physiol. 1977, 23, 73–7.
10. Roberts, B. Austr. J. Zool., 1977, 25, 233–41.
11. Vogel, W.; Masner, P.; Graf, O.; Dorn, S. Experientia 1979, 35, 1254–6.
12. Radwan, W.; Sehnal, F. Experientia 1974, 30, 615–8.
13. Cruickshank, P.A.; Mitt. Schweiz. Entomol. Ges. 1971, 44, 98–133.
14. Riddiford, L.M.; Ajami, A.M; Boake, Chr. J. Econ. Ent. 1975, 68, 46–8.
15. Cristodorescu, G.; Nosec, I.; Tacu, V. Arch. Roum. Path. Exp. Microbiol. 1978, 37, 47–53.
16. Cornwell P.B. The Cockroach, vol I, Hutchinson, London 1968, p. 48.
17. Chamberlain, W.F.; Becker, J.D. The Southwestern Entomologist 1977, 2, 179–182.

18. Clark, G.N. Chemical Times and Trends 1982, 5, 52-5.
19. Cochran, D.G. Pest Control, 1982, 50, 16-20.
20. Staal, G.B. In: Annual Review of Entomology 1975, 20, 417-460.
21. Edwards, J.P.; Menn, J.J. In: Chemie der Pflanzenschutz und Schadlingsbekampfungmittel. Springer Verlag New York 1981, 185-214.
22. Henrick, C.A. In Insecticide Mode of Action; Academic Press Inc., New York, 1982, Chapt. 11, pp 315-401.
23. Cruickshank, Ph.A.; Palmere R.M. Nature 1971, 233-488-g.
24. Henrick, C.A.; Willy, W.E.; Staal, G.B. Agr. Food Chem. 1976, 24, 207-18.
25. Henrick, C.A. US Patent 4,021,461, 1977.
26. Pallos, F.M.; Menn, J.J.; Letchworth, P.E.; Miaullis, J.B. Nature (London) 1971, 232, 486-
27. Schwartz, M.; Miller, R.W.; Wright, J.E.; Chamberlain, W.F.; Hopkins, D.E. J. Econ. Entomol. 1974, 67, 598 -
28. Edwards, J.P. Proc 8th Brit. Insectic., Fungic. Conf. 1975, 267-75.
29. Bilbie, I.; Nicolescu, G. Arch. Roum. Path. Exp. Microbiol. 1981, 40, 173-5.

RECEIVED November 23, 1984

Some Chemical Ecological Approaches to the Control of Stored-Product Insects and Mites

IZURU YAMAMOTO

Tokyo University of Agriculture, Setaga-ku, Tokyo 156, Japan

Three novel approaches to the control of insects are
described and evaluated. Each involves modification
of behavior.
Control of mating. A new type of sex pheromone was
found present in Callosobruchus chinensis, which is
not a conventional sex attractant, but rather induces
the male to extrude his genital organ and to effect
copulation. A female dummy bearing the pheromone
mimic, elicits copulation and ejaculation by the
male.
Control of oviposition. C. chinensis and C. maculatus develop a strategy to reduce competition among
larvae by first marking the beans and then eliminating the excess eggs, using the same marker at
accumulated doses. The marking pheromone is a mixture of lipids, and precoating the beans with certain
edible oils kills the deposited eggs and prevents the
injury.
Control of feeding. Many bean weevils Callosobruchus
spp. deposit eggs on kidney beans, but the penetrated
larvae do not develop. An unidentified fraction of
the bean was found as the growth inhibitor, but it is
not lectin.

If a crop is completely protected from pests, its yield may be
expected to double. This goal has been earnestly pursued in the
field to improve crop production. Lately, the need to place more
emphasis on protection from insects during post-harvest food
storage has been recognized. There are several reasons why this
area has been neglected: first, insufficient research has been
conducted; second, because loss evaluation is difficult and
seemingly no problem exists with the modern production technology,
there is little incentive; third, there is an assumption that food
storage is unnecessary under conditions where there is no food
surplus. We are more or less accustomed to handle storage problems
with our advanced technology, but in developing countries, post-

0097-6156/85/0276-0219$06.00/0
© 1985 American Chemical Society

harvest losses are about 30%, thus nullifying much of the effort
invested in field production strategies. The major reason for
these losses is insect pests. Until the present, pesticides have
played a major role in pest control. However, their negative
environmental impact is being severely criticized, making the concept of integrated pest management relatively more attractive.
Different control methods have been explored in storage insect
management, but examples of a truly integrated approach are few and
the assessment of the practicality of any of these new approaches
has not been completed. Considering pesticides for storage insect
control, particularly fumigants, the use of some of the existing
ones is decreasing because of problems of toxicity, residues and
resistance. The development of new pest control chemicals based on
new concepts is a desirable goal and may be rewarding. The
emerging countries in the tropics, having a favorable climatic condition for pest emergence, do not have the economic resources to
provide for proper storage space and/or fumigation. Therefore, a
simpler technology which can be utilized in such countries is
needed.

Pest Management with Ecochemicals

The present concepts of pest management have evolved from basic
knowledge of the chemical interactions between organisms.
Chemicals directly involved in the ecological relationship between
organisms are called ecochemicals, ecomones, or semiochemicals;
those acting between different species are called kairomones, allo-
mones, or allelochemics, while those acting within the same species
are called pheromones.
    Research on ecochemicals is most advanced in the case of
insects, thus providing the opportunity for its application through
pest management. Through evolutionary processes, insects have
acquired habits which are conditioned by pheromones and/or allomo-
nes. The reactions to these substances by insects differ from that
to poisonous insecticides in that it involves a normal and specific
response. In general, ecochemicals have been found to be effective
at very low levels because their direct effect is magnified through
behavioral responses. Moreover, these ecochemicals generally have
a simple chemical structure and often their analogs are available.
The present emphasis of utilizing these chemicals is to regulate
pest activities in order to prevent crop and food injury rather
than to kill pests. In the remainder of this report, research conducted at our laboratories will be described and discussed.

Control of Feeding Behavior

Insect preferences for certain types of food can be considered from
a chemical ecological point of view as follows: presence of
attractant, fixing factor oviposition-stimulant, and feeding stimulant; absence of repellent, oviposition deterrent, feeding
deterrent, nutritional defect, and growth-deterrent. Conversely,
the opposite is true for certain food types undisturbed by insects.

Attractants in cereal grains of Sitophilus zeamais. S. zeamais is
known as the rice weevil in Japan, but as the maize weevil in the
U.S., while S. oryzae is the small rice weevil in Japan, but the
rice weevil in the U.S. Both insects cause injury to rice, maize
and wheat, or to their processed products, thus being the most
important injurious insects to cereals. There is evidence that
olfactory stimulus is involved in the host selection. In warm
areas, the insects migrate during the harvest of rice, maize or
wheat into the fields from the store-houses, thereby causing post-
harvest injury after harvest. We began this study with the objective of elucidating chemical factors in grains involved in
host-finding behavior of S. zeamais.

     Using olfactometers, the following samples, fractions,
extracts, or compounds were shown to possess attractancy: 1.
polished rice, brown rice, maize and wheat; 2. ether, methanol,
acetone or hexane extracts of the above samples, among which the
ether extract showed the highest activity; 3. acidic and neutral
fractions of the ether extract. These observations suggested that
the composition of the attractive constituents is of a complex
rather than a simple nature. The attractants in rice and maize
grains were found to be a mixture of C_6 and C_7 carboxylic acids, γ-
lactones of C_3 hydroxycarboxylic acids, and phenylethanol. Of
these, hexanoic acid and 2-nonen-4-olide play the major role.
Similar attractants were also found in wheat. The attractancy of
related carboxylic acids was investigated extensively, however no
others were found to be attractive beyond the above-mentioned compounds. As for the lactones, several synthesized compounds were
found to be attractive. The above food attractant appears to be
important in understanding the host-finding behavior of the rice
weevil. Studies should be conducted to determine whether there is
a correlation between the concentration of attractants in grains
and observed susceptibility. Then, breeding for resistant
varieties could be conducted on a rational basis.

Attractant in Cheese of Tyrophagus putrescentiae. T. putrescentiae
(formerly T. dimidiatus) is a grain mite that feeds on stored
foodstuffs such as dairy products, grain powders, chocolate, spi-
ces, soybean paste and dried fish. The mite has been found to
prefer cheese, and is thus commonly called the cheese mite. A
question arises as to whether these foodstuffs have the same or
different factors that govern host-finding by the mite. For this,
an olfactometer bioassay was designed using cheddar cheese which is
easily available and possesses a strong attractancy as the source
material.

     Upon freeze drying the cheese, the attractancy was condensed in
the cold trap with the volatiles and further concentrated into a
neutral fraction, which was separated into 3 fractions by chromatography. Each fraction was assayed to determine its attractancy.
Fraction I gave no activity; fr. II gave 1/6 of the starting activity, and fr. III gave faint activity, but when fr. II was combined
with fr. I or III, the attractancy increased to 1/2; evidence of
synergism between fractions. By combining all of the fractions,
the original attractancy was restored. 8-Nonen-2-one of fr. II
showed some activity, but other methylketones in the same fraction

were not active. However, a mixture of 8-nonen-2-one with
heptan-2-one, octan-2-one and nonan-2-one showed strong activity, a
synergistic effect among methylketones. Fr. III contained lower
alcohols, but the attractancy of fr. III appeared to be due to
3-methylbutanol alone which further synergized the activity of the
mixture of the above-mentioned methylketones.

Methylketones are present in dairy products, formed from
triglycerides through β-ketoacids. 3-Methylbutanol originates from
L-leucine and is present in various fermented products. The presence of such common constituents is thought to be involved in host
finding for different kinds of foodstuffs by the mite.

Much less research has been done on food attractants than on
sex pheromones. More emphasis on the understanding of such allomo-
nes and their potential use for insect control is needed.

Growth inhibitor in kidney bean of bean weevils. A larva of the
azuki bean weevil, Callosobruchus chinensis, grows inside the azuki
bean and other beans, but dies at the first instar inside the kidney bean. Ishii (1) had presumed that this was due to a certain
growth inhibitor in the bean, and other workers had attempted to
grow resistant bean plants by grafting. Janzen et al. (2) claimed
that the cowpea weevil, C. maculatus, is not able to grow inside
the kidney bean due to the action of lectin. We have attempted to
show these same relationships with C. chinensis. The lectin fraction, purified by affinity chromatography, did not show any growth
inhibitory activity, however. A possibility that a trypsin inhibitor is responsible for the growth inhibition remains. Such a study
in combination with efforts to breed beans for resistance could
contribute to the attainment of a weevil-free bean.

Control of Mating Behavior

The use of sex attractants seems promising particularly with stored
product insects because they must live in a restricted space; this
has been well documented by Burkholder (4). Recently, a new type
of sex pheromone was found from C. chinensis by us.

Mating pheromones of C. chinensis. Two sex pheromones are
involved in the mating behavior of C. chinensis: a female sex
attractant, and a copulation release pheromone. The former
attracts the male to the female, but does not induce further
responses. The latter induces the male to extrude his genital
organ and to attempt copulation, and thus is named "erectin" (3).
Erectin is released by both the male and the female (more by the
female), but only the male responds to it.

The evidence for copulation release activity has been
demonstrated with several insects: Costelytra zealandica;
Trogoderma glabrum; Limonius canus; Tribolium confusum; Tenebrio
molitor, and Callosobruchus maculatus (F.) (4). However, erectin
of C. chinensis was the first to be identified, and it is involved
only in copulation release activity.

Chemically, erectin consists of two synergistically acting
fractions, neither having any activity. One is a mixture of
C_{26}-C_{35} hydrocarbons and the other is a dicarboxylic acid, named

"callosobruchusic acid", (E)-3,7-dimethyl-2-octene-1,8-dioic acid. (5).

If erectin can be made available in sufficient quantities for practical use, and if the copulation of males with female dummies bearing erectin can effect the lowering of the population density, this apparent reversal of the sterilized male technique could become a new control approach.

The first requisite for practical use was achieved by the synthesis of callosobruchusic acid, and the substitution of a complex hydrocarbon mixture by octadecane. Only the (E) form was active. Synthesis of the two optical forms was also achieved, but both forms showed the same activity as the natural callosobruchusic acid, thus making it impossible to assign the absolute configuration (5). The second requirement has been pursued by selecting different dummies. When an amount of erectin equivalent to that of one female was applied to a glass rod, the male attempted copulation, but did not ejaculate. However, the male did show the insertion and ejaculation behavior with a dummy of aluminum foil tube bearing erectin.

Control of Oviposition Behavior

Examples of the control of oviposition behavior with chemicals have been demonstrated in this laboratory with an oviposition stimulant for C. chinensis and C. maculatus and an oviposition regulator.

Oviposition stimulant from bean seed coat. Oviposition on kidney, cowpea and azuki beans by C. chinensis and C. maculatus is stimulated by at least two factors. This was shown using glass beads of different sizes treated with the extract of the bean seed coats as the oviposition substrate. C. maculatus oviposited only when a chemical stimulant was provided, whereas C. chinensis required adequate physical stimuli such as size, and the chemical stimulant played only a secondary role. Isolation of the chemical (5) is in progress. Such an oviposition stimulant with a suitable substrate could modify the oviposition behavior of these pest insects.

Oviposition regulator of C. chinensis and C. maculatus. There are two prominent phenomena that govern the oviposition behavior of the two weevils. First, under low density conditions, females oviposit evenly among beans. Second, under high density conditions where each bean holds many eggs, the egg distribution becomes random and only a few eggs hatch, leaving the rest to die. The bean surface becomes more shiny as oviposition progresses. An ether-soluble substance which imparts this shininess was shown to have marking activity. The weevil prefers lesser marked beans for further oviposition, thus resulting in an even distribution of the eggs. Moreover, as this substance increases to a level of 100 μg/bean, causing many eggs to be deposited, the substance shows ovicidal action, thus eliminating most of the eggs except those oviposited in earlier stages that had already hatched and penetrated into the beans. These insects thus have developed a strategy to reduce competition among larvae, and to maximally utilize the host beans by using the same substance at different levels.

This pheromone consists of triglycerides, fatty acids, and hydrocarbons, and each has some marking and ovicidal activity. The fatty acid composition of the triglycerides is similar to that of edible oils, and treatment of the azuki bean with certain edible oils (200 μg/bean) could protect the beans from injury by C. maculatus, C. chinensis, and Zabrotes subfaciatus.

Conclusion

As illustrated, new approaches to pest control are being developed that have the potentiality to lessen injuries by lowering the insect population without direct kill. One future direction may well be that of identifying ecochemicals of natural origin as a forerunner to the development of synthetic "pestistatics" which mimic the function of the former. The methods of crop and food protection will certainly change with the transformation of social consciousness and with progress of science and technology. Along with this trend, the form of chemicals now called pesticides will inevitably be altered.

Literature Cited

1. Ishii, S. Bull. Natl. Inst. Agric. Sci. Ser. C., No. 1. 1952, 185.
2. Janzen, D. H.; Juster, H. B.; Liener, I. E. Science 1976, 192, 795.
3. Tanaka, K.; Ohsawa, K.; Honda, H.; Yamamota, I. J. Pesticide Sci. 1981, 6, 75.
4. Yun-Tai Qi; Burkholder, W. E. J. Chem. Ecol. 1982, 8, 527.
5. Mori, K.; Ito, T.; Tanaka, K.; Honda, H.; Yamamoto, I. Tetrahedron 1983, 39, 2303.

RECEIVED December 14, 1984

Phytochemical Disruption of Insect Development and Behavior

WILLIAM S. BOWERS

Department of Entomology, New York State Agricultural Experiment Station, Cornell University, Geneva, NY 14456

Plants deploy many defensive strategies against attack by insects. Plant secondary chemicals with toxic actions against insects are well known and have often served as prototypic models for the development of optimized analogs subsequently reduced to commercial practice. Less well understood and investigated are those secondary chemicals targeted to more subtle disruption of pest specific biology. These include mimics of important developmental hormones and chemicals of communicative significance.

Competitive species living together in a restricted space eventually seek to exclude or prey upon each other. Plants and insects are prime examples of coexisting species locked in an eternal struggle for space and sustenance.

Plants, as the ultimate collectors of solar energy and the vegetative dominants, are well matched against insects, which are the principal animate life forms.

Insects are so successful because of their mobility, high reproductive potential, ability to exploit plants as a food resource, and to occupy so many ecological niches. Plants are essentially sessile and can be seen to produce flowers, nectar, pollen, and a variety of chemical attractants to induce insect cooperation in cross-pollination. However, in order to reduce the efficiency of insect predation upon them, plants also produce a host of structural, mechanical, and chemical defensive artifices. The most visible chemical defenses are poisons, but certain chemicals, not intrinsically toxic, are targeted to disrupt specific control systems in insects that regulate discrete aspects of insect physiology, biochemistry, and behavior. Hormones and pheromones are unique regulators of insect growth, development, reproduction, diapause, and behavior. Plant secondary chemicals focused on the disruption of insect endocrine and pheromone mediated processes can be visualized as important components of plant defensive mechanisms.

0097–6156/85/0276–0225$06.00/0
© 1985 American Chemical Society

Notwithstanding the continuing dynamics of evolutionary
processes, certain plant secondary chemicals employed as defensive
strategies against pests and pathogens can serve as prototypic
models leading to new chemical control agents and/or resources for
the genetic engineering of pest resistant crop cultivars. Various
plant derived toxicants have provided the initial chemistry leading
to successful commercial insecticides including the pyrethroids and
carbamates. Less well explored are those plant chemical defenses
targeted to more subtle interruption of insect-specific aspects of
development and behavior.

Insect Phytohormones

Phytojuvenoids. Wigglesworth (1) demonstrated that a hormone
secreted by the insect corpora allata was responsible for the
control of differentiation in immature insects and reproduction in
adult female insects. Williams (2) prepared an active extract of
this hormone from adult male cecropia moths and called it "juvenile
hormone". We were able to derive sufficient knowledge of the
chemistry of the juvenile hormone from the study of the active
cecropia extract to synthesize JH III (3). Seven years later its
presence as a natural hormone in the tobacco hornworm was confirmed
(4). Three other analogous juvenile hormones (JH 0, I, II) have
been found to occur only in lepidoptera (5, 6, 7) (Figure 1).
Juvenile hormone III is the principal juvenile hormone of insects
and has been demonstrated in all of the insect taxa investigated.
        In 1965 Slama and Williams (8) observed that contact with
certain paper products disrupted the metamorphosis of the linden
bug. The biological activity was identical in every respect to
that produced by the insect juvenile hormone. Thus the ultimate
nymphal stages were induced to molt into nymphal-adult intermediates or supernumerary nymphs which eventually died without
completing metamorphosis. They demonstrated that the active
agent(s) was present in the balsam fir tree from which the paper
had been manufactured. Bowers et al. (9) isolated and identified
the principal active compound in the balsam fir as a monocyclic
sesquiterpenoid ester and called it juvabione. This discovery
revealed a new plant defensive strategy focused on disruption of
the insect endocrine system and stimulated world-wide efforts to
investigate plants as a source of endocrine-active compounds.
Although Schmialek (10) had demonstrated the juvenile hormone
activity of farnesol, it was isolated and identified as a natural
product of mealworm feces rather than as the ubiquitous sesquiterpenoid alcohol found in plants. Though of low juvenile hormone
activity, farnesol may act as a protective secondary chemical in
certain plants. Dehydrojuvabione was identified as a second morphogenetic agent in a Czechoslovakian fir (11). It is about ten-
fold less active than juvabione, but sufficiently active to disrupt
insect metamorphosis at submicrogram levels.
        The two naturally occurring drug and insecticide adjuvants,
sesamin and sesamolin from sesame oil, were found to posses modest
juvenile hormone activity and served as models for the design of
the first aromatic juvenile hormone analogs (12, 13).

Numerous compounds with low to moderate juvenile hormone
activity have been found in plants, and their chemistry is summarized in Table I (8-22). Of particular note are the juvocimones,
found in the distillate of the essential oil of sweet basil, whose
biological activity against certain species is quite pronounced.
They are far more active than the natural juvenile hormones and
induce the ultimate morphogenetic effects against milkweed bugs at
dosages in the pico-gram range (22).

Phytoecdysteroids. Molting and differentiation in insects are
controlled by molting hormones which are polyhydroxylated steroids
called ecdysones. Ecdysone (Figure 2) was isolated by Butenandt
and Karlson (23), and the structure proposed by Karlson et al. (24)
was confirmed with a full elucidation of structure and absolute
configuration by Huber and Hoppe (25) through X-ray analysis. A
second hormone was characterized as 20-hydroxyecdysone by several
laboratories including Hampshire and Horn (26), Hocks and Wiechert
(27), and Kaplanis et al. (28). Several additional hormones and/or
metabolites with significant molting hormone activity have been
subsequently identified (Figure 2).
      In 1966 Nakanishii et al. (29) identified ponasterone A, a
polyhydroxysteroidal component of the plant Podocarpus nakaii
(Figure 3). Recognizing its similarity to the insect ecdysones he
was able to confirm that it possessed authentic molting hormone
activity. Later, this phytoecdysteroid was discovered to exist as
a natural ecdysone of several crustacea (30). Similarly the
phytoecdysteroid inokosterone (Figure 3) was shown to occur in the
crustacean Callinectes sapidus (31). The discovery of plant
steroids with molting hormone activity stimulated a widespread
search among plants for similar hormone mimics. Recently, 111
plant families have been shown to contain at least 69 phytoecdysteroids with molting hormone activity (32). It is especially
noteworthy that the natural insect molting hormone, 20-
hydroxyecdysone, is the most commonly encountered hormone in the
major plant divisions Pteridophyta, Gymnospermae, and Angiospermae.
In descending order the most abundant ecdysteroids are ponasterone
A, polypodine B, ecdysone, and pterosterone (32).
      Ponasterone A, a natural phytoecdysone, is reported to have
the highest hormonal activity in insects and to show the greatest
affinity for ecdysteroid receptors (33). Ecdysteroids and analogs
incorporated into artificial diets are shown to deter feeding in
many insects (34) as well as to inhibit growth and reproduction
(27-30). Galbraith and Horn (39) were the first to suggest that
the phytoecdysteroids might constitute a plant defensive strategy.
This thesis has been echoed by numerous investigators and is rein-
forced by their clear biological effects on a wide variety of
insects. In view of the widespread occurrence of phytoecdysones
and their acknowledged interference with insect molting and
metomorphosis, their role as plant defensive components seems
assured.

Anti-Juvenile Hormones. A plant defensive strategy targeted to
disruption of the endocrine regulation of the early larval stages
of metamorphosis was revealed with the discovery of anti-juvenile

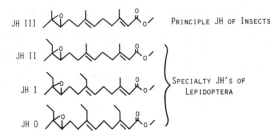

JH III — PRINCIPLE JH OF INSECTS

JH II —

JH I — SPECIALTY JH'S OF LEPIDOPTERA

JH 0 —

Figure 1.   The juvenile hormones of insects.

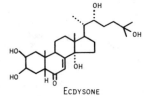
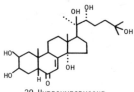

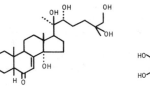
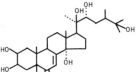

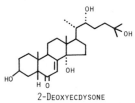

ECDYSONE   20-HYDROXYECDYSONE

20,26-DIHYDROXYECDYSONE   24-METHYL-20-HYDROXYECDYSONE

2-DEOXYECDYSONE

Figure 2.   The insect molting hormones, "ecdysones".

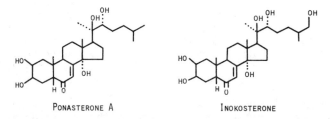

PONASTERONE A   INOKOSTERONE

Figure 3.   Representative phytoecdysones also found in
invertebrates.

Table I.   Phytochemicals with Juvenile Hormone Activity

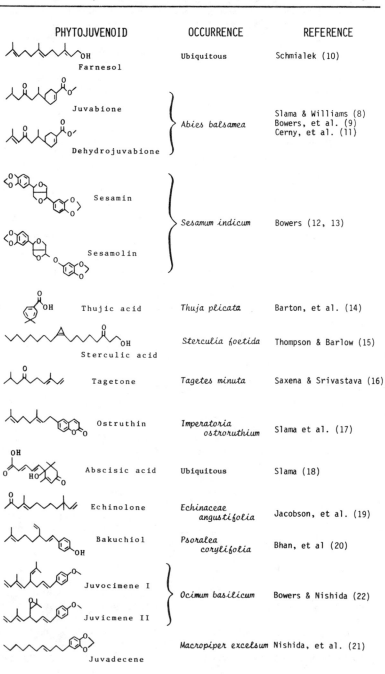

PHYTOJUVENOID	OCCURRENCE	REFERENCE
Farnesol	Ubiquitous	Schmialek (10)
Juvabione	*Abies balsamea*	Slama & Williams (8) Bowers, et al. (9) Cerny, et al. (11)
Dehydrojuvabione		
Sesamin	*Sesamum indicum*	Bowers (12, 13)
Sesamolin		
Thujic acid	*Thuja plicata*	Barton, et al. (14)
Sterculic acid	*Sterculia foetida*	Thompson & Barlow (15)
Tagetone	*Tagetes minuta*	Saxena & Srivastava (16)
Ostruthin	*Imperatoria ostroruthium*	Slama et al. (17)
Abscisic acid	Ubiquitous	Slama (18)
Echinolone	*Echinaceae angustifolia*	Jacobson, et al. (19)
Bakuchiol	*Psoralea corylifolia*	Bhan, et al (20)
Juvocimene I	*Ocimum basilicum*	Bowers & Nishida (22)
Juvicmene II		
Juvadecene	*Macropiper excelsum*	Nishida, et al. (21)

hormonal substances in the plant Ageratum houstonianum (40, 41).
Two simple substituted chromenes were isolated and identified from
Ageratum and called precocenes I and II (Figure 4). By contact,
feeding, or fumigation, the precocenes induced precocious metamorphosis in immature insects, sterilized adult females, induced
diapause in certain beetles, disrupted embryogenesis, and inhibited
sex pheromone production. All of these anti-juvenile hormonal
activities were completely reversed by simultaneous, combined, or
subsequent treatment with juvenile hormone III, demonstrating that
the antihormonal effects resulted from the absence of the juvenile
hormone rather than inhibition at a receptor site (40, 41). Work
in other laboratories demonstrated additional biological actions of
the precocenes including: specific inhibition of endocrine controlled processes during embryogenesis (42), reduction in mating of
males (43), effect on migratory behavior (44, 45), diminished
neurosecretory activity (46, 47), caste/morph determination (48,
49), and sex determination (50).
     Although paurometabolous insects were found to be most sensi-
tive to the precocenes, Kiguchi (51) obtained precocious
metamorphosis in the silkworm with precocene II, and Mathai and
Nair (52) were similarly successful with the armyworm.
     Cupp et al. (53) discovered that precocene II inhibited pupa-
tion in the yellow fever mosquito, and Tarrant and Cupp (54) found
that precocene induced precocious adults of Rhodnius prolixus and
stopped their feeding. The consequences of feeding inhibition
should be useful in preventing Chagas disease transmission. More
recently Azambuja et al. (55) demonstrated that the natural
precocene II was a powerful antifeedant for Rhodnius whereas the
more potent synthetic anti-hormonal analog ethoxy precocene [i.e.
7-ethoxy-6-methoxy-2,2-dimethyl chromene (56)] possessed very
little antifeedant effect. This result suggests that the antifeedant and antihormonal actions may depend on different
mechanisms. The precocenes have also been found to inhibit molting
(40, 41, 54, 57, 58) and that the administration of ecdysone over-
comes the ecdysial stasis (40, 57).
     Mode of action studies reveal that the corpora allata are
destroyed following exposure to the precocenes (59-61). Metabolism
studies indicated that precocene was rapidly oxidized and hydrated
to the 3,4-dihydrodiol and that the epoxide appeared to be an
activated intermediate (62). Recent evidence indicates that the
epoxide is an alkylating agent which destroys the corpora allata
terminating juvenile hormone production (63, 64).
     The precocenes appear to be multifunctional plant protectants
designed to affect many insect-specific endocrine mediated events
as well as feeding and communication activities.

Insect Phytopheromones

Many studies of the direct and immediate interactions occurring
between plants and insects, as insect predators attempt to feed on
plants, have been documented. Thus, research on repellants and
antifeedants has received much attention and will not be discussed
here. More subtle interactions of plant secondary chemicals interferring with basic insect communication systems have received much

less attention. Two instances in which plants have apparently
developed chemical defensive mechanisms interferring with or
simulating specific signals of insect communicative significance
are exemplified by the studies of aphid alarm pheromones and the
cockroach sex pheromones.

Cockroach Phytosexpheromonal Mimics. In 1971, Bowers and
Bodenstein (65) discovered that certain plants contained compounds
which initiated in the American cockroach behavioral responses
identical to those induced by the natural sex pheromone of the
virgin female. Male courtship display and copulatory attempts
occurred when males were exposed to needle and cone oil distillates
of spruce, Picea rubra, and fir, Abies siberica, Abies alba. The
active substance, on isolation and characterization, was found to
be D-bornyl acetate (Figure 5), a well-known constituent of conifer
trees. This discovery prompted a more widespread survey of plants
and resulted in the discovery of additional plant extracts that
also simulated the natural sex pheromone. In addition to the
conifers, active extracts were prepared from plants of the families
simarubareae, araliaceae, labiatidae, and compositae. Several of
these plants including Erigeron annus contained a $C_{15}H_{24}$ hydrocarbon with significant activity (65). From Erigeron annus Tahara et
al. (66) identified the active hydrocarbon as germacrene D (Figure
5). Although the plant compounds possessed significant sex
pheromonal activity, Bowers and Bodenstein (65) demonstrated that
crude extracts of the virgin female cockroach midgut [a rich source
of the pheromone (67)] were much more active than the purified
plant mimics. Thus, at no time was it suspected that any of the
plant compounds were identical with the natural pheromone. The
midgut extracts upon fractionation gave unweighable samples (<50
ug) which could be diluted one billion times and still elicit the
characteristic mating display from male cockroaches. Chemical
manipulation of active fractions indicated that the natural
pheromone was an unsaturated ketone. This was the first accurate
information on the chemistry of the cockroach pheromone since Day
and Whiting (68) disproved the pheromone structure assigned by
Jacobson et al. (69). The natural pheromone was finally characterized by Persoons et al. (70) as a derivative of germacrene D and
called it Periplanone B (Figure 5). Washio and Nishino (71) confirmed the authenticity of the behavioral responses to the
monoterpenoids by comparison of their electroantennogram responses
with the natural pheromone. Nishino et al. (72) reported that (+)-
trans-verbenyl acetate (Figure 5) also possessed significant sex
pheromone activity for the American cockroach and evaluated 67
synthetic analogs (73).
     The frequent occurrence of pheromonal mimics in plants is
disturbing in view of the oft-presumed specificity of pheromonal
chemicals. One might wonder whether their presence in plants has
communicative significance for cockroaches. A defensive strategy
based upon the possession of a sex pheromone mimic seems of dubious
value to a plant unless the stimulation to sexual activity over-
rides or depresses feeding activities. Alternatively, attraction
of omnivorous cockroaches might result in their destruction of
competing plants or parasites.

Aphid Alarm Pheromones. When aphids are attacked by predators they
produce droplets of secretion from their cornicles whose odor
initiates escape behavior in nearby siblings. The first alarm
pheromone was identified by Bowers et al. (74) for the rose, pea,
greenbug, and cotton aphids as trans-β-farnesene. The macrocyclic
hydrocarbon germacrene A was subsequently identified as the alarm
pheromone of the sweetclover and spotted alfalfa aphids (Figure 6)
(75, 76).

The repellant nature of the alarm pheromones makes them an
attractive resource for the development of plant protection
chemicals. Synthetic optimization efforts resulted in the preparation of several active analogs (77-79). However, the analogs were
significantly less active than the natural pheromones and field
trials were not attempted.

The induced mobility of aphids on exposure to trans-β-
farnesene was demonstrated to enhance the effectiveness of contact
insecticides by Griffiths and Pickett (80), and Gibson and Pickett
(81) made the exciting discovery that an insect resistant wild
potato, Solanum berthaultii, possessed glandular hairs on its
leaves which exude trans-β-farnesene. Although the glandular hairs
may also entrap aphids with sticky secretions, they found that the
air above S. berthaultii leaves contained the alarm pheromone in
concentrations sufficient to effectively repel aphids. This discovery is the first demonstration of a plant defensive strategy
employing the identical chemical messenger used by the insect. No
doubt future investigations will reveal similar results. With the
chemical basis of resistance established, direct chemical analytical techniques for the measurement and quantitation of the
pheromone will permit rapid selection of resistant varieties from
conventional breeding efforts. Genetic engineering technologies
should eventually allow for the incorporation and expression of
such resistance mechanisms in other plant species.

Summary

Historically, phytochemicals have been an important resource for
the development of pesticides, both for their direct use as well as
prototypes from which more efficient pest control chemicals may be
developed. Early efforts in natural product chemistry revealed
many potent toxins. Basic research into insect physiology and
biochemistry coupled with the study of insect/plant chemical interactions has revealed several subtle defensive strategies by plants
to limit insect predation. Chemical mimics of the insect juvenile
and molting hormones are found to be abundant in plants. Chemical
communication signals similar to or identical with that of the
natural insect pheromones have also been shown to be a part of
plant defensive strategies. Future research in the chemistry of
insect/plant interactions will establish a rational basis for the
systematic breeding of insect-resistant plants through conventional
efforts, and serve as an exciting resource for genetic engineering.

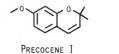

PRECOCENE I   II

Figure 4.   Antijuvenile hormones from Ageratum.

BORNYL ACETATE

VERBENYL ACETATE

GERMACRENE D

PERIPLANONE B

Figure 5.   Plant compounds evoking cockroach sex pheromone
activity and periplanone B the authentic sex pheromone of the
American cockroach.

(E)-β-FARNESENE   GERMACRENE A

Figure 6.   Aphid alarm pheromones.

Literature Cited

1. Wigglesworth, V. B. Quart. J. Mic. Sci. 1934, 77, 191-222.
2. Williams, C. M. Nature 1956, 178, 212-13.
3. Bowers, W. S.; Thompson, M. J.; Uebel, E. C. Life Sci. 1965, 4, 2323-31.
4. Judy, K. J.; Schooley, D. A.; Dunham, L. L.; Hall, M. S.; Bergot, B. J.; Siddall, J. B. Proc. Nat. Acad. Sci. USA 1973, 70, 1509-23.
5. Roller, H.; Dahm, K. H.; Sweeley, C. C.; Trost, B. M. Angew. Chem. 1967, 79, 190-1.
6. Meyer, A. S.; Schneiderman, H. A.; Hanzman, E.; Ko, J. H. Proc. Nat. Acad. Sci. USA 1968, 60, 856-60.
7. Bergot, B. J.; Jamieson, G. C.; Ratcliff, M. A.; Schooley, D. A. Science 1980, 210, 336-8.
8. Slama, K.; Williams, C. M. Proc. Nat. Acad. Sci. USA 1965, 54, 411-414.
9. Bowers, W. S.; Fales, H. M.; Thompson, M. J.; Uebel, E. C. Science 1966, 154, 1020-2.
10. Schmialek, P. Zeitschreft fur Naturforschung 1961, 16, 401-4.
11. Cerny, V.; Dolys, L.; Labler, L.; Sorm, F.; Slama, K. Coll. Czech. Chem. Commun. 1967, 32, 3926-33.
12. Bowers, W. S. Science 1968, 161, 895-7.
13. Bowers, W. S. Science 1969, 164, 323-5.
14. Barton, G. M.; MacDonald, B. F.; Sahota, T. S. Can. For. Serv. Bi-Month. Res. Notes 1972, 28, 22-3.
15. Thompson, S. N.; Barlow, J. S. Ann. Entomol. Soc. Am. 1973, 66, 797-800.
16. Saxena, B. P.; Srivastava, J. B. Experientia 1972, 28, 112-13.
17. Slama, K.; Romanuk, M.; Sorm, F. In "Insect Hormones and Bioanalogues"; Springer-Verlag: New York, 1974.
18. Slama, K. In "Herbivores: Their Interaction with Secondary Plant Metabolites"; Rosenthal, G. A.; Janzen, D. H., Eds.; Academic Press: New York, 1979.
19. Jacobson, M.; Redfern, R. E.; Mills, G. D. Lloydia 1975, 38, 473-6.
20. Bhan, P.; Soman, R.; Sukh, Dev. Agric. Biol. Chem. 1980, 44, 1483-7.
21. Nishida, R.; Bowers, W. S.; Evans, P. H. Arch. Insect. Biochem. Physiol. 1983, 1, 17-24.
22. Bowers, W. S.; Nishida, R. Science 1980, 209, 1030-2.
23. Butenandt, A.; Karlson, P. Z. Naturforschg. 1954, 9 (b), 389-91.
24. Karlson, P.; Hoffmeister, H.; Hummel, H.; Hocks, P.; Spiteller, G. Chem. Ber. 1965, 98, 2394-402.
25. Huber, R.; Hoppe, W. Chem. Ber. 1965, 98, 2403-24.
26. Hampshire, F.; Horn, D. H. S. Chem. Comm. 1966, 2, 37-8.
27. Hocks, P.; Weichert, R. Tetrahedian Lett. 1966, 26, 2989-93.
28. Kaplanis, J. N.; Thompson, M. J.; Yamamoto, R. T.; Robbins, W. E.; Louloudes, S. J. Steroids 1966, 8, 605-23.
29. Nakanishi, K.; Koreeda, M.; Dasaki, S.; Chang, M. L.; Hsu, H. Y. Chem. Comm. 1966, 915-17.
30. McCarthy, J. P. Steroids 1979, 34, 799-806.
31. Faux, A.; Horn, D. H. S.; Middleton, E. J.; Fales, H. M.; Lowe, M. E. Chem. Commun. 1969, 175-6.

32. Bergamasco, R.; Horn, D. H. S. In "Endocrinology of Insects"; Downer, G. H.; Laufer, H., Eds.; Alan R. Liss: New York, 1983; pp. 627-54.

33. Fristrom, J. W.; Yund, M. A. In "Progress in Ecdysone Research"; Hoffman, J. A., Ed.; Elsevier/North-Holland, 1980; pp. 349-62.

34. Jones, C. G.; Firn, R. D. J. Chem. Ecol. 1978, 4, 117-38.

35. Robbins, W. E.; Kaplanis, J. N.; Thompson, M. J.; Shortino, T. S.; Cohen, C. F.; Joyner, S. C. Science 1968, 161, 1158-60.

36. Robbins, W. E.; Kaplanis, J. N.; Thompson, M. J.; Shortino, T. S.; Joyner, S. C. Steroids 1970, 16, 105-25.

37. Singh, P.; Russell, G. B. J. Insect Physiol. 1980, 26, 139-42.

38. Singh, P.; Russell, G. B.; Fredricksen, S. Entomol. Exp. Appl. 1982, 32, 7-12.

39. Galbraith, M. N.; Horn, D. H. S. Chem. Commun. 1966, 905-6.

40. Bowers, W. S. In "The Juvenile Hormones"; Gilbert, L. I., Ed.; Plenum Press: New York, 1976; pp. 394-408.

41. Bowers, W. S.; Ohta, T.; Cleere, J. S.; Marsella, P. A. Science 1976, 193, 542-7.

42. Dorn, A. Gen. Comp. Endocr. 1982, 46, 47-52.

43. Walker, W. F. Physiol. Ent. 1978, 3, 147-55.

44. Rankin, M. A. J. Insect Physiol. 1979, 26, 67-74.

45. Rankin, S. M.; Rankin, M. A. Physiol. Ent. 1980, 5, 175-82.

46. Unnithan, G. C.; Nair, K. K.; Kooman, C. J. Experientia 1978, 34, 411-12.

47. Masner, P.; Bowers, W. S.; Kalin, M.; Muhle, T. Gen. Comp. Endocr. 1979, 37, 156-66.

48. MacKauer, M.; Nair, K. K.; Unnithan, G. C. Canad. J. Zool. 1979, 57, 856-9.

49. Hales, P. F.; Mittler, T. E. J. Insect Physiol. 1981, 27, 333-7.

50. Hales, D. F.; Mittler, T. E. Proc. Ninth Int. Symp. Comp. Endocr., University of Hong Kong Press, 1981.

51. Kiguchi, K. Vth Internat. Congr. Pesticide Chemistry, Kyoto, Japan, 1982.

52. Mathai, S.; Nair, V. S. K. Arch. Insect Biochem. Physiol. 1984, 1, 199-203.

53. Cupp, E. W.; Lok, J. B.; Bowers, W. S. Ent. Exp. Appl. 1977, 22, 23-8.

54. Tarrant, C.; Cupp, E. W. Trans. Roy. Soc. Trop. Med. Hyg. 1978, 72, 666-8.

55. Azambuja, P. D.; Bowers, W. S.; Ribeiro, M. C.; Garcia, E. S. Experientia 1982, 38, 1054-5.

56. Bowers, W. S. In "Natural Products and the Protection of Plants"; Marini-Bettolo, G. B., Ed.; Pontif. Acad. Sci., Scripta Varia 41, 1977.

57. Azambuja, P. D.; Garcia, E. S.; Ribeiro, J. M. C. Gen. Comp. Endocr. 1981, 45, 100-4.

58. Azambuja, P. D.; Garcia, E. S.; Furtado, A. F. C. R. Acad. Sci. Paris 1981, 292, 1173-5.

59. Bowers, W. S.; Martinez-Pardo, R. Science 1977, 197, 1369-71.

60. Unnithan, G. C.; Nair, K. K.; Bowers, W. S. J. Insect Physiol. 1977, 23, 1081-94.

61. Unnithan, G. C.; Nair, K. K.; Syed, A. Experientia 1980, 36, 135-6.
62. Ohta, T.; Kuhr, R. J.; Bowers, W. S. Agric. Food Chem. 1977, 25, 478-81.
63. Soderlund, D. M.; Messeguer, A.; Bowers, W. S. J. Ag. Food Chem. 1980, 28, 724-31.
64. Jennings, R. C.; Ottridge, A. P. Chem. Commun. 1979, 920-1.
65. Bowers, W. S.; Bodenstein, W. G. Nature 1971, 232, 259-61.
66. Tahara, S.; Yoshida, M.; Mizutani, J.; Kitamura, C.; Takahashi, S. Agri. Biol. Chem. 1975, 39, 1517-18.
67. Bodenstein, W. G. Ann. Entomol. Soc. Amer. 1970, 63, 336-7.
68. Day, A. C.; Whiting, M. C. Proc. Chem. Soc. (Lond.) 1964, 368.
69. Jacobsen, M.; Beroza, M.; Yamamoto, R. T. Science 1963, 139, 48-9.
70. Persoons, C. J.; Verwiel, P. E. J.; Ritter, F. J.; Talman, E.; Nooijen, P. J. F.; Noijen, W. J. Tetrahendron Lett. 1976, 24, 2055-8.
71. Washio, H.; Nishino, C. J. Insect Physiol. 1976, 22, 735-41.
72. Nishino, C.; Tobin, T. R.; Bowers, W. S. Appl. Ent. Zool. 1977, 12, 287-90.
73. Nishino, C.; Takayanagi, H. J. Chem. Ecol. 1981, 7, 853-65.
74. Bowers, W. S.; Nault, L. R.; Webb, R. E.; Dutky, S. R. Science 1972, 177, 1121-2.
75. Bowers, W. S.; Montgomery, M. E.; Nault, L. R.; Nielson, M. W. Science 1977, 196, 680-1.
76. Nishino, C.; Bowers, W. S.; Montgomery, M. E.; Nault, L. R.; Nielson, M. W. J. Chem. Ecol. 1977, 3, 349-57.
77. Nishino, C.; Montgomery, M. E.; Bowers, W. S.; Nault, L. R. Agr. Biol. Chem. 1976, 40, 2303-4.
78. Nishino, C.; Bowers, W. S. Appl. Ent. Zool. 1976, 11, 340-3.
79. Bowers, W. S.; Nishino, C. J. Insect Physiol. 1977, 23, 697-701.
80. Griffiths, D. C.; Pickett, J. A. Ent. Exp. & Appl. 1980, 27, 194-201.
81. Gibson, R. W.; Pickett, J. A. Nature 1983, 302, 608-9.

RECEIVED November 8, 1984

Proallatocidins

FRANCISCO CAMPS

Instituto Quimica Bio-Orgánica (C.S.I.C.), Jorge Girona Salgado, 18-26, 08034-Barcelona, Spain

The rationale for design of proallatocidins related to chromenic structure of precocenes is discussed. Stabilization against environmental conditions and insect metabolic pathways, as well as modification of transport properties, are some of the leads followed for the synthesis of more powerful insect growth regulators of this type. Some aspects of the chemistry of 3,4-epoxyprecocenes are also examined.

Precocenes I and II are natural products with a simple chromene structure (7-methoxy- and 6,7-dimethoxy-2,2-dimethylchromene) which were isolated from plant sources and exhibited powerful antijuvenile hormone activities in several types of insects (1).

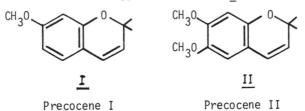

I   II

Precocene I   Precocene II

There is now strong evidence that precocenes act as proallatocidins, which are transformed by allatal mono-oxygenases into a highly reactive 3,4-epoxide. This epoxide is assumed to be the true cytoxic agent by selective alkylation of cellular elements of the insect corpora allata glands (2).

On the other hand, the recent discovery of antijuvenile hormone activity in several o-isopentenylphenols, compounds biogenetically related to chromenes, has led to the proposal of a direct alternative alkylation pathway, via rearrangement of precocenes to quinone methides (3).

For the above characteristics, precocenes were considered as potential lead compounds for the development of a new generation of insecticides. Accordingly, in the last years, extensive research has

been carried out in a variety of academic and industrial laboratories
for the design of more powerful insect growth regulators of this type,
which should simultaneously combine, among other features, greater
resistance to the peripheral detoxification in the insect and minimal
toxicity against non-target organisms, particularly vertebrates. In
the present communication, we summarize some of our efforts in this
area.

Chemical Stabilization of Precocene Structures

In view of the reported lability of the chromene skeleton under en-
vironmental-like conditions, i.e. acid promotes dimerization and
light causes rearrangement of chromene to quinone methide (4) one of
our first concerns was the chemical stabilization of the precocene
structures for its potential application in field trials.

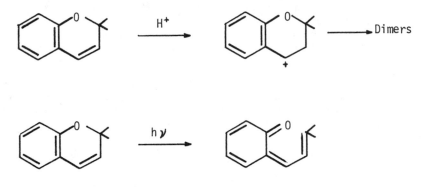

We anticipated that replacement of one vinyl hydrogen by fluorine in
the natural compound would reduce the reactivity of the chromene 3,4-
double bond without preventing the above enzymatic epoxidative bioactivation of taking place, due to the similitude of atomic radii of
fluorine and hydrogen. However, 3-fluoroprecocene analog III, prepared
by modification of known procedures for the synthesis of chromenes (5)
was inactive. Likewise, substitution of trifluoromethyl group for one
of the two gem-dimethyl groups at the C-2 site (6), to strenghten the
C-2 oxygen bond, precluding the above rearrangement, resulted in a
decrease of activity in the corresponding trifluoromethyl analog IV.

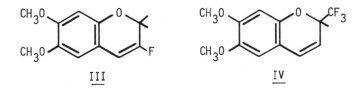

III   IV

Stabilization against Insect Metabolic Degradation

One of the conclusions from AJH structure activity studies is the importance of an unsubstituted double bond and a C-7 alkoxy substituent
in the chromene structure to elicit AJH activity. However, results of
precocene metabolism in different insect species revealed that cleavage at this substituent is one of the main detoxification mechanisms
observed. Accordingly, we expected that an increase of steric hin-
drance at this site might prevent the occurrence of such a cleavage
and, in this way, enhance the AJH activity. This requirement was ful-
filled by preparation of dihydrobenzodipyran derivatives V and VI, in
which a bulky pseudoterbutoxy substituent is present at C-7 of the
chromene structure, as part of a 2,2-dimethylchroman ring, linking
this site and C-6 or C-8, respectively in linear isomers V or angular
derivatives VI (7).

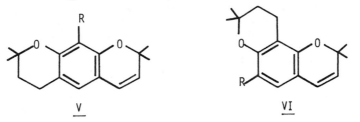

$$\underline{V} \qquad\qquad\qquad\qquad\qquad \underline{VI}$$

Preliminary results of antijuvenile hormone bioassays for precocious metamorphosis in immature stages of Oncopeltus fasciatus revealed that some of the non-linear derivatives (VI) showed higher activities and lower toxicities that those exhibited by natural precocenes
or 7-ethoxy-6-methoxy-2,2-dimethylchromene, one of the most active
synthetic analogs known. In addition, some of these compounds were
unexpectedly inactive in the sterilization bioassay with adults of
the same species (8). Further investigation of antijuvenile hormone
activities of selected compounds of this series in other insects is
now in progress.

Modification of transport properties

Although the differential sensitivity of hemi- and holometabolous
larvae to the precocenes remains unexplained, it has been shown that
the corpora allata of holometabolous species are sensitive to the precocenes in vitro (9). Likewise, it has been demonstrated that precocenes are rapidly sequestered by hemolymph proteins in several insects
preventing an effective amount of the precocenes from ever reaching
the corpora allata (10).

Consequently, we anticipated that to overcome these problems it
might be important to incorporate moieties in the precocene structure
to alter the transport properties of the natural compounds. For this
aim, we synthesized crown ether precocenes VII (11), in which C-6 and
C-7 of the chromene skeleton were incorporated into a 15-crown-5 or
18-crown-6 ether ring. We also prepared several chromene derivatives
VIII bearing polyoxyethylenated groups and sugar residues at the C-8
position (12).

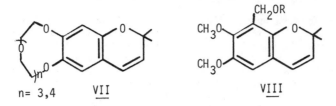

n= 3,4   <u>VII</u>   <u>VIII</u>

So far all these compounds were inactive in the standard contact
test but further investigation under in vitro conditions will be carried out in the near future.

Another strategy used to modify the transport properties of natural precocenes was to prepare analogues in which the double bond was
masked in such a way that might be liberated under mild oxidative conditions (13). It was thought that by an appropriate choice of these
protective groups, the AJH activity might be enhanced if the proalla-
tocidin structure could be liberated close to the corpora alata. For
this aim we have prepared in our laboratory (14) precosyl thiophenyl
ethers IX, in view of the usefulness of this protective group in organic synthesis, allowing the easy regeneration of the double bond
by oxidation to the corresponding sulfoxides and sulfones followed by

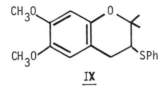

<u>IX</u>

elimination under very mild conditions (15). Biological studies of
these compounds are under way with different insects.

Chemical Studies of 3,4-epoxyprecocenes

Investigations of the in vivo and in vitro metabolism of the precocenes disclosed that 3,4-dihydroxyprecocenes were the most common metabolites. The abundance of these diols suggested that precocene must
have undergone epoxidation followed by hydration, leading to a constant isomer ratio of 70:30 trans cis isomers, (2,16). Recently, it
was demonstrated by incubation in vitro with Locusta migratoria corpora allata that 4-^{3}H-precocene I was metabolized stereospecifically
to (−)-trans-(3R,4S) and (+)-cis-(3R,4R) diols (17).

To prove the above hypothesis of precocene mode of action, it
was required to synthesize the corresponding 3,4-epoxyprecocenes and
to study its chemical reactivity. When this was accomplished in two
different laboratories (16,18) the lability of these epoxides towards
nucleophilic or electrophilic attack was confirmed. In each case the
chemical hydrolysis of these compounds gave the same isomer ratio of
diols observed in the enzymatic metabolic process.

We anticipated that stabilization of these type of epoxides
might be achieved by replacement of 7-alkoxy substituent by a fluoro-
alkoxy group. In fact, this was the case and 7-trifluoroethoxy-3,4-

epoxy precocene derivatives X exhibited higher stabilities than the corresponding non-fluorinated analogs (19), whithout loss of AJH activity in the olefinic precursors.

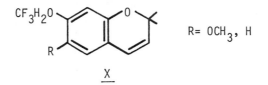

R= OCH$_3$, H

X

In searching for a relationship between antijuvenile hormone activities and epoxide chemical reactivity, we attempted to apply as a chemical probe the m-chloroperoxybenzoic-alkaline fluoride system, a reagent developed in this laboratory for preparation of acid labile epoxides (20). However, formation of hemiesters of 3,4-dihydroxy precocene, was the predominant reaction in the case of activated chromene structures.

Likewise, as it has been studied in the chromenes (21), we examined the possible relationship between precocene-like activity and ^{13}C chemical shifts of C-3 and C-4 in several 3,4-epoxides of active and inactive chromenes. In all cases observed, these chemical shifts differed too slightly, within the range of 0.5 ppm for C-3 and 0.7 ppm for C-4, to be of any diagnostic value.

Recently, several nucleophilic reagents have been used to establish the mode of action of the metabolites of polycyclic aromatic hydrocarbons (PAH). Among them, several phosphodiesters have been examined to clarify the possibility of reaction of PAH epoxides with the phosphate groups (P-alkylation) of nucleic acids (22). In this context we have studied the reaction of 3,4-epoxyprecocene II with dibenzyl phosphate under a variety of conditions. In all cases, instead of the formation of phenol or phosphotriesters observed with PAH epoxides, we obtained predominantly dimer XI. This compound was also the main component of the mixtures obtained by reaction of the above precocene epoxide with other acid catalysts, along with dimers XII and XII. Dimer XII was formed almost exclusively by thermal treatment. The structure and configuration for compound XII has been established by spectral and X-ray diffraction analyses (23).

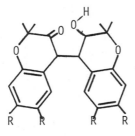

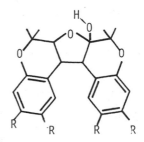

R= OCH$_3$

XI   XII

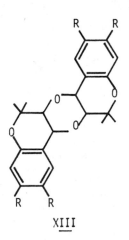

XIII

It is worth of note that the formation of a dimer, with different dioxane structure, in the treatment of 3,4-dihydroxyprecocene I with p-toluensulfonic acid has been recently reported (13). Further work, to study the reactivity of precocene epoxides with selected nucleophiles, which can shed light on the mode of action of these compounds, is in progress.

As a complement to the study of the chemistry of 3,4-epoxyprecocenes, we have also prepared the corresponding 2,2-dimethyl-3-chromanones by pyrolysis of hemiesters of 3,4-dihydroxyprecocenes. These chromanones might afford by enolization 3-hydroxyprecocenes, tautomers of the 3,4-epoxyprecocenes, with an enhanced reactivity towards nucleophiles at C-4. However, preliminary results of antijuvenile

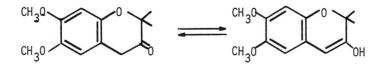

hormone activity of these chromanones were negative.

In short, although precocenes are a very valuable tool for carrying out chemical allatectomy in insect physiology studies, the future application of agents of the precocene type in insect control depends on overcoming two main problems of the natural precocenes, namely, its reported toxicity in vertebrates (24,25) and its insensitivity to holometabolous insects.

Acknowledgments

Financial support from Comisión Asesora de Investigación Científica y Técnica (Grant 1664-82) and Joint American-Spanish Committee for Scientific and Technological Cooperation (Grant 0394-11) is gratefully acknowledged.

Literature cited

1.   Bowers, W.S.; Ohta, T.; Cleere, J.S.; Marsella, P.A. Science, 1976, 193, 542–7.
2.   Pratt, G.E.; Jennings, R.C.; Hamnett, A.F.; Brooks, G.T. Nature 1980, 284, 320–3.
3.   Bowers, W.S.; Evans, P.E.; Marsella, P.A.; Soderlund, D.M.; Bettarini, F. Science, 1982, 217, 647–8.
4.   Merlini, L. In Advances in Heterocylic Chemistry; Katritzky, A.R.; Boulton, A.C., Eds. Vol. 18, Academic Press, New York, 1975, pp. 159–98.
5.   Camps, F.; Coll, J.; Messeguer, A.; Pericás, M.A. J. Heterocyclic Chem. 1980, 17, 1377–9.
6.   Camps, F.; Coll, J.; Messeguer, A.; Pericás, M.A. J. Heterocyclic Chem. 1980, 17, 207–8.
7.   Camps, F.; Colomina, O.; Coll, J.; Messeguer, A. Tetrahedron, 1982, 19– 2955–9.
8.   Bowers, W.S.   unpublished data.
9.   Bowers, W.S.; Feldlaufer, M.F. Gen. Comp. Endocrinol. 1982, 47, 120–4.
10. Soderlund, D.M.; Feldlaufer, M.F.; Takahashi, S.Y.; Bowers, W. A. In "Juvenile Hormone Biochemistry"; Pratt, G.E.; Brooks, G. T., eds. Elsevier–North Holland: Amsterdam, 1981; pp. 353–62.
11. Camps, F.; Coll, J.; Ricart, S. J. Heterocyclic Chem. 1983, 20, 249–50.
12. Camps, F.; Coll, J.; Ricart, S. J. Heterocyclic Chem., 1984, 21 745–8.
13. Anastasis, P.; Brown, P.E. J. Chem. Soc. Perkin Trans. 1983, 1431–7.
14. Ricart, S., unpublished results.
15. Thomas, P.; Whiting, D.A. Tetrahedron Lett. 1983, 24, 1099–102.
16. Soderlund, D.M.; Messeguer, A.; Bowers, W.S. J. Agric. Food Chem. 1980, 28, 724–31.
17. Hamnet, A.F.; Pratt, G.E. Life Sciences 1983, 32, 2747–53.
18. Jennings, R.C.; Ottridge, A.P. Chem. Commun. 1979, 920–1.
19. Camps, F., Coll, J.; Messeguer, A.; Pericás, M.A. Tetrahedron Lett. 1980, 21, 2361–4.
20. Camps, F.; Col, J.; Messeguer, A.; Pujol, F. Chem. Lett. 1983, 971–4.
21. Ottridge, A.P.; Jennings, R.C.; Brooks, G.T. In "Juvenile Hormone Biochemistry"; Pratt, G.E.; Brooks, G.T.; Eds. Elsevier–North Holland: Amsterdam, 1981; pp. 381–6.
22. Di Raddo, P.; Chan, T.H. Chem. Commun., 1983, 16–7.
23. Camps, F.; Coll, J.; Conchillo, A.; Messeguer, A.; Molins, E.; Miravitlles, C. J. Cryst. Spec. Res. in press.
24. Hsia, M.T.; Grossman, S.; Schrankel, K.R. Chem. Biol. Interactions 1981, 37, 265–77.
25. Halpin, R.A.; Vyas, K.P.; El-Naggar, S.F.; Jerina, D.M. Chem. Biol. Interactions 1984, 48, 297–315.

RECEIVED November 8, 1984

Propionate and Methyl Malonate Metabolism in Insects

GARY J. BLOMQUIST, PREMJIT P. HALARNKAR, and LAWRENCE A. DWYER

Department of Biochemistry, University of Nevada, Reno, NV 89557

Insects utilize propionate and methylmalonate in
the biosynthesis of ethyl branched juvenile
hormones and methyl branched cuticular hydrocarbons. The sources of propionate and methylmalonate in some insects appear to differ from
those in mammals. Succinate is the precursor of
propionate and methylmalonate in a termite, whereas
valine and probably other amino acids are the
sources of propionate and methylmalonate in several
other species. An unusual pathway for propionate
metabolism has been shown to occur in insects and
it may be related to the absence or low levels of
vitamin B_{12} found in many species. Propionate is
converted directly to acetate with carbon 1 of
propionate lost as CO_2, carbon 2 of propionate
becoming the methyl carbon of acetate and carbon 3
of propionate becoming the carboxyl carbon of
acetate. This pathway suggested the possibility
that 2-fluoropropionate might be selectively
metabolized in insects to the toxic 2-fluoro-
acetate. However, preliminary data indicate that
2-fluoropropionate is not toxic to the housefly or
the American cockroach.

Propionate serves several unique and important roles in insects.
It is used by some insects, in very small amounts, as a precursor
to homomevalonate which is an intermediate in the biosynthesis of
juvenile hormone (JH) II (1,2) and probably JH I and JH 0 as
well. Much larger amounts of propionate and methylmalonate are
needed for the biosynthesis of methyl branched hydrocarbons which
are major cuticular components in most of the approximately 100
insect species whose cuticular lipids have been examined (3-7).
Until recently, there was little information available on either
the source of propionate or its metabolism in insects. In
mammals vitamin B_{12} is a key cofactor in propionate and
methylmalonate metabolism (8-9). Recent observations that some
insect species lack or contain low levels of vitamin B_{12} (10)

0097–6156/85/0276–0245$06.00/0

raised questions as to how these insects perform or circumvent
such reactions. Evidence is presented indicating that insects
have a novel pathway for propionate metabolism as compared to
vertebrates. Sources of propionate and methylmalonate and their
utilization as substrates for JH and methyl branched hydrocarbon
biosynthesis are reviewed.

Sources of Propionate and Methylmalonate

In vertebrates, the major sources of propionate and methylmalonate are odd chain fatty acids and the amino acids
isoleucine, valine and methionine (8,9). In the termite
Zootermopsis angusticollis, which can incorporate propionate as
the methyl branch unit of mono- and dimethylalkanes (7), these
sources were considered unlikely because the diet of termites
presumably contains little fatty acid and Z. angusticollis
contains very small amounts of odd chain length fatty acids (Chu
and Blomquist, unpublished). Likewise, termites would be
expected to conserve essential amino acids such as the ones that
could serve as precursors to propionate and methylmalonate.
    A series of experiments were performed in which the in vivo
incorporation of $[1-^{14}C]-$, $[2,3-^{14}C]-$ and $[2,3-^{3}H]$succinates into
methyl branched alkanes were compared. $[1-^{14}C]$Succinate was preferentially incorporated into the normal alkanes and $[2,3-^{3}H]-$
succinate was preferentially incorporated into the methyl
branched alkanes (11). This and other evidence (11) suggested
that succinate could be a precursor to methylmalonate in this
termite. Direct evidence that this termite could utilize succinate as the precursor to methylmalonate was obtained by examining
the incorporation of $[2,3-^{13}C_2]$succinate into methylalkanes by
^{13}C-NMR. Carbons 2 and 3 of succinate were incorporated into the
branching methyl group(s) and the tertiary carbons(s) of mono-
and dimethylalkanes (12). These data indicate that succinate is
metabolized to methylmalonyl-CoA and then is incorporated in
place of malonyl-CoA at specific points during chain elongation.
    Recent experiments in our laboratory have utilized HPLC to
separate organic acids from homogenates of insect tissue.
Mitochondrial preparations from Z. angusticollis were apparently
able to metabolize $[2,3-^{14}C]$succinate to methylmalonate. Radioactivity was recovered in the fraction corresponding to
propionate. This indicates that this termite is able to convert
succinate to propionate, presumably via a methylmalonyl-CoA
intermediate. When succinate dehydrogenase was inhibited by
malonate during a mitochondrial incubation, the conversion of
succinate to propionate was increased, further indicating that
the mitochondrial pool of succinate can be used to form
methylmalonate and propionate. The general flow of carbon in
mammals is propionate to methylmalonate to succinate, which is
then metabolized by tricarboxylic acid cycle enzymes. Thus, the
flow of carbon in the termite appears reversed from that observed
in mammals.
    In other insects, including the housefly Musca domestica
(13,14) and the cockroach Periplaneta americana (15,16), studies
with both radioactive and stable isotopes clearly showed that
succinate was not a major precursor to the methyl branching unit.

In these two species, $[^3H]$valine was readily incorporated into the branched alkanes. A ^{13}C-NMR examination of the incorporation of $[3,4,5-^{13}C_3]$valine into the branched alkanes of the housefly (13) showed that carbons 3, 4, and 5 were incorporated intact (as determined by $^{13}C-^{13}C$ coupling) into the branching methyl carbon, tertiary carbon, and carbon adjacent to the tertiary carbon, respectively. Similar data were obtained when the incorporation of $[3,4,5-^{13}C_3]$valine into 3-methylpentacosane was examined in the American cockroach (17).

Whether or not propionate is produced as an obligate intermediate in the metabolism of valine to methylmalonyl-CoA has been a subject of controversy (18). Indeed, some textbooks still show that methylmalonic semialdehyde is directly oxidized to methylmalonyl-CoA (19). Using stable isotopes, Baretz and Tanaka (18) have presented convincing evidence that rats convert methylmalonic semialdehyde to propionyl-CoA, which is then carboxylated to form methylmalonyl-CoA. It appears that a similar pathway occurs in the housefly and the American cockroach (Figure 1). If methylmalonyl semialdehyde were converted directly to methylmalonyl-CoA, one of the ^{13}C labeled methyl groups from $[3,4,5-^{13}C_3]$-valine would become the free carboxyl carbon of methylmalonyl-CoA and would then be lost as CO_2 during incorporation into the alkyl chain. The observation that carbons 3, 4 and 5 of valine were incorporated intact indicates that this does not happen and that valine probably is metabolized via propionyl-CoA to methylmalonyl-CoA (Figure 1).

Vitamin B_{12} is a required cofactor for methylmalonyl-CoA mutase, which is involved in propionate catabolism in mammals. This fact and the differences observed in the precursors to propionate and methylmalonate among insect species (Table I) prompted an examination of a number of insects for vitamin B_{12} levels. The termite Z. angusticollis, which readily converts succinate to methylmalonate, has large amounts of vitamin B_{12}, whereas the American cockroach has low levels and the housefly does not have detectable amounts of vitamin B_{12} (10). Thus, both dietary considerations and levels of vitamin B_{12} may play a role in determining the precursors to propionyl and methylmalonyl derivatives in insects.

Metabolism of Propionate

It is well established that propionate can be utilized for the methyl branch unit in methyl branched hydrocarbons in insects (4-7), and recent data have shown that the methyl branches are inserted early during chain elongation rather than toward the end of the process (13,16). ^{13}C-NMR analysis demonstrated that propionates labeled with ^{13}C in either the 1, 2 or 3 positions are incorporated into the methyl branched alkanes of insect cuticular lipids. C-3 of propionate becomes the branching methyl carbon, C-2 becomes the tertiary carbon and C-1 the carbon adjacent to the tertiary carbon (13,16,17) in these methyl branched hydrocarbons.

Indirect evidence from studies with radioactive precursors suggested that in addition to labeling the methyl branch unit of 3-methyl and internal methyl branched hydrocarbons, propionate

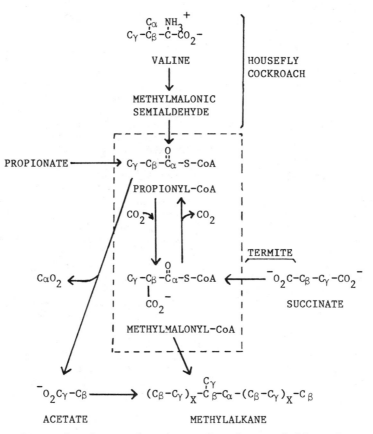

Figure 1. Proposed pathways for the metabolism of propionate and methylmalonate in selected insects.

Table I. Metabolism of Propionate and Methylmalonate in Insects

Insect	Order	Amounts of Vitamin B_{12}	Source of Propionate and Methylmalonate	Convert Propionate to Acetate	Subcellular Location
Cockroach (Adult) Periplaneta americana	Dictyoptera	Low (10)	Valine (Other a.a.?) (17)	Yes (17)	Mito (17)
Housefly (Adult) Musca domestica	Diptera	Not Detectable (10)	Valine (Other a.a.?) (14)	Yes[a]	Mito[a]
Termite Zootermopsis angusticollis	Isoptera	High (10)	Succinate (16)	Yes[a]	Mito[a]
Cabbage Looper (Larva) Trichoplusia ni	Lepidoptera	Not Detectable (10)	Valine (Other a.a.?) (20)	Yes[a]	Mito[a]

[a] Unpublished results, Halarnkar, P.P., Blomquist, G.J. and Heisler, C. R.

could be converted to acetate and subsequently label the straight chain portion of hydrocarbons (7,11). Direct evidence obtained by ^{13}C-NMR studies in the housefly (14) and the American cockroach (17) showed that [2-^{13}C]propionate labels hydrocarbons in the same positions as [2-^{13}C]acetate and that [3-^{13}C]propionate labels hydrocarbons in the same positions as [1-^{13}C]acetate. These data suggest that propionate is converted to acetate with carbon 1 lost as CO_2, carbon 3 becoming the carboxyl carbon of acetate and carbon 2 becoming the methyl carbon of acetate.

Propionate also serves as a precursor for juvenile hormone biosynthesis in Manduca sexta (1). The ethyl branch in JH II (and presumably those in JH I and JH 0) is derived from propionate which is incorporated into homomevalonate. The data presented by Schooley et al. (1) on the incorporation of [2-^{14}C]-propionate did not entirely fit the postulated pathways for JH biosynthesis. Radioactivity from [2-^{14}C]propionate was recovered in JH III, which lacks ethyl branches, and its degradation products, neither of which should have been labeled. Radioactivity was also found in degradation products of JH II from portions of the molecule that should not have been labeled by propionate. It was suggested (1) that other, unspecified, metabolic pathways for propionate could account for the labeling of JH III by [2-^{14}C]-propionate. On the other hand, [1-^{14}C]propionate did not label JH III, but did label JH II in the pattern expected.

These results could be explained if, as was suggested by the data of Dillwith et al. (14), there is a pathway in which propionate is directly converted to acetate with the loss of carbon 1 and the oxidation of carbon 3. With a pathway of this type, propionate labeled in carbon 1 would lose its label if it were converted to acetate prior to incorporation into other compounds. However, if propionate were labeled in carbon 2 or carbon 3 it would retain its label upon conversion to acetate. Therefore, any label incorporated into JH or hydrocarbon from propionate labeled in carbon 1 would have to be the result of propionate being utilized as an intact unit. Incorporated label from propionate labeled in carbon 2 or carbon 3 could result from direct incorporation of propionate or from conversion of propionate to acetate prior to being utilized for JH or hydrocarbon biosynthesis. Thus, the labeling patterns seen by Schooley, et al. (1) and Dillwith, et al. (14) could be the result of the retention of label from carbon 2 or carbon 3 of propionate, and the loss of label from carbon 1 of propionate during the conversion of propionate to acetate.

In vivo and in vitro studies in our laboratory have directly demonstrated that such a pathway does take place in insects and may represent a major pathway for the degradation of propionate. Following injection of [1-^{14}C]propionate or [2-^{14}C]-propionate, insects were killed, homogenized, organic acids extracted and separated by HPLC, and radioactivity in each fraction assayed by liquid scintillation counting. Radioactivity from [2-^{14}C]propionate was recovered in acetate as well as in citrate and succinate. In vivo and in vitro studies, as a function of time, indicated that propionate was first converted to acetate which subsequently labeled tricarboxylic acid intermediates. Radioactivity from [1-^{14}C]propionate was not found in acetate or any TCA intermediate. These results were essentially the same in all four of the insect species studied to date (Table I). The rate of conversion of propionate to acetate varied with insect species and occurred most rapidly in the housefly where about 50% of the injected $[2-^{14}C]$propionate was converted to acetate within 5 min.

Subcellular localization of the propionate to acetate pathway was examined in the housefly, termite, American cockroach and cabbage looper. In all four species the mitochondrial fraction had significant activity whereas the microsomal fraction (105,000g pellet) and soluble fraction (105,000g supernatant) did not convert propionate to acetate to any appreciable extent. These results were similar to results of work done in plants, which showed that propionate is metabolized to acetate (20,21) by mitochondrial preparations.

A pathway for converting propionate to acetate is not unique in biological systems. Plants, many of which apparently do not contain vitamin B_{12}, convert propionate directly to acetate via a 3-hydroxypropionate intermediate (22). The finding that many insects either do not have detectable levels of B_{12} or have very low levels (10) suggests that, like plants, insects have evolved an alternative route of propionate catabolism. The conversion of propionate to acetate may be a general pathway in insects, as even the termite Z. angusticollis, which has large amounts of vitamin B_{12}, has this pathway.

Prestwich and coworkers (23,24) have shown that by judiciously placing fluorines on selected positions of fatty acids and sterols; insects will metabolize the fluorinated precursor to the potent toxin 2-fluoroacetate (25). Because insects convert propionate to acetate, it was possible that they might convert 2-fluoropropionate to 2-fluoroacetate by the same pathway. However, preliminary experiments using houseflies and cockroaches indicated that 2-fluoropropionate was not readily converted to 2-fluoroacetate. Houseflies injected with 2-fluoropropionate (3 µg/insect) were not affected after 1.5 hr, whereas control insects injected with 2-fluoroacetate (2 µg/insect) were all dead within 0.5 hr. Similar results were obtained with the American cockroach.

Summary

Propionate is a key intermediate in JH and hydrocarbon biosynthesis in insects. It serves as a precursor for methyl branched hydrocarbons which in many insects are important compounds for communication and cuticular protection, and it is a precursor for juvenile hormone biosynthesis (JH 0, JH I and JH II). Sources of propionate have been shown to be succinate in a termite and certain amino acids such as valine in other species.

Insects have an unusual pathway for catabolizing propionate which may be related to the absence or low levels of vitamin B_{12} found in many species. The propionate to acetate pathway is present in all insects which have been studied, including the termite, which has high levels of vitamin B_{12}. The presence of this unusual metabolic pathway for propionate metabolism offered the potential for selectivity in developing insect control

agents. The compound 2-fluoropropionate seemed to be an ideal
candidate. Unfortunately, it was not toxic to the housefly or
American cockroach, presumably because it was not metabolized to
fluoroacetate at a sufficient rate. Nonetheless, exploitation of
unique metabolic pathways in insects offers the potential for
novel control techniques.

Acknowledgments

Supported in part by the Science and Education
Administration of the U.S. Department of Agriculture under Grant
83-CRCR-1-1210 from the Competitive Research Grants Office. A
contribution of the Nevada Agriculture Experiment Station.

Literature Cited

1. Schooley, D. A.; Judy, K. J.; Bergot, B. J.; Hall, M.
    S.; Siddall, J.B. Proc. Natl. Acad. Sci. USA 1973,
    70, 2921-5.
2. Tobe, S. S.; Feyereisen, R. In "Endocrinology of
    Insects"; Downer, R. G. H. and Laufer, H., Ed.; Alan
    R. Liss, Inc.: New York, 1983; pp. 161-78.
3. Blomquist, G. J.; Dillwith, J. W. In "Comprehensive
    Insect Physiology, Biochemistry and Pharmacology.
    Vol. 3. Integument, Respiration and Circulation";
    Kerkut, G. A. and Gilbert, L. I., Eds.; 1984; in
    press.
4. Howard, R. W.; Blomquist, G. J. Ann. Rev. Ent. 1982,
    27, 149-73.
5. Blomquist, G. J.; Jackson, L. L. Prog. Lipid Res.
    1979, 17, 319-45.
6. Blomquist, G. J.; Kearney, G. P. Arch. Biochem.
    Biophys. 1976, 173, 546-53.
7. Blomquist, G. J.; Howard, R. W.: McDaniel, C. A.
    Insect Biochem. 1979, 9, 371-74.
8. Rosenberg, L. E. In "The Metabolic Basis of Inhereted
    Disease"; Stanburg J. B., Wyngaarden J. B. and
    Fredrickson D. S., Eds.; McGraw-Hill Inc.: New York,
    1972; pp. 440-58.
9. Coates, M. E. In "The Vitamins"; Sebrell W. H., Jr.
    and Harris, R. S., Eds.; Academic Press: New York,
    1968; pp. 212-20.
10. Wakayama, E. J.; Dillwith, J. W.; Howard, R. W.;
    Blomquist, G. J. Insect Biochem. 1984, 14, 175-79
11. Chu, A. J.; Blomquist, G. J. Arch. Biochem. Biophys.
    1980, 201, 304-12.
12. Blomquist, G. J.; Chu, A. J.; Nelson, J. H.; Pomonis,
    J. G. Arch. Biochem. Biophys. 1980, 204, 648-50.
13. Dillwith, J. W.; Blomquist, G. J.; Nelson, D. R.
    Insect Biochem. 1981, 11, 247-53.
14. Dillwith, J. W.; Nelson, J. H.; Pomonis, P. G.;
    Nelson, D. R. and Blomquist, G. J. J. Biol. Chem.
    1982, 257, 11305-14.

15. Blomquist, G. J.; Major, M. A.; Lok, J. B. Biochem. Biophys. Res. Comm. 1975, 64, 43-50.
16. Dwyer, L. A.; Blomquist, G. J.; Nelson, J. H.; Pomonis, P. G. Biochim. Biophys. Acta 1981, 663, 536-44.
17. Halarnkar, P. P.; Nelson, J. H.; Heisler, C. R.; Blomquist, G. J. Arch. Biochem. Biophys. 1985, in press.
18. Baretz, B. H.; Tanaka, K. J. Biol. Chem. 1978, 253, 4203-13.
19. Rawn, J. D. "Biochemistry"; Harper and Row: New York, 1983; pp. 843-45.
20. de Renobales, M.; Blomquist, G. J. Insect Biochem. 1983, 13, 493-502.
21. Giovanelli, J.; Stumpf, P. K. J. Biol. Chem. 1958, 231, 411-26.
22. Hatch, M. D.; Stumpf, P. K. Arch. Biochem. Biophys. 1962, 96, 193-98.
23. Prestwich, G. D.; Melcer, M. E.; Plavcan, K. A. J. Agric. Food Chem. 1981, 29, 1023-27.
24. Prestwich, G. D.; Phirwa, S. Tetrahedr. L. 1983, 24, 2461-64.
25. Peters, R. A. "Carbon-Fluorine Compd: Chem., Biochem. Biol. Act." CIBA Found. Symp., 1971, 1972, 1-27.

RECEIVED November 8, 1984

Suicidal Destruction of Cytochrome P-450
In the Design of Inhibitors of Insect Juvenile Hormone Biosynthesis

RENÉ FEYEREISEN[1,2], DAN E. FARNSWORTH[1], KATHRYN S. PRICKETT[3], and PAUL R. ORTIZ DE MONTELLANO[3]

[1] Department of Entomology, Oregon State University, Corvallis, OR 97331
[2] Department of Agricultural Chemistry, Oregon State University, Corvallis, OR 97331
[3] Department of Pharmaceutical Chemistry, School of Pharmacy, University of California, San Francisco, CA 94143

The possibility of designing compounds which would act as selective suicide substrates for the cytochrome P-450 monooxygenase of insect corpora allata was investigated. Three types of assays were performed using the cockroach, Diploptera punctata: Inhibition of juvenile hormone (JH) biosynthesis by corpora allata incubated in vitro; in vitro inhibition of midgut microsomal cytochrome P-450 monooxygenases and in vivo inhibition of oocyte growth following topical application. Compounds containing an acetylenic function and structurally related to precocene or to the natural substrate of the allatal cytochrome P-450 (methyl farnesoate) were found to be better inhibitors of JH III biosynthesis in vitro than precocene II. These preliminary results suggest that the relatively lax substrate specificity of the allatal cytochrome P-450 monooxygenase may be exploited in the rational design of powerful inhibitors of insect juvenile hormone biosynthesis.

There is a growing need for new, safe and specific approaches for insect control which are compatible with Integrated Pest Management strategies. These approaches, based on a fundamental knowledge of insect biology should lead to compounds that induce pest-specific biochemical lesions or that regulate natural control mechanisms ([1], [2]). Considerable attention has been focussed during the last 15 years on compounds which interfere with the normal program of growth, metamorphosis and reproduction in insects. The practical utility of juvenile hormones and their synthetic analogues (juvenoids) will remain limited to certain special applications ([3]). Juvenoids are not perceived to have a major future impact on agricultural chemistry ([4], [5], [6]), in part because plant stress would be aggravated by insect pests blocked in the feeding larval stage. In contrast, anti-juvenile hormone agents, compounds which antagonize JH biosynthesis, release, transport, uptake or mode of

0097–6156/85/0276–0255$06.00/0

action, are more attractive for insect control strategies because
they should provoke rapid deficiencies in the JH-controlled
processes, with dramatic consequences such as precocious metamorphosis or inhibition of reproduction. Thus, anti-juvenile
hormone agents (sensu lato) would be more versatile than juvenoids,
because they would have a lesser need for critical timing in
application and a shorter response time (7, 8).

Two of the better known "anti-juvenile hormone" agents, precocenes (9) and fluoromevalonate (10) are inhibitors of JH
biosynthesis. The mode of action of fluoromevalonate at the
molecular level is unknown. Elucidation of the mode of action of
precocenes indicates that these plant chromene derivatives reach
the site of JH biosynthesis, the corpora allata (CA), where they
undergo a lethal epoxidation leading to extensive macromolecular
alkylation and ultimately cause cell death (11, 12). Bioactivation
of precocenes to the highly reactive precocene epoxide (13) in the
corpora allata is almost certainly catalyzed by methyl farnesoate
(MF) epoxidase (14), a cytochrome P-450 monooxygenase (15) which is
the last enzyme of the JH biosynthetic pathway (at least in locusts
and cockroaches).

Precocenes may not provide the new approach to insect control
originally expected ("4th generation" insecticides, 9) because they
are not active in some major groups of agricultural pests
(Lepidoptera) and because their mode of action (cytotoxicity) is
not compatible with environmental concerns. Indeed, precocene II
has been shown to be hepatotoxic and nephrotoxic in rats (16, 17).
However, research on precocenes has led to (at least) two important
conclusions: (1) compounds structurally unrelated to the JH
biosynthetic pathway can reach critical sites (e.g. epoxidase)
within the CA, and (2) such compounds can be catalytically
processed (e.g. epoxidized) by enzymes of JH biosynthesis. The lax
substrate specificity of methyl farnesoate epoxidase in the corpora
allata and its catalytic competence might be exploited in the
design of irreversible inhibitors of JH biosynthesis (Figure 1).

A number of inhibitors of methyl farnesoate epoxidase of
Blaberus giganteus corpora allata have been described (18). They
include typical cytochrome P-450 monooxygenase inhibitors such as
methylenedioxyphenyl compounds and substituted imidazoles. In
assays of JH III biosynthesis by Periplaneta americana CA in vitro
some methylenedioxyphenyl compounds were shown to inhibit hormone
production at moderate to high concentrations (19). Both
methylenedioxyphenyl compounds and terpenoid imidazoles have also
some anti-juvenile hormone activity in Lepidoptera (7, 20).

Another class of cytochrome P-450 inhibitors, compounds with a
monosubstituted acetylenic function, are well known for their
potential as insecticide synergists (21) and some have already been
reported to be active as JH biosynthesis inhibitors as well (19,
22). Ortiz de Montellano and Kunze (23) have shown that many
ethynyl substrates cause the destruction of rat hepatic cytochrome
P-450, when the prosthetic heme is alkylated during attempted
metabolism of the triple bond. Such suicide substrates must bind
to the enzyme and be catalytically acceptable thereby offering a
potential for selectivity. In fact, selectivity of suicide
substrates for particular molecular forms (isozymes) of hepatic

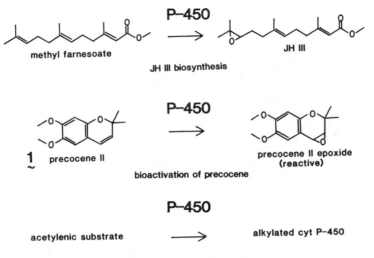

Figure 1. Three reactions catalyzed by methyl farnesoate epoxidase (a cytochrome P-450 monooxygenase) in the corpora allata.

microsomal cytochrome P-450 has been demonstrated (review in 24).
This is important if the ultimate goal is to find an inhibitor
selective for insect allatal methyl farnesoate epoxidase. Recent
indirect evidence suggests that with certain suicide substrates
such as 1-aminobenzotriazole or substituted phenyl 2-propynyl
ethers, self-catalyzed destruction of microsomal cytochrome P-450
is isozyme-selective in the housefly (25). We report in this paper
that acetylenic inhibitors of JH synthesis can be modelled on
either precocene or on the substrate of the allatal cytochrome
P-450 monooxygenase, methyl farnesoate. We also report the in vivo
activity of these compounds in the cockroach Diploptera punctata.

Materials and Methods

Diploptera punctata was maintained as described (26). Precisely
timed newly emerged females were removed from the colony and mating
was confirmed within 24 hours by the presence of a spermatophore.
       Precocene II (1) and 7-hydroxy-4-methylcoumarin were obtained
from Aldrich. Acetylenic MF analog (2) was synthesized according
to (27). CGA 167126 (3) was a gift from Ciba-Geigy. Insecticide
synergist (4) 2,4,5-trichlorophenyl 2-propynyl ether was obtained
from Hoffmann-La Roche. 7-Methoxy-4-methylcoumarin was obtained
from Calbiochem.
       In vitro activity of the corpora allata was measured by the
radiochemical assay of Pratt and Tobe (28). Glands were dissected
from 5-day old mated females (terminal oocyte length 1.3-1.5 mm)
and incubated in medium TC 199 without methionine (GIBCO) but with
[methyl-^{14}C]methionine (56.7 mCi/mmol, Amersham) at a final
concentration of 0.26-0.30 mM and Ficoll (2%, Sigma). Incubation
of single gland pairs was for 3 hours at 28°C in the presence or
absence of inhibitor. Six to 12 individual pairs of CA were
incubated at each concentration of inhibitor. Because the
inhibitors were supplied in ethanol, both control and experimental
incubation contained a constant 1.5% concentration of ethanol.
Biosynthesized JH III and methyl farnesoate from the incubation
medium and the glands were extracted, separated by TLC and assayed
by liquid scintillation counting (29).
       Activity of the inhibitors in vivo was assayed as described in
Feyereisen et al. (30). The compounds were topically applied to
the ventral surface of the thorax in 2 μl of acetone. Adult mated
females were treated on day 2 and the length of the terminal
oocytes was measured on day 5. Seven to 20 insects were used for
each dose of inhibitor. Growth of the oocytes during that period
was compared to the growth of oocytes from control insects treated
with acetone alone. In some experiments the insects were treated
sequentially with inhibitor and 200 μg hydroprene (ZR512) on day 2.
       Microsomal cytochrome P-450 monooxygenases: Midguts from 20
to 36-day-old adult females were dissected and collected in
Yeager's cockroach saline. The tissue was homogenized in 50 mM
MOPS (morpholinopropanesulfonate) buffer, pH 7.2, containing 1 mM
EDTA, 0.4 mM freshly prepared PMSF (phenylmethylsulfonyl fluoride)
and 10% sucrose. A microsomal fraction was prepared by sequential
centrifugation of the homogenate at 1,000 g for 15 min., 10,000 g
for 15 min. and 100,000 g for 65 min. The final "microsomal"

pellet was resuspended in 50 mM MOPS buffer, pH 7.2, containing
1 mM EDTA and 0.4 mM PMSF, to a concentration of 1 mg protein/ml.
Assays for aldrin epoxidation contained, in a final volume of
500 μl: 50 mM MOPS buffer, pH 7.2; 1 mM EDTA; 0.4 mM PMSF; 1.15 mM
NADP; 50 mM glucose-6-phosphate; 0.6 units glucose-6-phosphate
dehydrogenase, 2.3 mM NADH and 100 μg of microsomal protein. This
assay mixture was incubated for 5 min. at 30°C with or without the
inhibitor, before the reaction was started by the addition of
aldrin to a final concentration of 50 μM. The aldrin epoxidation
reaction was stopped after a 2 min. incubation at 30°C by the
addition of 3 ml hexane. The product (dieldrin) was analyzed and
quantified by electron-capture gas chromatography.

Assays for 7-methoxy-4-methylcoumarin O-demethylation
contained, in a final volume of 500 μl: 50 mM MOPS buffer, pH 7.2;
1 mM EDTA; 0.4 mM PMSF; 1.15 mM NADP; 50 mM glucose 6-phosphate;
0.6 units of glucose-6-phosphate dehydrogenase and 200 μg of microsomal protein. This assay mixture was incubated for 5 min. at 30°C
with or without the inhibitor, before the reaction was started by
the addition of 7-methoxy-4-methylcoumarin to a final concentration
of 200 μM. The O-demethylation reaction was stopped after 5 min.
incubation at 30°C by the addition of 330 μl of 8% (w/v) perchloric
acid. The mixture was neutralized with 170 μl of 17% (w/v)
K_2CO_3 and centrifuged. A 200 μl aliquot of the supernatant was
added to 2.5 ml of 0.1 M carbonate/bicarbonate buffer, pH 10.
7-Hydroxy-4-methylcoumarin was assayed fluorometrically at 30°C in
a Perkin-Elmer 650-10S instrument set at 370 nm, 3 nm slit for
excitation and 455 nm, 5 nm slit for emission.

Results

We chose adult female Diploptera punctata as the test organism
because the effects of precocene II on this insect have been
described in detail (30) and because the CA of this insect produce
exclusively JH III.   Precocene II is bioactivated by a monooxygenase in D. punctata CA, just as precocene I is bioactivated in
Locusta migratoria CA (11, 12). Indirect evidence for precocene II
bioactivation to a highly reactive 3,4-epoxide is not only provided
by electron microscopy which shows extensive damage to the CA cells
after in vivo or in vitro treatment (30). In addition, in vitro
studies show that radiolabelled precocene II is metabolized by D.
punctata CA to a mixture of cis- and trans-3,4-dihydrodiols which
can be extracted and analyzed by HPLC (Fig. 2, HPLC conditions:
ether/pentane 1/1 + 0.2% 2-propanol; 1.5 ml/min. 300 x 4 mm
Micropak Si 5). The ratio of precocene II cis- and trans-3,4-
dihydrodiols (36:64) is remarkably similar to the ratio of
dihydrodiols obtained in incubations of rat liver microsomes with
precocene II (16) or in incubations of L. migratoria CA with
precocene I (11).

Three acetylenic compounds (Figure 3) were compared to
precocene II (1) for their ability to inhibit JH biosynthesis in
vitro, to inhibit oocyte growth in vivo, and to inhibit midgut
microsomal (i.e. extra-allatal) cytochrome P-450 monooxygenases.
The compounds were chosen to represent an analog of the substrate
of the allatal monooxygenase, methyl farnesoate, with the

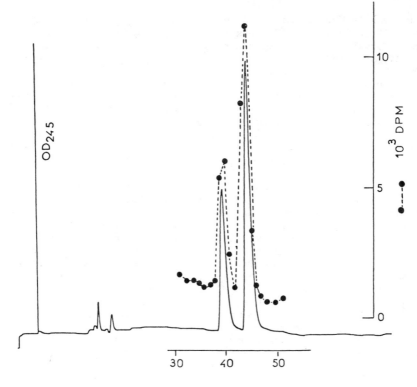

Figure 2. HPLC of precocene II cis- and trans-3,4-dihydrodiols obtained from in vitro incubation of Diploptera punctata corpora allata with ³H-precocene II.

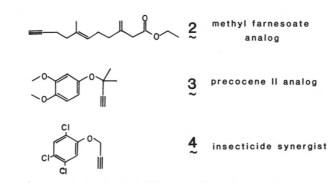

2 methyl farnesoate analog

3 precocene II analog

4 insecticide synergist

Figure 3. Acetylenic inhibitors of JH biosynthesis.

10,11-double bond replaced with an acetylenic function (2); a
precocene II analog with the 3,4-double bond replaced with an
acetylenic function (3); and a known inhibitor of insect cytochrome
P-450 monooxygenases, originally described in the literature (31)
as a carbamate insecticide synergist (4).

Inhibition of JH III biosynthesis in vitro. All three acetylenic
compounds were significantly better inhibitors of JH biosynthesis
than precocene II under similar incubation conditions (Fig. 4).
Compound 2 was the best inhibitor tested here, with an I_{50} of 16
μM, more than 25 times better than precocene II. Whether this
better performance as an inhibitor is due to a closer structural
analogy with the natural substrate of the epoxidase or whether it
is a consequence of the presumed difference in the mode of action
(irreversible inhibition of the epoxidase for 2; cytotoxicity of
precocene epoxide eventually resulting in decreased JH biosynthetic
rate for 1) is not presently known. The insecticide synergist (4)
proved to be about 7 times better than precocene II. Its activity
as an inhibitor of JH biosynthesis by D. punctata CA is similar to
that of 3 and the methylenedioxyphenyl analog of JH (Ro 20-3600).
In order to show that 4 was indeed acting as an inhibitor of methyl
farnesoate epoxidase, the ratio of intraglandular methyl farnesoate
(MF) to JH III biosynthesized was calculated at each dose of
inhibitor. A dose-dependent increase in the MF/JH III ratio was
observed (Fig. 5) resulting in a significant accumulation of methyl
farnesoate in CA inhibited by 4 when compared to control values.
This is consistent with the hypothesis that inhibition occurs at
the level of methyl farnesoate epoxidation. Accumulation of methyl
farnesoate was also observed in D. punctata CA inhibited with other
types of cytochrome P-450 monooxygenase inhibitors, such as
metyrapone and the methylenedioxyphenyl compound Ro 20-3600 (30).
In Periplaneta americana CA, no accumulation of methyl farnesoate
was noted with acetylenic inhibitors of JH III biosynthesis,
although this effect was observed with methylenedioxyphenyl
inhibitors (19).

Inhibition of midgut microsomal cytochrome P-450 monooxygenase
activities. We tested the acetylenic inhibitors of JH biosynthesis
on extra-allatal cytochrome P-450 monooxygenases to determine, in
an indirect manner, the relative selectivity of the compounds.
Aldrin epoxidation and 7-methoxy-4-methylcoumarin O-demethylation
were assayed following a 5 min. preincubation of midgut microsomes
in the presence of NADPH and various concentrations of the
inhibitors. The Table summarizes the I_{50} values obtained from
the dose-dependent enzyme inhibition curves. Compound 4 was, as
expected, the best inhibitor of microsomal monooxygenase activities
in vitro, but 2 was also a good inhibitor. In our experimental
conditions, both compounds inhibited up to 90% of epoxidation and
O-demethylation at the highest concentration tested (200 μM). In
contrast, 1 and 3 were only poor inhibitors of midgut microsomal
monooxygenases. Definitive evidence that the acetylenic compounds
inhibit these enzyme activities through a self-catalyzed
destruction of cytochrome P-450 is still lacking, but preliminary
data showed that inhibition of epoxidation and O-demethylation by

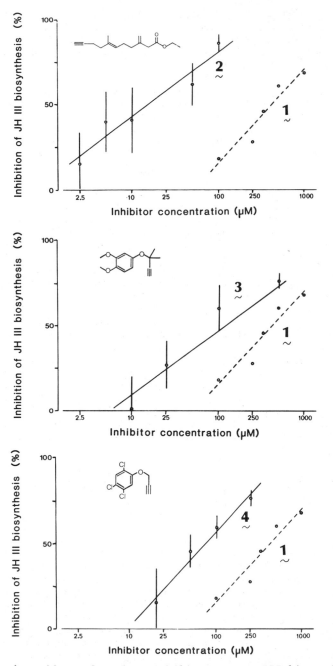

Figure 4. Effect of various inhibitors on JH III biosynthesis by <u>Diploptera</u> <u>punctata</u> corpora allata <u>in vitro</u>.

the acetylenic compounds was dependent on the time of preincubation and on the presence of NADPH during preincubation. In contrast, the low levels of inhibition by precocene were not dependent on the presence of NADPH and therefore may not involve the 3,4-epoxide.

Inhibition of midgut microsomal monooxygenase activities

	(1)	(2)	(3)	(4)
Aldrin epoxidation	> 200	8	> 200	5
7-methoxy-4-methyl-coumarin O-demethylation	175	12	80	12

Inhibition is expressed as I_{50} (μM). Compounds were tested at 5 concentrations ranging from 1 to 200 μM with 4 replicates   at each concentration.

In vivo activity of the inhibitors. In order to assess the potential "anti-juvenile hormone" effects of inhibitors of bio-synthesis, we tested the three acetylenic compounds as inhibitors of oocyte growth in mated, adult females. Oocyte growth in these insects is under the control of active corpora allata, and can be restored in allatectomized insects by topical application of a juvenile hormone agonist such as hydroprene (32). Figure 6 shows that 3 was not significantly better than 1 in this test. Compound 2 hardly showed any activity in vivo, even though it was the best inhibitor in vitro. This discrepancy may be attributed to pene-tration or metabolism problems, but is more likely to be caused by the JH agonist activity of this compound. The latter possibility is presently being investigated. Compound 4 was the most active compound in vivo. Oocyte growth was restored to normal levels in insects treated with 3 and 4 by a simultaneous treatment with hydroprene, thus suggesting that 3 and 4 did not inhibit oocyte growth by interfering with the mode of action of JH, (i.e., at the level of JH receptors or beyond). Because insects treated with 3 or 4 can be rescued with a JH agonist, and because no significant toxicity can be observed at the effective doses, we infer that inhibition of oocyte growth is indeed caused by the inhibition of JH synthesis leading to a decrease in JH titer.

Discussion

Inhibitors of JH biosynthesis may be found by random screening or by biochemical design. This pilot study shows that biochemical design is a useful tool and should be further exploited to search for potential new insect control agents. Our goal is to find compounds which, instead of destroying the corpora allata by a non-selective "shotgun attack" on the cell (12), would selectively destroy the methyl farnesoate epoxidase. This approach has also been taken by Brooks et al. (33) who have recently reported that some acetylenic MF analogs are very powerful inhibitors of JH biosynthesis in P. americana CA. Although we have not conclusively

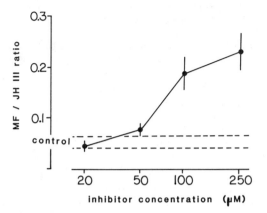

Figure 5. Accumulation of methyl farnesoate in corpora allata
inhibited by the acetylenic insecticide synergist (4).

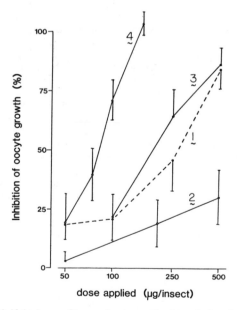

Figure 6. Inhibition of oocyte growth in adult female Diploptera
punctata over a 3-day period (day 2 to day 5).

established a suicidal interaction between the acetylenic compounds
and the enzyme, we can discuss the selectivity of the compounds,
their activity in vivo and prospects for the design of better
inhibitors of JH biosynthesis. Compounds 3 and 4 are equally
effective as inhibitors of JH biosynthesis in vitro, but 4 is a
much better inhibitor of midgut microsomal epoxidation and
O-demethylation. Also, 2 and 4 are equally effective as inhibitors
of midgut microsomal epoxidation and O-demethylation, but 2 is a
much better inhibitor of JH biosynthesis in vitro. This suggests
that 2 and 3 are probably more selective inhibitors of methyl
farnesoate epoxidase than 4 and that the activity of a compound as
JH biosynthesis inhibitor is not correlated to its activity as
inhibitor of other cytochrome P-450 monooxygenases. The in vitro
and in vivo activities of the three compounds are not in agreement,
and this stresses the extra-allatal fate of the compounds
(penetration, metabolism) as a major determinant of in vivo
activity. We believe that testing potential suicide substrates of
methyl farnesoate epoxidase using at least three criteria
(inhibition of JH biosynthesis in vitro; inhibition of extra-
allatal monooxygenase activities; inhibition of JH-dependent events
in vivo) is essential for the design of further compounds, and that
the information gained with few compounds in different tests is
superior to the information that would be obtained by running a
long list of compounds through a sophisticated test such as the
inhibition of JH biosynthesis in vitro. Indeed, the lack of in
vivo activity of Compound 2, if attributed to its JH agonist
activity, would rule out further testing of all similar compounds
suspected as JH agonists. Careful structure-activity relationships
analysis of analogs of 3 and 4 may explain why 3 is selective for
methyl farnesoate epoxidase and may guide the synthesis of more
powerful inhibitors. Our results suggest that suicide inhibitors
of cytochrome P-450 monooxygenases which are selective for the
allatal methyl farnesoate epoxidase can be rationally designed and
tested. If such compounds are also devoid of JH agonist activity,
they may become prototypes of new insect control agents.

Acknowledgments

This work was supported by NIH grants AI 19192 and RR 07079 (OSU)
and GM 25515 (UCSF). We are grateful to Dr. H. R. Waespe, Ciba-
Geigy Ltd. Basel, for the gift of CGA 167126. Oregon Agricultural
Experiment Station Technical Paper No. 7261.

Literature Cited

1. Geissbuhler, H.; Brenneisen, P.; Fischer, H. P. Science 1982,
   217, 505.
2. Hedin, P. A. J. Agr. Food Chem. 1982, 30, 201.
3. Staal, G. B. This volume.
4. Menn, J. J.; Henrick, C. A. Phil. Trans. R. Soc. Lond. B.
   1981, 295, 57.
5. Menn, J. J.; Henrick, C. A.; Staal, G. B. In "Regulation of
   Insect Development and Behavior"; Sehnal, F. et al. Eds;
   Technical University of Wroclaw, Wroclaw, 1981; pp. 735-748.

6.   Hammock, B. D.   Ibid.   pp. 841-852.
7.   Staal, G. B.; Henrick, C. A.; Bergot, B. J.; Cerf, D. C.;
     Edwards, J. P.; Kramer, S. J.   Ibid.   pp. 323-340.
8.   Kramer, S. J.; Staal, G. B.   In "Juvenile Hormone
     Biochemistry"; Pratt, G. E.; Brooks, G. T. Eds.; Elsevier,
     Amsterdam, 1981; pp. 425-437.
9.   Bowers, W. S.; Ohta, T.; Cleere, J. S.; Marsella, P. A.
     Science 1976, 193, 542.
10. Quistad, G. B.; Cerf, D. C.; Schooley, D. A.; Staal, G. B.
     Nature 1981, 289, 176.
11. Pratt, G. E.; Jennings, R. C.; Hamnett, A. F.; Brooks, G. T.
     Nature 1980, 284, 320.
12. Pratt, G. E.   In "Natural Products for Innovative Pest
     Management"; Whitehead, D. L.; Bowers, W. S. Eds.; Pergamon
     Press, Oxford 1983; pp. 323-355.
13. Hamnett, A. F.; Ottridge, A. P.; Pratt, G. E.; Jennings, R.
     C.; Stott, K. M.   Pestic. Sci. 1981, 12, 245.
14. Pratt, G. E.; Feyereisen, R.   1979. Unpublished results.
15. Feyereisen, R.; Pratt, G. E.; Hamnett, A. F.   Eur. J.
     Biochem. 1981, 118, 231.
16. Hsia, M. T. S.; Grossman, S. J.; Schrankel, K. R.   Chem.
     Biol. Interact. 1981, 37, 265.
17. Schrankel, K. R.; Grossman, S. J.; Hsia, M. T. S.   Toxicol.
     Let. 1982, 12, 95.
18. Hammock, B. D.; Mumby, S. M.   Pest. Biochem. Physiol. 1978,
     9, 39.
19. Pratt, G. E.; Finney, J. R.   In "Crop Protection Agents -
     Their Biological Evaluation"; McFarlane, N. R., Ed.   Academic
     Press, London 1977; pp. 113-132.
20. Kuwano, E.; Takeya, R.; Eto, M.   Agric. Biol. Chem. 1983, 47,
     921.
21. Wilkinson, C. F.   In "The Future for Insecticides: Needs and
     Prospects"; Metcalf, R. L.; McKelvey, J. J., Eds.   Wiley, New
     York, 1976; pp. 195-218.
22. Brooks, G. T.   Proc. Br. Crop Protect. Conf. Pests Dis. 1977,
     3, 731.
23. Ortiz de Montellano, P. R.; Kunze, K. L.   J. Biol. Chem.
     1980, 255, 5578.
24. Ortiz de Montellano, P. R.; Correia, M. A.   Ann. Rev.
     Pharmacol. Toxicol. 1983, 23, 481.
25. Feyereisen, R.; Langry, K. C.; Ortiz de Montellano, P. R.
     Insect Biochem. 1984, 14, 19.
26. Stay, B.; Coop, A.   J. Insect Physiol. 1973, 19, 147.
27. Prickett, K. S.; Ortiz de Montellano, P. R.   Unpublished
     results.
28. Pratt, G. E.; Tobe, S. S.   Life Sci. 1974, 14, 575.
29. Feyereisen, R.   Methods. Enzymol. vol. 111 (in press).
30. Feyereisen, R.; Johnson, G.; Koener, J.; Stay, B.; Tobe, S.
     S. J. Insect Physiol. 1981, 27, 855.
31. Barnes, J. R.; Fellig, J.   J. Econ. Entomol. 1969, 62, 86.
32. Tobe, S. S.; Stay, B.   Nature 1979, 281, 481.
33. Brooks, G. T.; Pratt, G. E.; Ottridge, A. P.; Mace, D. W.
     Proc. Br. Crop. Protect. Conf. Pests Dis. 1984, in press.

RECEIVED November 8, 1984

Detoxification Enzyme Relationships in Arthropods of Differing Feeding Strategies

CHRISTOPHER A. MULLIN

Pesticide Research Laboratory and Graduate Study Center, Department of Entomology, The Pennsylvania State University, University Park, PA 16802

Detoxification enzymes were compared in 36 arthropod species representing both chewing and sucking herbivores and their natural enemies. Enzymes studied include aldrin epoxidase (MFO), trans-epoxide hydrolase (trans-EH), cis-epoxide hydrolase (cis-EH), and 1-naphthyl acetate esterase. Major selectivities were found for MFO and EH. High MFO and trans-EH activities were consistently associated with herbivory, whereas entomophagous arthropods had a low trans-EH to cis-EH ratio. Phloem-sucking insects were different, exhibiting a low trans-EH to cis-EH ratio. Based on these distinct selectivities, EH may be an appropriate enzyme site for design of a broad-spectrum bioregulator of herbivorous pests that will have little impact on natural enemies.

Arthropods in different feeding niches tend to have contrasting susceptibilities to pesticides. Hence, lepidopteran larvicides are often chemically distinct from aphicides (1-3), and conventional pesticides with few exceptions exhibit greater lethalities for predators and parasites than the herbivorous pests they are targeted for (4). Knowledge of the defensive strategies arthropods use to selectively survive a toxicant exposure is necessary for successful design of chemical bioregulators that act to control pest populations, but have the appropriate safety for nontarget species. Although sequestration, penetration barriers and excretion are notable factors, metabolism and action at the target site are of greater importance in explaining the species variation in susceptibility to toxicants (3). Generally, enzymatic detoxification is the most direct and dependable way for an animal to survive a toxicant overexposure.

Metabolic transformation of lipophilic toxicants including pesticide and plant allelochemicals to excretable products usually proceeds by a series of enzymatic events to ultimately detoxify the chemical. Many of the initial reactions can generate intermediates

0097-6156/85/0276-0267$06.00/0
© 1985 American Chemical Society

that are more toxic than the parent xenobiotic. Included among
these are selective oxidations catalyzed by the cytochrome P-450
monooxygenases (mixed-function oxidase, MFO) exemplified by desulfuration at the phosphorus bond (P=S to P=O), thioether oxidations,
and epoxidation (5,6). For example, epoxidation of olefins and
arenes largely by MFO can produce reactive epoxides harmful to the
animal (7,8). The enzyme epoxide hydrolase (EH) catalyzes the
addition of water to the epoxide, thereby detoxifying it to a more
excretable 1,2-dihydroxy metabolite.

     Examples of an olefin to diol pathway include the metabolism
of aldrin, carbaryl, polycyclic aromatic hydrocarbons, and the
plant toxicants rotenone, pyrethrins, precocenes, and limonene to
the respective trans-diols (5-12). The high chemical reactivity of
many epoxides and the lack of an effective in vivo inhibitor of EH
often negates the isolation of epoxide intermediates from biological systems, and thus many remain putative. Nevertheless, some,
such as dieldrin, are refractory to hydration, and serve as useful
models in investigations of epoxide forming and degrading pathways
(9). Understanding the balance of activation and detoxification
enzymes such as epoxidase relative to EH available to an organism
will help define its adaptability to chemical stress.

     The well-known selectivities of some organophosphates may be
explained by the balance of enzymatic events. The reduced toxicity
of the insecticide malathion to mammals is largely the result of
rapid activation by desulfuration in the insect and the more rapid
detoxificaton by carboxylesterases and glutathione transferases in
the mammal (3). Design of new pest bioregulators should exploit
enhanced activation and decreased detoxification capabilities in
the targeted pests.

     Exploration of biochemical bases for pesticide selectivities
between chewing and sucking herbivores, and natural enemies has
lagged because of difficulty in rearing entomophages and the usually insufficient biomass available for enzyme assay. The typically
small natural enemy, especially parasitoids, precludes dissection
of specific organs where detoxification enzymes reside including
the midgut, fat body or malphighian tubules, but rather necessi-
tates use of whole body homogenates which may release factors that
impair enzyme measurements (13). Regardless, more sensitive and
rapid enzyme assays, and stabilizing additives including antioxi-
dants and inhibitors of proteinases and phenoloxidases now allow
the satisfactory in vitro study of detoxification enzymes within
whole body preparations of microarthropods (13-15). These techniques should aid in understanding the biochemical events responsible for chemical selectivities.

Enzyme Associations with Herbivore Status

Selectivities to synthetic pesticides may be explained, in part, by
preadaptations to toxic dietary chemicals. Leaf chewing pests,
phloem-sucking pests, and entomophagous natural enemies should have
very different exposures to dietary toxicants. Plant defensive
chemicals are thought to be allocated mostly to specialized organ-
elles or tissues of external structures, and only at low loadings

in vascular tissues (16,17). However, phloem loading and translocation of chemicals within plants is poorly understood (18,19).
Nevertheless, chewing herbivores such as lepidopteran larvae and
coleopterans expectantly consume higher loadings of plant toxicants
than phloem-sucking counterparts such as aphids. Thus, metabolic
adaptations to toxic chemicals should be better developed in chewing relative to sucking herbivores. This is indicated by the generally higher susceptibility of sucking herbivores to conventional
pesticides than chewing herbivores (1-3). Arthropod parasitoids
and predators, however, are usually exposed to plant toxicants via
their passive accumulation in nonessential tissues of the herbivorous host or prey (20). It may be expected that carnivores, because
of lowered encounter, would lack well-developed detoxification
fitness for plant allelochemicals. Comparison of the toxicological
bases that allow pestiferous (i.e., herbivory) and beneficial
(i.e., carnivory) activities to concur will assist our understanding of how to manage a realistic complex of crop arthropods.

   Herbivorous insects must contend with toxic phytochemicals,
many of which are epoxides or their olefinic precursors (21-23).
These phytochemicals often exhibit trans-geometry, or are higher
substituted epoxides and olefins, whereas animals preferably bio-
synthesize cis-olefins (Table I). Epoxidation of olefins, either
within the plant or the consuming insect, would produce reactive
epoxides that may undergo detoxification by an appropriate epoxide-
metabolizing enzyme. Use of a suitable model substrate for plant-
derived epoxides would expedite biochemical associations between
plants and animals. Trans-β-ethylstyrene oxide is an excellent
substrate for several EHs, and mimics the epoxides known to or
potentially derived from phenylpropenoids (24) and -butenoids (27)
of wide occurrence in the plant kingdom (Figure 1).

   Numerous investigations have demonstrated the association of
an insect MFO epoxidase with increased encounter with plant allelochemicals (11,28). This cytochrome P-450 dependent activation
reaction is obviously enhanced in many herbivorous pests. Hence,
it is of interest to explore the role of EH detoxification in
arthropod herbivory.

Table I. Propensity for Trans- and Cis-Olefin Biosynthesis in
Animals and Plants[a]

Chemical group	Animal	Plant
Trans-olefins	Infrequent	Common
Examples	Pheromones	Fatty acids
	Prostaglandins	Cinnamic acids
		Chalcones, stilbenes
		Carotenoids
		Phenylpropenoids
	Fumaric acid and sphingosine in both	
Cis-olefins	Predominant	Common
Examples	Fatty acids in both	

[a] References (24-26)

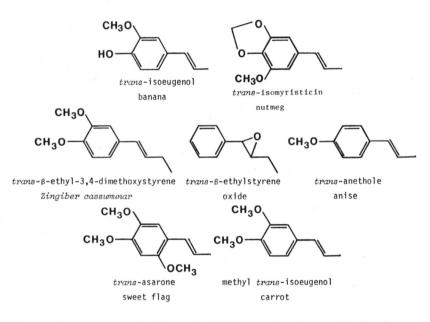

Figure 1. Analogy of <u>trans</u>-β-ethylstyrene oxide with common plant phenylpropenoids and –butenoids.

Enzyme profiles were compared in a few leaf chewing and piercing-sucking herbivores, as well as carnivorous arthropods (Table II).   Aldrin epoxidase, trans-β-ethylstyrene oxide hydrolase, cis-stilbene oxide hydrolase, and 1-naphthyl acetate esterase were measured using established methods (14,29).   Elevated epoxidase was associated with herbivores, however the most distinct group difference was with trans-EH.   Chewing herbivores consistently had higher trans-EH than carnivores (p<0.05); in addition the ratio of trans-EH to cis-EH was higher in chewing herbivores than either the sucking herbivores or carnivores.   Cis-EH was less variable between trophic groups presumably since the substrate is selective for EHs acting on animal-derived epoxides, and thus would represent detoxification events common to all animals.   Cis-EH, because of the low variability, serves as a useful base for EH ratios which appear more reliable than absolute enzyme levels in comparing species of different life stages, ages, and from which a differing tissue source was utilized.   The similar levels of general esterase between arthropods of alternative feeding strategies may be explained in part by incorporation of digestive enzymes common to all arthropods into the esterase measurement (14,29).

Phloem-sucking aphids, by contrast, had lower epoxidase, trans-EH and a trans-EH to cis-EH ratio than leaf chewing insects that often cohabit the same host plants (Table II).   Comparison of highly polyphagous aphids with the oleander aphid, a specialist on hosts from Asclepiadaceae and Apocynaceae, indicates that higher epoxidase and trans-EH are nevertheless associated with increasing encounter with phytochemicals.   These results support the view that phloem-feeding insects retain minimal enzyme capabilities to deal with plant allelochemicals because of their low loading in phloem relative to external plant tissues.

Contrary to phloem-feeding aphids, the large milkweed bug ingests large concentrations of toxic cardenolides from its food (20), and the elevated levels of MFO and EH in this piercing-sucking arthropod (Table II) may reflect the allelochemical richness of milkweed seeds.   Although the absolute enzyme levels of the milkweed bug are similar to the chewing herbivores, the low EH ratio is indicative of the piercing-sucking group, and emphasizes the potential of this comparative index in biochemical ecology.   Moreover, enzyme ratios will promote comparisons between the whole body burden of activity and that of an enzyme-rich tissue such as the midgut since tissue-to-tissue differences in activity profiles will be less than that of absolute enzyme levels (3).   Pooled whole body values from many individuals of differing generations and life stages should be a superior population assessment than an optimal in vitro activity that neglects the total population dynamics for interacting with natural toxicants.   Hence, comparisons in Table II were based on a similar number of whole body and midgut preparations for each feeding group.

The association of trans-EH with herbivory was explored further by comparing epoxide hydrolase in 36 species of macro- and microarthropods.   The trans- and the cis-EH activities for each species were plotted (Figure 2) relative to suitable isolines of ratios of activities.   Immediately apparent is the distinct clus-

TABLE II. Lipophile Mobilizing Enzymes in Arthropods of Differing Feeding Strategies[a]

Group / Species	Tissue	Enzyme Activity (pmol/min – mg protein)					Host Range
		Aldrin Epoxidase	Epoxide Hydrolase trans	cis	$\dfrac{trans}{cis}$	Esterase (X 10^{-3})	
Chewing Herbivores							
Two-spotted spider mite[b]	WB	1.44	1710	117	14.6	389	G
Orange tortrix[b]	WB,MG	26.8	1340	536	2.50	593	G
Mexican bean beetle	MG	4.16	780	352	2.22	94	S
Piercing-sucking herbivores							
Green peach aphid	WB	1.36	177	817	0.22	433	G
Potato aphid	WB	2.01	220	1032	0.21	171	G
Oleander aphid	WB	0.09	38	66	0.57	226	S
Large milkweed bug	MG	31.2	1090	909	1.20	118	S
Carnivores							
Amblyseius fallacis[b]	WB	0.27	310	431	0.72	318	G
Oncophanes americanus[b]	WB	0.85	407	727	0.56	307	G
Pediobius foveolatus	WB	0.67	198	415	0.48	83	S
Convergent lady beetle	MG	--	767	1390	0.55	231	G

[a] Activities in adult stages unless indicated otherwise. Abbreviations: WB = whole body; MG = midgut; G = generalist; S = specialist.

[b] Composite values for preparations from both adult and immature stages as cited (14,29).

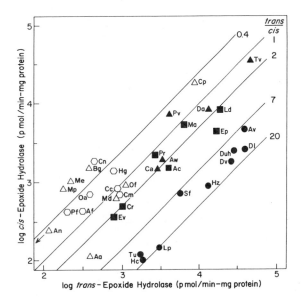

Figure 2. Relationship of <u>trans</u>- and <u>cis</u>-epoxide hydrolase activities
with arthropod feeding specializations. Actively feeding adults or
last instar larvae (Aa, Ac, Cr, Da, Hc, Hz, Lp, Ma, Md, Pr, Sf only)
were surveyed. Midgut preparations of the macroarthropods or whole
body preparations of the microarthropods (Aa, Af, An, Me, Mp, Oa, Pf,
Tu) were assayed. Epoxide hydrolase activities for <u>trans</u>-β-ethyl-
styrene oxide and <u>cis</u>-stilbene oxide are plotted relative to isolines
of trans/cis ratios with ● for generalist, ■ for oligophagous, and
▲ for specialist chewing herbivores, △ for saprophagous and sucking
insects, and ○ for entomophagous arthropods. Aa = <u>Aedes</u> <u>aegypti</u>
(Linn.), Ac = <u>Argyrotaenia</u> <u>citrana</u> (Fernald), Af = <u>Amblyseius</u> <u>fallacis</u>
(Garman), An = <u>Aphis</u> <u>nerii</u> Fonscholombe, Av = <u>Acalymma</u> <u>vittata</u>
(Fabricius), Aw = <u>Altica</u> <u>woodsi</u> (Isely), Bg = <u>Blattella</u> <u>germanica</u>
(Linn.), Ca = <u>Crioceris</u> <u>asparagi</u> (Linn.), Cc = <u>Chrysopa</u> <u>carnea</u>
Stephens, Cm = <u>Coleomegilla</u> <u>maculata</u> (De Geer), Cn = <u>Coccinella</u>
<u>novemnotata</u> (Herbst), Cp = <u>Chauliognathus</u> <u>pennsylvanicus</u> (De Geer),
Cr = <u>Choristoneura</u> <u>rosaceana</u> (Harris), Da = <u>Delia</u> <u>antiqua</u> (Meigen),
Dl = <u>Diabrotica</u> <u>longicornis</u> (Say), Duh = <u>D.</u> <u>undecimpunctata</u> <u>howardi</u>
(Barber), Dv = <u>D.</u> <u>virgifera</u> Le Conte, Ep = <u>Epicauta</u> <u>pennsylvanica</u>
(De Geer), Ev = <u>Epilachna</u> <u>varivestis</u> Mulsant, Hc = <u>Hyphantria</u> <u>cunea</u>
(Drury), Hg = <u>Hippodamia</u> <u>convergens</u> Guerin-Meneville, Hz = <u>Heliothis</u>
<u>zea</u> (Boddie), Ld = <u>Leptinotarsa</u> <u>decemlineata</u> (Say), Lp = <u>Lymantria</u>
<u>dispar</u> (Linnaeus), Ma = <u>Malacosoma</u> <u>americanum</u> (Fabricius), Md =
<u>Musca</u> <u>domestica</u> (Linn.), Me = <u>Macrosiphum</u> <u>euphorbiae</u> (Thomas), Mp =
<u>Myzus</u> <u>persicae</u> (Sulzer), Oa = <u>Oncophanes</u> <u>americanus</u> (Weed), Of =
<u>Oncopeltus</u> <u>fasciatus</u> (Dallas), Pf = <u>Pediobius</u> <u>foveolatus</u> (Crawford),
Pr = <u>Pieris</u> <u>rapae</u> (Linn.), Pv = <u>Plagiodera</u> <u>versicolora</u> (Laicharting),
Sf = <u>Spodoptera</u> <u>frugiperda</u> (J. E. Smith), Tu = <u>Tetranychus</u> <u>urticae</u>
Koch, and Tv = <u>Trirhabda</u> <u>virgata</u> (LeConte). Adapted with permission
from Ref 21. Copyright 1984, <u>Experientia</u>.

tering of chewing herbivores in the region of both high trans-EH and high trans/cis ratios, whereas entomophagous, sucking, and more specialized feeders group at low trans-EH and at low ratios of activities. Indeed, the 20 chewing herbivorous pests of economic importance surveyed here had on the average a 30-fold higher trans-EH and a 13-fold higher EH ratio than the four sucking herbivores (Table III), whereas the latter group and the beneficial arthropods had similar EH profiles.

The spectrum of EH activities and ratios were much less variable in entomphagous relative to phytophagous arthropods (Table II, Figure 2), probably since dietary and endogenously formed epoxides would be more equivalent in the food of the latter. To further explore the situation of the herbivore, gut EH levels from adult leaf feeding beetles of the Chrysomelidae were examined relative to encounter with plant allelochemicals as estimated by host plant range (Figure 3). Ratios of trans/cis EH correlated well (r>0.92) with either number of plant families or plant genera consumed. This strongly implicates trans-EH in the detoxification of plant-derived epoxides.

Synergists of Plant Toxicants as Selective Pest Bioregulators

The clear association of a cytochrome P-450 epoxidase and trans-EH with herbivory may have important consequences for integrated pest management (IPM), since these enzymes will determine, in part, the ability of phytophagous arthropods to be pests. Epoxidase action will activate many phytochemical defenses and thus be disadvantageous to the herbivore unless a concurrent detoxification event such as EH is available. Hence, the well-developed olefin to diol pathway of chewing crop pests may be countered by selective inhibition of trans-EH. Potent in vitro inhibitors of this enzyme are known for mammals although they appear less efficacious for arthropods (15,30). The appropriate structural modification of chalcone or flavonoid derivatives may lead to selective herbivore bioregulators that have little impact on entomophagous arthropods important in IPM. Perhaps the similar enzymologies of phloem-feeding aphids and carnivores (Tables II, III) may limit use of identical chemical strategies for aphid control unless ecological selectivity via systemic applications is implemented.

Inhibitors of a herbivore's ability to detoxify plant toxicants may expectedly be more slow-acting than neurotoxicants such as organophosphates. Nevertheless, securance of synergists of dietary toxicants that complement activation reactions and inhibit detoxification reactions in the targeted pests will have potential for IPM, since they should be compatible or moreover, augment plant antibiosis, natural enemies and concurrently used pesticides.

Relevance of Detoxification Dissimilarities to In Vivo Toxicosis

Recently much work has been devoted in understanding the anti-juvenile hormone action of plant chromenes from Ageratum spp. (10, 31,32). The allatocidal activity of precocenes is apparently due to a balance of decreased detoxification in peripheral tissues and

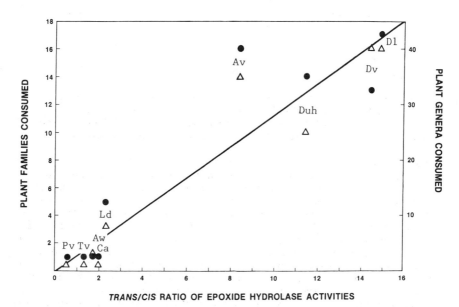

Figure 3. Association of trans-epoxide hydrolase with arthropod herbivory. Trans/cis ratios of epoxide hydrolase in midguts of adult chrysomelid beetles are plotted relative to number of plant families (●) or genera (△) found within the host range of each species. See Figure 2 for details.

TABLE III.   Epoxide Hydrolase in Pest Relative to Beneficial
Arthropods of Economic Importance[a]

Economic Group	Trans	Cis	Trans/cis
Chewing Herbivores n = 20	20.7^a	3.0^a	10.2^a
Sucking Herbivores n = 4	0.7^b	0.8^b	0.8^b
Beneficials n = 7	1^b	1^b	1^b

[a] Measured in gut preparations except for acarines, aphids and parasitoids, where whole body preparations used. Groups with same letter are not significantly different at $p < 0.01$.

enhanced oxidative bioactivation to a cytotoxic epoxide in the corpora allata (CA). It is interesting that most of the insects sensitive to precocene, and thus expected to exhibit enhanced epoxidase in the CA but decreased peripheral detoxification to diols, glucosides, etc., are sucking species and not holometabolous chewing herbivores (10,32). Inhibitors of appropriate detoxification enzymes should synergize precocene action.

A CA epoxidase perhaps identical to the precocene epoxidase biosynthesizes insect juvenile hormones (JH) from the analogous inactive olefinic precursor, and the enzyme activity appears higher in precocene-sensitive species (32). Subsequent detoxification of JH occurs primarily by EHs and esterases in peripheral tissues, and preliminary information does not indicate major differences for JH degradation routes between chewing and sucking herbivores, or insect carnivores (33,34). More study of the role of detoxification in regulating the action of JH in target tissues is required.

Esterases for trans-pyrethroids are well-developed in piercing-sucking herbivores such as the milkweed bug (35) and green-peach aphid (36), and may explain, in part, the general higher tolerance of these organisms over chewing herbivores for pyrethroids. Carnivorous lacewing larvae, moreover, are less suscepible than herbivores to cis-isomers of pyrethroids such as deltamethrin and permethrin apparently since they retain esterases that detoxify the cis-isomers faster than the less insecticidal trans-isomers (37).

Other enzyme-feeding guild associations may be important to insect control. β-glucosidases are more active in herbivores, particularly the piercing-sucking types, than carnivores, and are probably responsible for the selective activation of glucosidic juvenogens in some hemipteran insects (38,39). In this regard, it is interesting that glucoside formation for precocene metabolites appears to be retarded in sucking relative to chewing herbivores (40), and suggests the involvement of β-glucosidases. In contrast, glutathione transferases for arylhalides appear to be higher in chewing relative to sucking herbivores (41).

In summary, results of investigations on comparative detoxification in arthropods may help explain the frequent increases in sucking pests when control measures for the major chewing pests are successful, and should aid the more effective use of selective chemistry in the management of crop pests. Indeed, even toxicant sensitivity at nerve target sites may be partly regulated by detoxification such as is indicated for dieldrin (42) and pyrethroids (43). Certainly, inhibition of detoxification activities essential to an organism is a viable strategy for biorational control (44,45); indeed the highly effective action of organophosphates and carbamates in animals is based on impediment of acetylcholine detoxification at the nerve target.

Acknowledgments

The technical aid of E. M. Reeves and R. J. Steighner, and the support of the National Science Foundation (grant BSR-8306008) are much appreciated. This is paper No. 7054 of The Pennsylvania Agric. Exp. Stn. journal series.

Literature Cited

1. Saito, T. Residue Rev. 1969, 25, 175-186
2. Weiden, M. H. J. Bull. Wld. Hlth. Org. 1971, 44, 203-213.
3. Hollingworth, R. M. In "Insecticide Biochemistry and Physiology"; Wilkinson, C. F., Ed.; Plenum: New York, 1976; pp. 431-506.
4. Croft, B. A.; Brown, A. W. A. Ann. Rev. Entomol. 1975, 20, 285-335.
5. Kulkarni, A. P.; Hodgson, E. Ann. Rev. Pharmacol. Toxicol. 1984, 24, 19-42.
6. Casida, J. E. In "Pesticide Chemistry: Human Welfare and the Environment, Mode of Action, Metabolism and Toxicology"; Matsunaka, S.; Hutson, D. H.; Murphy, S. D., Eds.; Pergamon: New York, 1983; Vol. III, pp. 239-246.
7. Conney, A. H. Cancer Res. 1982, 42, 4875-4917.
8. Manson, M. M. Br. J. Ind. Med. 1980, 37, 317-336.
9. Brooks, G. T. Gen. Pharmac. 1977, 8, 221-226.
10. Bowers, W. S. Ent. Exp. Appl. 1982, 31, 3-14.
11. Brattsten, L. B. In "Plant Resistance to Insects"; Hedin, P. A., Ed.; ACS SYMPOSIUM SERIES No. 208, American Chemical Society: Washington, D. C., 1983; pp. 173-195.
12. Dowd, P. F.; Smith, C. M.; Sparks, T. C. Insect Biochem. 1983, 13, 453-468.
13. Wilkinson, C. F. In "Xenobiotic Metabolism: In Vitro Methods"; Paulson, G. D.; Frear, D. S.; Marks, E. P., Eds.; ACS SYMPOSIUM SERIES No. 97, American Chemical Society: Washington, D.C., 1979; pp. 249-284.
14. Croft, B. A.; Mullin, C. A. Environ. Entomol. 1984, 13, in press.
15. Mullin, C. A.; Matsumura, F.; Croft, B. A. Comp. Biochem. Physiol. 1984, 79C, 85-92.

16. McKey, D. In "Herbivores - Their Interaction with Secondary Plant Metabolites"; Rosenthal, G. A.; Janzen, D. H., Eds.; Academic: New York, 1979; pp. 55-133.
17. Waller, G. R.; Nowacki, E. K. "Alkaloid Biology and Metabolism in Plants"; Plenum: New York, 1978; pp. 121-141.
18. Giaquinta, R. T., this volume.
19. Chamberlain, K.; Burrell, M. M.; Butcher, D. N.; White, J. C. Pestic. Sci. 1984, 15, 1-8.
20. Duffey, S. S. Ann. Rev. Entomol. 1980, 25, 447-477.
21. Mullin, C. A.; Croft, B. A. Experientia 1984, 40, 176-178.
22. Cross, A. D. Quart. Rev., Chem. Soc. Lond. 1960, 14, 317-335.
23. Kolattukudy, P. E. Science 1980, 208, 990-1000.
24. Manitto, P. "Biosynthesis of Natural Products"; Ellis Horwood Ltd.: Chichester, England, 1981; pp. 287-429.
25. Bell, E. A.; Charlwood, B. V. "Secondary Plant Products"; Springer-Verlag: New York, 1980.
26. Harwood, J. L. In "The Biochemistry of Plants"; Stumpf, P. K., Ed.; Academic: New York, 1980. Vol. IV, pp. 1-55.
27. Tuntiwachwuttikul, P.; Pancharoen, O.; Jaipetch, T.; Reutrakul, V. Phytochemistry 1981, 20, 1164-1165.
28. Yu, S. J. Pestic. Biochem. Physiol. 1983, 19, 330-336.
29. Mullin, C. A.; Croft, B. A.; Strickler, K.; Matsumura, F.; Miller, J. R. Science 1982, 217, 1270-1272.
30. Mullin, C. A.; Hammock, B. D. Arch. Biochem. Biophys. 1982, 216, 423-439.
31. Bowers, W. S., this volume.
32. Pratt, G. E. In "Natural Products for Innovative Pest Management"; Whitehead, D. L.; Bowers, W. S., Eds.; Pergamon: New York, 1983; pp. 323-355.
33. Hammock, B. D.; Quistad, G. B. In "Progress in Pesticide Biochemistry"; Hutson, D. H.; Roberts, T. R., Eds.; Wiley: New York, 1981; Vol. I, pp. 1-83.
34. Ajami, A. M.; Riddiford, L. M. J. Insect Physiol. 1973, 19, 635-645.
35. Jao, L. T.; Casida, J. E. Pestic. Biochem. Physiol. 1974, 4, 465-472.
36. Devonshire, A. L.; Moores, G. D. Pestic. Biochem. Physiol. 1982, 18, 235-246.
37. Casida, J. E.; Gammon, D. W.; Glickman, A. H.; Lawrence, L. J. Ann. Rev. Pharmacol. Toxicol. 1983, 23, 413-438.
38. Robinson, D. Biochem. J. 1956, 63, 39-44.
39. Slama, K.; Wimmer, Z.; Romanuk, M. Hoppe-Seyler's Z. Physiol. Chem. 1978, 359, 1407-1412.
40. Bergot, B. J.; Judy, K. J.; Schooley, D. A.; Tsai, L. W. Pestic. Biochem. Physiol. 1980, 13, 95-104.
41. Cohen, A. J.; Smith, J. N.; Turbert, H. Biochem. J. 1964, 90, 457-464.
42. Brooks, G. T. In "Insect Neurobiology and Pesticide Action"; Society of Chemical Industry: London, 1980; pp. 41-55.
43. Soderlund, D. M.; Hessney, C. W.; Helmuth, D. W. Pestic. Biochem. Physiol. 1983, 20, 161-168.
44. Hedin, P. A. J. Agric. Food Chem. 1982, 30, 201-215.
45. Menn, J. J.; Henrick, C. A. Phil. Trans. R. Soc. Lond. 1981, 295B, 57-71.

RECEIVED November 8, 1984

δ-Endotoxin of *Bacillus thuringiensis* var. *israelensis*
Broad-Spectrum Toxicity and Neural Response Elicited in Mice and Insects

R. MICHAEL ROE[1], PETER Y. K. CHEUNG[2], BRUCE D. HAMMOCK[2], DAN BUSTER[2], and A. RANDALL ALFORD[3]

[1]Department of Entomology, North Carolina State University, Raleigh, NC 27695-7613
[2]Department of Entomology, University of California, Davis, CA 95616
[3]Department of Entomology, University of Maine, Orono, ME 04469

The alkaline-dissolved Bacillus thuringiensis israelensis (BTI) δ-endotoxin when introduced by injection was biologically active against a wide spectrum of host animals including insects from four orders and mice. The LD$_{50}$ for dissolved BTI δ-endotoxin in mice was 1.31 PPM and in Trichoplusia ni (Lepidoptera: Noctuidae) 3.71 PPM. Neuromuscular effects like heart cessation, lost coordination, tremor, and paralysis were observed in test animals. Using the appearance of lactate dehydrogenase in insect hemolymph post-injection as a cytosolic marker, we found that dissolved BTI δ-endotoxin was cytotoxic. In vivo recordings of activity in the ventral nerve cord post-injection indicated that dissolved BTI δ-endotoxin at the T. ni LD$_{50}$ elicited hyperexcitability and then nerve death as was also the case for the organophosphate, methamidophos. The cytotoxin phospholipase-A$_2$ when injected at its LD$_{50}$ elicited no neural response. BTI poisoning was also temperature dependent while BTI cytotoxicity was not. Proteins at 24, 27, 35, 49 and 68K daltons were resolved from the dissolved BTI δ-endotoxin. These were introduced in various combinations by injection and ingestion into mice and insects and compared to the alkaline-dissolved Bacillus thuringiensis kurstaki δ-endotoxin.

Within the sporangium of the bacterium Bacillus thuringiensis (BT) is synthesized a parasporal, proteinaceous crystal (1-2) that has found widespread use as a biological control agent (3). This crystal is commonly referred to as the "δ-endotoxin" as suggested by Heimpel (4). The taxonomy of BT is based on the serology of the flagellar H antigen (5), and 29 subspecies and 26 serotypes have been identified (6). The δ-endotoxins from the

0097-6156/85/0276-0279$06.00/0

majority of the serotypes are toxic when ingested by more than
182 species of insects, particularly in the economically
important order, Lepidoptera (3). The majority of the research
has centered on Bacillus thuringiensis subspecies kurstaki (BTK)
because of its larvicidal activity against major agricultural
pests in the order Lepidoptera. The serotype H14, Bacillus
thuringiensis subspecies israelensis (BTI), differs from the
other serotypes, however, by being highly toxic to members of the
insect order Diptera (7-8) and yet has little known toxicity to
Lepidoptera. The prospect of employing BTI to control
mosquitoes, blackflies, or other medically important insect pests
has stimulated great interest in elucidating the molecular basis
for its mode of action (3). All of the BT δ-endotoxins are also
of special interest because they appear to be highly selective
against insects and seem to pose no health risks to humans or
livestock.

The δ-endotoxin of BTK upon ingestion by larval Lepidoptera
is quickly activated by high gut pH and gut proteolytic activity
(9-11); gut epithelial cells swell, vacuoles form, and then the
cells separate from the basement membrane and each other
ultimately disrupting the gut-hemocoel barrier (12-15). Similar
observations in mosquito larvae fed BTI (16) led to the general
acceptance that the BT insecticidal activity was directed
against, if not restricted to the gut epithelium of the host
(12-13). An initial symptom of BTK poisoning is gut paralysis
(17) and it was hypothesized that an increase in the hemolymph pH
from the leakage of alkaline gut contents (17) or the influx of
K$^+$ into the hemocoel caused this paralysis (18-19). These
studies led to the discovery that BTK digests applied to the
ventral nerve cord of the cockroach, Periplaneta americana
(Orthoptera: Blattidae) caused excitation and then nerve blockage
(20-21) which appeared to be presynaptic in origin (20). Other
studies have shown that the alkaline-dissolved BTI δ-endotoxin is
cytotoxic to a number of different cell lines from insects and
mammals (3,22-24) and has a high affinity for specific
phospholipids in the plasma membrane (25). Thus, the objective
of this study is to assess the toxicity of alkaline-dissolved BTI
introduced into insects and mice by feeding and injection and to
assess the role of cytotoxicity and neurotoxicity in mortality
when dissolved δ-endotoxin is injected into insects.

SDS-PAGE Analysis of BTK and BTI δ-Endotoxin

BTK and BTI strain IFC-1 were provided by Biochem Products - US
Division (Salsbury Labs., Inc.). BTI was also isolated from a
commercial preparation provided by Sandoz Inc., cultured on GYS
medium (26). BTK and BTI toxin was prepared in an analogous
manner. Spores and crystals were separated from cell debris
by repeated washing with water and centrifugation. BTI crystals
were subsequently separated from spores by Renografin density
gradient centrifugation (27) and BTK crystals by discontinuous
sucrose gradient centrifugation (3). Crystals were then
dissolved by incubation for 3 h in 0.5% Na$_2$CO$_3$ (pH 11.0) and
dialyzed into 0.025 M sodium phosphate (pH 8.0) for storage at

-60°C. BTI (Sandoz) was further purified by DEAE-Cellulose
(DE-52, Whatman, 30 ml) in 0.025 M Tris-HCL (pH 8.00), eluted
with a 10 h, 100 ml, 0.0-0.5 M NaCl linear gradient; by acid
precipitation where the DEAE elutant was dialyzed into 0.05 M
sodium acetate (pH 4.5) and the percipitate removed by
centrifugation; and by Sephadex G-75 super fine (Pharmacia) gel
permeation chromatography (95 x 1.3 cm i.d. column) in 0.025 M
sodium phosphate (pH 8.0). Alkaline-dissolved and partially
purified δ-endotoxin was analyzed by 12.5% SDS-PAGE (28), stained
with Coomassie brilliant blue (Figure 1).

Incubation of BTK and BTI crystals in 0.5% Na_2CO_3 (pH 11.0)
solubilized a number of protein components (Figure 1). For BTK,
there were a number of proteins at 64K daltons and higher (Track
1). The standard procedure of washing BTK crystals with 1 M NaCl
before solubilization removes endogenous proteinases and results
in an enriched 130K dalton protein as the predominant component.
This procedure had little effect on the immunoreactivity of the
solubilized crystal or its toxicity in our studies. For BTI
(Sandoz, Track 2) there were proteins at 24, 27, 35, 49 and 68K
daltons. Of the lower molecular weight Sandoz BTI components,
the 27K proteins were the predominent component in the Salsbury
BTI (Track 8). Differences in the protein profile of the
δ-endotoxin from different BT varieties have been reported
previously (11,30-31). All alkaline-solubilized δ-endotoxin of
BTI (Sandoz) adsorbed to DE-52 and eluted in one peak (Track 3)
with an apparent concentration of the 68K component. The acid
precipitate (Track 4) was enriched with the 35K component which
was re-solubilized only at high pH. Because of its limited
solubility, the acid precipitate could not be bioassayed in later
studies. The soluble fraction was enriched with the 24K and 27K•
components (Track 5). Gel permeation chromatography enriched the
27 and 24K proteins (Tracks 6 and 7, respectively).

δ-Endotoxin Toxicity in Mice and Insects

BTI and BTK alkaline-dissolved and partially purified
δ-endotoxins were injected and/or fed to insects of 6 orders and
to mice (Tables I and II). The δ-endotoxin was injected in 0.15
M NaCl, 0.05 M Na_2HPO_4, and 0.02 M KH_2PO_4 at pH 7.2 into the
insect hemocoel or intraperitoneally into mice. In feeding
experiments, the toxin was dissolved in 5% sucrose and force-fed
in 2 μl volumes, to Trichoplusia ni and Heliothis zea
(Lepidoptera: Noctuidae). Aedes aegypti (Diptera: Culicidae)
larvae fed for a standard incubation period in water containing
BT toxin preparations; adults were given rectal injections (32).
Mice were also given BT by gavage. Data collected were subjected
to Probit analysis (33).

The alkaline-dissolved δ-endotoxin of BTI (Sandoz, Track 2)
was toxic by injection to all animals tested except Tenebrio
molitor (Coleoptera: Tenebrionidae) (Table I). Mice,
Trichoplusia ni, and Periplaneta americana were the most
sensitive, the LD_{50} being 1.3, 3.7 and 4.4 PPM, respectively.
The susceptibility to BTI poisoning also varied significantly
within a single insect family (the Noctuidae) with the LD_{50}

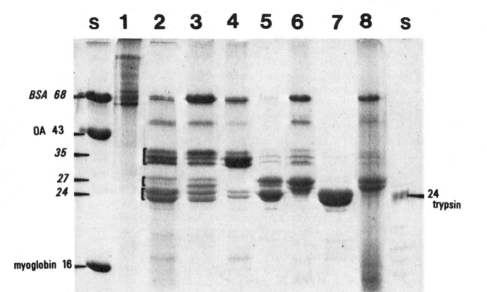

Figure 1. SDS-PAGE analysis of alkaline-dissolved Bacillus
thuringiensis subspecies kurstaki (BTK) and israelensis (BTI)
δ-endotoxin at 25 μg per track: (1) BTK δ-endotoxin from
Biochem Products - US Division (Salsbury Labs., Inc.), (2) BTI
δ-endotoxin from Sandoz Inc., (3) BTI (Sandoz) δ-endotoxin
purified by DEAE-anion exchange chromatography, (4) percipitate
formed after dialysis of BTI (Sandoz) δ-endotoxin into pH 4.5
sodium acetate buffer, (5) soluble fraction after dialysis of
BTI (Sandoz) δ-endotoxin into pH 4.5 sodium acetate buffer, (6)
BTI (Sandoz) δ-endotoxin purified by Sephadex G-75 gel
filtration chromatography at R_f 1.35, (7) at R_f 1.58, and (8)
BTI strain IFC-1 δ-endotoxin from Biochem Products - US
Division (Salsbury Labs., Inc.). S, molecular weights as
indicated X1000 for bovine serum albumin (BSA), ovalbumin (OA),
trypsin, and myoglobin. Reproduced with permission from Ref.
29. Copyright 1984, Academic Press, Inc.

Table I. Injected toxicity of alkaline-solubilized BTI
(Sandoz, Track 2) and BTK (Salsbury, Track 1) δ-endotoxin.

Animal - A or L*	24h BTI LD$_{50}$+	24h BTK LD$_{50}$+
Aedes aegypti - A	11.6 + 2.2	>900
(Diptera: Culicidae)	(3.16)	
Musca domesticus - A	10.9 + 2.2	>150
(Diptera: Muscidae)	(2.17)	
Trichoplusia ni - L	3.71 + 0.32	>130
(Lepidoptera: Noctuidae)	(3.22)	
Heliothis zea - L	73.6 + 3.0	>100
(Lepidoptera: Noctuidae)	(19.23)	
Tenebrio molitor - L	>100	>100
(Coleoptera: Tenebrionidae)		
Oncopeltus fasciatus - A	27.7 + 7.0	>300
(Hemiptera: Lygaeidae)	(1.96)	
Periplaneta americana - A	4.42 + 0.36	>20
(Orthoptera: Blattidae)	(7.04)	
Swiss-Webster Mice	1.31 + 0.23	>30
	(4.47)	

*Adult or Larva.

+PPM or mg/kg body weight + 1 S.D. with slope of probit
analysis in parenthesis.

ranging from 3.7 PPM for T. ni to 73.6 PPM for Heliothis zea.
The basis for this difference is unknown but these species
differences could be useful in the elucidation of the mechanism
for toxicity. BTI toxicity by injection also was not peculiar to
the Sandoz strain but was also noted for the IFC-1 strain from
Salsbury (Table III). By contrast the alkaline-dissolved
δ-endotoxin of BTK (Track 1) showed no toxicity when injected
into the same species (Table I). Obvious fundamental differences
exist between the dissolved δ-endotoxin of BTI and BTK.

When the purified, parasporal crystal of BTI (Sandoz) was
fed to A. aegypti larvae, the LC$_{50}$ was 2.95 + 0.59 ng/ml.
Alkaline-solubilization decreased the toxicity significantly to
an LC$_{50}$ of 2.29 + 0.06 µg/ml and rectal injections in adults
produced a LD$_{50}$ of 54.5 + 3.1 PPM (Table II). These findings
were consistent with previous work (34). Dissolved BTI
δ-endotoxin when fed to Lepidoptera and mice as expected (3) was
not toxic (Table II). The BTI IFC-1 strain was also similar to
the Sandoz strain in that both were toxic when fed to A. aegypti
(Table III). BTK dissolved δ-endotoxin when given orally was
toxic only to the lepidopteran, T. ni (Table II). The results
from all of these feeding experiments were consistent with
previous reports that BTK when fed to lepidopterans is active
while BTI is toxic to only certain dipterans (3) and is
supportive evidence that the BTI and BTK preparations used in our
studies were similar to preparations previously used by other
investigators. Furthermore, cross-contamination between BTI and

BTK as determined by ELISA was less than 0.01%, the detectable
limit of the assay (35).

The combined 24, 27, and 35K components of BTI (Sandoz)
δ-endotoxin (Figure 1, Tracks 2 and 3) had an equivalent toxicity

Table II. Oral toxicity of alkaline-solubilized
BTI (Sandoz, Track 2) and BTK (Salsbury, Track 1)
δ-endotoxin.

Animal	24h BTI LD$_{50}$+	24h BTK LD$_{50}$+
A. aegypti		
Larva	2.29 + 0.06* (12.41)	>40*
Adult (ENEMA)	54.5 + 3.1 (6.37)	>900
T. ni Larva	>50	2.30 + 0.27 (2.85)
H. zea Larva	>35	>35
Swiss-Webster Mice	>30	>30

+PPM or mg/kg body weight + 1 S.D. with slope of
probit analysis in parenthesis.

*μg/ml of water in which larvae were incubated.

Table III. Toxicity of partially purified alkaline-
solubilized BTI δ-endotoxin fed and injected into A.
aegypti and T. ni, respectively.

Toxin	A. aegypti 24h LC$_{50}$*	T. ni 24h LD$_{50}$+
Alkaline dissolved, Sandoz (Track 2)	2.29 + 0.06 (12.4)	3.71 + 0.32 (3.22)
DEAE (Track 3)	1.91 + 0.30 (2.92)	1.96 + 0.64 (2.16)
pH 4.5 soluble (Track 5, 27 & 24K)	2.02 + 0.54 (2.06)	1.95 + 0.14 (4.32)
G-75, R$_f$ 1.35 (Track 6, 27K)	2.71 + 0.12 (5.72)	3.54 + 0.50 (3.28)
G-75, R$_f$ 1.53 (Track 7, 24K)	>20	2.97 + 0.18 (7.00)
Alkaline dissolved, Salsbury (Track 8, 27K)	5.78 + 0.42 (4.76)	6.09 + 0.46 (4.67)

*μg/ml + 1 S.D. with slope of probit analysis in
parenthesis. Larvae were incubated in water with BTI
δ-endotoxin added.

+PPM or mg/kg body weight + 1 S.D. with slope of probit
analysis in parenthesis. Larvae were injected.

before and after DEAE (Table III for both A. aegypti and T. ni),
even though there appeared to be a concentration of the 68K component in this purification step. The 24 and 27K component
individually (Figure 1, Tracks 7 and 6, respectively) also had an
equivalent toxicity when injected into T. ni (Table III) but the
24K component (Figure 1, Track 7) was not toxic when fed to A.
aegypti whereas the 27K component was toxic (Table III). This
inactivity cannot be explained by the absence of the 35 and 68K
components in Track 7 (Figure 1) because these same components
are also absent in Track 5 and yet Track 5 retained oral toxicity
to A. aegypti (Table III). In fact the absence of the 68 and 35K
components (Figure 1, Tracks 5 and 7) likewise did not affect the
T. ni activity (Table III). Thus it appears that at least the
27K proteins are necessary for A. aegypti oral toxicity while
both the 24 and 27K components can impart toxicity to T. ni when
injected. The 27K component was also toxic in both A. aegypti
and T. ni regardless of the source (Track 6 and Track 8, Figure 1
and Table III).

Neural Toxicity of BTI δ-Endotoxin

The injection of alkaline-dissolved BTI δ-endotoxin led to a
number of immediate neuromuscular effects (Table IV) including,

Table IV. Symptoms elicited after the injection of alkaline-
dissolved BTI δ-endotoxin (Track 2) into mice and insects

Time Post-Injection	Trichoplusia ni (5 PPM)	Periplaneta americana (6 PPM)	Swiss-Webster Mice (1.5 PPM)
0-1h	Mouth palpation of injection site Increased wandering Heart arrest Abdominal paralysis List side to side when crawling Total paralysis & flaccidity	Loss of motor activity	Ruffled fur Lost alertness Not inquisitive Reduced responsiveness Slow in righting themselves Breathing shallow Lost activity in hind legs
20-24h	Localized black-ening of the body Total blackening of the body No response to head stimulation	In Survivors Failure to right themselves Tremor	In Survivors, Constipation Dead animals with a pinched waist

in the insects tested, listing from side to side when crawling,
heart arrest, paralysis, and tremors. These symptoms were
observed for both the Sandoz and Salsbury BTI δ-endotoxin and
were also observed for partially purified BTI δ-endotoxin (Table
III). The symptoms observed in mice were somewhat similar to
those observed in botulism poisoning (a neurotoxin), which
included a loss of alertness, shallow breathing, and in some
cases lost activity in the hind legs. Dead mice had a pinched
waist, a sign of diaphram arrest. The symptoms observed in
insects following the injection of BTI δ-endotoxin were clearly
different from those following the ingestion of BTK δ-endotoxin.
When T. ni were fed BTK δ-endotoxin, there was regurgitation
within 15 min, a total cessation of feeding until death, and no
overt neuromuscular anomalies. The injection of alkaline-
dissolved BTK δ-endotoxin produced no obvious adverse effects in
insects or mice.

   A number of other lines of evidence also suggested that
there may be another mode-of-action for BTI poisoning by
injection other than its known, general cytolytic activity
(3,22-24). Using the appearance of cytosolic lactate
dehydrogenase (LDH) in insect hemolymph post-injection as a
marker for cytotoxicity (36-37), we found that dissolved BTI
δ-endotoxin was a potent cytotoxin. When T. ni, however, were
injected with dissolved BTI δ-endotoxin and then incubated at 28,
15, and 9°, there was an increase in the LD_{50} with a decrease in
temperature (Figure 2) but the LDH levels at 3.5 PPM BTI were
unaffected by temperature.

   A pharmocological study of ventral nerve cord function in T.
ni also suggested that alkaline-dissolved BTI δ-endotoxin was
affecting the insect nervous system as a nerve poison (Figures 3
and 4). After 7-60 min post-injection of alkaline-dissolved BTI
δ-endotoxin at the LD_{50} concentration of 3.7 PPM, the ventral
nerve cord of T. ni exhibited spontaneous-high frequency
discharges. This was followed by a reduced baseline activity and
sensitivity to sensory stimulation (S, Figure 4) at 24 h
post-treatment. By 2-90 min post-injection of methamidophos,
there also was spontaneous-high frequency discharge which lasted
20 min - 6 h and was followed by a reduced baseline activity and
sensitivity to sensory stimulation (S, Figure 4) by 24 h. The
response to methamidophos was slightly more rapid and sustained
than with BTI toxin. Studies also suggested that the primary
site of action for BTI might be the peripheral nervous system and
that the mode-of-action of BTI and methamidophos on the insect
nervous system were probably different. When the peripheral
nervous system is severed from the T. ni ventral nerve cord, the
methamidophos application still elicits spontaneous discharge,
but there is no response to BTI toxin. When the cytotoxin,
phospholipase-A_2, is injected into T. ni at its LD_{50} of 35 PPM,
there is no spontaneous-high frequency discharge in the ventral
nerve cord and no reduction in the stimulus-response (S, Figure
4) as was the case for BTI.

   A 25K dalton component was isolated at relatively high
purity as determined by SDS-PAGE (Figure 5, Track 3) from
alkaline-dissolved BTI δ-endotoxin (Figure 5, Track 1). The

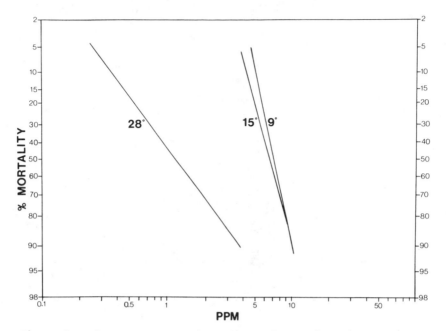

Figure 2.   The temperature dependency of alkaline-dissolved BTI δ-endotoxin injected into <u>Trichoplusia</u> ni.

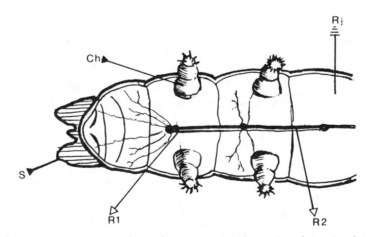

Figure 3.   Neurophysiological preparation of <u>Trichoplusia</u> <u>ni</u>. Head, thorax and gut are removed.   Tungsten electrodes were placed into the hemocoel along side abdominal ganglion VIII (at R1), the ventral nerve cord (at R2) and the abdominal wall (ground, R_i).   Injections of alkaline-dissolved BTI δ-endotoxin, methamidophos and phospholipase-A_2 were into the second pair of abdominal prolegs (Ch).   Mechanical sensory stimulation with a glass probe was at the anal proleg (S). Activity in the ventral nerve cord was monitored through 24 h post-treatment (<u>38-40</u>) (see Figure 4).

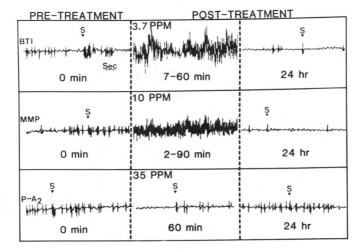

Figure 4. Time dependency of nervous activity in the ventral nerve cord of <u>Trichoplusia ni</u> injected with 3.7 PPM alkaline-dissolved BTI δ-endotoxin (Sandoz), with 10 PPM methamidophos (MMP) and with 35 PPM phospholipase-A$_2$ (P-A$_2$). Mechanical sensory stimulation is given at arrow <u>S</u>. The control response was the same as the recording for P-A$_2$. BTI and P-A$_2$ were injected into <u>T</u>. <u>ni</u> at their respective LD$_{50}$.

S   1   2   3   S

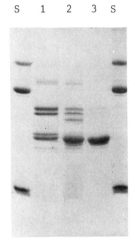

Figure 5. SDS-Page analysis of alkaline-dissolved <u>Bacillus</u> <u>thuringiensis</u> <u>israelensis</u> (BTI) δ-endotoxin from Sandoz Inc. at 25 μg per track: (1) BTI δ-endotoxin as prepared in Figure 1 (Track 2), (2) soluble fraction after dialysis of BTI δ-endotoxin into pH 4.5 sodium acetate buffer, and (3) 25K component from BTI δ-endotoxin after pH 4.5 percipitation and DEAE-anion exchange chromatography. S, molecular weight markers from top to bottom bovine serum albumin (68K daltons), ovalbumin (43K), and myoglobin (16K).

supernatant (Figure 5, Track 2) after pH 4.5 percipitation of the
crude endotoxin was further purified by DEAE-Cellulose (DE-52,
Whatman, 30 ml) in 0.025 M Tris-HCl (pH 8.00), eluted with a 48
h, 400 ml, 0.0-0.2 M NaCl linear gradient (Figure 5, Track 3).
The resulting 25K dalton component (Figure 5, Track 3) at 0.85
μg/ml produced 50% cell lysis in a 1% human red blood cell
solution (25°C, 15 min) and was also hemolytic against sheep and
rabbit red blood cells. Hemolytic activity increased for the 25K
component (Figure 5, Track 3) after a 15 sec incubation at 70°C
and was inactivated at 90°C as was also the case for the crude
toxin (Figure 5, Track 1). At 25 μg/ml, however, the 25K protein
was not toxic orally to A. aegypti even after 15 sec heat
treatments at 45, 70 or 90°C nor was it lethal when injected into
T. ni larvae at 4.6 PPM (the 48 h $LD_{50} > 4.6$ PPM). Injections
did disrupt larval-pupal metamorphosis later in development. The
crude toxin (Figure 5, Track 1), however, was toxic to A. aegypti
and T. ni at 45 and 70° but not after the 90° treatment. So the
25K component is cytotoxic but does not have either oral toxicity
to A. aegypti or injected toxicity to T. ni.

Conclusions

It appears from our studies and reports in the literature that the
parasporal crystal of BTI has gut-toxicity when fed to the
mosquito (16), both in vitro and in vivo cytotoxicity (3,22-24)
and in vivo neurotoxicity. Injected toxicity occurred in a
number of insect species. The crude alkaline-dissolved
δ-endotoxin of BTI when injected into T. ni was strongly
cytotoxic and at its LD_{50} also neurotoxic. This was unlike the
cytotoxin, phospholipase-A_2 which demonstrated no neurotoxicity
at its LD_{50}. The toxicity of BTI δ-endotoxin injected was also
temperature-dependent while its cytolytic activity was unchanged
in the same temperature range. A 25K component isolated from BTI
(Figure 5, Track 3) was also found to be cytolytic but when
injected had no toxicity.
    Until each of the components from the alkaline-dissolved
δ-endotoxin of BTI (Figure 1, Track 2) can be purified and
separately tested for cytotoxicity and neurotoxicity, the
interrelationships of these modes-of-action to the many poly-
peptides found in the δ-endotoxin of BTI will be in question.
The variations obtained with different BTI δ-endotoxin
preparations and BTI strains will magnify the complexity of the
problem. Nevertheless, a number of lines of evidence now exists
to suggest that injected toxicity and neurotoxicity are not
necessary a function of general cytolytic activity. The evidence
is clear that there are also cytolytic components that
demonstrate no gut toxicity. The finding of species difference
within a single family of Lepidoptera to the susceptibility of
BTI δ-endotoxin poisoning by injection will be an important tool
for studying mode-of-action in the future. Reports on the nerve
blocking action from digests of the δ-endotoxin of BTK (20-21),
suggest that a neurotoxic element may be common to the action of
other members of the BT complex.

Acknowledgments

Use of trade names in this publication does not imply endorsement
of the products named or criticism of similar ones not mentioned.
This paper is Number 9592 of the Journal Series of the North
Carolina Agriculture Research Service, Raleigh, North Carolina
27695.
    R. M. Roe was supported in part by the Department of Health
and Human Services, National Service Award 1 F32 GM09223-02 from
the National Institute of General Sciences; and B. D. Hammock by
NIEHS Research Career Development Award 5 K04 ES00107-05.
Partial support for this research was provided by NIEHS Grant
ES02710-04, the University of California General Fund for
Mosquito Research, and the respective state agricultural experiment stations. The assistance of Dr. Charles L. Judson and Ms.
Mary Ann Montague for their mosquito injections of BTI and BTK is
most appreciated. Also the assistance of Kenji Ota, Rafael del
Vecchio, Jim Ottea, and Terry Hanzlik in insect rearing is grate-
fully acknowledged.

Literature Cited

1. Somerville, H. J. <u>Trends Biochem. Sci.</u> 1978, 3, 108-10.
2. Bulla, L. A. Jr.; Bechtel, D. B.; Kramer, K. J.; Shethna, Y.
   I.; Aronson, A. I.; Fitz-James, P. C. <u>C. R. C. Crit.</u>
   <u>Rev. Microbiol.</u> 1980, 8, 147-204.
3. Thomas, W. E.; Ellar, D. J. <u>J. Cell Sci.</u> 1983, 60, 181-97.
4. Heimpel, A. M. <u>Ann. Rev. Entomol.</u> 1967, 12, 287-322.
5. de Barjac, H.; Bonnefoi, A. <u>Entomophaga.</u> 1962, 7, 5-31.
6. de Barjac, H., personal communication.
7. Goldberg, L. J.; Margalit, J. <u>Mosq. News.</u> 1977, 37, 355-8.
8. de Barjac, H. <u>C. R. Hebd Séanc. Acad. Sci. Paris Serie D.</u>
   1978, 286, 797-800.
9. Bulla, L. A. Jr.; Kramer, K. J.; Cox, D. J.; Jones, B. L.;
   Davidson, L. I.; Lookhart, G. L. <u>J. Biol. Chem.</u> 1981, 256,
   3000-4.
10. Nickerson, K. W. <u>Biotechnol. Bioeng.</u> 1980, 22, 1305-33.
11. Tyrell, D. J.; Bulla, L. A. Jr.; Andrews, R. E. Jr.; Kramer,
    K. J.; Davidson, L. I.; Nordin, P. <u>J. Bacteriol.</u> 1981,
    145, 1052-62.
12. Endo, Y.; Nishiitsutsuji-Uwo, J. <u>J. Invertebr. Pathol.</u>
    1980, 36, 90-103.
13. Percy, J.; Fast, P. G. <u>J. Invertebr. Pathol.</u> 1983, 41,
    86-98.
14. Sutter, G. R.; Raun, E. S. <u>J. Invertebr. Pathol.</u> 1967, 9,
    90-103.
15. Ebersold, H. R.; Luethy, P.; Mueller, M. <u>Bull. Soc. Ent.</u>
    <u>Suisse</u> 1977, 50, 269-76.
16. de Barjac, H. <u>C. R. Hebd. Séanc. Acad. Sci. Paris Serie D.</u>
    1978, 286, 1629-32.
17. Heimpel, A. M.; Angus, T. A. <u>J. Insect Pathol.</u> 1959, 1,
    152-70.
18. Angus, T. A. <u>J. Invertebr. Pathol.</u> 1968, 11, 145-6.
19. Ramakrishnan, N. <u>J. Invertebr. Pathol.</u> 1968, 10, 449-50.

20. Cooksey, K. E.; Donninger, C.; Norris, J. R.; Shankland, D.
     J. Invertebr. Pathol. 1969, 13, 461-2.
21. Aronson, J. N.; Crowder, L. A. SIP 16th Annual Meeting
     Abstract. 1983.
22. Murphy, D. W.; Sohi, S. S.; Fast, P. G. Science. 1976, 194,
     954-6.
23. Nishiitsutsuji-Uwo, J.; Endo, Y.; Himeno, M. J. Invertebr.
     Pathol. 1979, 34, 267-75.
24. Johnson, D. E. J. Invertebr. Pathol. 1981, 38, 94-101.
25. Thomas, W. E.; Ellar, D. J. FEBS Letters. 1983, 154,
     362-8.
26. Nickerson, K. W.; Bulla, L. A. Jr. Appl. Microbiol. 1974,
     28, 124-8.
27. Sharpe, E. S.; Nickerson, K. W.; Bulla, L. A. Jr.; Aronson,
     J. N. Appl. Microbiol. 1975, 30, 1052-3.
28. Laemmli, U. K. Nature. 1970, 227, 680-5.
29. Cheung, P. Y. K.; Roe, R. M.; Hammock, B. D.; Judson, C. L.;
     Montague, M. A. Pestic. Biochem. Physiol. 1984, 21, In
     Press.
30. Yamamoto, T.; Iizuka, T; Aronson, J. N. SIP 16th Annual
     Meeting Abstract. 1983.
31. Calabrese, D. M.; Nickerson, K. W. Can. J. Microbiol.
     1980, 26, 1006-10.
32. Spielman, A.; Wong, J. Biol. Bull. 1974, 147, 433-42.
33. Finney, D. J. "Probit Analysis"; Cambridge Univ. Press:
     Great Britian, 1971; pp. 1-333.
34. Klowden, M. J.; Held, G. A.; Bulla, L. A. Jr. Appl.
     Environ. Microbiol. 1983, 46, 312-5.
35. Wie, S. I.; Andrews, R. E. Jr.; Hammock, B. D.; Faust, R.
     M.; Bulla, L. A. Jr. Appl. Environ. Microbiol. 1982, 43,
     891-4.
36. Bergmeyer, H. -U.; Bernt, E. Meth. Enzymatic Anal. 1974, 2,
     574-9.
37. Wing, K. D.; Sparks, T. C.; Lovell, V. M.; Levinson, S. O.;
     Hammock, B. D. Insect Biochem. 1981, 11, 473-85.
38. Gammon, D. W. Pestic. Biochem. Physiol. 1977, 7, 1-7.
39. Miller, T.; Kennedy, J. M. Pestic. Biochem. Physiol. 1973,
     3, 370-83.
40. Narahashi, T. Adv. Insect Physiol. 1971, 8, 1-93.

RECEIVED November 15, 1984

Bioassay of Anti Juvenile Hormone Compounds: An Alternative Approach

THOMAS C. SPARKS[1], R. MICHAEL ROE[2], ADRIAN BUEHLER[3], and
BRUCE D. HAMMOCK[4]

[1] Department of Entomology, Louisiana Agricultural Experiment Station, Louisiana State
University Agricultural Center, Baton Rouge, LA 70803
[2] Department of Entomology, North Carolina State University, Raleigh, NC 27695
[3] Abteilung fuer Zoophysiologie, Universitaet Bern, Erlachstrasse 9a, CH-3012 Bern, Switzerland
[4] Department of Entomology, University of California, Davis, CA 95616

During the last larval stadium of the cabbage looper,
Trichoplusia ni, Hubner, the prepupal burst of juvenile
hormone (JH) stimulates the appearance of juvenile
hormone esterase (JHE) which, in turn, degrades the
JH. Disruption of this prepupal burst of JH by anti-
juvenile hormones (AJHs) such as fluoromevalonolactone,
result in a variety of teratogenic effects including
delayed tanning and/or pupation, malformed larvae and
reduced JHE activity. Although general toxicants and
esterase inhibitors may also cause malformed larvae,
there is usually no delay in tanning and/or pupation.
These observations provide the basis for a simple,
rapid AJH bioassay using an economically important pest
insect (T. ni). A simple 'key' was devised to help
facilitate the use of this bioassay.

At the heart of any search for bioactive molecules is the need
for effective bioassays. Several bioassays have been developed
for the identification of compounds with anti-juvenile hormone
(AJH) activity. The most common of these AJH bioassays involves
the treatment of young larvae or nymphs with the potential AJH
by incorporation into the diet or contact application, and then
waiting for several days (or, in some cases, weeks) for precocious
development (or other AJH response) to occur (1,2). Alternatively,
AJH activity can be determined using in vitro assays such as corpora
allata cultures or epidermal cell cultures to monitor for inhibition
of juvenile hormone (JH) biosynthesis (3-6) or blockage of the
JH induced inhibition of pupal commitment (7), respectively.
    Although the above assays have all proven useful, there are
a number of disadvantages associated with each of them. For example,
the contact and feeding bioassays can require a large quantity
of compound, compared to a typical topical bioassay, and it usually
takes several days before the AJH effects are observed. The in
vitro assays are tedious, require special skills in microsurgery,
dedicated sterile facilities, are unsuitable for screening large
numbers of compounds, detect a limited number of possible mechanisms,

0097-6156/85/0276-0293$06.00/0

and in vitro activity may not translate into in vivo activity due
to in vivo metabolism (8). Finally, the insects most commonly
used for AJH bioassays, the tobacco hornworm, Manduca sexta (L.),
and the large milkweed bug, Oncopeltus fasciatus (Dallas), are
of limited economic importance, and both species seem to be
hypersensitive such that compounds active on them (e.g. ETB,
precocene) display little activity on major pest insects, especially
among the Lepidoptera (2,9,10).
      The cabbage looper, Trichoplusia ni Hubner, is a pest of a
wide variety of agricultural crops (11-13) and represents a major
family of insect pests, the Noctuidae, which also includes the
corn earworm, Heliothis zea (Boddie), the tobacco budworm, Heliothis
virescens (F.), the soybean looper, Pseudoplusia includens (Walker),
the beet armyworm, Spodoptera exigua (Hubner), the fall armyworm,
Spodoptera frugiperda (J. E. Smith) and others. There is also
a growing body of knowledge concerning the endocrinology of T. ni
(14,15). Recent studies on JH and JH esterase (JHE) regulation
in wandering last stadium larvae and prepupae of T. ni (16-18)
suggested an alternative approach to bioassaying compounds for
AJH activity that circumvents some of the disadvantages associated
with other AJH bioassays. From a more fundamental point of view,
this report illustrates how key xenobiotics can be used to dissect
developmental processes during critical phases of insect development.

The Bioassay

The bioassay presented herein is based on what appears to be a
consistent set of teratogenic morphological effects that are
produced in response to disruptions of the JH mediated regulation
of development during the late last stadium of T. ni. To facilitate
the use of this bioassay, a simple key has been devised and is
presented below along with hypotheses for the physiological basis
for each step.
      Since the responses observed in this bioassay (and others)
can be the result of a variety of effects, not only on the endocrine
system, but also on the nervous and other systems (19), a selected
group of JHs, AJHs and pesticides (Table I) were used in developing
this bioassay in an effort to test and eliminate responses due
to non-AJH compounds. Likewise, it is important in any successful
bioassay to simply and rapidly eliminate inactive and inappropriate
compounds (i.e. non-AJHs). Thus, the first steps in our bioassay
attempt to exclude non-AJHs and inactive compounds, leaving the
later steps for the confirmation of AJH activity to those few
compounds that are most likely AJHs.

The Insect and Assays. Last (fifth) stadium (L5) larvae of T. ni
were used throughout. Larvae were reared on a 14L:10D photoperiod
(lights on at 5 AM) at 27°±2°C, and staged as described previously
(31). Larvae were treated with the desired compounds either by
injection along the mid-dorsal line of abdominal segments 1 & 2
(1 μl in distilled water), or by topical treatment (1-2 μl in
acetone or ethanol) on the dorsum of the last thoracic segment.
Controls received acetone or ethanol (topically, 1-2 μl), or water
(injected, 1 μl). Obviously, the method of treatment will be a

function of the solubility of the test compound and/or its suspected ability to penetrate the insect cuticle. Injection is easily accomplished using either a sharpened 10 µl Hamilton syringe or a finely-drawn glass capillary tube. The capillary tube has the advantage of practically eliminating any bleeding after injection. Treated larvae were examined for behavioral and developmental changes or abnormalities.

A pooled (3-5 larvae) hemolymph sample was collected by clipping either anal or thoracic legs of the larvae. JHE activity in the hemolymph was monitored as described previously (32,33) using JH III (H^3 at C_{10}, 11 Ci/mmol., New England Nuclear) as the substrate (5×10^{-6} M). All assays were run in duplicate on at least two (usually three or more) occasions.

The Key. The bioassay is divided into 5 steps which form a key for the identification of AJH activity. At each step examples are provided of compounds that exhibit the possible responses. Since these compounds were used to develop the key, they do not necessarily follow the same logical progression that one would see if the bioassay was being used to evaluate their AJH activity.

Identification of AJH Activity

1. Are there larval-pupal intermediates or malformed larvae?
   Treatment: L5D3 (last stadium, day 3) larvae treated ca. 9 AM; 200 nmol./larva.
   Expected Result: A majority (>50%) of the treated larvae become tanned, malformed larvae without displaying any obvious toxic response (see explanation below). Pupation/tanning in controls generally occurs on L5D4 ca. 3 PM (ca. 30 hr. posttreatment). Results are expressed as the percentage of larvae displaying the above effect relative to the controls (solvent only).
   A) YES - GO TO 2. Examples: FMev, EPPAT, DEF (Table I).
   B) NO - Increase dose and/or change the time of treatment or mode of application and try again; otherwise STOP. Examples: Epofenonane, hydroprene, ETB, precocene II, ethoxy-precocene, DFP, pipernoyl butoxide, chlordimeform, TFT, carbaryl (Table I).

Explanation: The appearance of tanned, malformed larvae following treatment with an exogenous compound can be the result of AJH activity or a response on the part of the insect to an external stress such as an insecticide. Since a variety of molecular structures are likely to be tested for AJH activity, it is important to eliminate compounds that are general toxicants at the beginning of the bioassay.

If the compound is causing the insect to be stressed or intoxicated, the insect larva (in this case a prepupa) may have trouble casting the old cuticle, yet be capable of tanning. Thus, the insect becomes a tanned larva or what appears to be a half pupa/half larva (due, in part, to incomplete ecdysis). For example, larvae treated with paraoxon (200 nmol./larva; data not included in Table I due to high mortality) caused the formation of tanned larvae in all larvae treated. However, all of these larvae had obviously been adversely affected (intoxicated) by the paraoxon

Table I. Compounds[1] used to Develop and Test the AJH Bioassay, and Their Effect on T. ni.

Compound	Dose used for key (nmol.)	Previously tested for AJH activity	% Malformed Larvae[2]	T_{50} for tanning (hrs.)[3]	JHE Activity (% Control)[4]	
					L5D3	L5D1
Used in the Development of the AJH Bioassay						
1. Juvenile hormone I	200	--	0	4.0+	304±83	105±23
2. Juvenile hormone III	200	20	0	0.0	164±54	112±9
3. Epofenonane: 1-(4'-ethyl-phenoxy)-6,7-epoxy-3-ethyl-7-methylnonane	200	--	0	1.5-	445±226	110±31
4. Hydroprene: ethyl (2E,4E)-3,7,11-trimethyl-2,4-dodeca-dienoate	200	9, 21	0	1.5+	365±32	100±25
5. FMev: tetrahydro-4-fluoro-methyl-4-hydroxy-2H-pyran-2-one	200	4, 9, 18, 22-26	100	7.8-	45±21	99±7
6. ETB: ethyl 4-(2-(tert-butyl-carbonyloxy)butoxy)benzoate	200	2, 7, 9 22, 24, 26 28	0	1.0+	365±139	87±26
7. Precocene II: 6,7-dimethoxy-2,2-dimethylchromene	200	1, 2, 3, 10 (for review), 28	0	0.1-	96±19	87±25

Table 1. Con't.

Compound	Dose used for key (nmol.)	Previously tested for AJH activity	% Malformed Larvae[2]	T_{50} for tanning (hrs.)[3]	JHE Activity (% Control)[4] L5D3	L5D1
8. Ethoxy-precocene: 7-ethoxy-6-methoxy-2,2-dimethylchromene	200	10, (for review), 29	0	1.1-	87±14	91±10
9. EPPAT: O-ethyl S-phenyl phosphoramidothiolate	200	--	72	1.5-	17±20	11±8
10. DFP: O,O-diisopropyl phosphorofluoridate	200	--	3	0.8-	98±7	105±15
11. DEF: S,S,S-tri-n-butyl-phosphorotrithiolate	200	--	71	1.8-	34±13	81±22
12. Piperonyl butoxide: 3,4-methylenedioxy-6-propylbenzyl n-butyl diethylglycol ether	200	2, 3, 9, 20	0	0.1-	111±29	94±27
13. Chlordimeform: N,N-dimethyl-N'-[2-methyl-4-chlorophenyl]-formamidine	200	--	0	4.4-	50±19	98±15
14. TFT: 1,1,1-trifluorotetra-decan-2-one	200	--	0	0.8-	122±28	89±24

Continued on next page

Table 1. Con't.

Compound	Dose used for key (nmol.)	Previously tested for AJH activity	% Malformed Larvae[2]	T_{50} for tanning (hrs.)[3]	JHE Activity (% Control)[4] L5D3	L5D1
15. Carbaryl: 1-naphthyl N-methylcarbamate	200	--	7	0.8-	111 ± 21	89 ± 5

Evaluated Using the AJH Bioassay

Compound	Dose used for key (nmol.)	Previously tested for AJH activity	% Malformed Larvae[2]	T_{50} for tanning (hrs.)[3]	JHE Activity (% Control)[4] L5D3	L5D1
16. Methyl compactin: methyl 7-[1,2,6,7,8,8a-hexahydro-2-methyl-8-(2-methylbutyryloxy)naphthylenyl-1]-3,5-hydroxy-heptanate	100	5, 6, 29	0	1.3-	--	--
17. L-643,049-01K01: 7-[1(S),2(S), 6(R),7,8(S),8a(R)-hexahydro-2,6-dimethyl-8-(2,2-diethyl-butyryloxy)-naphthalenyl-1]-3(R),5(R)-hydroxyheptanic acid sodium salt	100[5]	--	89	5.7-	57 ± 9	103 ± 17
18. L-643,049-00H03: 6(R)-[2-[8(S)-(2,2-diethylbutyryloxy)-2(S),6(R)-diemthyl-1,2,6,7,8,8a(R)-hexahydronaphthyl-1(S)]-ethyl]-4(R)-hydroxy-3,4,5,6-tetrahydro-2H-pyran-2-one	56[5]	--	10	1.7-	--	--

Table I. Con't.

Compound	Dose used for key (nmol.)	Previously tested for AJH activity	% Malformed Larvae[2]	T$_{50}$ for tanning (hrs.)[3]	JHE Activity (% Control)[4] L5D3	L5D1
19. L-643,737-00S03: 6(R)-[2-[8(S)-(2-ethyl-2-methylbutyryl-oxy)-2(S),6(R)-diemthyl-1,2,6,7,8,8a(R)-hexahydronaphthyl-1(S)lethyl]-4(R)-hydroxy-3,4,5,6-tetrahydro-2H-pyran-2-one	58[5]	--	83	3.8-	68±41	124±52

1. Sources of the compounds were as follows; 1,2 Calbiochem-Berhing; 3, P. Masner, Dr. Maag Ltd.; 4-6, G. Quistad, D. Schooley and G. Staal, Zoecon Corporation; 7,8, Sigma; 9, P. Magee, Chevron Chemical Company; 10,11, Aldrich; 11,12,13,15, Chem Service; 14, synthesiszed as described in (30); 16-19, R. Dybas, Merck, Sharp & Dohme.

2. Tanned and/or malformed larvae due to incomplete ecdysis or other causes.

3. Time for 50% of the treated larvae to tan and/or pupate relative (+ = before; - = after) to the controls.

4. JHE activity measured *in vitro* from larvae treated on L5D3 or L5D1. Values are means ± their standard devations.

5. Compound was injected. All others were applied topically.

treatment (i.e. flaccid, hemolymph accumulation in the abdomen,
failure to migrate to the top of the rearing containers, dehydration). Likewise, chlordimeform modified the behavior of the larvae
in the form of abnormal spinning and wandering behavior. Thus,
neither of these compounds (at the dose tested) would be considered
as a potential AJH. It should be remembered, however, that for
some insect species the effective dose for a particular AJH may
be near the toxic dose for that compound (see 22). One disadvantage
of any in vivo bioassay, including this one, is that they will
not detect the AJH activity of toxic molecules which, in an in
vitro system, would function as AJHs.

     As with the insecticides, FMev (a highly active AJH for the
Lepidoptera; 22) also causes the formation of tanned larvae. While
these tanned larvae may also be the result of general teratogenic
or stress effects, the removal of the corpora allata (site of JH
biosynthesis and release) also causes similar effects (34). Although
the presence of tanned larvae seems to consistently correlate with
AJH activity, it is imperative that the formation of tanned larvae
not be used as the sole criterion for the determination of AJH
activity. Rather, it is a single marker that is only of significance when coupled with Step 2 (below).

2. Is there a delay in time of pupation?
   Treatment: L5D3 larvae (9 AM); 200 nmol./larva.
   Expected Result: Treated larvae display a distinct (ca. 4 hr.)
   delay in the time of pupation/tanning. Results are expressed
   as the advancement '+' or delay '-' (in hrs) in the time of
   tanning/pupation relative to the controls (solvent only).
   Controls usually pupate/tan about 30 hr. posttreatment (L5D4,
   ca. 3 PM).
   A) YES - Compound may possess AJH activity - GO TO 3 for
      verification. Examples: FMev, chlordimeform (Table I).
   B) NO - STOP. Examples: JH I, JH III, epofenonane, hydroprene,
      ETB, precocene II, ethoxy-precocene, EPPAT, DFP, DEF,
      piperonyl butoxide, TFT, carbaryl (Table I).

Explanation: In many lepidopterans, including T. ni, the presence
of JH in the prepupa seems to accelerate the time of ecdysis to
the pupa (18,35-38). Conversely, a reduction in the JH titer due
to an AJH causes a delay in the time of ecdysis to the pupae and/or
the tanning process (18,37). This particular effect can be
prevented, in part, by the coapplication of JH I (18). Thus, a
delay in the time of tanning/pupation seems to be related to the
ability of a compound to block JH biosynthesis/release or action,
and hence act as an AJH.

                    Verification of AJH Activity

3. Is there reduced JHE activity in L5D4 larvae?
   Treatment: L5D3 (9 AM) larvae; 200 nmol./larva.
   Expected Result: Larvae are assayed 24 hr. posttreatment (L5D4,
   9 AM, larvae are now prepupae) for JHE activity. A compound
   with AJH activity should cause the level of JHE activity to
   be lower than normal. Results are expressed as a percentage
   of the JHE activity present in larvae treated with only the
   solvent.

A) YES - GO TO 4. Examples: FMev, EPPAT, chlordimeform, DEF (Table I).
B) NO - STOP. Not likely to be an AJH for T. ni. Examples: JH I, JH III, epofenonane, hydroprene, ETB, precocene II, ethoxy-precocene, DFP, piperonyl butoxide, TFT, carbaryl (Table I).

Explanation: Hemolymph JHE activity in late last stadium larvae of T. ni is directly regulated by the JH titer (16-18). Thus, any compound that affects JH biosynthesis /release or action at JH receptor sites will cause a change in the prepupal peak of JHE activity. Compounds with JH activity stimulate JHE production, often to greater than normal levels (e.g. JH I, JH III, epofenonane, hydroprene and ETB; Table I). For an AJH, a reduction in the JHE activity is the expected outcome. However, a reduction in JHE activity can also be the result of an inhibitor that interacts directly with the JHE (see below).

4. Normal JHE activity in L5D1 larvae?
   Treatment: L5D1 larvae (3 PM); 200 nmol./larva.
   Expected Result: Treated larvae are assayed for JHE activity 1-2 hr. posttreatment. An AJH should have no effect on JHE activity in this short of a time period. Results are expressed as a percentage of the normal (control) JHE activity.
   A) YES - Compound is an AJH. GO TO 5 for determination of mode of action. Example: FMev (Table I).
   B) NO - STOP. Compound is a general toxicant and/or JHE inhibitor, and is not an AJH. Example: EPPAT (Table I).

Explanation: Step 4 insures that the potential AJH is not merely acting as a JHE inhibitor. Since there is always the possibility of bioactivation, an in vivo test is used. AJHs will have no effect on JHE activity in the time-frame used in this step (1 hour) while this amount of time should be more than sufficient for JHE inhibitors to act (e.g. EPPAT; 19,39,40). If the compound under consideration makes it through this part of the bioassay key, then it has some AJH activity.

AJH Classification (Optional)

5. Does the AJH block the juvenoid induced increase in JHE activity?
   Treatment: L5D3 (9 AM) larvae; selected doses of the AJH (100 nmol./larva is a good starting place) are coapplied with the juvenoid, epofenonane (100 nmol./larva).
   Expected Result: Larvae are assayed for JHE activity on L5D4 (9 AM). The JHE induction should not be affected if the AJH acts by blocking JH biosynthesis. Results are expressed as a percentage of the JHE activity induced by epofenonane (100 nmol./larva) alone.
   A) YES - AJH is probably a suboptimal JH or a JH receptor antagonist. Example: ETB (Figure 1).
   B) NO - AJH probably functions by directly blocking JH bio-synthesis and/or release. Example: FMev.

Explanation: This section is included in an effort to provide a
means to gain some insights into the mode of action of the test
compound. Epofenonane appears to be a JH agonist for T. ni that
is somewhat less active than JH I (16,26). Unlike JH I, epofenonane
is not susceptible to ester hydrolysis, which eliminates the problem
of the JHE affecting the level of its own induction. Like epofenonane, the JH agonist-antagonist ETB was able to induce the
appearance of the hemolymph JHE in T. ni, but only at much higher
dosages (Figure 1). However, at lower doses ETB caused a dose
dependent reduction in the JHE activity induced by epofenonane
in T. ni (Figure 1). Since other tests demonstrated that ETB was
not a direct JHE inhibitor (26), it appeared that ETB was acting
as a JH receptor antagonist (26). Unlike ETB, FMev appeared to
have no effect on the JHE induction by epofenonane (41). This
observation is consistent with other studies that show it to function as an inhibitor of JH biosynthesis (22,23). Thus, this assay
can potentially provide information on the mode of action of an
AJH. Running similar tests in ligated abdomens would further insure
that the action being observed was probably due to competition
for JH receptors (probably the fat body), and not due to other
interactions with the brain, corpora allata, etc.

Using the Key to Test for AJH Activity. Besides the compounds
used to develop the key, methyl compactin and three of its
analogues (Table I) were evaluated for AJH activity (through step
4). Of the 4 compounds tested, both L-643,049-01K01 and L-643,737-
00S03 (tested at 100 and 58 nmol./larva, respectively; injected)
caused the formation of tanned malformed larvae (89 and 83%,
respectively; Step 1) and a large delay (ca. 4-6 hr.) in the time
of tanning/ pupation (Step 2) (Table I). Thus, both L-643,049-01K01
and L-643,737-00S03 appeared to possess AJH activity and were tested
further.
     Treatment (100 and 58 nmol./larva, respectively; injected)
of L5D3 larvae with either of these compounds resulted in a 30-40%
decrease in the prepupal JHE activity peak (Step 3), while L5D1
larvae treated (as above) with either compound possessed normal
levels of JHE activity one hour posttreatment (Step 4) (Table I).
These results confirm that both L-643,049-01K01 and L-643,737-00S03
are AJHs for T. ni.

Discussion

The bioassay-key presented here provides a relatively simple
approach to identifying compounds with AJH activity. Obviously,
the sensitivity of the assay is, in part, a function of the number
and magnitude of the dosages used. In using the bioassay and key,
it is best to start with as high of a dose as possible to rapidly
eliminate the inactive compounds. In our assays 200 nmol./larva
is typically used as the starting dose. In doing so, however,
it should be kept in mind that JH agonists/antagonists, such as
ETB (2,9,26), may only display AJH activity over a narrow dose
range which may be far below our 200 nmol./larva starting dose.
At this high dose only JH like activity may be seen. For example,
it has been demonstrated that ETB can antagonize the JH action
of the juvenoid epofenonane in T. ni (26, Figure 1) and yet be

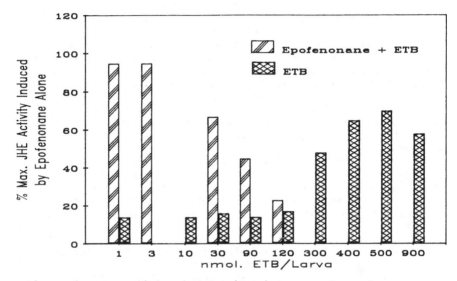

Figure 1: JHE activity in L5D3 (9 AM) larvae of T. ni treated with epofenonane (100 nmol./larva) and selected doses of ETB (1-120 nmol./larva) compared to larvae treated with ETB (1-900 nmol./larva) alone. JHE activity is expressed as a percentage of the JHE activity produced by epofenonane (100 nmol./larva) alone. Data is adapted from (26).

devoid of AJH activity at doses tested in both the classical AJH
bioassays as well as the one presented here (41 and see above).
Likewise, ETB displays JH-like activity at high doses (> 200 nmol.;
18,26, Figure 1). In L5D3 larvae of T. ni, JH activity in the
whole animal is manifested by normal wandering and prepupal activities leading to apparently normal pupae from which, however, there
is no adult emergence (18). Thus, if apparently normal pupation
occurs, but no adult eclosion is forthcoming, then it is possible
that the compound, at the dose being tested, is acting as a JH.

      With any new procedure, there is always the concern for its
proper use. In our hands, the results of the first two steps of
the bioassay, which easily can be combined into a single test,
have provided a very consistent test of AJH activity that has,
to date, always been confirmed by the later steps in the bioassay.
However, care must be taken in the evaluation of compounds that
display toxic effects. As already mentioned, pesticides like
paraoxon can produce tanned larvae and this tanning process can
be very greatly delayed relative to the controls (solvent only).
However, since the larvae that display these effects were behaving
in an abnormal/intoxicated manner (i.e. flaccid, lack of activity
when probed, hemolymph accumulation, failure to spin, etc.), the
results should be discounted. As a safety measure, however, it
is important to test all suspected AJHs as described in steps 3
and 4. This process will insure that compounds that are general
toxicants will not be mistakenly assigned as AJHs. Finally, it
is suggested that appropriate controls (solvent only) be run at
all times since there can be some day to day variation in the exact
time of ecdysis to the pupa and/or tanning. If at all possible,
a standard such as FMev should also be tested occasionally to insure
that the assay is working properly.
      Unlike some of the more classic AJH bioassays, the one
presented here allows for the rapid (ca. 36 hrs) elimination of
inactive compounds, saving the more intensive labor for only the
more promising compounds. An additional advantage is that only
a relatively small amount of material is needed for the bioassay.
Since the test organism, T. ni, is a major economic pest insect
with a broad host range, it can be assumed that, unlike M. sexta
or O. faciatus, it has a rather well developed detoxification
systems. Thus, compounds active on T. ni are more likely to be
active on other economically important insect pests.
      The bioassay presented here for T. ni is based on a growing
body of knowledge that has been accumulated during the last decade.
While few other insects have been so well studied in terms of
endocrine regulation of metamorphosis, it may be possible to adapt
this bioassay-key to other less studied, but economically more
important insect pests, such as H. virescens, S. frugiperda and
P. includens. Likewise, this bioassay should be easily adapted
to M. sexta for which there is a wealth of endocrinological
information (42,43).
      Compactin exhibits AJH activity in several insects (5,6,29)
and seems to function as an inhibitor of 3-hydroxyl-3-methylglut-
aryl-CoA reductase in both the rat (44,45) and M. sexta (5). Although the methyl ester of compactin was inactive as an AJH in
our bioassay, two of its analogs (L-643,049-01K01 and L-643,737-

00S03) did display AJH activity. Thus, for T. ni, analogs of
compactin would appear to be good candidates for continued
exploration as potential AJHs.

Acknowledgments

We thank B. Bondy for assistance in rearing the insects, and G.
Quistad, D. Schooley, G. Staal (Zoecon Corp.), P. Masner (Dr. R.
Maag Ltd.), P. Magee (Chevron Chemical Co.), and R. A. Dybas (Merck,
Sharp and Dohme Research Laboratories) for providing many of the
compounds used in this study. This research was supported, in
part, by Merck, Sharp and Dohme, the Louisiana Agricultural Experi-
ment Station, Basic Research Grant No. 32-003 from Louisiana State
University (to TCS), North Carolina State University, Department of
Health and Human Resources, National Service Award (to RMR) F32
GM09223-02 from the National Institute of General Medical Science,
and Grant No. 2 Ro 1 ES02710-04 and Research Career Development
Award 5 K04 ES00107-05 (to BDH) from the National Institutes of
Environmental Health Sciences.

Literature Cited

1. Bowers, W. S. In "The Juvenile Hormones"; Gilbert, L. I. Ed.;
   Plenum Press: New York, 1976; pp. 394-408.
2. Staal, G. B. In "Natural Products and the Protection of Plants";
   Marini-Bettolo, G. B. Ed.; Elsevier: New York, 1977; pp. 353-83.
3. Hammock, B. D.; Mumby, S. M. Pestic. Biochem. Physiol. 1978,
   9, 39-47.
4. Kramer, S. J.; Staal, G. B. In "Juvenile Hormone Biochemistry";
   Pratt, G. E.; Brooks, G. T., Eds.; Elsevier/North Holland
   Biomedical Press: New York, 1981; pp. 425-37.
5. Monger, D. J.; Lim, W. A.; Kezdy, F. J.; Law, J. H. Biochem.
   Biophys. Res. Commun. 1982, 105, 1374-80.
6. Edwards, J. P.; Price, N. R. Insect Biochem. 1983, 13, 185-9.
7. Riddiford, L. M.; Roseland, C. R.; Thalberg, S.; Curtis, A. T.
   J. Insect Physiol. 1983, 29, 281-6.
8. Bowers, W. S.; Feldlaufer, M. F. Gen. Comp. Endocrinol. 1982,
   47, 120-4.
9. Staal, G. B.; Henrick, C. A.; Bergot, B. J.; Cerf, D. C.;
   Edwards, J. P.; Kramer, S. J. In "Regulation of Insect
   Development and Behavior"; Sehnal, F.; Zabza, A.; Menn, J. J.;
   Cymborowski, B. Eds.; Wroclaw Technical University Press:
   Wroclaw, Poland, 1981; pp. 324-40.
10. Bowers, W. S. In "Insecticide Mode of Action"; Coats, J. R. Ed.;
    Academic Press: New York, 1982; pp. 403-27.
11. Metcalf, C. L.; Flint, W. P.; Metcalf, R. L. "Destructive
    and Useful Insects"; McGraw-Hill Book Co.: New York, 1962;
    pp. 664-5.
12. Chalfant, R. B. J. Econ. Entomol. 1969, 62, 1343-4.
13. Harris, C. R.; Svec, H. J.; Chapman, R. A. J. Econ. Entomol.
    1978, 71, 642-644.
14. Hammock, B. D.; Jones, D.; Jones, G.; Rudnicka, M.; Sparks,
    T. C.; Wing, K. D. In "Regulation of Insect Development and
    Behavior"; Sehnal, F.; Zabza, A.; Menn, J. J.; Cymborowski,
    B. Eds.; Wroclaw Technical University Press: Wroclaw, Poland,
    1981; pp. 219-35.

15. Hammock, B. D. In "Comprehensive Insect Physiology, Biochemistry and Pharmacology"; Kerkut, G. A.; Gilbert, L. I. Eds.; Pergamon Press: New York, 1984; Vol. 7, In Press.
16. Sparks, T. C.; Hammock, B. D. J. Insect Physiol. 1979, 25, 551-60.
17. Jones, G.; Hammock, B. D. J. Insect Physiol. 1983, 29, 471-5.
18. Sparks, T. C. J. Insect Physiol. 1984, 30, 225-34.
19. Sparks, T. C.; Hammock, B. D. Pestic. Biochem. Physiol. 1980, 14, 290-302.
20. Pratt, G. E.; Finney, J. R. In "Crop Protection Agents - Their Biological Evaluation"; McFarlane, N. R. Ed.; Academic Press: New York, 1976; pp. 113-32.
21. Edwards, J. P.; Bergot, B. J.; Staal, G. B. J. Insect Physiol. 1983, 29, 83-9.
22. Quistad, G. B.; Cerf, D. C.; Schooley, D. A.; Staal, G. B. Nature. 1981, 289, 176-7.
23. Quistad, G. B.; Staiger, L. E.; Cerf, D. C. J. Agric. Food Chem. 1982, 30, 1151-4.
24. Beckage, N. E.; Riddiford, L. M. J. Insect Physiol. 1983, 29, 633-7.
25. Farag, A. I.; Varjas, L. Ent. exp. & appl. 1983, 34, 65-70.
26. Sparks, T. C.; Wing, K. D.; Hammock, B. D. Life Sciences. 1979, 25, 445-50.
27. Kiguchi, K.; Mori, T.; Akai, H. J. Insect Physiol. 1984, 30, 499-506.
28. Hammock, B. D.; Kuwano, E.; Ketterman, A.; Scheffrahn, R. H.; Thompson, S. N.; Sallume, D. J. Agric. Food Chem. 1978, 26, 166-70.
29. Hiruma, K.; Yagi, S.; Endo, A. Appl. Ent. Zool. 1983, 18, 111-5.
30. Hammock, B. D.; Wing, K. D.; McLaughlin, J.; Lovell, V. M.; Sparks, T. C. Pestic. Biochem. Physiol. 1982, 17, 76-88.
31. Sparks, T. C.; Willis, W. S.; Shorey, H. H.; Hammock, B. D. J. Insect Physiol. 1979, 25, 125-32.
32. Hammock, B. D.; Sparks, T. C. Analyt. Biochem. 1977, 82, 573-9.
33. Hammock, B. D.; Roe, R. M. Methods in Enzymology, 1984, In Press.
34. Jones, G.; Hammock, B. D. J. Insect Physiol. 1985, 31, In Press.
35. Cymborowski, B.; Stolarz, G. J. Insect Physiol. 1979, 25, 939-42.
36. Hiruma, K. Gen. Comp. Endocr. 1980, 41, 392-9.
37. Sieber, R.; Benz, G. Physiol. Ent. 1980, 5, 283-90.
38. Safranek, L.; Cymborowski, B.; Williams, C. M. Biol. Bull. mar. biol. Lab. Woods Hole 1980, 158, 248-56.
39. Abdel-Aal, Y. A. I.; Roe, R. M.; Hammock, B. D. Pestic. Biochem. Physiol. 1984, 21, 232-41.
40. Prestwich, G. D.; Eng, W. -S.; Roe, R. M.; Hammock, B. D. Arch. Biochem. Biophys. 1984, 228, 639-45.
41. Sparks, T. C.; Wing, K. D.; Hammock, B. D. Unpublished data.
42. Granger, N. A.; Bollenbacher, W. E. In "Metamorphosis, A Problem in Developmental Biology"; Gilbert, L. I.; Frieden, E. Eds.; Plenum Press: New York, 1981; pp. 105-37.
43. Sparks, T. C.; Hammock, B. D.; Riddiford, L. M. Insect Biochem. 1983, 13, 529-41.
44. Endo, A.; Kuroda, M.; Tanzawa, K. FEBS Letters 1976, 72, 323-6.
45. Endo, A. TIBS 1981, 6, 10-13.

RECEIVED November 8, 1984

Applications of Immunoassay to Paraquat and Other Pesticides

J. M. VAN EMON, J. N. SEIBER, and B. D. HAMMOCK

Department of Environmental Toxicology, University of California, Davis, CA 95616

The enzyme-linked immunosorbent assay (ELISA) is a
rapid immunochemical procedure which can be used for
trace analysis. We have applied the procedure to
paraquat and other compounds difficult to analyze by
the more classical methods. The immunoassay for
paraquat shows the practicality of the method for
fortified and actual residue samples, and is being
compared with a gas chromatography procedure. Our
work with the ELISA illustrates that the
immunochemical technology can be used to solve
problems encountered in pesticide residue analysis.

It has been stated that progress in pesticide analysis will no
longer be made in search of the proverbial zero residue level of
detectability, but rather will lie in devising methods of greater
selectivity for the positive identification of nanogram quantities
of pesticide residues (1). We could add to this statement the
need for assays of greatly reduced cost and increased speed. Many
residue procedures are too expensive to be used routinely in
regulatory procedures or, perhaps of greater importance, to be
employed effectively in optimizing pesticide usage and monitoring
worker health and safety. Immunochemical methods of analysis offer
many advantages, including sensitivity, specificity, and speed of
analysis (2). Compounds which are most difficult to analyze by
classical procedures due to high polarity or low volatility are
frequently amenable to analysis by immunochemistry. We are now
investigating enzyme-linked immunosorbent assays (ELISAs) rather
than radioimmunoassays (RIAs) for pesticide residue work. ELISAs
are quicker, cheaper, and safer than RIAs as radioactivity is not
used. However, as all immunoassays function on the principle of
mass action, the same immunochemical tools can be used to devise a
number of different assay procedures.

General ELISA Methodology

Immunoassays are physical rather than biological assays; they
possess the specificity and sensitivity of bioassays with the speed

0097–6156/85/0276–0307$06.00/0

and precision of physical assays. Specific antibodies are raised
in an experimental animal in response to a large foreign molecule
(antigen). Most pesticide molecules are not large enough to
stimulate the immune system, and must be conjugated to a large
molecule such as a protein. Antibodies against the pesticide are
then obtained using this pesticide-protein conjugate. A small
molecule which becomes immunogenic after attachment to the large
carrier molecule is termed a hapten. The pesticide may already
have a functionality, such as -OH, -SH, -COOH, or $-NH_2$, useful
for conjugation to the carrier, but frequently a derivative of the
pesticide possessing such functionality must first be synthesized
before conjugation can occur. In either case antibodies can be
obtained which are directed against the original pesticide.

Serum antibodies are heterogeneous molecules of varying
antigen specificity and affinity. Within this polyclonal
population, there are probably antibodies present which recognize
the carrier protein, but the contribution of these antibodies to
assay binding can be eliminated by using a different carrier
(2). Monoclonal antibody technology could be used to select
the clone with the highest recognition to hapten. Although
polyclonal antibodies are adequate for most ELISAs, monoclonal
antibodies could be developed against pesticide haptens yielding an
analytical reagent that is physically, chemically, and
immunologically homogeneous. As monoclonal antibodies would be in
a virtually unlimited supply, ELISAs could be easily standardized
as several laboratories would be using the same antibody clone.
Although a very small amount of antibody is needed per ELISA,
having a large supply of monoclonals could alleviate fears of
eventually exhausting one's supply. Hammock and Mumma (2)
discuss some of the advantages of hybridoma technology. Yet, it is
important to consider that, in some cases, polyclonals will be
superior to monoclonal antibodies. Unless one is developing a
sophisticated system for avoiding the separation step in pesticide
immunoassay, it will be a rare situation where production of
monoclonal antibodies for pesticide residue analysis can be
justified on a purely scientific basis. However, as immunoassay of
pesticides moves into the private sector, there will be compelling
administrative and legal pressures to employ monoclonal antibodies.
For these reasons they may dominate the field in a few years.

The ELISA is based on the fact that antigen or antibody can be
attached to a solid-phase support while retaining immunological
activity, and that either antigen or antibody can be linked to an
enzyme with the complex retaining both immunological and enzymatic
activity. A variety of enzymes, including alkaline phosphatase,
horseradish peroxidase, and glucose oxidase have been linked to
antibodies and antigens. This method has been used successfully
for detection of either antigen or antibody (3-4), and it has
been used by us for the detection of nucleotides, insecticides,
surfactants and a variety of other compounds.

To perform a microplate ELISA for the measurement of antigen,
plates sensitized with the specific antigen are incubated with a
mixture of reference antibody and the test sample. If antigen is
present in the test solution, it combines with the reference
antibody which cannot then react with the sensitized plate. The
amount of antibody attached to the solid phase is then indicated by

an enzyme labelled anti-immunoglobulin conjugate and enzyme substrate. There is a proportional relationship between the amount of inhibition of substrate converted to products in the test sample and to the amount of antigen in the test system. We routinely run samples in plastic plates containing 50 wells with which all of the procedures can be executed very rapidly. The end point of the assay is the bright yellow color of p-nitrophenol, an end product from the alkaline phosphatase-mediated hydrolysis of p-nitrophenyl phosphate; this product can be visually estimated for semi-quantitative answers or rapidly and precisely measured in an inexpensive colorimeter for quantitation.

Immunoassays can be designed to analyze parent compounds and metabolites separately or as a group. Ercegovich (5) obtained antibodies against parathion which also detected its metabolite p-nitrophenol. Other workers were successful in raising antibodies to DDT and malathion metabolites (6-7). Although the specificity of immunoassays are usually very high there is no guarantee against cross reactivity. Just as good chromatographic techniques require controls, so do immunoassays. Fortunately immunoassays are quickly and easily performed so that the necessary controls can be run to check for interferences. The sensitivity and selectivity of immunoassays can also greatly reduce the cost of analysis by minimizing the amount of sample preparation.

Examples of Pesticide Immunoassays

We will first review several of the immunoassays developed in this laboratory as they illustrate some of the advantages of immunoassay in pesticide residue analysis. Then we will move on to a more detailed discussion of our recent analytical work with the herbicide paraquat.

Diflubenzuron. The benzoylphenyl urea insect growth regulators, for example, pose a formidable residue analysis problem. The compounds are nonvolatile and thus must be derivatized for GC analysis by a rather arduous chemical procedure. The immunoassay developed in this laboratory is much more sensitive and reproducible at a fraction of the cost and can be used to analyze the more difficult matrices such as milk. For instance, a sensitivity of 1 ppb is routinely obtained when milk is added directly to the assay (8). A series of partition steps can also be added to further clean diflubenzuron milk extracts yielding a sensitivity in the low ppt range (8). However this increase in sensitivity may not be needed since methods in current use provide a detection limit of only 10-50 ppb.

An impressive aspect of immunoassays is their specificity. One immunoassay for diflubenzuron can distinguish it from the very closely related BAY SIR 8514 as well as a variety of other closely related materials (9). High resolution HPLC columns can resolve these compounds but the analysis is slow and expensive. The ELISA can distinguish these materials when applied directly, or less specific assays can be used as a highly selective detector when used as an adjunct to HPLC.

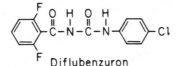

Diflubenzuron   BAY SIR 8514

Triton X and N. Surfactants have many industrial applications
and are found in such diverse products as pesticide formulations and
cosmetics. These nonionic compounds are difficult to extract, clean
up, and analyze and, consequently, no sensitive method existed for
their analysis. An ELISA was developed in this laboratory which
distinguishes between the nonionic surfactants Triton X and Triton N
(10). These compounds, mixtures of ethoxylates of varying length
of 4-(1,1,3,3-tetramethylbutyl)phenol and 4-nonylphenol, give
numerous overlapping peaks upon chromatographic analysis and have
almost identical UV and IR spectra. An ELISA has been developed for
class selective detection of the Triton X series; the antibody
detects the 4-(1,1,3,3-tetramethylbutyl)phenyl moiety and does not
distinguish among molecules with ethoxylated side chains of varying
lengths.

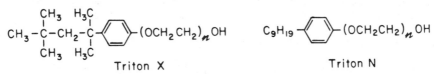

Triton X   Triton N

S-Bioallethrin. The pyrethroid S-bioallethrin (1R,3R,4'S) and
its inactive isomer (1S,3S,4'R) can be readily distinguished by
another ELISA procedure, illustrating the assay's ability to
determine chirality at the residue level (Figure 1). Antibodies
were raised against S-bioallethrin using the allethrin hemisuccinate
conjugated to various proteins (11-12).

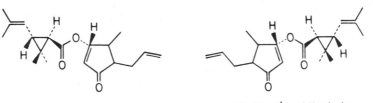

S- Bioallethrin   1S,3S,4'R Allethrin

Pyrethroids, as well as some carbamate and organophosphate
insecticides, are marketed as isomer mixtures, each isomer having a
different degree of activity. It can be expected that environmental
degradation and metabolism will occur preferentially with some
isomers. Thus the ability to distinguish between optical isomers at
the residue level may be of critical importance in monitoring the
safety of treated substances.

Paraquat

During a recent year, over 950,000 pounds of paraquat dichloride

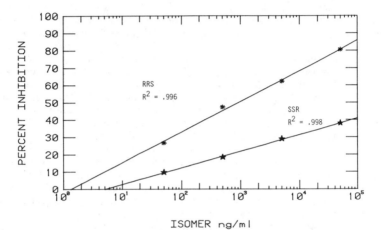

Figure 1. Standard curves for optical isomers of S-bioallethrin using ELISA.

(1,1'-dimethyl-4,4'-bipyridinium dichloride) was used in California agriculture (13). Much of this herbicide was used on cotton, primarily as a harvest aid. The value of this cotton production is estimated to exceed one billion dollars (14). Although it is extensively used, little is known regarding the long term chronic effects of paraquat exposure.

Paraquat has a number of biological effects, but its main biochemical action is apparently as a potent redox uncoupler. Regardless of the route of administration, symptoms of paraquat poisoning are centered in the lungs leading to fibrosis and pneumonia (15). Submicrogram quantities of paraquat deposited in the lung can cause fibrotic lesions (16) which could lead to asthma and emphysema symptoms in chronically exposed individuals (17). Unfortunately, the conventional methods of analysis of paraquat are too laborious and/or insensitive to handle the large sample load generated by rigorous studies needed to evaluate human exposure.

Immunochemical procedures were developed to overcome this difficulty. Factori and Hunter (18) reported a radioimmunoassay for paraquat, and Niewola et al., (19) reported an ELISA for estimating paraquat in serum. This ELISA is similar to ours but uses a different enzyme, incubation temperature and time, and shows a high degree of cross reactivity with ethyl paraquat. Methods similar to those previously reported were used to generate paraquat haptens for our study. N-Methyl-4-(4-pyridyl)pyridinium bromide was reacted with ethyl 5-bromovalerate forming a paraquat analogue capable of conjugating with a protein. New Zealand white rabbits were injected with a paraquat-protein conjugate and antibodies capable of recognizing paraquat were obtained. The antibodies are quite selective showing no cross reactivity to compounds structurally related to paraquat (e.g. ethyl paraquat, N,N'-dimethyl-4,4'-bipiperidine) or compounds used in conjunction with paraquat (e.g. diquat).

Using the selective antibody for paraquat, environmental samples can be analyzed with little or no cleanup. Sample throughput can be measured in samples per hour rather than days per sample. This was a remarkable discovery as paraquat is notorious for requiring extensive sample preparation. The limit of detection (2 ng/ml) in the ELISA procedure is also much lower than the colorimetric procedure, which has a sensitivity of 200 ng/ml (20), and lower than the GC procedure (Figure 2) (21). Perhaps the greatest advantage is the speed of the ELISA enabling large numbers of samples to be analyzed without requiring an extensive cleanup procedure, thus reducing analytical time.

Gas-liquid chromatography following reduction of paraquat to the mono- and diunsaturated derivatives (21) is of adequate sensitivity for most work when N-selective detectors are employed. Seiber and Woodrow (22) modified this method for assaying paraquat in air samples. The method is time consuming and labor intensive, involving acid extraction and many concentration and evaporation steps. The maximum sample output per analyst per day is 6-8 with no duplicates. The reported recovery efficiency was 75% (22), although an efficiency closer to 50% is frequently encountered in practice. A modified acid extraction combined with analysis by the ELISA provides recoveries of 75% (Figure 3). This

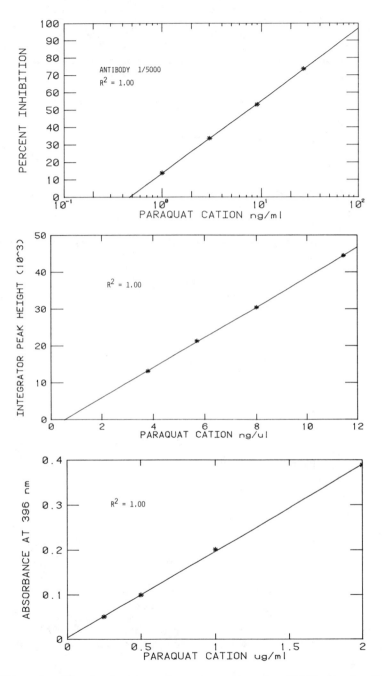

Figure 2. Standard curves for paraquat using ELISA (top), GC (middle), and colorimetry (bottom).

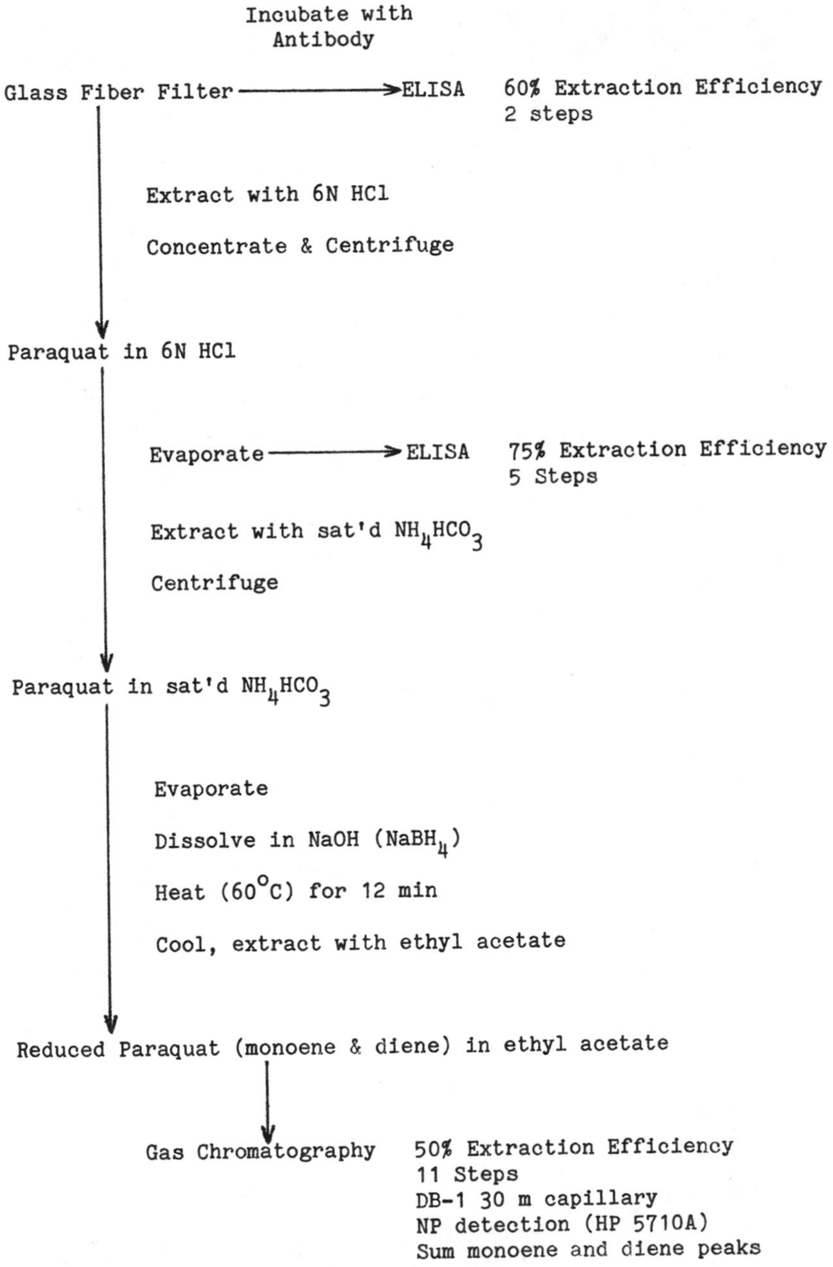

                              Incubate with
                                Antibody

Glass Fiber Filter ──────────►ELISA   60% Extraction Efficiency
         │   2 steps
         │
         │   Extract with 6N HCl
         │
         │   Concentrate & Centrifuge
         │
         ▼
Paraquat in 6N HCl

                   Evaporate ────────►ELISA   75% Extraction Efficiency
         │   5 Steps
         │
         │   Extract with sat'd NH_4HCO_3
         │
         │   Centrifuge
         │
         ▼
Paraquat in sat'd NH_4HCO_3

                   Evaporate
         │
         │   Dissolve in NaOH ($NaBH_4$)
         │
         │   Heat ($60^{\circ}C$) for 12 min
         │
         │   Cool, extract with ethyl acetate
         │
         ▼
Reduced Paraquat (monoene & diene) in ethyl acetate
                   │
                   │
                   ▼
         Gas Chromatography   50% Extraction Efficiency
                               11 Steps
                               DB-1 30 m capillary
                               NP detection (HP 5710A)
                               Sum monoene and diene peaks

Figure 3. Comparison of sample preparation steps for analysis
of paraquat on air filters using ELISA and gas chromatography.

illustrates the successful adaptation of an extraction technique
intended for use with GC determination to an immunochemical method
of analysis. Alternatively, the antibody itself can be used to
extract paraquat from a glass fiber filter, eliminating the
time-intensive extraction procedure. Using the specific antibodies
for extraction with subsequent detection by ELISA, 50 samples can be
analyzed in triplicate in one day with a routine extraction
efficiency of 60% (Figure 3). However, this method uses a rather
large amount of antibody making the modified acid extraction
procedure preferable for analyzing a large sample load.

     This rapid sample-processing capability makes it feasible to
measure exposure of field workers literally on the same day the
exposures occur. In order to test this capability air sampling was
conducted in the San Joaquin valley during a paraquat application on
cotton, and the samples are now being analyzed using both ELISA and
GC methods to compare the two techniques in terms of speed,
precision (by calculating the percent coefficient of variation), and
accuracy (by comparing results from fortified samples using both
techniques). Preliminary results indicate that the ELISA compares
well with GC literature values in these three parameters and
additionally has the anticipated greater sample throughput
capability.

Conclusions

It is probable that in certain situations immunochemical methods
will provide distinct advantages over conventional analytical
methods. However, it is unlikely that immunochemical methods will
completely replace current, established analytical methods of
pesticide analysis (5). This is in spite of the fact that
chemical classes currently assayed by immunochemical techniques in
clinical analytical labs contain the same type of functional groups
as many pesticides.

     ELISA could potentially be used advantageously in many types of
exposure and monitoring situations, for paraquat and other
pesticides amenable to ELISA analysis. An obvious use of ELISA is
the detection of pesticide residue levels in plant and animal
tissues, or food extracts. Biological specimens such as plasma and
urine currently analyzed by RIA seem particularly amenable to
analysis by ELISA. Portable field kits could be developed to
determine safe worker re-entry times into treated fields.
Environmental samples such as soil, water, and air, can be analyzed
by the ELISA. Pesticide conjugates have been proposed for skin
testing of individuals suspected of sensitivity to pesticides
(6); the ELISA could be used to detect specific antibodies in
these individuals and aid in exposure studies.

     Antibodies have been raised against representative compounds
from the major classes of pesticides. Although the ELISA will be
useful for individual analysis of a wide variety of compounds, if
one needed to analyze several different compounds simultaneously in
one matrix immunoassay may not be the method of choice, due to the
large amount of controls and standards needed. However, it could be
successfully used for the rapid screening of a large number of
samples for the presence of specific types of pesticides and for
confirmatory tests (5). The work reported here with paraquat,

allethrin, diflubenzuron, Triton X and Triton N provides evidence of
the ELISA's ability to distinguish closely related compounds. The
ELISA promises to be a good supplement to current methods of residue
analysis.

Literature Cited

1. Zweig, G. In "Essays in Toxicology"; Blood F.R. Ed.; Academic
    Press: New York, 1970; Chap. 3.
2. Hammock, B. D.; Mumma R. O. In "Recent Advances in Pesticide
    Analytical Methodology"; Harvey J.; Zweig, G., Eds.; ACS
    SYMPOSIUM SERIES No. 136, American Chemical Society:
    Washington, D.C., 1980; Chap. 18.
3. Voller, A.; Bidwell, D.E.; Bartlett, A. Bull World Health
    Org. 1976, 53, 55-65.
4. Van Weeman, B.K.; Schuurs, A.H.W.M. FEBS Letters 1971, 15,
    232-6.
5. Ercegovich, C.D. In "Pesticide Identification at the Residue
    Level"; Gould, R.F., Ed.; ADVANCES IN CHEMISTRY SERIES No. 104,
    American Chemical Society: Washington, D.C., 1971; Chap. 11.
6. Centeno, E.R.; Johnson, W.J.; Shehon, A.H. Int. Arch.
    Allergy Appl. Imm. 1970, 37, 1-13.
7. Haas, G.J.; Guardia, E.J. Proc. Soc. Expt. Biol. Med.
    1968, 129, 546-51.
8. Wie, S.I.; Hammock, B.D. J. Agric. Food Chem. 1984,
    Accepted.
9. Wie, S.I.; Hammock, B.D. J. Agric. Food Chem. 1982,
    30, 949-57.
10. Wie, S.I.; Hammock, B.D. Anal. Biochem. 1982, 125, 168-176.
11. Wing, K.D.; Hammock, B.D.; Wuster, D.A. J. Agric. Food
    Chem. 1978, 26, 1328-33.
12. Wing, K.D.; Hammock, B.D. Experientia. 1979, 35, 1619-20.
13. "Pesticide Use Report by Commodity." California Department of
    Food and Agriculture, 1983.
14. "California Agriculture." California Department of Food and
    Agriculture, 1982.
15. Smith, P.; Heath, D. "Paraquat" CRC Crit Revs. Toxicol. 4,
    411-45.
16. Zavala, D.C.; Rhodes, M.L. Chest 1978, 74, 418-20.
17. Maddy, K.T. "Human Health Problems with the Herbicide Paraquat
    in California 1965 through 1974." California Department of Food
    and Agriculture, 1975.
18. Fatori, D.; Hunter, W.M. Clinic Chimica Acta 1980, 100,
    81-90.
19. Niewola, Z.; Walsh, S.T.; Davies, G.E. Int. J. Immuno-
    pharmac. 1983, 5, 211-18.
20. Lott, P.F.; Lott, J.W. J. Chrom. Sci. 1978, 16, 390-5.
21. van Dijk, A.; Ebberink, R.; deGroot, G.; Maes, R.A.A.; Douze,
    J.M.C.; van Heyst, A.N.P. J. Analy. Tox. 1977, 1, 151-54.
22. Seiber, J.N.; Woodrow, J.E. Arch. Environm. Contam. 1981,
    10, 133-49.

RECEIVED November 8, 1984

A Direct Correlation Between Bioassay and ELISA for
the Bacillus thuringiensis var. israelensis δ-endotoxin.

Peter Y. K. Cheung and Bruce D. Hammock

Department of Entomology
University of California
Davis, California 95616

A partially purified Bacillus thuringiensis var.
israelensis (Bti) δ-endotoxin was used to immunize
rabbits. The antisera obtained have an improved specificity towards the mosquito larvacidal activity of
the toxin, as opposed to antiserum raised when the
whole crystal was used as immunogen. Using a two
step/indirect ELISA (enzyme linked immunosorbent
assay) procedure developed in our laboratory, fourteen
experimental formulations were tested, and the results
were compared with bioassays. An average of 69.1
international units ± 20% c.v. was found to associate
with each ug of toxin detected by the ELISA. Our data
indicate that when toxin specific antisera are
available, immunoassays can be used to predict the
biological activity of Bti samples with reasonable
accuracy.

RECEIVED December 17, 1984

Euplectrus plathypenae Parasitization of Trichoplusia ni. Effect
on Weight Gain, Ecdysteroid Titer and Molting.

Thomas A. Coudron and Thomas J. Kelly [1]

U. S. Department of Agriculture
Biological Control of Insects Research
Laboratory, P. O. Box A
Columbia, MO 65205

Euplectrus spp. (Hymenoptera: Eulophidae) are gre-
garious ectoparasites of several lepidopteran species
that are pests to agricultural crops. Parasitization
of Trichoplusia ni larvae by Euplectrus plathypenae
results in an inhibition of the larval larval molting
process in the host. After parasitization the host
temporarily refrains from eating and lags behind the
synchronized non-parasitized larvae in weight gain.
However the parasitized larvae does resume eating and
continues to increase its body weight prior to the
development of the parasite larvae. There is no sign
of new cuticle formation in parasitized larvae. A
prolonged association between the parasite and host is
not necessary to elicit an effect on the host. Larvae
which are immediately separated from the parasitic egg
will fail to molt into the next instar. The inhibition of host molt may be related to the absence of
peak in the ecdysteroid titer in the hemolymph which
occurs 20 hrs before ecdysis in non-parasitized T. ni.
This unique type of host development control appears
to be distinct from the paralyzing effect of many
Hymenoptera venoms and the algogenic effects of some
Hemiptera salivary toxins.

RECEIVED February 11, 1985

---

[1] Current address: Insect Reproduction Laboratory, BARC-E,
Bldg. 306, Beltsville, MD 20705

CONTROL OF PESTS
WITH NATURAL PRODUCTS

Role of Natural Product Chemistry

J. R. PLIMMER

Agrochemicals and Residues Section, Joint FAO/IAEA Division, International Atomic Energy Agency, Vienna, Austria

The rapidly expanding growth of knowledge of natural product structures now provide clearer understanding of biochemical mechanisms. This has made possible "biorational" approaches to the design of pest control agents. Natural products may be of potential value in pest control in several ways. They may be a source of structures for screening. They may possess activity that is applicable to pest control directly or after structural modification of the original structure. Finally, the recognition of their function in nature may suggest new approaches to pest control. However, their practical application may be limited by economics. Resistant plants are important in managing insect pests and their resistance may arise from many factors. Some plants contain insecticidal principles that may be exploited. Compounds that modify insect behaviour are not directly lethal, but may be valuable in pest control. However, their efficacy may be difficult and costly to determine.

During the latter half of the nineteenth century and the early decades of the twentieth century, the groundwork of structural organic chemistry was laid. In 1828, the theory that a "vital force" was utilized in plants and animals to elaborate natural products was disproved, when Wöhler announced that urea could be formed by heating ammonium cyanate. The investigation of the theoretical basis of organic chemistry received its stimulus not only from the contemporary spirit of philosophical inquiry, but also from the prospects of a practical outcome. The industrial revolution had as its basis the availability of energy in the form of coal. Coal gas was used for lighting early in the nineteenth century. Coal tar was recognized as a source of organic chemicals, and W.H. Perkin's discovery of the first coal tar dye in 1856 occurred when he attempted preparation of quinine by oxidation of aniline. Subsequently, synthetic dyestuffs rapidly replaced those from natural sources. The ability to

0097–6156/85/0276–0323$06.00/0
© 1985 American Chemical Society

produce a variety of synthetic chemicals from coal tar and
subsequently from petroleum feedstocks has greatly influenced the
development of the chemical industry.

The prevalence of structural types based on aromatic or
heteroaromatic nuclei among chemicals available from coal tar or
petroleum contrasts with the structural types that represent the
major groups of natural products. Chemicals from natural sources
are a rich source of structural diversity and reflect the
complexity of biological systems in which the processes of
biosynthesis and the biological roles of molecules have undergone
continual change and modification.

Natural product chemistry remains a challenging branch of
study. Progress in both theory and practice of organic chemistry
has hastened the process of elucidation of molecular structure.
The paucity of starting material and relatively primitive
technology had, in some cases, meant that almost a century was
required to establish details of molecular structure. Molecules
of even moderate complexity presented a major challenge until
modern instrumental techniques simplified the problems of
separation, characterization, and stereochemistry. Within the
last three decades, it has become possible to derive structural
information from a few nanograms of material and the elucidation
of structure and synthesis of biological polymers has now become
almost routine. In addition, a better understanding of
conformational principles and the influence of stereochemistry on
reactivity of molecules now contributes greatly to rapid
elucidation of structure.

Biorational Approaches

Such information has led to explosive growth in the understanding
of biochemical processes. Knowledge of metabolism, biosynthetic
processes, neurochemistry, regulatory mechanisms, and many other
aspects of plant, animal and insect biochemistry has provided a
basis on which the mode of action of a pesticide may often be
more clearly understood. The exploitation of biological
information can lead to the synthesis of a new molecule designed
to act at a particular site or block a key step in a biochemical
process.

The potential value of this approach has generated
increasing interest in the functioning of insect neurosecretory
systems. That selective chemical control methods might emerge
from an increased understanding of insect biochemistry became
apparent with the discovery of the insect juvenile hormones.
Investigation of secretions from the corpora allata led to the
identification of insect juvenile hormones and provided a
stimulus for the study of the molecular basis of insect
physiology (1, 2, 3). Juvenile hormone analogues are proving
very valuable for selective control of insects. A potentially
more attractive method appears to be the use of chemicals that
would block juvenile hormone production, because such compounds
would shorten the lifetime of the larvae, the life stage of the
insect which inflicts most damage on the host. Screening of

plant extracts led to the discovery that precocenes (from the
plant <u>Ageratum</u> <u>houstonianium</u>) possessed this property (4), but
the activity is considered to result from cytotoxicity and
degeneration of the <u>corpora</u> <u>allata</u> (5, 6) and not from specific
interference with biosynthetic routes.

Investigation of the biosynthetic pathways involved in the
production of juvenile hormones (tetrahydro-4-fluoromethyl-4-
hydroxy-2H-pyran-2-one) in the <u>corpora</u> <u>allata</u> revealed the
involvement of mevalonolactone as an intermediate and led to the
synthesis of potential blocking agents with the subsequent
discovery of the activity of fluoromeralonolactone as an
inhibitor of juvenile hormone biosynthesis (7). Exploitation of
these findings has continued in a search for more potent
compounds of practical value (8). Such discoveries gave rise to
the term "biorational" to describe synthetic insecticides that
are designed by consideration of specific biochemical targets.

These developments indicate potential new modes of selective
insecticidal action. Through a better understanding of the
physiological processes, the basis of screening can be broadened
and the increased knowledge of the diverse biochemical pathways
suggests new approaches to the design of pesticides.

Natural Products in Pest Control

The utilization of natural products in pest control may be
considered from a number of standpoints. First, the variety of
structural types provides a rich source of compounds or models
for conventional screening programmes. Second, consideration of
the known biological activity of a natural product may lead to
its application in pest management, either directly or after
structural modification. Third, the recognition and
understanding of the function of a chemical in nature may reveal
potential approaches to pest management.

For control of some pest problems, materials from natural
sources have proved extremely successful. In particular, many
plant, human, and animal diseases are controlled by antibiotics
produced by microorganisms. Insecticides from plant sources have
been used effectively for many years, but the scale on which they
have been used does not compare with that of the synthetic
organic insecticides introduced in the years following World War
II.

Application of natural products to pest control is subject
to similar considerations to those that affect synthetic
compounds. Elucidation of structure may be a relatively routine
matter, but synthesis may present difficulties, particularly
where there are several stereochemical possibilities. There are
often great differences in biological activity among different
stereoisomers and the presence of inactive isomers in a product
may be undesirable. However, such technical problems can be
overcome as in the case of the synthetic pyrethroids where
specific stereoisomers of high purity are now produced on a
commercial scale.

Screening for biological activity also presents a major

problem. Biological testing for many types of activity is
important. In screening for insecticidal activity, the number of
compounds passing through research and development steps from
synthesis to market per commercial product was 14,500 in 1979
compared with 1,800 in 1956 (9). The number of compounds that
must be screened will rise considerably as the end of the century
approaches. Thus, screening must be rapid or modified procedures
must be devised. The problem of testing for types of biological
activity other than direct toxicity to the target often presents
additional complexity.

If the type of activity to be examined is not directly
lethal, i.e. control of the pest or suppression of pest
population is achieved through effects on behaviour, development
or reproduction, the problem of biological testing may become
much more difficult, as may also the problem of demonstrating
efficacy. It is in these areas in which the spectrum of
biological activity of natural products is most likely to be
concentrated. The interface between host and pest organisms is
modified by a variety of chemical interactions. To understand
these complex interrelationships is a difficult goal, but the
exploitation of "natural products" for the benefit of agriculture
will depend on progress towards it.

Economic Aspects

A major problem in achieving this goal is the question of
resources. The U.S. loses a considerable percentage of its
agricultural production to pests and also pays heavily to control
pests. Such losses have been estimated at about $10 billion
annually (10). However, the benefit of reducing losses cannot be
measured completely in monetary terms. Factors such as improved
environmental quality, conservation and utilization of land and
water resources are difficult to include in the arithmetic of
potential benefits. The supremacy of pesticide chemicals as
control agents has influenced the process of controlling pests by
emphasizing the profitability of a market as a major criterion
for development. Such a criterion is useful, but limited in its
applicability. The elimination of the screwworm fly from the
southern United States provides an example of a successful
alternative approach to control of a major pest. Benefits to the
farmer and consumer are a measure of the degree of success of
this programme.

The development of individual natural product structures as
pesticides will be subject to the same economic factors that
affect synthetic pesticides. Natural products do not differ from
compounds synthesized in the laboratory, but they may, as
products of biological processes, be more readily degraded than
many man-made structures. Although the potential for facile
degradation may have favourable implications for environmental
safety, there is little justification for the assumption a
priori, that because a compound is a natural product, it
possesses no undesirable toxicological properties. Toxicological
tests must be performed for both natural and man-made compounds
before registration as pesticides.

The cost of developing and registering a pesticide for the
U.S. market is about $20 million (11). Therefore, the compounds
developed must have a large potential market. For an
insecticide, this implies broad spectrum activity and use on
several major crops. Worldwide, only a few major industrial
corporations possess adequate resources to undertake such
activity. Thus, it will be difficult to realise the potential of
natural products for pest control without recourse to a broader
institutional framework.

The biological activity associated with natural products is
often species-specific. For example, suppression of insect
population by mass-trapping or by mating disruption relies on
technology that is specific for the target insect. Such specific
pest control methods have the merit that they do not adversely
affect beneficial insects, but they are rarely appealing to the
major commercial investor. The specificity of such methods and
their low potential for environmental pollution appears to offer
great promise and, to stimulate their development, there has been
an appropriate response at the political level. Simplified
toxicity and other testing procedures have been proposed by the
U.S. Environmental Protection Agency for biochemicals to
facilitate registration (12).

Although the regulatory agencies have indicated their
interest in stimulating new methods of pest control as
alternatives to the use of conventional pesticides, the problems
of economic and scientific development of such techniques
remain. There is often much disparity in progress in different
fields of science. Breakthroughs often arise when new techniques
or ideas are transferred from one field of endeavour to another.
Such a process of cross fertilization is hindered by rigid
academic separation of basic disciplines. The information
required for successful pest control does not reside solely in
the hands of the chemist, but the cooperation of several
specialists is essential. This applies particularly to methods
of insect control that involve modification of behaviour through
the use of attractants, repellents or resistant plants.

Resistant Plants and Ecological Chemistry

Natural product chemistry has evolved through the stages of
structural elucidation to the examination of the function of
natural compounds within ecological systems. Ecological
chemistry is a growing field of science in which plant-insect,
plant-plant, and other interorganismal relationships can be
examined in terms of the effects of chemicals on biological
functions, thus providing a more fundamental appreciation of
plant resistance and the interrelationship between pests and
agricultural crops.

There is a chemical basis for the interdependence of insect
and plant life within an ecosystem. Although the influence of
herbivores, diseases, etc. and the morphology of insects and
plants are also important, the nutritional status of the plant
hosts significantly affects the survival of an insect community

and its vigor. Despite similarities in the major constituents of
plants, insects are highly selective in their feeding and other
behaviours, such as oviposition, that occur in association with a
specific host. In this respect, secondary plant products are
extremely important and many thousands of secondary plant
compounds have been elaborated during the course of plant
evolution. Swain (13) discussed the occurrence and variation of
secondary compounds in terms of plant and insect coevolution.
Some of the controversy concerning the role of insects in
coevolution has been summarized in an article by Feeny (14). In
terms of agricultural pest control, he suggests that research
should concentrate on the diversity of compounds in plants that
are toxic at the metabolic level (antibiosis) rather than at the
behavioural level (nonpreference resistance). Feeny also
discusses the concept of "apparency", which is the susceptibility
of plants or plant tissue to be discovered by insects.
Agricultural practices often favour the "apparency" of plants
when monocultures extending over large areas present an
attractive host for insects. Thus, additional defences are
needed over and above those required in a natural community.
Defences include crop protection chemicals and the cultivation of
plants with a high degree of insect resistance. The reduction of
"apparency", e.g. by crop rotation, would reduce the need for
defensive measures and might improve agronomic quality by
favouring plants in which considerable metabolic resources were
not expended on synthesis of defensive compounds.

     Resistant crop plants are the subject of intensive
programmes of research. Study of the relationships among
chemical composition, variety and insect resistance has, in many
instances, made it possible to correlate observed resistance to
insect attack with plant chemistry. However, resistance may be
conferred by factors other than chemical composition, nor is
plant resistance always compatible with agronomic quality. This
problem and the problem of discovering new sources of resistant
germplasm are major limitations of the technique. Although the
host-plant resistance method of control utilizes mechanisms that
have evolved naturally, the selection of agronomically valuable
species for cultivation by man interferes substantially with the
evolutionary process. The mechanisms of resistance have been
summarized (15) as nonpreference resistance, antibiotic
resistance, and tolerance. In the case of nonpreference
resistance, the plant is less suitable for food, oviposition or
shelter and resistance may involve chemical or physical factors.
The chemical basis of resistance is governed by the complex of
compounds that affect insect behaviour, such as feeding
deterrents, oviposition deterrents, attractants, etc. in contrast
to antibiotic resistance where the presence or absence of
compounds that affect the development or survival of the insect
is involved. Tolerance to insects is the ability of the plant to
survive attack because pest damage is repairable or does not
impair the vigour of the plant. Knipling (16) discusses the
effect of the types of resistance on insect population and
considers that antibiotic resistance should have the greatest

effect. High mortality should cause a decline in insect
populations. This will, of course, depend on the rate of
increase of insect populations and the level of antibiotic
resistance, among other factors.

Experimental evaluation of effects on insect populations
presents a major source of difficulty and it was emphasized in
the preceding discussion that a total ecosystem study would be
required to determine the effectiveness of such control measures
and that small plots are inadequate. Resistant plant varieties
are advantageous because they reduce numbers of insects, but from
the point of view of the chemist seeking leads to structures of
high biological activity, the study may be disappointing
particularly if a number of factors are involved. In studies of
resistance, tests for insecticidal activity per se may yield only
limited information. Although pyrethroids, rotenoids, nicotine,
ryanodine, and a number of other plant components are
insecticidal, it is important in a screening programme to
recognize compounds that may affect insect development, such as
the precocenes (4), ecdysones (17), juvenile hormone mimics (17),
as well as compounds that affect insect behaviour, such as the
tannins, azadirachtin, and a number of terpenoids which act as
insect antifeedants.

Chemical factors are also involved in the resistance of
plants to disease and in the competitive ability of a plant to
survive within a community of plants. Plant stress may also
generate a chemical response giving rise to compounds known as
the phytoalexins, the nature of which will depend on the
chemistry of the host plant (18, 19). Such response to injury or
infection is of great interest because it has stimulated
investigations of the nature of the bioregulatory processes
involved.

This is illustrated in the case of grapevines where the
exploitation of the defences of the plant against fungal attack
by vine downy mildew (Plasmopara viticola) and Botrytis cinerea
has been studied (20). A group of phytoalexins, the viniferins,
which may be formed by oligomerization of resveratrol (a
trihydroxy-trans-stilbene) were found only in infected or injured
vineleaves and had moderate antifungal activity in vitro. Pryce
(21) also discussed the formation of antifungal compounds in rice
leaves when the rice was treated with
2,2-dichloro-3,3-dimethylcyclopropane carboxylic acid. Responses
that are elicited when the natural defences of the plant are
stimulated suggest new approaches to the control of diseases or
pests.

Natural Products as Leads to New Pesticides

Natural products have provided leads to new pesticides.
Knowledge of the biological activity of many plants has been
recorded in the ancient pharmacopoeias of India and China and is
preserved in the folklore of most nations. The practical
application of much of this knowledge has proved difficult
because it has been accumulated throughout the ages and it is

based on a mixture of experience, tradition, and magic. Natural products continue to be sold as drugs or as insecticides in many parts of the world and the market place is evidence of their value.

Nicotine is an example of a natural insecticide that has been in use for many years. Pyrethrin, a crude mixture of natural pyrethroids, was used by Caucasian tribes as an insecticide before 1800. The extract of pyrethrum from chrysanthemum cinaeriaefolum contains pyrethrins, cinerins and jasmolins. Although this product was a valuable insecticide when few effective insecticides were available, its use in agriculture has been limited by its photochemical instability. Replacement of such naturally occurring insecticides by compounds synthesized from petrochemicals was favoured by wartime conditions.

Synthetic insecticide development and official action to approve their use have received additional stimuli because supplies of imported insecticides have been restricted from time to time by political or economic factors (22). Thus, attempts to develop pyrethroids were continued and many successful synthetic analogues have been produced since the elucidation of their structure. Some of the earliest of these were synthesized as soon as the nature of the active materials had been established (23). In the forefront of research on synthetic pyrethroids have been Elliott and his co-workers. As a result of their sustained effort, the pyrethroids have now achieved important position in agriculture. The major structural features necessary for their activity have been elucidated (24), and there are now many related compounds commercially available.

The new pyrethroid insecticides have required many years for successful development, notwithstanding the structural information available in 1924. The environmental persistence of the new pyrethroids is generally lower than that of the organochlorine insecticides, and the structure is capable of extensive modification with retention of activity. Fortunately, they were introduced into agricultural use at a time when heavy reliance on organochlorine insecticides was no longer feasible and there was an urgent need for new insecticides.

This group is perhaps the most outstanding example of natural products that have provided leads for modification. Other groups of naturally occurring insecticides have been investigated, but without the same spectacular degree of success. Quassia, rotenones, sabadilla, ryania, mamey, and several other plant products have been used as insecticides. The chemistry of the major groups is known, but many plant principles remain to be explored (25). Screening of natural products presents a problem and raises the issue of suitable bioassay in terms of insect species, the part of the plant that should be examined and the appropriate tests for activity.

The number of discoveries of new insecticides based on natural product structures as prototypes is limited, although insecticidal activity is widely distributed in nature. For example, carbamates occur in nature and it has long been known that phyostigmine, an alkaloid from the calabar bean, is toxic to

mammals. However, many other lines of thought might have led to
the inclusion of carbamates in an insecticide screening
programme, assuming that they were not to be included on a random
basis.

The powerful action of some natural products on the central
nervous system may suggest their potential value in insect
control. If their mammalian toxicity can be reduced by
structural modification they may be useful insecticides. Cartap,
a rice insecticide, is a bisthiocarbamate. Its structure was
based on that of the natural product, nereistoxin, a neurotoxin
isolated from shellfish, and it is likely that an <u>in vivo</u>
metabolic conversion of cartap to nereistoxin or related compound
is responsible for its activity (11).

$$(CH_3)_2\ NH\ CH(CH_2SCONH_2)_2 \xrightarrow[\text{metabolism}]{} (CH_3)_2\ NH\ CH\ \begin{matrix} CH_2S \\ CH_2S \end{matrix}$$

            cartap   nereistoxin.

As screening techniques become more sophisticated, the range
of compounds that are potentially valuable as leads should
increase. As, an example, the antijuvenile hormone action of the
precocenes has been cited. Their spectrum of activity is
limited, but the study of their action has revealed potential
biochemical target sites.

Naturally occurring insecticides are subject to fluctuations
in production and availability and, in developed countries, they
can rarely compete effectively with manufactured products. All
new structures must face the same searching toxicological
examination whether they are natural or synthetic in origin and
their economic potential must, therefore, be equally as
attractive. Although such factors may appear discouraging,
chemicals that possess unusual modes of action such as
behaviour-modifying chemicals have found an acceptable place in
pest management.

Behavioural Compounds

Many interactions between insects and insects or between plants
and insects are mediated by chemicals and in the section on plant
resistance, the role of nonpreference resistance was discussed.

There has been considerable progress in isolating and
identifying compounds that affect insect behaviour. A recent
review contains more than 800 references to the literature and
lists over 300 compounds as insect attractants, attractant
pheromones and related compounds together with the corresponding
insect species (26). Other pheromones (trail pheromones, hair
pencil secretions, etc.), feeding deterrents, oviposition
deterrents, an other types of behavioural compounds are not
included in this comprehensive review, but the volume of
information on behavioural compounds is constantly being expanded
as new compounds are identified.

There are many ways in which behavioural compounds may be

used in pest management systems. Potential application of
pheromones and attractants and the status of research was
recently summarized (10). Antifeedants have also received much
attention, particularly extracts of the neem tree (Melia
azdirachta). Volatile attractants, arrestants, and stimulants
from plants also appear promising as topcs for future resarch.

In practice, the predominant application of insect
pheromones and attractants is for detection and survey of
infestations. Pheromone traps are an extremely valuable
component of many pest management systems and provide information
on an area-wide basis that permits timely application of control
measures.

For suppression or control of insect populations, attractant
pheromones or attractants may be used in several ways: (a) in
combination with baits, toxicants, sterilants or pathogens;   (b)
mass-trapping;   or (c) to disrupt mating.

The application of pheromones to area-wide suppression of
population by permeating the air with pheromone to disrupt mating
communication has been used in the management of several major
pest insects.

There has been considerable research and some commercial
interest in this approach. There are, however, major practical
problems in the way of successful development. It has a major
advantage that only a few grammes of pheromone per hectare is
needed for application. Since, pheromones are generally degraded
rapidly, environmental burden is minute compared with that
typical of many insecticides. The problem of pheromone
formulation has been the subject of much research because they
are usually labile and volatile compounds (27). One of the major
difficulties in the use of pheromones (or other nonlethal methods
of control) is the difficulty of demonstrating efficacy (28).
Extremely large plot sizes are usually necessary to acquire
sufficient information, particularly if it becomes necessary to
measure the impact of the treatment on mating and population.
Research on the gypsy moth over a number of years has
demonstrated the difficulties and defined the limits of the
technique for this species (29). Although the technique suffers
from the disadvantage that considerable knowledge of the
population dynamics and mating behaviour of the target species
are essential prerequisites for application, limited successes
are encouraging continued trials and development.

The use of behaviour-modifying compounds offers great
promise for integrated systems of pest management, but it
requires substantial knowledge of the target species.
Modification of structure of behaviour-modifying compounds for
population suppression is a potential approach and, in some
cases, analogues are of value in mating disruption. However, the
response of an insect to pheromones depends on very precise
structural requirements and even slight modification of structure
normally reduces the biological activity by several orders of
magnitude.

Investigation of the physiology of pheromone production has

shown   that in the female corn earworm moth (Heliothis zea)
pheromone production is controlled by a brain hormone (30).   The
stimulation of pheromone production by this hormone is an
important stage in the reporductive process.   Such key events in
the reproductive process suggest new potential for insecticidal
modes of action.

Conclusion

Natural products are a rich source of structural variety.   They
may provide source materials for screening programmes to detect
biological activity and natural products that possess known
activity may serve as templates for structural modification.   An
outstanding example of the optimization of structure to improve
potency and stability is the synthesis and successful
commercialization of pyrethroids and their analogues.
    Bacteria present great potential as sources of biologically
active chemicals.   Developments in biotechnology should greatly
increase the feasibility of manufacturing new bacterial products
in quantity.   As a future approach to insect control, the
prospect of transferring the capability to produce selective
insect toxins from bacteria to plants may lie within the
potential of new genetic engineering techniques.
    Such developments in biology as the elucidation of
bioregulatory processes and the discovery of new receptor sites
and receptors have been instrumental in the development of
biorational pesticides capable of blocking essential steps in
metabolic processes that are peculiar to the organism or
organisms that are to be controlled.
    Natural products are the key to understanding ecological
systems.   There has been rapid progress in elucidation of the
chemistry of primary and secondary plant components and in the
growth of knowledge of insect biochemistry.   For its application
to pest control, parallel developments in physiology and
behavioural studies are essential for revealing the underlying
complex of chemically mediated interactions among organisms.
    Biorational methods of control based on altered insect-plant
relationships are being effectively used, if we extend this term
to describe application of host plant resistance.   Such control
methods have been developed empirically; but the rational basis
for further development will be the outcome of new genetic
studies.
    Success in deciphering the structure and function of natural
molecules has dramatically changed the approach to biological
problems.   The structure and functions of biological
macromolecules, particularly nucleic acids and proteins are
rapidly becoming known and the ability to engineer or alter their
subunits in living organisms provides potential technology for
altering biological outputs.   Thus, the nutritive value of a
plant protein may be improved by changing the sequence of amino
acids.   The insertion of genes into plants and their subsequent
expression is a concept that promises a great future
development.   However, the question of resistance to pests by

plants is complicated by the contribution of many factors, each affected by different genes. The evolution of host-plant preferences and particularly, the adaptation of insect behaviour, are important factors which require more study and Dethier (31) has argued that this is important at the genetic level. The behaviour of insects at the feeding and reproductive stages is critically important in determining the pest status of a species. Although chemicals that modify behaviour currently have a role in pest control, their application on an area-wide scale in attempts to suppress insect populations by using pheromones to alter mating behaviours, may meet with little success unless the biochemical and behavioural aspects of the system are well understood.

One of the most important roles of natural products in pest control may be their role in stimulating thought. Pest control may be conceived as an activity directed towards changing the existing conditions in an ecological system for the benefit of man. A better understanding of the chemically-mediated interactions among organisms within the system should make it possible to devise ways in which minimal changes yield maximum benefits. However, this cannot be achieved by the chemist alone, even though new molecular biology provides an explanation of biological phenomena in chemical terms. Behaviour, population dynamics and other factors associated with whole organisms and communities are also essential components of the cooperative study required to achieve control of pests in an ecologically sound and effective manner.

Literature Cited

1. Roller, H.; Dahm, K.H.; Sweeney, C.C.; Trost, B.M. Angew. Chem. Int. Ed. 1967, 6, 179.
2. Meyer, A.S.; Schneiderman, H.A.; Hanzmann, E. Proc. nat. Acad. Sci. U.S.A. 1968, 60, 853.
3. Judy, K.J.; Schooley, D.A.; Dunham, L.L.; Hall, M.S.; Bergot, B.J.; Siddall, J.B. Proc. nat. Acad. Sci. U.S.A. 1973, 70, 1509.
4. Bowers, W.S. In "The Juvenile Hormones"; Gilbert, L.I., Ed.; Plenum, New York, 1976, pp. 394-408.
5. Brooks, G.T.; Pratt, G.E.; Jennings, R.C., Nature 1979, 281, 570.
6. Pratt, G.E.; Jennings, R.C.; Hamnett, A.F.; Brooks, G.T. Nature 1980, 284, 320.
7. Quistad, G.B.; Cerf, D.C.; Schooley, D.A.; Staal, G.B. Nature 1981, 289, 176.
8. Kramer, S.J.; Baker, F.C.; Miller, C.A.; Cerf, D.C.; Schooley, D.A.; Menn, J.J. In "Human Welfare and the Environment", Eds. J. Miyamoto et al, Pergamon, New York, 1983, pp. 177-182.
9. Menn, J.J.; Henrick, C.A. Phil. Trans. Roy. Soc. Lond. 1981, B 295, 57.
10. Klassen, W.; Ridgway, R.L.; Inscoe, M. In "Insect Suppression with Controlled Release Pheromone Systems", Vol.

I, Kydoneius, A.F., Beroza, M., Eds.; CRC Press, Boca Raton, FL, 1982, pp. 13-130.

11. Casida, J.E., In "Advances in Pesticide Science" Part I, H. Geissbühler, Ed.; Pergamon, Oxford, 1979, pp. 45-53.
12. EPA "Approvals for Pyrethrum Substitutes Stepped Up", Chemistry and Engineering News, 1979, Oct. 1.
13. Swain, T. Proc. 15th Ann. Cong. Entomol. 1976, pp. 249-256.
14. Feeny, P. In "Natural Products for Innovative Pest Management", Whitehead, D.L., Bowers, W.S., Eds.; Pergamon, Oxford, 1983, pp. 167-185.
15. Painter, R.H. "Insect Resistance in Crop Plots", Univ. Press of Kansas: Lawrence and London, 1968, 520 pp.
16. Knipling, E.F., The Basic Principles of Insect Population Suppression and Management, U.S. Department of Agriculture, Agriculture Handbook No. 512, 1979.
17. Jacobson, M.; Crosby, D.G., "Naturally Occurring Insecticides", Dekker: New York, 1971.
18. Bowers, W.S.; Nishida, R. Science 1980, 193, 542.
19. Kuc, J.; Caruso, F.L. In "Host Plant Resistance to Pests", Hedin, P.A., Ed.; Am. Chem. Soc. Symp. Series 1977, 62, pp. 90-114.
20. Cruickshank, I.A.M. In "Natural Products and the Protection of Plants", Marino Bettolo G.B., Ed.; Elsevier: Amsterdam, 1976, pp. 509-561.
21. Pryce, R.J., In "Natural Products for Innovative Pest Management", Whitehead, D.L., Bowers, W.S., Eds.; Pergamon: Oxford, 1983, pp. 73-90.
22. "Pesticide Registration Data: Proposed Data Requirements Part III", U.S. Environmental Protection Agency 1982, 40 CFR 158, 53192-53202, EPA 79.
23. Staudinger, L.; Ruzicka, L. Helv. chim. Acta. 1924, 7, 177,201, 212, 236, 245, 377, 390, 406, 442, 448.
24. Elliott, M.; Janes, N.F. Chem. Soc. Rev. 1978, 7, 473.
25. Crosby, D.G., In "Minor Insecticides of Plant Origin", Jacobson, M., Crosby, D.G., Eds.; Dekker: New York, 1971, pp. 177-239.
26. Inscoe, M.N., In "In Insect Suppression with Controlled Release Pheromone Systems", Vol. II, Kydoneius, A.F., Beroza, M., Eds.; CRC Press: Boca Raton, FL, 1982, pp. 201-295.
27. Kydoneius, A.F.; Beroza, M., Eds.; "Insect Suppression with Controlled Rlease Systems", Vols. I & II; CRC Press: Boca Raton, FL, 1982.
28. Roelofs, W.L., Ed.; "Establishing Efficacy of Sex Attractants and Disruptants for Insect Control"; Entomol. Soc. America: College Park, MD, 1979.
29. Plimmer, J.R.; Leonhardt, B.A.; Webb, R.E. In "Insect Pheromone Technology", Leondardt, B.A., Beroze, M. Eds.; Am. Chem. Soc. Symp. Series, 1982, 190, pp. 231-242.
30. Raina, A.K.; Klun, J.A. Science 1984, 225, 531.
31. Dethier, V.G. Proc. 4th Insect/Host Plant Symp. Ent. exp. and appl. 1978, 24,559.

RECEIVED November 23, 1984

Do Plants "Psychomanipulate" Insects?

L. L. MURDOCK, G. BROOKHART, R. S. EDGECOMB, T. F. LONG, and L. SUDLOW

Department of Entomology, Purdue University, West Lafayette, IN 47907

The hypothesis that plants attain a measure of defense
against herbivory by "psychomanipulation" of insect
herbivores is presented. In brief, the psychomanipulation hypothesis asserts that certain secondary plant
chemicals imbibed in sublethal doses by insects feeding
on plant parts interfere with information processing in
the insectan central nervous system and thereby modify
insect-herbivore behavior in ways that reduce or
prevent further herbivory. Behavioral changes could
include the evocation of unadaptive behaviors,
inhibition of hunger, onset of narcosis, etc. Evidence
will be presented that numerous plant substances do
have "psychomanipulative" effects at sublethal levels
and that the pyschomanipulation mechanism might operate
in nature.

One important defensive strategy of plants is the synthesis and
accumulation of chemicals in their tissues that discourage or stop
feeding by herbivores. The mechanisms by which these chemicals
affect potential herbivores are currently being elucidated. Two
general modes of action seem to be widely accepted: (i) deterrency
and (ii) toxicity. Deterrents are gustatory stimulants that
negatively affect the insect's biting, feeding, and maintenance of
feeding (1). Deterrency involves an action of the protectant
chemical on the sensory nervous system of the unadapted herbivore.
The message sent to the insect central nervous system is a series of
action potentials from a specific sensory neuron or a certain
patterned output from a set of sensory neurons (2,3). It is
interpreted as "danger" or "no good" and its effect is to cause
feeding to cease (2,3). Toxicity, on the other hand, disrupts the
functioning of physiological and biochemical systems within the
herbivore. Toxic action in the insect may cause sickness or
weakness. The insect may not grow, mature, or reproduce normally
and perhaps die prematurely.
     Deterrents and toxins differ in their sites and modes of

0097-6156/85/0276-0337$06.00/0

action. Deterrents act on the peripheral nervous system, chiefly on
the chemosensory apparatus, resulting in a behavioral change in the
insect. In contrast, toxins disrupt cellular, biochemical or
physiological processes in the insect. The hypothesis presented
here suggests a third general mode of action for secondary plant
compounds. It begins with the premises (a) that the insect central
nervous system (CNS) is an excellent target for plant defensive
chemicals -- at least as suitable as the chemosensory portion of the
peripheral nervous system; (b) that the importance of many plant
chemicals is not their lethality but rather the changes they evoke
in the insect herbivore's behavior by interfering with central
nervous function. We suggest the term "psychomanipulation" to
describe the behavior-modifying effects of such chemicals acting in
the CNS. Our purpose in this paper is to define the
psychomanipulation hypothesis, provide evidence in support of it,
and suggest how potential psychomanipulants might be studied. The
primary question is how plant chemicals modify insect behavior
through cellular and molecular mechanisms. New answers to that
question might open approaches to the discovery of novel insect
control agents.

The Psychomanipulation Scenario

Imagine a hungry insect alighting on the leaf of a non-host plant
that accumulates a psychomanipulant. Because the leaf contains
sufficient positive stimuli and insufficient deterrents, feeding
begins. Some of the imbibed behavior-modifying chemical escapes the
detoxication machinery in the gut wall, enters circulation, and
penetrates the blood-brain barrier into the CNS. There it reacts
with particular neuronal types to excite, inhibit, modulate, or
otherwise affect normal patterns of activity. This modified
neuronal activity is manifested as a change in the insect's behavior
such that it no longer consumes plant material, or at least consumes
less than it would otherwise.

The Insect CNS as Target. The insect central nervous system is composed of a dorsal and rostral supraesophageal ganglion or brain,
which communicates via circumesophageal connectives to the first of
a chain of ventrally located ganglia (fused to different degrees in
different species). Sensory information from the environment enters
ganglia of the CNS via the axons of sensory neurons, whose cell
bodies are located in the periphery of the organism. The sensory
axons project into the neuropil and form synapses with specific
classes of neurons. Information is transmitted between neurons in
most instances by means of chemical transmitter substances released
at presynaptic terminals. These transmitter molecules diffuse
across the narrow gap (synaptic cleft) between the presynaptic cell
and the postsynaptic cell and interact with specific receptors
exposed on the surface of the latter. The transmitter-receptor
interaction alters the level of excitability of the postsynaptic
cell. Excitation or inhibition of that neuron can have behavioral
consequences if the change in its excitability affects other neurons
to which it communicates. Motor neurons are the immediate
determinants of movement and communicate directly with muscle fibers

via chemical transmission. Motor neuron cell bodies are located in the periphery of CNS ganglia. The dendrites of these neurons are often elaborate arborizations located within the central regions of the ganglion, collecting information from other neurons. The nature and intensity of that information determines whether the motor neuron becomes active and fires action potentials. Interneurons reside completely within the CNS. They coordinate the activities of sensory and motor neurons and subserve all major integrative and higher functions of the nervous system. Evidence is accumulating that insects also make use of modulatory neurons, neurons whose activities change the quality of information passing through synapses, or modify the spontaneous activity of receptive neurons or muscle cells (4,5). The best-characterized modulatory neuron is the Dumeti cell, which innervates the extensor tibiae muscle in the migratory locust. This cell, whose unpaired soma is located in the midline of the metathoracic ganglion, releases octopamine from its terminals. Neurons of this type may be widespread in insects and some of them may be interneurons (cf. 5).

The range of neurochemical messenger substances used by insects is remarkably similar to those in mammals. In insects, acetylcholine seems to be the neurotransmitter released by most types of sensory neurons and L-glutamic acid apparently serves as the transmitter of motor neurons innervating skeletal muscles (4). Peripheral inhibitory neurons may release gamma-aminobutyric acid (GABA) at their terminals. In a few instances there is evidence that aromatic biogenic amines serve as neuromessengers in the periphery, e.g., octopamine is the modulator substance released by Dumeti neurons referred to above (4), and the light organs of the firefly also seem to be controlled by neurons releasing octopamine (6). The salivary glands of various insects, including the locust, Schistocerca gregaria (7), the cockroach, Nauphoeta cineria (8), and the moth, Manduca sexta (9) may be innervated by catecholaminergic nerve fibers. A few muscles appear to be innervated by neurons that release neuroactive peptides, notable examples being the cockroach hindgut, where proctolin is released (10), and the coxal depressor muscles innervated by the Ds motoneuron in Periplaneta americana (11). All of these neuromessengers as well as others undoubtedly also serve to communicate information between neurons within the insect central nervous system.

Plant Protection by Behavior Modification. Behavior is generally equated with the movement of an organism or body part and the cessation of such movements (12). To be classified as behavior, movement requires the participation and direction of the CNS. The intimate involvement of the CNS in the coordination of muscles and body parts results in behaviors ranging from the fundamental (e.g., walking and grooming) to the complex (e.g., navigation and learning). Even relatively simple behaviors such as jumping or grasping food, require the participation of neurons on many levels.

Since behavior depends on the proper functioning of every neuron involved in its expression, disruption of any participating neuron could have severe consequences. These consequences, however, may be important without being obvious. For example, greatly increased locomotor activity might be detrimental beause the insect

is spending far less time feeding and concurrently expending food
reserves as unnecessary activity. Other behavioral changes might
cause an increase in the response time needed for escape, thus increasing the chances for that insect to end up as another's meal.
By affecting neurons involved in specific behaviors, plant psychomanipulants could negatively affect insect behavior and thus aid in
the defense of the plant.

Action in the CNS of Plant-derived Compounds

Since communication between neurons by means of chemical neuromessenger substances is basic to the functioning of the insect CNS,
and since the CNS is the wellspring of all complex and patterned
insect behavior, interference with chemical communication within the
CNS should prevent behaviors from appearing when they normally
would, or evoke behaviors when they normally would not.
Psychomanipulant chemicals reaching the insect CNS could disrupt
chemical communication by acting as receptor agonists, receptor
antagonists, synaptic clearance antagonists, neuromessenger
releasers, and neuromessenger supply suppressants. A wide variety
of plant secondary substances fitting these categories could
interfere with insect neurochemical messenger systems and effect
behavioral changes. Our purpose here is to mention a few examples.

Receptor Agonists. By mimicking the natural chemical messenger at
its receptors, psychomanipulants could cause spurious excitation or
inhibition in the CNS. Two examples of receptor agonists are
nicotine found in the leaves of the tobacco plant,

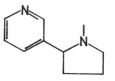

NICOTINE

Nicotiana tabacum, and lobeline, present in the dried leaves and
tops of the herb, Lobelia inflata (13). Both nicotine and lobeline
act as agonists on a specific type of acetylcholine receptor, the
nicotinic cholinergic receptor. In mammals, nicotinic cholinergic
receptors mediate cholinergic neurotransmission in skeletal muscles,
autonomic ganglia, and the central nervous system. At these sites
nicotine's action has two phases, excitation and depression. Insect
central nervous tissues are rich in nicotinic cholinergic receptors
(cf. 14), but their role in behavior is not well understood.
     Among the known agonists of muscarinic cholinergic receptors
isolated from plants are pilocarpine, which is isolated from
leaflets of South American shrubs of the genus Pilocarpus, and
arecoline, isolated from the seeds of Areca catechu, the betel nut.
Cholinergic receptors with pharmacological properties closely
resembling those of vertebrate muscarinic receptors are found in the
insect CNS (14). It is interesting that muscarinic agents appear to

possess less killingpower than nicotine agonists (15). Many other plant-derived cholinergic agonists are known, including the alkaloids cytisine and matrine.

There are naturally-occurring opioid agonists from plants, such as morphine, which owes its dramatic psychopharmacological effects in mammals to an interaction with receptors for enkephalins and endorphins. As evidence increases that insects and other arthropods utilize neuroactive peptides as neuromessengers (16), it becomes more likely that some of the plant opiates are defensive chemicals having a "psychomanipulant" type of mechanism.

Receptor Antagonists. By blocking neuromessenger receptors, receptor antagonists may evoke inappropriate behaviors through inhibition or disinhibition of a neuron or group of neurons. GABA receptors in arthropods and mammals are blocked by two chemicals of plant origin, bicucculine, and picrotoxin (17). Scopolamine, common in many Solanaceae (including the famous deadly nightshade, Atropa belladonna), is an excellent blocking agent for muscarinic cholinergic receptors, as is closely-related atropine (17). One of the more notorious plant chemicals is curare (used for poison arrows), which is prepared from a variety of species of Strychnos (18). The active compound common to all preparations of curare is D-tubocurarine. Another antagonist (nicotinic) of plant origin is β-erythroidine, from the seeds and other plant parts of several Erythrina species (13). One of the more specific antagonists of α-2 adrenergic receptors is yohimbine, the principal alkaloid in extracts of the bark of Pausinystalia yohimbe (19). Yohimbine also blocks one type of octopaminergic receptor in locusts (20).

Synaptic Clearance Antagonists. By preventing the removal of naturally-released transmitter from the region of its receptors, the effect of the neuromesssenger on the receiving cell will be prolonged and intensified. There are three principal routes by which neuromessengers are removed from the synaptic cleft: (i) enzymatic destruction of the transmitter (e.g., acetylcholine (ACh) which is hydrolyzed in the synaptic cleft by acetylcholinesterase); (ii) uptake into pre- and post- synaptic cells by membrane-associated pumps that have substantial specificty for molecules they will carry; (iii) diffusion away from the cleft.

The classical synaptic clearance antagonist is the cholin-esterase inhibitor physostigmine, from the calabar bean, Physostigma

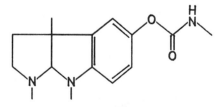

PHYSOSTIGMINE

venenosum, a West African perennial (13). Physostigmine reversibly inhibits acetylcholinesterase and thereby permits the buildup of

abnormally-high levels of ACh in the synaptic cleft. This results
in prolonged and intensified stimulation of the postsynaptic cell,
that can lead to dysfunction of the physiological process controlled
by the cholinergic input. Lesser known, but possibly significant as
a psychomanipulant, is guvacine (21), from betel nuts, which blocks
uptake of GABA in mammalian CNS preparations.

Cocaine is a powerful CNS stimulant as well as local anesthetic
(13). It is isolated from the leaves of Erythrolon coca, a tree
indigenous to Peru and Bolivia. Cocaine effectively blocks the
uptake of catecholamines into presynaptic neurons and thus promotes
the activity of synapses (both central and peripheral) involving
these amines (22).

Neuromessenger Releasers. Normal CNS functioning could be disrupted
by evoking premature or continued release of neuromessengers from
presynaptic stores. Although D(+)-amphetamine, a well-known psycho-
stimulant and appetite depressant, does not occur in plants, a
closely-related substance, DL-cathinone, does occur. It is found in
the leaves of the khat shrub, Catha edulis, which grows in East
Africa and in the Arab peninsula. Cathinone evokes the release of
norepinephrine from central and peripheral presynaptic stores and
has cardiovascular and appetite depressing effects similar to D-
amphetamine (23).

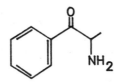

CATHINONE

Neuromessenger Supply Suppressants. By reducing the availability of
neuromessengers from presynatic stores, neuromessenger supply suppressants could alter the physiology and biochemistry of the presynaptic neuron such that these stores are depleted. The reduction
of neuromessenger supply could be caused by the drug reserpine,
isolated from the Indian plant, Rauwolfia serpentina (13). The

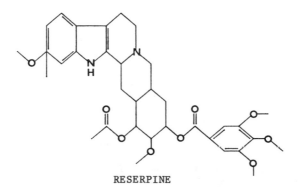

RESERPINE

effect of reserpine is the depletion of stores of aromatic biogenic amine transmitters, including dopamine, norepinephrine, 5-hydroxytryptamine, and octopamine. In insects, the effects of reserpine are relatively slow in onset and can persist for many days (24). Reserpine is active in both insects and mammals. In mammals, reserpine can reduce blood pressure and produce a state of sedation in which the individual is indifferent to environmental stimuli (25).

Others. There are other plant substances which could act as psychomanipulants via mechanisms that do not fit neatly into the above categories. Some could enter and disrupt the normal processes operating in a neuron. These would include substances like caffeine and theobromine, both of which inhibit intraneuronal phosphodiesterase, thereby affecting the excitability of neurons.

Presynaptic supply of transmitter could be affected by the dietary content of transmitter precursors. L-DOPA, the precursor of dopamine, a prominent insect CNS transmitter, occurs in several species of plants. L-5-Hydroxytryptophan, which could serve as a precursor of the indoleamine transmitter 5-hydroxytryptamine, also occurs at high levels in some plant materials. Although the consequences of altering levels of transmitters via diet are not clear, such alterations have been shown to occur with several transmitters in mammals (26) and could alter the behavior of an insect herbivore.

Candidate Psychomanipulants of Plant Origin

Physostigmine. There is evidence that cholinergic drugs at sublethal levels can indeed modify insect behavior. Adult male Locusta migratoria that have passed the imaginal molt at least three weeks earlier fly with a wing-beat frequency of 23 Hz. The elegant work of D.M. Wilson (27) established that the regular wing beating in locusts is the result of a central pattern generator located in the thoracic ganglia. While the pattern generator is capable of producing the motor program for flight in the absence of sensory feedback from the wings and thorax, such feedback can, nevertheless, influence the frequency with which the wings are flapped. Wind detectors on the head can also modify the output of the central pattern generator (28).

When adult L. migratoria were injected with 5 μg/g body weight of physostigmine salicylate, they exhibited a period of hyperactivity lasting several minutes, followed by a 2-3 hour period of hyperresponsiveness following handling or mechanical stimulation (29). Thereafter they appeared outwardly normal with little or no mortality at 24 hours. This sublethal dose of physostigmine had a dramatic effect on wingbeating frequency (Fig. 1). During the early test intervals many individuals only exhibited brief periods of flight with visibly reduced frequency. At 4 hours postinjection, the frequency of those individuals that exhibited sustained flights fell to about 16 Hz, returning slowly thereafter toward the control value.

The available data cannot demonstrate that slowing the wing-beat frequency following injection of physostigmine is solely or even partly due to elevated levels of acetylcholine at

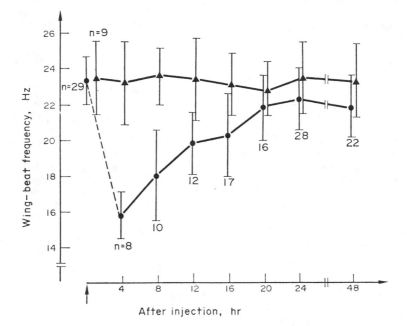

Figure 1. Wing-beat frequency of <u>Locusta</u> adults before (arrow on abscissa) and after injection of 5 μl/g H_2O (▲—▲) or 5 μg/g physostigmine (●—●). Reproduced with permission from Ref. 29. Copyright 1973, Pergamon Press.

sensory-central synapses in the CNS. It is possible that some or
all of the effect of physostigmine on the central pattern generator
is a result of acetylcholine accumulation at other types of central
cholinergic synapses, synapses that participate in the pattern
generation mechanism. However, this example does demonstrate that a
naturally-occurring anticholinesterase agent acting at sublethal
levels could significantly affect a crucial behavior. A locust able
to fly only a short distance or to fly only with reduced speed is
much more susceptible to predation.

Reserpine. As mentioned earlier, the Rauwolfia alkaloid reserpine
is noted for its ability to deplete aromatic biogenic amines in
nervous tissue of mammals and insects. One of the more remarkable
effects of reserpine in humans is to render individuals indifferent
to environmental stimuli. Reserpine appears to have a similar
effect on insects, although information is still relatively scanty.
In the cockroach, Periplaneta americana, reserpine at 50 µg/g causes
strong and long-lasting depletion of the aromatic biogenic amines
dopamine, octopamine, and 5-hydroxytryptamine (24). Numerous
authors have noted that reserpine has tranquilizing effects on
insects, e.g., the ant Formica rufa (30) and P. americana (24,31).
In our own experiments with adult Phormia regina, feeding reserpine
mixed with powdered milk and sucrose or injecting it at doses of 100
µg/g body weight and higher seemed to induce a state of lethargy and
inactivity.
    To obtain clearer understanding of the behavioral effects of
reserpine and their possible relationship to levels of aromatic
biogenic amines in the insect central nervous system, we injected
unfed, 2-day old adult P. regina with 2 µg/fly of reserpine.
Because of the insolubility of reserpine, we first dissolved it in a
small volume of dimethylacetamide (DMAD) and mixed the DMAD with
corn oil in a 1:10 ratio. Two groups of control flies were used,
one untreated and the other injected with 300 nl of the DMAD/corn
oil vehicle. The flies were held for one day with water available
ad lib., but no food. To determine the effectiveness of the drug,
two behaviors were observed: (i) meal size; (ii) proboscis-extension
responses to tarsal stimulation. Meal size was determined by
weighing the flies (plus the applicator sticks attached to them as
handles) before feeding and after allowing them to feed for 30
minutes on 1 M sucrose. The net weight gain was taken as the meal
size. Proboscis-extension responsiveness was estimated as the mean
acceptance threshold (M.A.T., the concentration of sucrose to which
the average fly in the population will respond) and was determined
following the up-and-down procedure described by Thompson (32).
Brains were dissected from control and reserpinized flies and
assayed for octopamine and dopamine by high performance liquid
chromatography (HPLC). The column used was a Brownlee RP-18 5 µ
particle size cartridge (250 mm x 2.1 mm i.d.). The amines were
eluted isocratically with 0.1 M KH_2PO_4 containing 2% methanol and
0.3 mM heptanesulfonic acid, sodium salt; mobile phase pH was 3.05
(33). The flow rate was 0.6 ml/min. Octopamine and dopamine were
detected and quantitated using an RC-3 electrochemical detector
(Bioanalytical Systems, W. Lafayette, Indiana) equipped with a
glassy carbon electrode operated at +0.95 V relative to a Ag/AgCl

reference electrode. Under these conditions the retention time of octopamine was 8 min and dopamine was 16 min.

Vehicle-injected flies exhibited a mildly elevated M.A.T. and a moderate meal size; concomitantly, there was a decline in brain octopamine content but no change in brain dopamine levels (Table I). Reserpine treated flies exhibited a 10-fold rise in threshold relative to the vehicle-injected controls and a large increase in meal size. These flies were hyperphagic in that they consumed their own body weight in 1 M sucrose, more than double that of the un-treated controls and 60 percent higher than the solvent controls. Brain levels of octopamine in the group injected with reserpine were significantly depleted relative to both untreated and solvent controls.

Clearly, injection of adult blowflies with reserpine caused them to become less responsive to food stimuli, while at the same time inducing them to eat more when they were offered 1 M sucrose, a highly stimulating food. Similar observations have been made with blowflies fed on reserpine (34). While it is not possible to ascribe the behavioral effects of reserpine on blowflies solely to actions in the CNS, it seems quite likely that the CNS plays a major role, particularly in light of the demonstrated capacity for reserpine to deplete CNS aromatic biogenic amines.

Both the meal size increase and the decreased responsiveness to tarsal stimulation with sucrose caused by reserpine can be explained as a requirement for increased intensity of sensory information to

TABLE I.   EFFECTS OF RESERPINE ON BLOWFLY FEEDING BEHAVIOR AND BRAIN BIOGENIC AMINES

	UNTREATED CONTROL	SOLVENT CONTROL	RESERPINE (2 μg/fly)
M.A.T. (mM)	5.2^a*	24^b	250^c
MEAL SIZE (mg)	$18^a \pm 6.3$ (10)	$25^b \pm 6.4$ (10)	$40^c \pm 5.8$ (10)
OCTOPAMINE (pmol/brain)	$7.6^a \pm 1.8$ (6)	$4.8^b \pm 1.9$ (6)	$2.4^c \pm 0.44$ (6)
DOPAMINE (6) (pmol/brain)	$2.8^a \pm 0.85$ (6)	$3.6^b \pm 1.7$ (6)	$0.94^c \pm 0.24$

*Values within a row followed by different letters are significantly different ($p \leq 0.05$).

initiate and terminate feeding. The higher M.A.T. reflects that higher concentrations of sucrose are needed to evoke proboscis-extension responses in P. regina adults. Higher concentrations of

sucrose evoke higher-frequency responses in the tarsal sugar
receptors (35).   In other words, more intense sensory information
must reach the amine-depleted CNS before the behavior ensues.
        Sensory information is also crucial to terminate a meal.   As
the fly initiates consumption, the crop begins filling.   As it
expands, the crop stimulates stretch receptors in a nerve net
associated with the crop and abdominal wall (36).   If the inhibitory
sensory information is not forthcoming, hyperphagia ensues (37-39).
According to our hypothesis, in the reserpine-affected CNS more
intense sensory stimuli is required in order to bring the meal to an
end.   The necessary extra stimuli would be forthcoming as the crop
continues to expand due to continued feeding.   This would impose
additional stretch on the abdominal stretch receptors, which would
respond by firing at higher and higher frequencies.   At some point
the firing rate of these receptors becomes sufficiently high to
trigger the termination of feeding in the CNS.
        These observations, combined with those in the literature,
support the concept that reserpine, if consumed by insects, could
exert a profound and long-lasting effect on insect behavior, an
effect which would decrease the insect's mobility and fitness to
respond to environmental stimuli, and to predators.   That effect,
probably due largely to a central action of the substance, could
result in plant protection by "psychomanipulation".

L-canavanine and L-canaline.   Secondary plant substances may evoke
particular types of stereotyped insect behavior via actions within
the CNS.   An excellent example is the work of Kammer, Dahlman, and
Rosenthal (40), who observed that injection of adult Manduca sexta
with L-canavanine and L-canaline led, within minutes, to sustained
flights lasting many hours.   The site of action of L-canavanine and
L-canaline was believed to be the CNS.   This produced continuous
motor output, which became less coordinated with time.   L-canavanine
and L-canaline are two of some 260 non-protein amino acids accumulated by various plants (41).   If an unadapted insect acquired

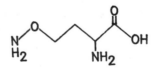

CANALINE

sufficient L-canaline or L-canavanine to exhibit the above
behavioral symptoms, the ensuing uncontrolled locomotion would not
only likely carry the insect away from its host plant, it would also
probably attract the attention of a predator.   The molecular mode of
action of L-canavanine and L-canaline in adult M. sexta remains to
be determined.   Kammer et al., however, noted that L-canaline, which
was more potent in evoking flight behavior than L-canavanine,
structurally resembles the inhibitory transmitter GABA (40).
Notably, injection of the formamidine insectostat,
N-demethylchlordimeform (DCDM), into adult M. sexta and Heliothis
zea also caused prolonged, uncoordinated flights (42).

Nectar of the Yellow Kowhai. The yellow kowhai (Sophora microphylla
Ait.) occurs in the North and South Islands of New Zealand. Its
yellow blossoms, which appear from July to October (late winter
through early spring), secrete abundant nectar and are very
attractive to honey bees. Bees only collect nectar from the
flowers, ignoring the pollen. New Zealand apiarists, who maintained
hives in the vicinity of kowhai groves, observed that hives failed
to build up normally in the spring. These failures could not be
attributed to the presence of Nosema or to pollen deficiency. In
some cases field bee mortalities were observed in the vicinity of
the hives. In seeking the cause of the poor hive development,
Clinch, Palmer-Jones, and Forster (43) observed that nectar from S.
microphylla had a narcotic effect on the bees. When fed 10 µl of S.
microphylla nectar, about 70 percent of the 70 bees tested became
narcotized within 30 min. When the treated bees were held in an
incubator at 30 deg. C., they all recovered within 4 hours with low
24-hour mortality (3-13%). The 24 hour mortality was much higher if
the treated bees were held at 20 deg. C. (74%) instead of 30 deg. C.
Clinch and coworkers (43) pointed out that seeds of S. microphylla
contain as major alkaloids methylcytisine, matrine, and smaller
amounts of cytisine and suggested that the observed narcosis and
mortality could be due to alkaloids reaching the nectar.

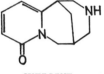

CYTISINE

Our own observations with pure cytisine support this. Worker
bees taken directly from a colony were held approximately 5 hours
without food, then offered cytisine in 40% aqueous sucrose. They
were fed by hand using a micropipette and allowed to consume volumes
of 10 µl. Cytisine at 10 µg/ 1 acted quickly: the bees fed avidly
on the solution, but within one to two minutes reared back, made
vigorous cleaning movements of their proboscises, fell over, made
wild leg movements, and exhibited poorly coordinated flight. Within
a few minutes they lay motionless on the bottom of the cage.
Similar, but less dramatic, symptoms were observed with bees offered
1 µg cystisine/µl. Bees fed 0.1 µg cytisine/µl were affected by
cytisine, but immobilization of all 10 treated bees was not attained
until 5-6 hours after treatment. The following day, 5 of the 10
bees treated with 1.0 µg had recovered and were maintained alive for
3 days by feeding them aqueous sucrose. While further observations
are needed, it is evident that honey bees are susceptible to oral
doses of cytisine as low as 1-10 µg. It is interesting to note that
narcotic or poisonous nectar is not an isolated phenomenon. Honey
bees are also poisoned by the nectar of the karaka tree
(Cornyocarpus laevigata J.R. et G. Forst.), which induces weakness
and inability to fly, as well as mortality (44).
    These observations raise the question: why do certain plants
produce narcotic or toxic nectar? In the case of S. microphylla the

explanation may be relatively simple. As noted by Clinch et al.
(40), the honey bee is not an effective pollinator of S. microphylla
even though it avidly collects nectar. Rather, Clinch et al. (40)
reported that birds consume large quantities of kowhai nectar, and
that they may be the most important pollinators. Thus, psychomanipulants in nectar could function as a defense of the plant
against the honey bee and other non-pollinating nectar thieves.
This hypothesis is consistent with the observations that birds seem
unharmed by kowhai nectar and that preparations of kowhai nectars
are not toxic when fed to white mice (40).

Summary

The psychomanipulation hypothesis predicts that plants obtain a
measure of protection from insect herbivores by accumulating
secondary substances which act in the CNS to change the herbivore's
behavior in ways that reduce or prevent further herbivory.
Secondary plant substances, acting at specific neural sites, could
modify insect behavior, e.g., by interfering with central
integrative processes, suppressing appetite, blocking learning or
memory, or distorting vision, without killing the herbivore.
Evidence that substances can act as psychomanipulants has been
presented here, but further research in the area is necessary. To
definitively test this hypothesis it is essential to obtain
quantitative estimates of the individual psychomanipulative
substances occurring in plants, to demonstrate that they necessarily
and sufficiently account for the change in behavior, and finally to
show that psychomanipulation is the result of an action of these
substances within the insect CNS. The first two criteria can be met
using currently available methods. Locating the site of action
within the CNS is more difficult and will require innovative
application of combined behavioral, neurophysiological and
neuropharmacological techniques.
     Insects may be continually probing beyond their host ranges,
thereby encountering new chemicals with the potential to disrupt
their CNS information processing. Some of these chemicals may also
have the potential to be insecticidal. When insecticidal compounds
are involved, we would suggest that a phase of psychomanipulation
will always precede lethal intoxication, as the following
illustrates: Imagine a wild Nicotiana plant, with leaves containing
substantial levels of nicotine, a classic natural insecticide. The
nicotine may be significant to the plant not so much because it has
the capacity to kill unadapted insects, but because it can modify
their behavior after they have eaten small amounts of the plant's
tissue. That behavior modification, e.g., increased excitability
and locomotory activity, could increase the likelihood that the
insect departs the plant. Only when that insect were refractory and
persisted in feeding on the plant would it receive a killing dose.
     The chemical-ecological literature abounds with descriptions of
effects of secondary plant substances on insects, vaguely lumped
under the heading "toxic". If we persist in being satisfied with
"toxicity" as a description cum explanation of the effect of plant
substances on insects, we shall remain very much in the dark as to
the action of these important compounds. The only way we can

possibly discover subtle neurophysiological and neurochemical
effects of secondary compounds is to actively search for them.

Literature Cited

1. Hsiao, T. In "Insect and Mite Nutrition"; Rodriguez, J. G.,
    Ed.; North-Holland: Amsterdam, 1972; p. 225-40.
2. Dethier, V. G. Entomol. Exp. & Appl. 1982, 31, 49-56.
3. Schoonhoven, L. M. Proc. Kon. Nedar Akad. (Amsterdam) Ser. C
    1977, 80(4), 341-50.
4. Florey, E. Fed. Proc. 1967, 26, 1164-78.
5. Evans, P. D. Adv. Insect Physiol. 1980, 15, 317-473.
6. Carlson, A. D. Adv. Insect Physiol. 1969, 6, 51-97.
7. Klemm, N. Comp. Biochem. Physiol. 1972, 43A, 207-11.
8. Bland, K. P.; House, C. R.; Ginsbourg, B. L.; Lazlo, I.
    Nature: New biology 1973, 244, 26-7.
9. Robertson, H. A. J. Exp. Biol. 1975, 63, 413-19.
10. Brown, B. E.; Starratt, A. N. J. Insect Physiol. 1975, 21,
    1879-81.
11. O'Shea, M.; Bishop, C. A. J. Neurosci. 1982, 2, 1242-51.
12. Hoyle, G. Adv. Insect Physiol. 1970, 7, 349-444.
13. Taylor, C. In: "The Pharmacological Basis of Therapeutics,
    Sixth Ed."; Gilman, A. G.; Goodman, L. S.; Gilman, A., Eds.;
    MacMillan: New York, 1980; Chap. 6,10.
14. Sattelle, D. B. Adv. Insect Physiol. 1980, 15, 215-315.
15. Dudai, Y.; Nahum-Zvi, S.; Haim-Granot, N. Comp. Biochem.
    Physiol. 1980, 65C, 135-8.
16. Greenberg, M. J.; Price, D. A. Ann. Rev. Physiol. 1983, 45,
    271-88.
17. Windholz, M.; Budavari, S.; Blunetti, R. F.; Otterbein, E. S.,
    Eds. "The Merck Index, Tenth Ed."; Merck & Co.: Rahway, NJ,
    1983; p. 1209.
18. Cooper, J. R.; Bloom, F. E.; Roth R. H. "The Biochemical Basis
    of Neuropharmacology, Fourth Ed."; Oxford University: New
    York, 1982.
19. Goldberg, M. R.; Robertson, D. Pharmacol. Rev. 1983, 35,
    143-80.
20. Evans, P. D. J. Physiol. 1981, 99-122.
21. Johnston, G. A. R. Ann. Rev. Pharmacol. Toxicol. 1978, 18,
    269-89.
22. Nielsen, J. A.; Chapin, D. S.; Moore, K. E. Life Sciences
    1983, 33, 1899-907.
23. Kalix, P. Psychopharmacology 1981, 74, 269-70.
24. Weiner, N. In "The Pharmacological Basis of Therapeutics,
    Sixth Ed."; Gilman, A. G.; Goodman, A. G.; Gilman, A., Eds;
    MacMillan: New York, 1980; Chap. 9.
25. Sloley, D.; Owen, M. Insect Biochem. 1988, 12, 469-76.
26. Fernstrom, J. D.; Wurtman, R. J. Adv. Biochem Psychopharmacol.
    1974, 11, 133-42.
27. Wilson, D. M. J. Exp. Biol. 1961, 38, 471-90.
28. Weis-Fogh, T. Phil. Trans. R. Soc. 1956, B239, 553-84.
29. Kutsch, W.; Murdock, L. L. J. Insect Physiol. 1973, 19,
    1519-25.

30. Kostowski, W.; Beck, J.; Mesaroy, J. J. Pharm. Pharmacol. 1965, 17, 253-55.
31. Frontali, N. J. Insect Physiol. 1968, 14, 881-6.
32. Thompson, A. J. Can. J. Zool. 1977, 53, 451-5.
33. Hopkins, T. L. Personal communication.
34. Green, J.; Murdock, L. L., unpublished data.
35. Shiraishi, A.; Tanabe; Y. J. Comp. Physiol. 1974, 92, 161-79.
36. Gelperin, A. Z. Vergl. Physiol. 1971, 72, 17-31.
37. Gelperin, A. Annu. Rev. Entomol. 1971, 16, 365-78.
38. Dethier, V. G.; Gelperin, A. J. Exp. Biol. 1967, 47, 191-200.
39. Nuñez, J. A. Naturwissenschatter 1964, 17, 419.
40. Kammer, A.; Dahlman, D. L.; Rosenthal, G. A. J. Exp. Biol. 1978, 75, 123-9.
41. Rosenthal, G. A.; Bell, A. E. In: "Herbivores: Their Interaction with Secondary Plant Metabolites"; Rosenthal, G. A.; Janzen, D. H., Eds.; Academic: New York, 1979; chap. 9.
42. Lund, A.; Hollingworth, R. M.; Murdock, L. L. In "Advances in Pesticide Science"; Geissbuehler, H., Ed.; Pergamon: Oxford, 1979; Part 3, p. 465.
43. Clinch, P. G.; Palmer-Jones, T.; Forster, I. W. N.Z.J. Agric. Res. 1972, 15, 194-201.
44. Palmer-Jones, T.; Line, L. J. S. N.Z.J. Agric. Res. 1963, 5, 433-6.

RECEIVED February 5, 1985

Protein Hydrolysate Volatiles as Insect Attractants

KENT E. MATSUMOTO, RON G. BUTTERY, ROBERT A. FLATH, T. RICHARD MON, and
ROY TERANISHI

Biocommunication Chemistry Research Unit, Western Regional Research Center, Agricultural
Research Service, U.S. Department of Agriculture, Albany, CA 94710

Hydrolysed protein preparations have been used to
attract various insects. The general subject of
insect attractant use both in nature and by man is
introduced, with particular reference to the Tephritid
family of fruit flies. The work of the Biocommunication Chemistry Research Unit on the identification
of the active attractant compounds in the hydrolysed
corn protein, Nu-Lure Insect Bait (NLIB) is discussed.
Different isolates have been obtained by running
simultaneous steam distillation-extractions (SDE)
under vacuum and atmospheric pressure and under basic
and acidic conditions. Chemical fractionation of
these isolates has also been accomplished. Chemical
identification by gas chromatography/mass spectrometry
(gc/ms) is discussed.

Metcalf and Metcalf (1) have quoted Rachel Carson (2) as describing
attractants as "new, imaginative, and creative approaches to the
problem of sharing our earth with other creatures". Many of these
attractants are natural products. We would like to discuss: 1) the
use of attractants in pest control, 2) the economic importance of
the Tephritid family of fruit flies, which is the focus of our
research, and the use of attractants in its control, 3) the
approach the Biocommunication Chemistry Research Unit is taking
toward finding new attractants, our progress to date, and our
plans, and 4) some problems we have encountered in our research.
We will touch only in passing upon the very large and important
fields of pheromones and kairomones as they are covered in more
detail in the chapters by Tumlinson (3) and Klun (4).

Uses of Attractants

Attractants were defined by Dethier (5) as "chemicals that cause
insects to make oriented movements toward the source". He
differentiated them from arrestants which "cause insects to
aggregate". For the purposes of this paper, we will not be this

specific in our terminology, because, in many cases, the biological observations necessary to make this differentiation have not been made.

Insect control professionals have used attractants as tools to monitor populations as part of integrated pest management (IPM) programs and as means of selectively reducing populations by luring individuals to traps, poisons, or even chemosterilants (6). Some work is currently being done on the attraction of natural enemies as well (7). (Please note that examples shown and references cited are meant to be illustrative rather than exhaustive). But for what do insects use attractants? It is postulated that they are used for finding shelter, mates, oviposition sites, and food (1, 8). It is here that the biologist becomes indispensible to the chemist. Only from his observations can chemists know where to look for naturally occuring attractants. It is also important to note that not all attractants are chemical. Visual (9-11) and auditory (12) clues also play very important roles in the behavior of some insects. Even with chemicals, whereas one tends to think of attractants as volatile materials acting over some distance, some are active only over very short distances or induce the appropriate behavior only upon contact (13).

What are some examples of these attractants? (Figure 1) For mate finding, the insect-produced pheromones are the primary examples. However, environmental factors may also play an important role in the effectiveness of attractants. In the southern pine beetle, Dendroctonus frontalis, alpha-pinene released from an attacked tree is necessary along with the endogenously produced frontalin in order to attract males for mating (14).

For species whose larvae are specialist feeders, finding suitable plants for oviposition is of great importance. Corn earworm moths, Heliothis armigera, will oviposit on twine impregnated with an extract of corn silk (15). The rice stemborer, Chilo plejadellus, female will be attracted to and oviposit near a component of rice plants identified as p-methylacetophenone (16). Some of these oviposition attractants are contact materials and, thus, are probably of no use in practical applications. This is the case for many of the butterflies of the Nymphalid family. The Indian butterfly, Papilio demoleus, seems to require some non-volatile component in citrus leaves to induce oviposition, although it seems to be attracted, at least partially, to the odor of the leaves (17).

Food finding is another area in which attractants play a major role in insect behavior. This is true for some larvae as well as adults. Saxena (18) has shown that the larvae of P. demoleus are attracted to the leaves of citrus and cotton plants by their odor. However, it seems that most chemicals which positively affect insect feeding are gustatory stimulants such as sinigrin for the cabbage butterfly, Pieris brassicae (19-21). This insect feeds on plants of the family Cruciferae whose members are unique in having high concentrations of this and other mustard oil glycosides. Sinigrin, or, more likely, a volatile relative, will also attract the adults of the diamondback moth, Plutella maculipennis (22), and the cabbage root fly, Delia brassicae (23). Leek volatiles will attract the Ichneumid Diadromus pulchellus, a parasitoid of the pupae of the leek moth Acrolepiopsis assectella (24).

However, the use of natural attractants is best known when the adults are host feeders: bran and molasses for grasshoppers, Melanoplus spp. and Camnula spp. (25); syrups for ants, Iridomyrmex humilis and Tapinoma sessile (26); and peanut butter for the imported fire ant, Solenopsis saevissima richteri (27). From some natural attractants of this type discreet chemical attractants have been isolated (Figure 2): sotolone for ants, house flies, and cockroaches (28); geraniol and eugenol for the Japanese beetle, Popillia japonica (29); methyl eugenol for the oriental fruit fly, Dacus dorsalis (30); carbon dioxide and lactic acid for mosquitoes, Aedes aegypti (31-32); and 3-hexen-1-ol and 2-hexen-1-ol for the silkworm, Bombyx mori (33). These last two compounds belong to the "green odour complex" (34), and electrophysiology work has shown the existence of "green" receptors in this species (35-36) and other species (37-38).

There have also been many synthetic attractants (Figure 3) that have been classified as food lures (39): p-acetoxyphenethyl methyl ketone (cue-lure) (40) for the melon fly, Dacus curcurbitae; tertiary-butyl 2-methyl-4-chlorocyclohexanecarboxylate (trimedlure) (41) for the Mediterranean fruit fly, Ceratitis capitata; phenethyl propionate (42), which, along with eugenol, is used for the Japanese beetle; propyl 1,4-benzodioxan-2-carboxylate (amlure) (43) for the European chafer, Amphimallon majalis; ethyl 3-isobutyl-2,2-dimethylcyclopropanecarboxylate (44) for the coconut rhinoceros beetle, Orcytes rhinoceros; and heptyl butyrate (45) for yellow jackets of the genus Vespula. Wasp traps containing pentyl valerate are even available in retail garden shops. Although these have been called food lures, cue-lure, trimedlure, and methyl eugenol attract mainly males. They have also been referred to as parapheromones (46). A synthetic attractant for a predator has also been found: cyclohexyl phenylacetate (47) will attract the checkered flower beetle, Trichodes ornatus.

Tephritid Attractants

Our search for attractants is focused on the Tephritid family of fruit flies which includes species that are of economic importance in Europe, Asia, Australia, and the Americas. It is estimated that the olive fly, Dacus oleae, causes ten percent fruit drop in European olives. Of the infested fruit remaining on the trees, 25 percent of the flesh is destroyed (48). A conservative estimate of the annual cost of the recent Medfly infestation in California, not including capital outlays, is $59 million for chemical controls, $38 million for quarantine and fumigation, and $260 million in crop losses (49). It is estimated that 70% of the susceptible fruit in Egypt is infested by the Medfly (50) and a $50 million control program has been started there.

Because of their very large potential for crop damage and for economic losses to the export market, the Animal and Plant Health Inspection Service (APHIS), as well as the Agricultural Research Service (ARS), is very concerned with the control of Tephritid species. APHIS has a very active program of traps to spot infestations of the Mediterranean fruit fly, the oriental fruit

ATTRACTANTS	USE	SPECIES
 α-PINENE (FROM TREES)	MATE FINDING	DENDROCTONUS FRONTALIS
CORN SILK EXTRACT	OVIPOSITION	HELIOTHIS ARMIGERA
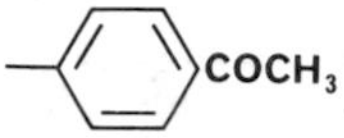 4'-METHYLACETOPHENONE (FROM RICE STALKS)	OVIPOSITION	CHILO PLEJADELLUS
CITRUS LEAF	OVIPOSITION (NON-VOLATILE)	PAPILIO DEMOLEUS
CITRUS LEAF	FOOD (VOLATILE)	PAPILIO DEMOLEUS (LARVAE)

Figure 1.   Attractants for mate finding, oviposition and food.

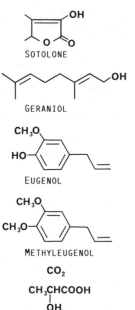

SOTOLONE   ANTS
                               HOUSEFLIES
                               COCKROACHES

GERANIOL

EUGENOL   JAPANESE BEETLE

METHYLEUGENOL   ORIENTAL FRUIT FLY

CO_2

$CH_3CHCOOH$
$\quad\quad|$
$\quad\ OH$
LACTIC ACID   MOSQUITOES

$CH_3CH_2CH=CHCH_2CH_2OH$
3-HEXEN-1-OL

                               SILKWORM
$CH_3CH_2CH_2CH=CHCH_2OH$
2-HEXEN-1-OL

Figure 2.   Compounds isolated from food sources.

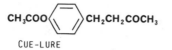

CUE-LURE

DACUS CUCURBITAE

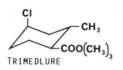

TRIMEDLURE

CERATITIS CAPITATA

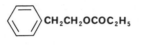

POPILLA JAPONICA

AMLURE

AMPHIMALLON MAJALIS

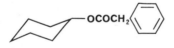

ORCYTES RHINOCEROS

$CH_3(CH_2)_5CH_2OCOC_3H_7$-n

VESPA SPP.

TRICHODES ORNATUS

Figure 3. Synthetic Attractants.

fly, the melon fly, the Mexican fruit fly (<u>Anastrepha</u> <u>ludens</u>), and the Caribbean fruit fly (<u>A</u>. <u>suspensa</u>). The California State Department of Food and Agriculture is involved in a trapping program to halt the spread of the apple maggot (<u>Rhagoletis pommenella</u>). The ARS has personnel in Honolulu and Hilo (Hawaii), Weslaco (Texas), Beltsville (Maryland), and Miami and Gainesville (Florida), as well as our facility in Albany (California) working on these problems. Attractants are a prominent part of the work at this laboratory.

Historically, shortly after the turn of the century, an eight year old girl noticed that flies were attracted to kerosene that her mother had daubed on a hitching post to keep ants away from the jam she was cooling at the top of the post. On further investigation, the girl's father found that these flies were Medflies which were indeed attracted to the kerosene and not to the jam (<u>51</u>). Later, Howlett (<u>52</u>) heard a neighbor complaining that he was being bothered by flies attracted to the oil of citronella he was using as a mosquito repellant. Howlett identified these flies as male <u>Dacus</u> <u>zonatus</u> and <u>D</u>. <u>doralis</u> and later identified methyl eugenol as the active component of the citronella oil (<u>30</u>). Now, trimedlure is used for surveying for the Medfly, cue-lure for the melon fly, methyl eugenol for the Oriental fruit fly, and protein hydrolysates for the Mexfly and Caribfly. In fact, protein hydrolysates can be used for many species of fruit flies. Protein hydrolysates also differ significantly from the synthetic lures in that they will attract female flies, especially gravid ones, as well as males. The sex pheromones of the Medfly and the olive fruit fly have been identified (<u>53</u>-<u>55</u>), synthesized (<u>53</u>-<u>56</u>) and tested in the field (<u>57</u>-<u>59</u>). Methyl eugenol, mixed with an insecticide and applied to cardboard pieces (<u>60</u>) or spot sprayed (<u>61</u>), has been used in male-annihilation eradication programs. Protein hydrolysates mixed with an insecticide have been used in many Medfly (<u>62</u>-<u>65</u>) and Mexfly (<u>64</u>, <u>66</u>) eradication projects, including recent ones in California (<u>61</u>).

Experimental Approach, Results, and Discussion

The Biocommunication Chemistry Research Unit, WRRC, USDA, has developed a program to find other attractants for members of the Tephitid family allowing for the development of new lures, improvement of current IPM programs, imparting species selectivity to programs, and possible replacement of currently used baits. These are in addition to finding attractants for pests currently uncontrolled in this way.

We are examining those commonly used baits, the protein hydrolysates. Initial studies have used the corn gluten hydrolysate commonly known as PIB-7 or, now, as Nu-Lure Insect Bait (NLIB). This material was used in the recent successful Medfly eradication program in California. Since the fruit flies are probably attracted to the volatile emanations from the bait, we have used equipment and techniques previously developed by members of our group for flavor research. For example, a modified Likens-Nickerson simultaneous steam distillation-extraction head was developed by Flath and Forrey (<u>67</u>). Also, there is a 90 liter pilot plant system (68) that can be used for isolating enough volatile material for chemical and biological assays. This is very important because without sufficient quantities of material, replicated biological testing and proper chemical fractionation, yielding sufficient material for bioassays, would be impossible. Conventional vapor trapping techniques yield only 10^{-6} to 10^{-3} grams of material. Using a laboratory scale, 12 liter, simultaneous steam distillation-extraction system 10^{-4} to 10^{-2} grams can be collected. The 90 liter pilot plant system will yield 10^{-2} to 1 gram. Recent work has shown that the figures for these last two methods can be increased by an order of magnitude. Thus, using the equipment and techniques already developed by our group, it is possible to produce material in sufficient quantity to conduct both the biological and chemical investigations.

Our first separation method involved running the simultaneous steam distillation extraction under 100 mm vacuum in order to minimize heat effects. This was followed by extraction under atmospheric pressure in order to get more complete recovery. This atmospheric extraction was run for 10 days, using a fresh batch of solvent each day (68-69). Approximately 10 times as much material was collected each day at atmospheric pressure as was collected under vacuum. Since Schultz, et. al. (70) showed that many non-water-soluble alcohols, esters, aldehydes, and ketones can be recovered by this system in less than 3 hours, the collection of a large amount of material after 10 days is indicative of a very complex and probably dynamic system. Gas chromatograms for these extracts (68) and some compound identifications (69) have been reported. (Other reports on the identification of volatiles from protein hydrolysates are given in references 71-75). Preliminary results have shown that the vacuum extracts are more attractive for the Medfly than the atmospheric ones.

Next, in agreement with the work of Bateman and Morton (76) on the Queensland fruit fly, we found that increasing the pH of the hydrolysate from its normal 4.2 to 8.5 to 9 increases the attractancy of the bait for our test species. Gas chromatography (Figure 4) and gc/mass spectrometry (Finnigan/Incos model 4500X GC/MS with OV-101 fused silica capillary column) on the isolated volatiles of pH 4.2 and pH 9 hydrolysate show considerable differences.

The major volatile components of the pH 4.2 bait are 2-methylpropanal, 3-methylbutanal, 2-methylbutanal, furfural, 3-(methylthio)propanal, acetylfuran, phenylacetaldehyde, and acetophenone.

The major volatile components of the pH 9 bait are 2-methylpropanal, 3-methylbutanal, 2-methylbutanal, methylpyrazine, dimethylpyrazine, ethylpyrazine, 2,3-dimethylpyrazine, ethylmethylpyrazine, trimethylpyrazine, dimethylethylpyrazine, and diethylpyrazine.

The volatiles of the pH 4.2 bait are dominated by aldehydes and ketones while those of the pH 9 bait are dominated by pyrazines. Figure 5 shows the structures of some of these pyrazines. It is interesting to note that some of these pyrazines have also been found in the rectal gland secretion of the male melon fly (77).

We have separated the volatiles from the pH 9 bait into basic and neutral fractions by extraction with acid, followed by neturalization of the acid extract (Figure 6).

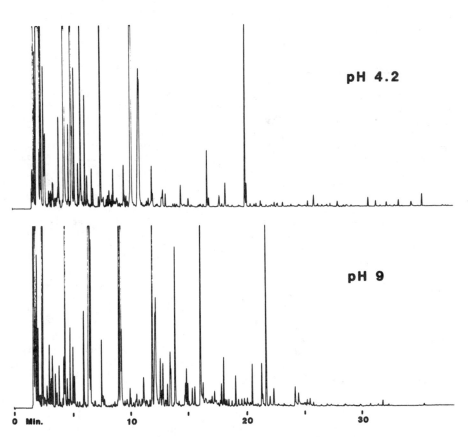

Figure 4.   Gas chromatograms of volatiles isolated from NLIB;
top chromatogram, pH 4.2; bottom chromatogram, pH 9; 30 m x 0.25
mm fused silica DB-1 column, temperature programmed: 50°C for
0.1 min., then 4°C/min. to 220°C, held at 220°C for 20 min.

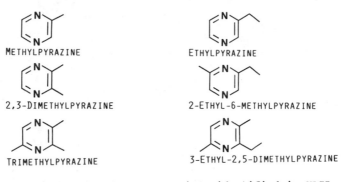

Figure 5.   Structures of some pyrazines identified in NLIB.

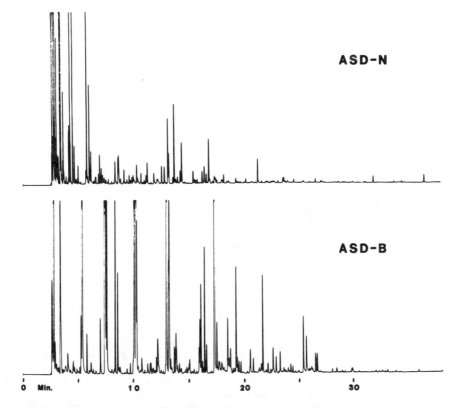

Figure 6.   Gas chromatograms of volatiles isolated from pH 9
NLIB; top chromatogram, neutral fraction; bottom chromatogram,
basic fraction; 60 m x 0.32 mm fused silica DB-1 column,
temperature programmed: 50°C for 0.1 min., then 4°C/min. to
230°C, held at 230°C for 10 min.

The major components of the neutral fraction are 2-methyl-propanal, 2-butanone, 3-methylbutanal, 2-methylbutanal, dimethyl disulfide, and 3-methylpentanal.

The major components of the basic fraction are 3-methylbutanal, methylpyrazine, dimethylpyrazine, ethylpyrazine, 2-ethyl-6-methyl-pyrazine, trimethylpyrazine, and 3-ethyl-2,5-dimethylpyrazine. (The aldehyde was probably carried over during the extractions because of its solubility in both water and organic solvents.) Preliminary results show that this fraction has some attraction for the male oriental fruit fly.

We are now further bioassaying these fractions and will further separate components in order to isolate the active materials. Our very able entomologist-cooperators and their insect test species are listed in Table I. We are also setting up facilities to study the neurophysiology and ethology of these insects. This will greatly expand the types of assays in our research program.

Table I
Entomologist-Cooperators

Roy T. Cunningham ARS, Hilo, HI	Mediterranean Fruit Fly Oriental Fruit Fly Melon Fly
William G. Hart ARS, Weslaco, TX	Mexican Fruit Fly
Peter Landolt Dennis Howard ARS, Miami, FL	Caribbean Fruit Fly
Shmuel Gothilf Volcani Institute Bet-Dagan, Israel	Mediterranean Fruit Fly

What are some of the problems that researchers on attractants might face? One is the possibility that mixtures are necessary for activity. Their components may even need to be in specific ratios. This is something that has become very apparent in pheromone work (78). In our investigations of feeding attractants, this may also be true (79). As most of us know, people often find the odor of a complex natural oil more attractive than that of its most characteristic single component. However, current chemical analytical methodology is designed to separate materials into single components, and we will have to find ways of determining how to combine materials most efficiently to obtain the best activity. A further complication when working with feeding attractants is that the combination may have to include non-volatile arrestants such as carbohydrates or peptides with the volatile materials in order to be useful.

Another problem with bioassays is that many possible attractants can act as repellents at higher concentrations. Protocols must be designed to take this into acount, even though the concentration ranges over which the attractancy and/or repellancy occur may vary substantially from compound to compound. Putting a compound into a test at too high a concentration can swamp out even the standards, much in the same way that a pheromone is used in a confusion technique. Also insects may be attracted to a compound only until it reaches a certain concentration and then stop coming. If this is the case, insects may be attracted to a region around a trap but never enter the trap. How these problems are handled will depend on whether the bioassay is of the olfactometer type or the field type.

Another problem is the lack of knowledge about insect behavior. Although an insect is attracted to a host plant, it may be for reasons other than food. For instance, although an attractant is isolated from a food source such as a fruit, it may actually be a signal for the formation of a lek, that is, a congregation of males for the purpose of attracting females. If this is so, then the material may be effective in attracting insects to an area but may be ineffective in inducing them to enter into, or land on, a trap.

Conclusions

We have found that: 1) volatiles from protein hydrolysates will attract fruit flies of various species, both male and female, 2) that volatiles from protein hydrolysates which had been brought to pH 9 are more attractive than volatiles from protein hydrolysate as it comes at pH 4.2, 3) that isolates from protein hydrolysates are attractive to various species of fruit flies, and 4) that the basic isolate is attractive to male oriental fruit flies. We are continuing to pursue vigorously the isolation, identification, and biological testing of the active components of this mixture.

I would also like to reemphasize the usefulness of attractants research in finding new ways of controlling insect pests; its efficacy has been shown. We must now refine the tools we have, as is the case with the protein hydrolysates, and expand the range of pests than can be selectively controlled by this technique.

Acknowledgements

I would again like to acknowlege our entomologist--cooperators: Roy Cunningham, Shmuel Gothilf, Bill Hart, Dennis Howard, and Peter Landolt; I would also like to thank W. G. Schultz for helpful discussions and C. Caulins for the leading references on the history of Tephritid attractants.

Reference to a company and/or product by name is only for purposes of information and does not imply approval or recommendation of the product by the Department of Agriculture to the exclusion of others which may also be suitable.

364   BIOREGULATORS FOR PEST CONTROL

Literature Cited

1. Metcalf, R. L.; Metcalf, R. A. In "Introduction to Insect
    Pest Management"; Metcalf, R. L.; Luckman, W. H., Eds.; John
    Wiley and Sons: New York, 1975; p. 275-6.
2. Carson, R. "Silent Spring"; Hoghton Mifflin: Boston, 1962;
    p. 296.
3. Tumlinson, J. H. In this volume.
4. Klun, J. A. In this volume.
5. Dethier, V. G.; Browne, L. B.; Smith, C. N. J. Econ. Entomol.
    1960, 53, 134-6.
6. Metcalf, R. L.; Metcalf, R. A. Loc. cit.; pp. 285 ff.
7. Davis, H. G.; George, D. A.; McDonough, L. M.; Tamaki, G.;
    Buditt, Jr., A. K. J. Econ. Entomol. 1983, 76, 674-5.
8. Il'ichev, A. L. Nauchnye Doklady Vysshei Shkoly, Biologi-
    cheskie Nauki 1983, 9, 48-52. Rev. Appl. Entomol. Series A
    1984, 72:1494.
9. Prokopy, R. J.; Owens, E. D. Ann. Rev. Entomol. 1983, 28,
    337-64.
10. Nakagawa, S.; Prokopy, R. J.; Wong, T. T. Y.; Ziegler, J. R.;
    Mitchell, S. M.; Urago, T.; Harris, E. J. Ent. exp. appl.
    1978, 24, 193-8.
11. Kirk, W. D. J. Ecolog. Entomol. 1984, 9, 35-41.
12. Burk, T; Webb, J. C. Ann. Entomol. Soc. Amer. 1983, 76,
    678-82.
13. Yamamoto, I. In this volume.
14. Metcalf, R. L.; Metcalf, R. A. Loc. cit.; pp. 288.
15. ibid.; p. 283.
16. Saito, T.; Munakata, K. In "Control of Insect Behavior by
    Natural Products"; Wood, D. L.; Silverstein, R. M.; Nakajima,
    M., Eds; Academic Press: New York, 1970; p. 225-36.
17. Saxena, K. N.; Goyal, S. Ent. exp. appl. 1978, 24, 1-10.
18. Saxena, K. N.; Khatter, P.; Goyal, S. Experientia 1977, 33,
    1312-13.
19. David, W. A. L.; Gardiner, B. O. Entomol. exp. appl. 1966,
    9, 247-55.
20. Hanson, F. E. In "Herbivorous Insects: Host-Seeking Behavior
    and Mechanisms:; Ahmad, S., Ed.; Academic Press: New York,
    1983; p. 12.
21. Ma, W. C. Meded. Landbouwhogesch. Wageningen 1972, 72-11,
    1-162.
22. Gupta, P. D.; Thorsteinson, A. J. Entomol. exp. appl. 1960,
    3, 305-14.
23. Finch, S. Entomol. exp. appl. 1978, 24, 350-9.
24. Lecomte, C.; Thibout, E. Entomophaga 1983, 28, 217-26. Rev.
    Appl. Entomol. Series A. 1984, 72: 272.
25. Metcalf, C. L.; Flint, W. P.; Metcalf, R. L. "Destructive
    and Useful Insects"; McGraw-Hill: New York, 1962; pp. 323,
    468-9.
26. ibid.; p. 324, 897-8.
27. ibid.; p. 897.
28. Tokitomo, Y.; Kobayashi, A.; Yamanishi, T.; Muraki, S. Proc.
    Japan Acad. Ser. B 1980, 56, 457-462.

29. Metcalf, C. L.; Flint, W. P.; Metcalf, R. L., Loc. cit., pp.
     394, 751.
30. Howlett, F. M. Bull. Entomol. Res. 1915, 6, 297-305ff.
31. Acree, Jr., F; Turner, R. B.; Gouck, H. C.; Beroza, M.;
     Smith, N. Science 1968, 161, 1346-7.
32. Davis, E. E. J. Insect Physiol. 1984, 30, 211-5.
33. Watanabe, T. Nature 1958, 182, 325-6.
34. Visser, J. H.; Ave, D. A. Ent. exp. appl. 1978, 24, 738-49.
35. Morita, H.; Yamashita, S. J. Exp. Biol. 1961, 38, 851-61.
36. Priesner, E. Ann. Zool. Ecol. anim., 1979, 11, 533-546.
37. Ma, W. C.; Visser, J. H. Ent. exp. appl. 1978, 24, 520-533.
38. Visser, J. H. Ent. exp. appl. 1979, 25, 86-97.
39. Metcalf, R. L.; Metcalf, R. A. Loc. cit., p. 284.
40. Alexander, B. H.; Beroza, M.; Oda, T. A.; Steiner, L. F.;
     Miyashita, D. H.; Mitchell, W. C. J. Agr. Food Chem. 1962,
     10, 270-6.
41. McGovern, T. P.; Beroza, M.; Ohinata, K.; Miyashita, D.;
     Steiner, L. F. J. Econ. Entomol. 1966, 59, 1450-5.
42. Ladd, Jr., T. L.; McGovern, T. P.; Beroza, M.; Buriff, C. R.;
     Klein, M. G. J. Econ. Entomol. 1976, 69, 468-70.
43. McGovern, T. P.; Fiori, B.; Beroza, M.; Ingangi, J. C. J.
     Econ. Entomol. 1970, 63, 168-71.
44. Barber, I. A.; McGovern, T. P.; Beroza, M.; Hoyt, C. P.;
     Walker, A. J. Econ. Entomol. 1971, 64, 1041-4.
45. Davis, H. G.; Eddy, G. W.; McGovern, T. P., Beroza, M. J.
     Econ. Entomol. 1969, 62, 1245.
46. Chambers, D. L. In "Chemical Control of Insect Behavior";
     Shorey, H. H.; McKelvey, Jr., J. J., Eds.; John Wiley and
     Sons: New York; 1977; p. 331.
47. Davis, H. G.; Eves, J. D.; McDonough, L. M. Environ.
     Entomol. 1979, 8, 147-9.
48. Michelakis, S. E.; Neuenschwander, P. In "Fruit Flies of
     Economic Importance"; Cavalloro, R., Ed.; A. A. Balkema:
     Rotterdam; 1983; p. 603-11.
49. Burk, T.; Calkins, C. O. Florida Ent. 1983, 66, 3-18.
50. Schultz, W. G., private communication.
51. Severin, H. H. P.; Severin, H. C. J. Econ. Entomol. 1913, 6,
     347-51.
52. Howlett, F. M. Trans. Entomol. Soc. Lond. 1912, 60, 412-8.
53. Jacobson, M.; Ohinata, K.; Chambers, D. L.; Jones, W. A.;
     Fujimoto, M. S. J. Med. Chem. 1973, 16, 248-51.
54. Baker, R.; Herbert, R.; Howse, P. E.; Jones, O. T.; Francke,
     W.; Reith, W. J. Chem. Soc., Chem. Comm. 1980, 52-3.
55. Baker, R.; Herbert, R. H.; Parton, A. H. J. Chem. Soc.,
     Chem. Comm., 1982, 601-3.
56. Kay, I. T.; Williams, E. G. Tetrahedron Lett. 1983, 24,
     5915-8.
57. Zumreoglu, A. In "Fruit Flies of Economic Importance";
     Cavalloro, R., Ed.; A. A. Balkema: Rotterdam; 1983; p. 495-9.
58. Jones, O. T.; Lisk, J. C.; Howse, P. E.; Baker, R.; Ramos,
     P. ibid.; p. 500-5.
59. Mazomenos, B. E.; Haniotakis, G. E.; Ioannou, A.; Spanakis,
     I.; Kozirakis, E. ibid.; p. 506-12.

60.   Steiner, L. F.; Hart, W. G.; Harris, E. J.; Cunningham, R. T.;
      Ohinata, K.; Kamakahi, D. C.   J. Econ. Entomol. 1970, 63,
      131-5.
61.   Dowell, R.   Department of Entomology Seminar, University of
      California, Berkeley; 30 January 1984.
62.   Steiner, L. F.   J. Econ. Entomol. 1952, 45, 838-48.
63.   Stephenson, B. C.; McClung, B. B.   Bull. Entomol. Soc. Amer.
      1966, 12, 374.
64.   Hart, W. G.; Ingle, S.; Reed, D.; Flitters, N.   J. Econ.
      Entomol. 1967, 60, 1264-5.
65.   Steiner, L. F.   Proc. 1st Int. Citrus Symp. 1968, 2, 881-7.
66.   Lopez-D., F.; Chambers, D. L.; Sanchez-R., M.; Kamasaki, H.
      J. Econ. Entomol. 1969, 62, 1255-7.
67.   Flath, R. A.; Forrey, R. R.   J. Agric. Food Chem. 1977, 25,
      103-9.
68.   Teranishi, R.; Buttery, R. G.; Mon, T. R.   In "Isolation and
      Characterization of Biologically Active Natural Products";
      Acree, T., Ed.; Walter DeGruyter: Berlin; in press.
69.   Buttery, R. G.; Ling, L. C.; Teranishi, R.; Mon, T. R.   J.
      Agric. Food Chem. 1983, 31, 689-92.
70.   Schultz, T. H.; Flath, R. A.; Mon, T. R.; Eggling, S. B.;
      Teranishi, R.   J. Agric. Food Chem. 1977, 25, 446-9.
71.   Manley, C. H.; Fagerson, I. S.   J. Food Sci. 1970, 18, 340-2.
72.   Manley, C. H.; Fagerson, I. S.   J. Food Sci. 1970, 35, 286-91.
73.   Markh, A. T.; Vinnikova, L. G.   Appl. Biochem. Microbiol.
      (Eng. Trans.) 1973, 9, 774-8.
74.   Withycombe, D. A.; Mookherjee, B. D.; Hruza, A.   In
      "Analaysis of Food and Beverages"; Charalambous, G., Ed.;
      Academic Press: New York; 1978; p. 81-94
75.   Morton, T. C.; Bateman, M. A.   Aust. J. Agric. Res. 1981, 32,
      905-16.
76.   Bateman, M. A.; Morton, T. C.   Aust. J. Agric. Res. 1981, 32,
      883-903.
77.   Baker, R.; Herbert, R. H.; Lomer, R. A.   Experientia 1982,
      38, 232-3.
78.   Roelofs, W.   In "Chemical Ecology: Odour Communication in
      Animals"; Ritter, F. J., Ed.; Elsevier: New York; 1979, p.
      159-68.
79.   Yamamoto, I.; Yamamoto, R.   In "Control of Insect Behavior by
      Natural Products"; Wood, D. L., Silverstein, R. M., Nakajima,
      M., Eds; Academic Press: New York; 1970; p. 331-45.

RECEIVED November 8, 1984

Beetles: Pheromonal Chemists par Excellence

J. H. TUMLINSON

Insect Attractants, Behavior, and Basic Biology Research Laboratory, Agricultural Research Service, U.S. Department of Agriculture, Gainesville, FL 32604

Pheromones of insect species in the order Coleoptera are characterized by considerable structural diversity. Unlike the lepidopterous sex pheromones, which are nearly all fatty acid derivatives, coleopterous sex pheromone structures range in complexity from the relatively simple 3,5-tetradecadienoic acid of the black carpet beetle to the tricyclic terpenoid, lineatin, of the striped ambrosia beetle. While the sex pheromones of many beetles consist of mixtures of compounds that act synergistically to elicit a behavioral response, other Coleoptera species appear to use only a single compound for chemical communication between the sexes. In the latter case the compound usually has at least one chiral center and chirality plays a major role in determining pheromone specificity. Some interesting relationships within groups of coleopterous species, based on subtle differences in structures or mixtures, have been unraveled in the last few years. These and the variety of structures comprising coleopterous pheromones provide a challenging opportunity for the natural products chemist.

There are more species of Coleoptera than of any other order of insects. It has been estimated that over 350,000 species of beetles have been described, and this order represents about half of the total number of species (1). Thus it is somewhat surprising to discover from recent surveys of the literature (2,3) that pheromones have been identified for only about 50 to 75 species of Coleoptera. In contrast, sex pheromones or sex attractants (most of the latter discovered by screening procedures) are available for over 500 species of Lepidoptera. Furthermore, over half of the known coleopterous pheromones have been isolated and identified from species in the family Scolytidae.

As previously noted (4,5), coleopterous pheromones are characterized by great diversity and complexity in both the chemical
structures and the composition of blends. Some species, particularly in the family Dermestidae, utilize very simple fatty acid-
like molecules similar to the lepidopterous pheromones. In contrast, Trypodendron lineatum (Olivier) uses a tricyclic acetal
(6). There are a variety of other coleopterous pheromone structures of intermediate complexity. Additionally, the phenomenon of
chirality introduces even more complexity and diversity into these
pheromone structures. Silverstein (5) lists nine possible categories of response by insects to enantiomers. So far we know of
examples for five of these categories. Thus an insect may produce
a single enantiomer and the other enantiomer may be active,
inactive, or inhibitory. Alternatively, the insect may produce
both enantiomers and respond optimally to the natural ratio of the
two enantiomers or more strongly to one than the other. For a
more detailed discussion of chirality and pheromone structures see
Silverstein (7).

It would seem that many of the coleopterous species could
achieve a high degree of specificity in their pheromone signals
with single components, given the many diverse structures available and the added degree of structural specificity provided by
differences in enantiomeric or stereoisomeric forms of a molecule. In fact, many closely related species use different enantiomers or stereoisomers to achieve specificity in their chemical
signal. However, most of the coleopterous pheromones identified
thus far are multicomponent and in the few cases where pheromones
have been identified for several closely related species in a
genus, we usually find that the different species are using blends
of a few compounds and achieving specificity by varying ratios,
number or types of components, and the chirality of individual
components.

Despite the general trend toward multicomponent pheromones,
several single component coleopterous pheromones have been isolated and identified. In many of these instances it appears that
the pheromone identification is incomplete, although a single compound may be attractive and may in fact be a very effective lure
for field trapping tests. No doubt further work will turn up more
components and more effective pheromones in many instances. However, there are cases in which very careful and thorough investigations have yielded a single compound with apparently the total
activity of the original extract or of the live insects. There is
still the possibility in many of these instances that the bioassay
did not measure all behavior associated with pheromone communication. Thus, more thorough studies of the behavior may indicate
the presence of more pheromone components. Additionally, where a
species is isolated from its native habitat and other related
species, the role of other components in providing specificity to
the signal may not be evident.

Therefore, as a general rule we should probably always expect
that the complete pheromone will consist of two or more compounds.
However, it is possible that a few species may utilize single component pheromones, particularly if the structure is sufficiently
unique and if chirality provides sufficient diversity to allow

specificity of the signal in the context of the associated environment.

Consideration of the environment in which a species lives and reproduces is very important, although often neglected in studying the chemical communications system of that species. In many cases it is very difficult to monitor the chemical isolation and purification with field bioassays and thus laboratory bioassays are often used. Obviously a laboratory bioassay is quite artificial in that many environmental factors that affect the insect's behavior are eliminated. In fact, bioassays conducted in the field may not measure all the behavioral components either. In either case a pheromone identification should be followed by an investigation of the response of other species to synthetic compounds, blends, isomers, enantiomers, etc., and comparison of the results with the natural interspecific responses or behavior. Additionally, odors from host plants or pheromones of other species may affect the response of members of a species to their pheromones.

For a variety of reasons, some of which have already been discussed, we can often learn more about a communication system by studying the pheromones and related behavior of several members of a genus or family than by an in-depth study of a single species. The interspecific interactions are often quite subtle but very important. For example, the importance of enantiomers or stereoisomers may only become evident after investigating the responses of several closely related species to a pheromone or pheromone blend.

There have been several good reviews -- particularly of the chemistry and behavior associated with bark beetle pheromones -- in the last few years (6, 9-12). Many aspects of my discussion here are discussed in more detail in one or more of these reviews. In the remainder of this chapter, I intend to illustrate with selected examples the points that I discussed above.

Scolytidae

Bark beetles are of great economic importance, which is one of the reasons more research has been done on the pheromones of the Scolytidae than on those of any other family of Coleoptera. Their pheromone systems also seem to be typical of the Coleoptera in that while there is considerable diversity in pheromone structure within this family, there also seems to be a pattern of structures, particularly within a genus. The first pheromone identified from a coleopterous species was from Ips paraconfusus Lanier (then I. confusus) by Silverstein et al. (13). Three compounds -- ipsenol (I), ipsdienol (II), and cis-verbenol (III) -- were identified. These compounds are synergistic, the combination of all three being required for activity with little or no attraction being elicited by any one of them alone.

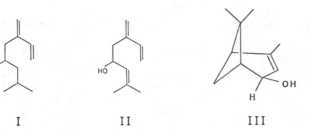

I   II   III

Each of these compounds is chiral and thus can exist in either
of two enantiomeric forms. Thus, the chemical signal can be
varied by (1) using different ratios of these three compounds, (2)
using different blends employing one, two, or three of these compounds, and (3) imposing the element of chirality on each of the
other two methods. The I. paraconfusus pheromone was identified
as (S)(-) ipsenol, predominantly (S)(+) ipsdienol, and (S)(+)
cis-verbenol (13). Subsequently the I. pini pheromone was
identified as (R)(-) ipsdienol (14), and as little as 3% of the
(S)(+) enantiomer completely interrupted I. pini response to the
(R)(-). Furthermore, while I. pini in the western U. S. produces
and responds to (R)(-) ipsdienol, I. pini in New York produces a
65:35 ratio of (+):(-) ipsdienol and responds more strongly to the
racemic synthetic ipsdienol in the field (15). A survey of the
literature on Ips pheromones suggests that most if not all Ips
species utilize one or more of these three components in various
combinations and of different chiralities to achieve the necessary
specificity in their chemical signal.

The aggregation pheromone of Dendroctonus brevicomis LeConte
consists of a synergistic mixture of exo-brevicomin (IV) produced
by females, myrcene (V) from the host tree, and frontalin (VI)
produced by males which respond to compounds IV and V. The mixture of all three compounds then attracts males and females (10).
Subsequently, verbenone (VII) and trans-verbenol (VIII) are produced and deter further attack (11). Additionally verbenone
deters the response of I. paraconfusus which utilizes the same
host tree species in the same area (16,17). Thus the two species
may be found colonizing different parts of the same tree. Again,
as in the Ips pheromones, chirality plays a role. The naturally
occurring enantiomers of exo-brevicomin and frontalin in combination with myrcene elicit a greater response from both male and
female D. brevicomis than do the other enantiomers.

There are many other scolytid species that utilize these and
similar compounds to effect intraspecific communication. Francke
et al. (18) have discovered several spiroketals as active components of scolytid pheromones. Additionally, several parasites and
predators utilize the pheromones produced by their scolytid prey
as kairomones. For example, the predatory beetle, Temnochila
chlorodia, responds specifically to exo-brevicomin produced by
female D. brevicomis (19). This same phenomenon has been
demonstrated recently in Lepidoptera; the egg parasite
Trichogramma sp. uses the Heliothis zea (Boddie) sex pheromone to
locate the H. zea eggs (20).

Thus, to quote from Birch (<u>10</u>), "Not only should the chemical communication system within a species be intimately known in order to manipulate behavior effectively, but the effect of manipulating one species on the complex of other species in that environment should also be understood."

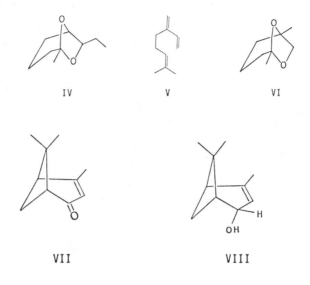

IV   V   VI

VII   VIII

Curculionidae

Relatively few pheromones have been identified from species in this family. The first curculionid pheromone identified, that of the boll weevil, <u>Anthonomus grandis</u> Boheman, is terpenoid in character like those of the Scolytidae. Two terpene alcohols (IX, X) and two aldehydes (XI, XII) were isolated and identified from male weevils and from frass (<u>21</u>). All four of these compounds are required to elicit optimum attraction.

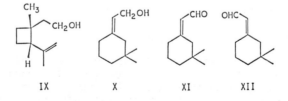

IX   X   XI   XII

Produced by the male, this pheromone acts as an aggregating pheromone in the field in the spring, but in midseason appears to act more like a true sex pheromone in that it primarily attracts females. Also, as in the bark beetles, components of the essential oils of the host plant appear to enhance the attractiveness of this pheromone (<u>22</u>).

Grandisol (IX) and the corresponding aldehyde, grandisal, also have been isolated and identified as components of the pheromone of two other curculionids, Pissodes strobi (Peck), the white pine weevil, and P. approximatus Hopkins, the northern pine weevil (23). The pecan weevil, Curculio caryae (Horn), also has been reported to respond to the synthetic boll weevil pheromone blend and to compound X alone (24), and Hedin et al. (25) isolated compound X from an active extract of female pecan weevils. Additionally, the New Guinea sugarcane weevil, Rhabdoscelus obscurus (Boisduval), was reported to respond to the synthetic boll weevil blend (26). Thus these compounds and similar ones may serve as pheromone components in several curculionid species from different genera. Interestingly, females of a cerambycid species Dectes texanus texanus LeConte have been reported to be attracted by these compounds (27). When we consider that the tricyclic acetal lineatin, a scolytid pheromone, has the same carbon skeleton as grandisol, it seems possible that this type of structure may be fairly widespread among coleopterous pheromones.

Very recently (R*,S*)-4-methyl-5-hydroxy-3-heptanone has been identified as the major component of the aggregation pheromone of the rice weevil, Sitophilus oryzae L., and the maize weevil, S. zeamais Motsch. (28). This compound is obviously very different in structure from the previously identified curculionid pheromones. In fact, it is very similar to the pheromone of two other stored-products pests, the drugstore beetle and the cigarette beetle, which are in the family Anobiidae (see later). Undoubtedly, we will find more diversity in structure as more curculionid pheromones are identified.

Since so few curculionid pheromones have been identified, we do not yet know what role chirality will play or the importance of blends in mediating behavior within this family. Obviously, since the boll weevil response to individual components or pairs of components is almost nil, the blend is important. However, the boll weevil is essentially an introduced pest, having migrated into the U. S. in about the second decade of this century from Central America. Thus interactions related to chemical communications between the boll weevil and other closely related species have not been studied. Much more research is needed in this area before we will begin to understand how this insect and related species interact in the ecosystem.

Scarabaeidae

The existence of chirality in pheromone molecules has been recognized since 1966, but as Silverstein (7) explains, most of us ignored it because the insects responded to the synthesized racemic compounds. Thus the insects' response to the chiral pheromones identified in earlier work appeared to fall into the first category described by Silverstein (5), i.e., the insect produced and responded to a single enantiomer and the other enantiomer was inactive. Furthermore, the paucity of natural pheromone obtainable from the insects makes it difficult, and in most cases impossible, to determine the stereochemistry of the natural material.

It was inevitable that eventually we would be forced to consider effects of chirality on pheromone activity. The importance of geometrical isomerism in determining the activity of lepidopterous pheromones was well recognized. However, it was still somewhat surprising when we discovered that the synthesized racemic Japanese beetle, <u>Popillia japonica</u> Newman, pheromone was inactive even though we had purified both the natural (active) pheromone and the synthesized pheromone to greater than 99%, and the two were indistinguishable by every chromatographic and spectroscopic analytical method available (<u>29</u>). It was only when the (<u>R</u>) and (<u>S</u>) enantiomers of XIII were synthesized (<u>30</u>) that we discovered that the pure (<u>R</u>)(−) enantiomer of XIII was very attractive to males and the (<u>S</u>)(+) was an inhibitor that, even when present in only a few percent, significantly reduced the activity of the (<u>R</u>)(−).

There are two other factors worth noting in regard to the Japanese beetle pheromone. A single compound apparently possesses all the activity of the pheromone. This is even more surprising when we note that the material obtained from females contained about 15% of the (<u>E</u>) isomer and 3% of the analog with a saturated side chain. However, these other two compounds had no effect on the activity of the (<u>R</u>)(<u>Z</u>)-isomer when added in the appropriate ratios whether the racemic or the pure enantiomeric forms were used.

Since the Japanese beetle is an imported pest, it remains to be seen whether these other isomers and/or enantiomers have a more significant role in interactions among closely related species in its native habitat. Only one other sex attractant has been identified for a scarab species. Phenol attracts males of the grass grub beetle, <u>Costelytra zealandica</u> (White) (<u>31</u>.). Thus, we do not know what types of compounds are likely to be most prevalent among the pheromones of the Scarabaeidae. However, it is worth noting that attractants derived from plants appear to synergize the response of both male and female Japanese beetles to the pheromone (<u>32</u>, <u>33</u>), and thus it is similar to many other coleopterous species in this respect.

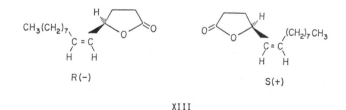

XIII

Chrysomelidae

In this very large coleopterous family the sex pheromone of only two species have been identified. <u>Diabrotica virgifera virgifera</u> LeConte, the western corn rootworm (WCR), produces and responds to

(R,R)-8-methyl-2-decanol propanoate (<u>34</u>). Sonnet et al. (<u>35</u>)
synthesized the four stereoisomers of this pheromone in high
stereoisomeric purity (Table I). Comparison of the captures of
WCR males in the field with traps baited with the individual
steroisomers and with the racemic mixture demonstrated that the
males respond strongly to the (R,R)-isomer and to a lesser extent
to the (2S,8R)-isomer. The other two isomers are inactive; i.e.,
they neither attract nor inhibit the response of WCR males, nor do
they synergize the activity of the (R,R)-isomer (<u>36</u>). The
(R,R)-isomer appears to be as attractive as an equal amount of the
natural pheromone in volatiles collected from females. Thus it is
possible that the WCR is utilizing a single component pheromone.

Table I. Purity of the synthesized stereoisomers of 8-methyl-2-
          decanol propanoate that were tested in the field for
          attractiveness to several species of <u>Diabrotica</u>.

Isomer	Gross purity (%)	Isomeric composition (%)			
		2S,8S	2R,8S	2S,8R	2R,8R
2S,8S	99.3	98.8	0.4	0.8	–
2R,8S	99.1	0.9	99.3	–	0.8
2S,8R	98.3	1.9	–	97.4	0.7
2R,8R	95.3	–	1.9	0.3	97.8
2RS,8RS	99+				

    Evidence from studies of the response of several closely
related <u>Diabrotica</u> species to traps in the field baited with WCR
female volatiles, the racemic synthetic pheromone, and the four
stereoisomers suggests that a number of closely related species
may be relying on the stereochemistry of this molecule to effect
specific communication. The Mexican corn rootworm (MCR), <u>D</u>. <u>v</u>.
<u>zea</u> Krysan & Smith responded to the same stereoisomers as the
WCR. This is not too surprising since the MCR also responds to
volatiles collected from WCR females, and these two taxa are
subspecies that have been reported to interbreed in those areas
where their geographical ranges abut (<u>37</u>). In other field tests
we found that the northern corn rootworm (NCR), <u>D</u>. <u>barberi</u> Smith &
Lawrence, responded to the racemic material at low concentrations
(optimum 0.1 μg on cotton wicks), but as the dose was increased
to 1 μg, the NCR response diminished rapidly and at 10 μg it
was almost nil. Conversely, in the same test the WCR response
continued to increase to 100 μg, the highest dose tested. This
behavior was explained when captures of NCR males in traps baited
with the stereoisomers was investigated. NCR males were captured
only in traps baited with the (2R,8R)-isomer at the 1 μg dose.

Furthermore, NCR male response to the (2R,8R)-isomer was reduced to the level of the blank when an equal amount of the (2S,8R)--isomer was added to the (2R,8R). The (2S,8S)-isomer also significantly reduced response to the (2R,8R), but not as much as (2S,8R), while (2R,8S) neither enhanced nor diminished response to (2R,8R) (38).

Two other species in this group, D. porracea Harold, which is sympatric with the MCR but not with WCR, and D. longicornis (Say), sympatric with WCR and NCR, respond to the racemic synthetic pheromone, but only to the (2S,8R)-isomer (36,38).

These field trapping studies with the stereoisomers have provided strong circumstantial evidence that the WCR female produces only the (2R,8R)-isomer, although WCR males respond to both (2R,8R) and (2S,8R). These results are summarized in Table II. Since the NCR males, which are strongly inhibited by (2S,8R), respond to volatiles from WCR females this suggests that these volatiles do not contain the inhibitory isomer. Furthermore, D. porracea responds to (2S,8R) and to the racemic pheromone and thus is not inhibited by the other isomers. Therefore the fact that D. porracea does not respond to WCR female volatiles also indicates that these volatiles do not contain (2S,8R). At this time we have discerned no obvious reason why WCR males respond to an isomer their females do not produce. This is even more curious when we consider that the (2S,8R)-isomer may be the pheromone of a sympatric species, D. longicornis.

Table II.   Attractiveness of volatiles collected from virgin
            Diabrotica virgifera virgifera females, synthesized
            racemic, (2R,8R)-, and (2S,8R)-8-methyl-2-decanol pro-
            panoate to males of 4 Diabrotica species in field tests.

Species[a]	D. virgifera volatiles	8-Methyl-2-decyl propanoate		
		Racemic	2R,8R	2S,8R
WCR D. v. virgifera	++[b]	++	++	+
MCR D. v. zea	++	++	++	+
D. porracea	0[b]	+	0	+
NCR D. barberi	++	+-[c]	++	--[b]

[a]WCR = western corn rootworm, MCR = Mexican corn rootworm, NCR = northern corn rootworm.
[b]+ Indicates attraction; ++ indicates strong attraction; -- indicates strong inhibition; 0 indicates no activity.
[c]The racemic material is attractive to NCR males at low concentrations (ca. 0.1 µg) but inhibitory at higher concentrations (ca. 10 µg).

Diabrotica cristata (Harris), a nonpest species taxonomically classified in the virgifera species group, has apparently developed a slightly different mechanism for species isolation. The male D. cristata respond to 8-methyl-2-decanol acetate. However, when the four isomers of the acetate were tested, D. cristata males responded only to the (2S,8R)-isomer (39). Thus both structural diversity and stereoisomerism are being used by this group of insects to achieve specificity in their chemical signals. No evidence of multicomponent pheromones in this group has yet been discovered.

The pheromone of the southern corn rootworm (SCR), D. undecimpunctata howardii Barber, was identified as (R)-10-methyl-2-tridecanone (40). Since it has only one asymmetric carbon, it can exist in only two enantiomeric forms. Synthesis of both the (R)- and (S)-enantiomers by Sonnet (41) in high enantiomeric purity again provided material for field tests. SCR males respond only to the (R)-enantiomer in the field. Although extensive field tests have not yet been conducted, preliminary evidence suggests that other closely related species may respond to this pheromone. Since fewer isomers are available, the number of specific signals that can be formulated is reduced correspondingly. Thus it might be necessary for this group to employ more components, although no evidence is available to indicate that this occurs.

Obviously, much more research on pheromone communication in this genus and in other chrysomelid species is needed. However, it seems possible that these Diabrotica species are achieving specificity in their signals through the use of stereoisomers as single component pheromones.

Other Families

The only other family for which pheromones have been identified for several species is Dermestidae. Approximately 11 dermestid pheromones have been identified thus far, and they are of two general types. Trogoderma species utilize 14-methyl-8-hexadecenal and the analogous alcohol and carboxylic acid methyl esters. Apparently all of the species studied thus far respond to the (R)-enantiomer of these compounds (42). Other dermestid phero-mones identified thus far are fatty acids or methyl esters of fatty acids, an example being the black carpet beetle, Attagenus megatoma (Fabricius), pheromone (E,Z)-3,5-tetradecadienoic acid (43).

Only two Anobiidae pheromones have been identified thus far. The drugstore beetle, Stegobium paniceum (L.), pheromone was identified as 2,3-dihydro-2,3,5-trimethyl-6-(1-methyl-2-oxo-butyl)-4H-pyran-4-one (XIV) (44) and the cigarette beetle, Lasioderma serricorne F., as 4,6-dimethyl-7-hydroxy-nonan-3-one (XV) (45). The similarity of these structures is worth noting.

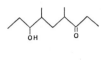

XIV   XV

Examples of structures identified from species in other
families illustrate even further the diversity of coleopterous
pheromone chemistry. The only Bruchidae sex attractant pheromone
known so far is methyl (E)-2,4,5-tetradecatrienoate from
<u>Acanthoscelides</u> <u>obtectus</u> (Say). The allenic nature of the
olefinic bonds in this molecule imparts chirality to the
structure. The pheromone of the Tenebrionidae species <u>Triboleum</u>
<u>castaneum</u> Herbst has been reported as 4,8-dimethyl decanal which
also has two chiral centers (<u>46</u>-<u>48</u>)). The lesser grain borer,
<u>Rhyzopertha</u> <u>dominica</u> (Fabricius), a member of the Bostrichidae
family, uses a mixture of esters as its pheromone,
(<u>S</u>)-1-methylbutyl (<u>E</u>)-2-methyl-2-pentenoate and (<u>S</u>)-1-methylbutyl
(<u>E</u>)-2,4-dimethylpentenoate (<u>49</u>).
     Very recently seven macrolide lactones have been isolated and
identified from volatiles collected from males of five species of
grain beetles in the family Cucujidae. The number of lactones
produced by each of these species varies from two to five,
although four of the species use only two components as their
pheromone and the fifth uses only three. Thus the rusty grain
beetle, <u>Cryptolestes</u> <u>ferrugineus</u> (Stephens), uses a synergistic
blend of (<u>E</u>,<u>E</u>)-4,8-dimethyl-4,8-decadien-10-olide and
(3<u>Z</u>,11<u>S</u>)-3-dodecen-11-olide (<u>50</u>); the flat grain beetle, <u>C</u>.
<u>pusillus</u> (Schönerr), uses (<u>Z</u>)-3-dodecenolide and
(<u>Z</u>)-5-tetradecen-13-olide (<u>51</u>); and the flour mill beetle, <u>C</u>.
<u>turcicus</u> (Grouvelle), uses (<u>Z</u>,<u>Z</u>)-5,8-tetradecadien-13-olide and
(<u>Z</u>)-5-tetradecen-13-olide (<u>52</u>). <u>Oryzaephilus</u> <u>mercator</u> (Fauvel)
uses a blend of (3<u>Z</u>,11<u>R</u>)-3-dodecen-11-olide and
(<u>Z</u>,<u>Z</u>)-3,6-dodecadien-11-olide, while <u>O</u>. <u>surinamensis</u> (L) uses a
three component blend of (<u>Z</u>,<u>Z</u>)-3,6-dodecadien-11-olide,
(<u>Z</u>,<u>Z</u>)-3,6-dodecadienolide, and (<u>Z</u>,<u>Z</u>)-5,8-tetradecadien-13-olide
(<u>53</u>). The chirality of only one of these male produced pheromones
has been determined, i.e. (<u>Z</u>)-3-dodecen-11-olide. However, it
appears that here again both multicomponent blends and chirality
are being used to achieve specificity in the pheromone signals
(<u>54</u>).
     In at least some cases where single component pheromones have
been identified the pheromone identification probably can be
labeled incomplete. As more research is conducted on the behavior
of these species and their interactions with other species,
particularly those in the same family and genus, more compounds
will undoubtedly be identified that are active in mediating the
behavior of these species. At this time I can only reiterate that
considerably more research is needed on coleopterous pheromones
before we can begin to understand these complex interactive
systems. For natural products chemists this should be a fruitful
area for investigation for some time.

Literature Cited

1. Pesson, P.   "The World of Insects," McGraw-Hill: New York, 1959.
2 Klassen, W.; Ridgway, R. L; Inscoe, M. N.   In "Insect Suppression with Controlled Release Pheromone Systems"; Kydonieus, A. F.; Beroza, M., Eds.; CRC Press, Inc.: New York, 1982; Vol. I.
3. Inscoe, M. N.   In "Insect Suppression with Controlled Release Pheromone Systems; Kydonieus, A. F.; Beroza, M., Eds.; CRC Press, Inc.: New York, 1982; Vol. II.
4. Silverstein, R. M.   In "Chemical Control of Insect Behavior: Theory and Application"; Shorey, H. H.; McKelvey, J. J. Jr., Eds.; Wiley: New York, 1977.
5. Silverstein, R. M.   In "Chemical Ecology: Odour Communication in Animals;" Ritter, F. J., Ed.; Elsevier/North Holland Biomedical Press: Amsterdam, 1979.
6. Borden, J. H.; Handley, J. R.; Johnston, B. D.; MacConnel, J. G.; Silverstein, R. M.; Slessor, K. N.; Swigar, A. A.; Wong, D. T. W.   J. Chem. Ecol. 1979, 5: 681-9.
7. Silverstein, R. M.   In "Semiochemistry: Flavors and Pheromones"; Acree; Soderland, Eds.; ACS SYMPOSIUM SERIES, American Chemical Society: Washington, D. C., 1984; (in press).
8. Borden, J. H.   In "Chemical Control of Insect Behavior: Theory and Application"; Shorey, H. H.; McKelvey, J. J. Jr., Eds.; Wiley: New York, 1977.
9. Birch, M. C.   Am. Scientist 1978, 66, 409-19.
10. Birch, M. C.   In "Chemical Ecology of Insects;" Bell, W. J.; Cardé, R. T., Eds.; Chapman & Hall: London, 1984.
11. Wood, D. L.   Ann. Rev. Entomol. 1982, 27, 411-46.
12. Vité, J. P.; Francke, W.   Naturwissenschaften 1976, 63, 550-5.
13. Silverstein, R. M.; Rodin, J. O.; Wood, D. L.   Science 1966, 154, 509-10.
14. Birch, M. C.; Light, D. M.; Wood, D. L.; Browne, L. E.; Silverstein, R. M.; Bergot, B. J.; Ohloff, G.; West, J. R.; Young, J. C.   J. Chem. Ecol. 1980, 6, 703-17.
15. Lanier, G. N.; Claesson, A.; Stewart, T.; Piston, J. J.; Silverstein, R. M.   J. Chem. Ecol. 1980, 6, 677-87.
16. Byers, J. A.; Wood, D. L.   J. Chem. Ecol. 1980, 6, 149-64.
17. Byers, J. A.; Wood, D. L.   J. Chem. Ecol. 1981, 7, 9-18.
18. Francke, W.; Hindorf, G.; Reith, W.   Naturwissenschaften 1979, 66, 618-9.
19. Bedard, W. D.; Wood, D. L.; Tilden, P. E.; Lindahl, K. Q., Jr.; Silverstein, R. M.; Rodin, J. O.   J. Chem. Ecol. 1980, 6, 625-41.
20. Lewis, W. J.; Nordlund, D. A.; Gueldner, R. C.; Teal, P. E. A.; Tumlinson, J. H.   J. Chem. Ecol. 1982, 8, 1323-31.
21. Tumlinson, J. H.; Hardee, D. D.; Gueldner, R. C.; Thompson, A. C.; Hedin, P. A.; Minyard, J. P.   Science 1969, 166, 1010-12.
22. McKibben, G. H.; Mitchell, E. B.; Scott, W. P.; Hedin, P. A.   Environ. Entomol. 1977, 6, 804-6.
23. Booth, D. C.; Phillips, T. W.; Claesson, A.; Silverstein, R. M.; Lanier, G. M.; West, J. R.   J. Chem. Ecol. 1983, 9, 1-12.

24. Polles, S. G.; Payne, J. A.; Jones, R. L. Pecan South 1977, 4, 26-8.
25. Hedin, P. A.; Payne, J. A.; Carpenter, T. L.; Neel, W. Environ. Entomol. 1979, 8, 521-3.
26. Chang, V. C. S.; Curtis, G. A. Environ. Entomol. 1972, 1, 476.
27. Patrick, D. R. J. Ga. Entomol. Soc. 1974, 9, 17.
28. Schmuff, N. R.; Phillips, J. K.; Burkholder, W. E.; Fales, H. M.; Chen, C. W.; Roller, P. P.; Ma, M. Tetrahedron Lett. 1984, 25, 1533-4.
29. Tumlinson, J. H.; Klein, M. G.; Doolittle, R. E.; Ladd, T. L.; Proveaux, A. T. Science 1977, 197, 789-92.
30. Doolittle, R. E.; Tumlinson, J. H.; Proveaux, A. T.; Heath, R. R. J. Chem. Ecol. 1980, 6, 473-85.
31. Henzell, R. f.; Lowe, M. D. Science 1970, 168, 1005.
32. Klein, M. G.; Tumlinson, J. H.; Ladd, T. L., Jr.; Doolittle, R. E. J. Chem. Ecol. 1981, 7, 1-7.
33. Ladd, T. L.; Klein, M. G.; Tumlinson, J. H. J. Econ. Entomol. 1981, 74, 665-7.
34. Guss, P. L.; Tumlinson, J. H.; Sonnet, P. E.; Proveaux, A. T. J. Chem. Ecol. 1982, 8, 545-56.
35. Sonnet, P. E.; Carney, R. L.; Henrick, C. A. J. Chem. Ecol. 1985, (in press).
36. Guss, P. L.; Sonnet, P. E.; Carney, R. L.; Branson, T. F.; Tumlinson, J. H. J. Chem. Ecol. 1984, 10, 1123-1131.
37. Krysan, J. L.; Smith, R. F.; Branson, T. F.; Guss, P. L. Ann. Entomol. Soc. Am. 1980, 73, 123-30.
38. Guss, P. L.; Sonnet, P. E.; Carney, R. L.; Tumlinson, J. H.; Wilkin, P. J., J. Chem. Ecol. 1985, 11, 000-00.
39. Guss, P. L.; Carney, R. L.; Sonnet, P. E.; Tumlinson, J. H. Environ. Entomol. 1983, 12, 1296-7.
40. Guss, P. L.; Tumlinson, J. H.; Sonnet, P. E.; McLaughlin, J. R. J. Chem. Ecol. 1983, 9, 1363-75.
41. Sonnet, Philip E. J. Org. Chem. 1982, 47, 3793-6.
42. Silverstein, R. M.; Cassidy, R. F.; Burkholder, W. E.; Shapas, T. J.; Levinson, H. Z.; Levinson, A. R.; Mori, K. J. Chem. Ecol. 1980, 6, 911-17.
43. Silverstein, R. M.; Rodin, J. O.; Burkholder, W. E.; Gorman, J. E. Science 1967, 157, 85.
44. Kuwahara, Y.; Fukami, H.; Howard, R.; Ishi, S.; Matsumura, F.; Burkholder, W. E. Tetrahedron 1978, 34, 1769-74.
45. Chuman, T.; Kohno, M.; Kato, K.; Noguchi, M. Tetrahedron Lett. 1979, 2361.
46. Suzuki, T. Agric. Biol. Chem. 1980, 44, 2519-20.
47. Suzuki, T. Agric. Biol. Chem. 1981, 45, 1357-63.
48. Suzuki, T. Agric. Biol. Chem. 1981, 45, 2641-3.
49. Williams, H. J.; Silverstein, R. M.; Burkholder, W. E.; Khorramshahi, A. J. Chem. Ecol. 1981, 7, 759-80.
50. Wong, J. W.; Verigin, V.; Oehlschlager, A. C.; Borden, J. H.; Pierce, H. D., Jr.; Pierce, A. M.; Chong, L. J. Chem. Ecol. 1983, 9, 451-74.
51. Millar, J. G.; Pierce, H. D., Jr.; Pierce, A. M.; Oehlschlager, A. C.; Borden, J. H.; Barak, A. V., J. Chem. Ecol. 1985, 11, 000-00.

52. Millar, J. G.; Pierce, H. D., Jr.; Pierce, A. M.;
    Oehlschlager, A. C.; borden, J. H., J. Chem. Ecol. 1985, 11,
    000-00.
53. Pierce, A. M.; Pierce, H. D., Jr.; Millar, J. G.; Borden, J.
    H.; Oehlschlager, A. C. Proc. Third Intern. Working Conf. on
    Stored Product Entomology, 1983 (in press).
54. Pierce, H. D., Jr.; Pierce, A. M.; Millar, J. G.; Wong, J. W.;
    Verigin, V. G.; Oehlschlager, A. C.; Borden, J. H. Proc.
    Third Intern. Working Conf. on Stored Product Entomology, 1983
    (in press).

RECEIVED November 15, 1984

Sexual Messages of Moths: Chemical Themes Are Known and New Research Challenges Arise

JEROME A. KLUN

Organic Chemical Synthesis Laboratory, Beltsville Agricultural Research Center, Agricultural Research Service, U.S. Department of Agriculture, Beltsville, MD 20705

Female moths produce chemical signals to attract and sexually stimulate mates, and research over the past two decades has defined the intricate chemical nature of these messages. The females employ fairly simple aliphatic compounds in their communications; the major variables are carbon-chain length, olefinic site and geometry, functionality, and optical form, and various combinations of compounds involving these variables constitute the chemical message. Attainment of a fundamental understanding of these chemical themes has brought this field to a threshold of new research challenges. There is now a need to understand the intermediary metabolic processes that regulate composition of the molecular message, the endogenous factors that regulate timely onset of biosynthesis and release of the chemical message, the nature of the olfactory chemoreceptor, and the molecular basis of perception and transduction of chemical stimuli into electrophysiological events in the sensory neurons.

Entomologists have been fascinated with molecular messages or "sex scents" of moths for a long time. Near the turn of the 20th century, researchers provided vivid descriptions of the amorous visitations of male moths to sites containing captive virgin female moths (1, 2). However, the famous French entomologist, Jean-Henri Fabre, had difficulty proving that the behavior exhibited by the male moths was triggered by an airborne scent emanating from the virgin females, and he underrated the importance of olfactory perception. In 1900, while studying the peacock moth (Saturnia pyri), Fabre formed the following hypothesis (1):

"Physical science is to-day preparing to give us wireless telegraphy, by means of the Hertzian waves. Can the Great Peacocks have anticipated our efforts in this direction? In order to set the surrounding air in motion and to inform pretenders miles away, can the newly-hatched bride have at her disposal electric or

magnetic waves, which one sort of screen would arrest and
another let through? In a word, does she, in her own
manner, employ a kind of wireless telegraphy? I see
nothing impossible in this: insects are accustomed to
invent things quite as wonderful."

Fabre tested this hypothesis and subsequently rejected the idea
that electromagnetic radiation was involved in the sexual communication systems of moths; however he remained skeptical of the
involvement of the sense of smell. Sixty years later the case for
smell was finally settled when German chemists, led by A. Butenandt
(3-5), were the first to isolate, identify, and synthesize an
insect sex pheromone, that of the female silkworm (Bombyx mori).
They found that the compound responsible for eliciting sexual
behavior from the male moth was (E,Z)-10,12-hexadecadien-1-ol.
This identification was no trivial accomplishment inasmuch as it
required extraction of 500,000 silkworm females (most moths produce
only nanogram amounts of pheromone per female) and it was achieved
in the days before the advent of gas-liquid chromatography and
other micro-analytical tools. Seven years later, in 1966, the
identity of a sex pheromone from a second species of moth, the
cabbage looper (Trichoplusia ni), was announced by R. S. Berger
(6). By 1970, pheromonal components had been identified from three
additional species of moths; (Z)-9-tetradecen-1-ol acetate (7) from
the fall armyworm (Spodoptera fugiperda) and (Z)-11-tetradecen-1-ol
acetate (8, 9) from the redbanded leafroller (Agyrotaenia
velutinana) and the European corn borer (Ostrinia nubilalis).
Discovery of the same compound in the leafroller and corn borer was
perplexing at the time because it was thought that the sex
pheromone of a species should be a single compound with a structure
unique to each species. This naïveté was short-lived and an
explosive growth period in the field of moth sex pheromone
chemistry ensued.

Virtually all of the earliest pheromone identifications were
chemically incomplete and sometimes inaccurate owing to inadequa-
cies of early analytical techniques and instrumentation and methods
of behavioral assay. However, over a period of 25 years much of
the chemistry of pheromones and the insect behavior they evoke was
defined with reasonable completeness and a clear picture of the
chemical themes used by the moths in their sexual communications
has emerged through the collective findings of scientists
throughout the world. We now know that moth sex pheromone
chemistry involves relatively simple straight carbon-chain
compounds (Figure 1); chain length, functionality, sites and
geometry of double bonds, specific proportions of compounds, and
permutations of these variables constitute the essence of moths'
sexual messages. Optical form is seemingly relegated to a minor
role in moth sex pheromone chemistry. Only a handful of the
several hundred identified moth pheromones (10, 11) involve
asymmetry of the epoxide moiety (12, 13), methyl branching (14,
15), or an ester of a secondary alcohol (16).

The chemical picture is deceptively simple until one appre-
ciates the fact that each species may use very specific permutations and/or proportions of compounds in the array of Figure 1. It
then becomes apparent that moths have evolved an elegant and

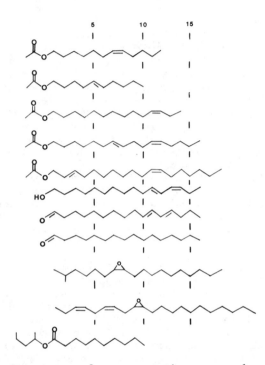

Figure 1.   Structures of representative compounds secreted by female moths as sex pheromones.

specific signaling system that is based upon multi-dimensional sets
of simple chemical variables.

Development of synthetic replicas of the sets of compounds
identified as sex pheromone components has constituted a significant challenge to synthetic organic chemists, because the requirements for pheromonal activity include stringent chemical and
isomeric purity that was not easily attained by traditional
synthetic organic chemical methods. These challenges were met
admirably by ingenious and imaginative organic chemists (17, 18).

Our recent study of the female sex pheromones of two closely
related moths, the pickleworm (Diaphania nitidalis) (19) and
melonworm (D. hyalinata) (20), provides an example of how trace
components and mixtures of compounds with different functionality
make up a moth's pheromonal signal. Figure 2 shows gas
chromatograms obtained by direct analysis of heptane extracts of
sex pheromone glands (located at the tip of the female abdomen) of
the two species. The chromatograms are arranged as opposing images
to allow comparison of the profile of compounds produced by each
species. Ten compounds were identified from the melonworm and
seven were found in the pickleworm. The major chromatographic
peaks represent ca. 50 ng compound. Consistent with the chemical
theme depicted in Figure 1, all of the compounds were C_{16}
straight-chain compounds with aldehyde, alcohol, or acetate functionality and with geometrical forms of mono-enes and conjugated
dienes. Figure 2 indicates that both species shared similar pheromonal components, but yet the complement of compounds from each was
unique. This quality makes for the species specificity of pheromonal signals among moths. In behavioral assays using male moths
in a wind tunnel, the specificity of male response could be easily
demonstrated. As an example, when a proportional amount of
melonworm pheromone component 6 [(E,E)-10,12-hexadecadienal] was
added to a synthetic mixture of pickleworm compounds the mixture
was rendered totally unattractive to pickleworm males.

Trace components in a set of compounds produced by a female
often are essential for the expression of biological activity of a
sex pheromone and the pickleworm pheromone exemplifies this quality.
Figure 2 shows that component 4 constituted only 0.1% of the total
mixture of compounds produced by the pickleworm female. Bioassays
showed, however, that if this compound was deleted from the set of
compounds produced by the female, males were behaviorally
unresponsive to the mixture despite the fact that all other
components of the pheromone complement were present. Similarly,
when component 2 [(Z)-11-hexadecenal], also a trace component (ca.,
3%), was deleted from the set of pickleworm compounds, the tendency
of males to fly upwind to the stimulus was reduced by 70% relative
to the response elicited by the complete set of compounds.

The development of our understanding of the chemical themes of
moth sex pheromones has been directly linked to spectacular
advancements in the technologies of analytical chemistry such as
open tubular capillary chromatography (OTCC), combined OTCC-mass
spectroscopy, and microchemical characterization techniques.
Surely, substances like component 4 of the pickleworm pheromone
would never have been discovered without such analytical
capability. Combined with advancements in analytical chemistry,
increasingly objective and effective behavioral and

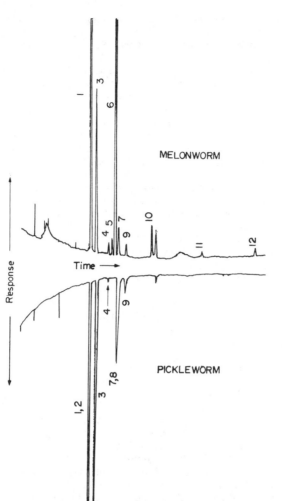

Figure 2. Chromatograms obtained by direct analysis of heptane extracts of the sex pheromone glands of pickleworm and melonworm females on a 60 m x 0.25 mm (ID) DB-1 column (J & W Scientific, Inc., Rancho Cordova, CA 95670). 1. (E)-11-hexadecenal; 2. (Z)-11-hexadecenal; 3. hexadecanal; 4. (E,Z)-10,12-hexadecadienal; 5. (Z,Z)-10,12-hexadecadienal, 6. (E,E)-10,12-hexadecadienal; 7. (E)-11-hexadecen-1-ol; 8. (Z)-11-hexadecen-1-ol; 9. hexadecan-1-ol; 10. (E,E)-10,12-hexadecadien-1-ol; 11. (E)-11-hexadecen-1-ol acetate; 12. (E,E)-10,12-hexadecadien-1-ol acetate. Unnumbered chromatographic peaks are either solvent contaminants or components that were not observed consistently in replicated analyses. The geometrical proportion of 11-hexadecenal in pickleworm extract was 96:4 (E:Z).

electrophysiological bioassays (21-24) have been developed, and
continue to be refined, that allow for better understanding of the
biological relevance of the various chemicals that influence the
behavior of insects.

The moth sex pheromones are among the most biologically potent
chemicals known and we have developed an in-depth understanding of
the chemical themes involved. The enthusiasm of researchers in the
field of insect sex pheromone chemistry and behavior is not only
directed to the expansion of fundamental understanding of this
phenomenal biological system but also to how this knowledge may
eventually be applied practically to relieve some of the many pest
insect problems that confront us. So far, our knowledge of insect
sex pheromone systems has had limited impact agriculturally;
however, fundamentally important research challenges lie ahead in
this area of biology that, if met successfully, may yield
unimagined and effective approaches to pest-insect manipulation.
There are many promising avenues of research.

Little is known of the biosynthetic pathways by which the
pheromones are produced. Preliminary probes in this area (25, 26)
indicate that mono-unsaturated acetate moth sex pheromones may
arise via a biosynthetic pathway involving Δ-11-desaturases that
operate on long chain saturated fatty acyl moieties of
glycerolipids to generate an olefinic site and enzymes that shorten
or lengthen fatty acyl chain length by 2 carbons. A tentative
outline for the biosynthetic origin of a major component of the sex
pheromone of the cabbage looper, (Z)-7-dodecen-1-ol acetate, is
shown in Figure 3. Much remains to be done, however, to expand our
understanding of the intermediary metabolic pathways that regulate
the qualitative and quantitative composition of the precise sets of
compounds that we know make up most pheromonal messages. As
example, the factors that regulate production of the specific
mixtures of geometrical isomers, used so frequently by the moths,
are undefined as is the biosynthetic origin of pheromones having
skipped and conjugated diene functionality.

Related to the issue of the biosynthetic origin of the pheromones, Ashok Raina and I recently discovered that female sex pheromone production in at least one species of moth is regulated by a
neurohormonal peptide that is produced in the female brain (27).
We observed that when adult females of the corn earworm moth
(Heliothis zea) were ligated between the head and thorax so that
normal hemolymph circulation between the brain and the rest of the
body was interrupted, they did not produce sex pheromone. However,
the ligated females could be stimulated to produce pheromone, in
amounts not different from that produced by normal females, by an
injection of a saline extract of the female brain. Our studies
with this nocturnal moth indicate that the neurohormone is stored
in the brain and released into the hemolymph in the scotophase
causing timely production of pheromone by the female.

Adding to the puzzlement of this system, we found that pheromone production could also be induced in the glands of ligated corn
earworm females by injection of brain extracts of the male corn
earworm and extracts of females from four different families of
moths (Pyralidae, Lymantriidae, Geometridae, and Psychidae) (27).
The sex pheromones of females representing these families have
little or no chemical similarity to the corn earworm sex pheromone

and the function of the neurohormone in the male corn earworm is
unknown. Thus, the hormone may be widely distributed in the order
but there is no obvious connection between its occurrence and the
physiological functions it may have in these assorted insects.
Elucidation of the chemistry and physiological modes of action and
functional significance of this neuroendocrine substance may not
only expand our knowledge of endogenous factors that regulate
pheromone biosynthesis in moths but also serve as a stepping stone
to the understanding of the mechanisms by which the brain exerts
its influence upon the vital life processes of the insects.

Another promising avenue for research that is unquestionably
at the frontier of science concerns development of an understanding
of processes involved in olfactory perception. We know from our
research on the sex pheromones that male moths possess a phenomenal
ability to perceive and respond behaviorally to incredibly minute
quantities of pheromonal compounds in the air. The chemoreceptive
sensitivity of this detector system is certainly unrivaled by any
organic-chemical detector system devised by man.

We know that the insects' olfactory chemoreceptor system is
located on the antennae, and considerable work has been done to describe the general morphological and fine structure features of
moth antennae (28). A typical moth antenna is often filamentous and
multisegmented. The ventral surface of each segment is heavily
clothed with hair-like projections called sensilla (Figure 4) that
have proprioceptive, mechanoreceptive, or chemoreceptive sensory
functions, according to morphological and electrophysiological
studies. Sensilla responsible for pheromone detection are usually
the most numerous on the antennae of male moths (29). Microscopic
studies of the chemoreceptive sensilla indicate that they are porous
extensions of the insect's cuticle and each sensillum often houses
one to five sensory neuron cells. Each sensory cell has dendrites
that project from the cell body to positions near the pores of the
sensillum and each cell sends an axon, without synapse, to a central junction in the brain (30). Electrophysiological studies (30,
31) show that these sensory cells send electrical impulses to the
brain when the pheromone molecules are moved over the antenna. It
is by these sensory-cell action potential impulses that the brain
receives information about its environment and then, ostensibly,
orchestrates appropriate physiological-behavioral responses
beneficial to the organism's survival.

Despite this seemingly straightforward description of pheromone
perception, little is actually known of the mechanistic details of
the process. The exact location and biochemical nature of the receptor site is unknown. Because most pheromones are specific
mixtures of compounds, it is probable that the organization of the
receptor system may involve an array of several receptors, each
designed to receive certain components of the mixture, that operate
in concert (32). Bestmann and Vostrowsky (33) proposed a
hypothetical model of a pheromone receptor based upon study of the
electrophysiological activity of sex pheromone analogs and the
influence of pheromone-like compounds on temperature-dependent
phase transitions of synthetic dipalmitoyl lecithin micelles. The
model depicts the receptor as being composed of 4 or more protein
subunits surrounded by a crystalline lipid matrix and located on
the dendrites of the sensory cells. According to their

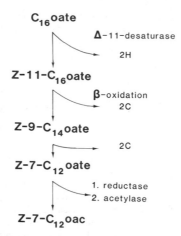

C_{16}oate

Δ-11-desaturase
⟶ 2H

Z-11-C_{16}oate

β-oxidation
⟶ 2C

Z-9-C_{14}oate

⟶ 2C

Z-7-C_{12}oate

1. reductase
2. acetylase

Z-7-C_{12}oac

Figure 3. Probable biosynthetic origin of a major pheromonal component of the cabbage looper (<u>Trichoplusia ni</u>) involving acyl moiety of triacylglycerols.

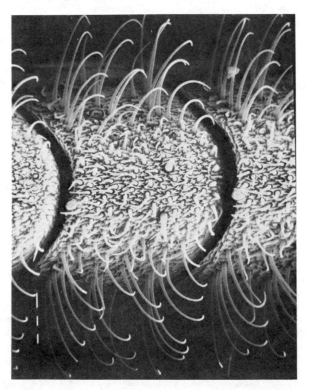

Figure 4. Scanning electron photomicrograph of a segment of the male antenna of the tobacco budworm moth (<u>Heliothis virescens</u>). The hair-like projections on the antenna are sensory sensilla responsible for detection of female sex pheromone. Photomicrograph by Michael Blackburn, Entomology Dept., U of MD, College Park.

speculations, pheromone molecules were thought to impinge on the sensilla, diffuse through the pores of the sensilla, and then interact with the dendrite-based receptor, causing the opening of ion channels via a conformational change in the receptor protein-lipid matrix. Their model did not include any consideration of how the receptor might be cleared of the pheromone and the ion gate returned to its original closed conformation. On the other hand, Vogt and Riddiford (33) have focused their attention upon proteins of the sensilla and specifically on a protein that binds pheromone and an esterase. They propose a molecular model in which pheromone reception may involve a dynamic kinetic equilibrium in which the pheromone (an ester) can follow 3 possible pathways in the reception process: sensillum pore to binding protein to receptor to the esterase; pore to binding protein to esterase; and pore to esterase. They suggest that this shiftable-equilibria model system can explain the broad stimulus-response range exhibited by some species of moths. The idea here is that as pheromone concentration increases, proportionally larger amounts of pheromone impinging on the sensillum are shunted directly to the esterase and thereby saturation of the receptor is prevented. A combination of these hypothetical models might begin to better approximate the biochemical events involved in the chemoreceptive process. There is an obvious need to expand our knowledge of the pheromone receptor system.

We certainly know much about the chemical stimuli that the receptor system is designed to receive. It is time to begin the search for substances that might bind, antagonize, or debilitate the receptor. Discovery of such substances would prove useful in localization of the receptor site and in elucidation of the biological chemistry of olfactory perception. It is very possible that the chemistry of substances that will antagonize the receptor will bear little resemblance to the chemistry of the natural product it is designed to receive. Precedents for this idea may be found in the neurosciences. As examples, opiates are not peptides, but they block endorphin receptors (34); and avermectins, macrocyclic lactones derived from an actinomycete (Streptomyces avermitillis), do not resemble the neurotransmitter, γ-aminobutyric acid, but they nevertheless block receptors for that compound (35). Intuitively, one is inclined to think that many of the techniques and technologies that have been developed and applied in the neurosciences (36) may have utility in probing the nature of the pheromonal olfactory receptor system. Although the moth pheromones have little structural similarity to the molecular messengers of the neurosciences, acetylcholine, etc., the machinery by which they are perceived may involve variations on the nervous system receptor theme rather than an entirely novel mechanism.

The momentum of research on the identity and functional significance of the molecular messengers in insects (neurotransmitters, pheromones, and neurohormones) and their receptors will undoubtedly build and, as it does, so too will the probability that novel, efficacious, and environmentally sound methods of pest-insect control will emerge. For those of us involved with studies of insect sex pheromones, the avenues for research are clear. We are

no longer content to ask, "What is the chemistry of these molecular
messengers?" We must now also ask, "How are they produced and how
are they perceived?"

Acknowledgments

I thank M. Inscoe, J. Neal, J. Oliver, A. Raina, R. Ridgway,
M. Schwarz, and D. Warthen, for their constructive comments as I
prepared this paper.

Literature Cited

1. Fabre, J.-H. In "Souvenirs Entomologiques"; Librairie Ch.
    Delagrave; Paris, France, 1900; pp. 339-60. Translated by
    A. Teixeiro de Mattos, in "The Life of the Caterpillar"; Dodd,
    Mead, and Co.; New York, 1916; pp. 246-78.
2. Lintner, J. A. West N.Y. Hort. Soc. Proc. 1882, 27, 52-66.
3. Butenandt, A.; Beckman, R.; Stamm, D.; Hecher, E. Z.
    Naturforsch. 1959, 14b, 283-4.
4. Butenandt, A.; Hecker, E.; Hopp, M.; Koch, W. Justus Liebigs
    Ann. Chem. 1962, 658, 39-64.
5. Truscheit, E.; Eiter, K. Ibid. 1962, 658, 65-90.
6. Berger, R. S. Ann. Entomol. Soc. Amer. 1966, 59, 767-71.
7. Sekul, A. A.; Sparks, A. N. J. Econ. Entomol. 1967, 60,
    1270-2.
8. Roelofs, W. L.; Arn, H. Nature 1968, 219, 513.
9. Klun, J. A.; Brindley, T. A. J. Econ. Entomol. 1970, 63,
    779-80.
10. Klassen, W.; Ridgway, R. L.; Inscoe, M. In "Insect
    Suppression with Controlled Release Pheromone Systems";
    Kydonieus, A. F.; Beroza; M., Eds. CRC Press, Inc. Boca
    Raton, Florida, 1982; Vol. 1, p. 13.
11. Baker, R.; Bradshaw, J. W. S. Aliphatic Rel. Nat. Prod. Chem.
    1983, 3, 66-106.
12. Iwaki, S.; Marumo, S.; Saito, T.; Yamada, M.; Katagiri, K.
    J. Amer. Chem. Soc. 1974, 96, 7842-4.
13. Hill, A. S.; Kovalev, B. G.; Nikolaeva, L. N.; Roelofs, W. L.
    J. Chem. Ecol. 1982, 8, 383-96.
14. Sugie, H.; Tamaki, Y.; Sato, R.; Kumakura, N. Appl. Ent.
    Zool. 1984, 19, 323-30.
15. Tamaki. Y.; Noguchi, H.; Horiike, M.; Hirano, C. Appl. Ent.
    Zool. 1984, 19, 245-51.
16. Leonhardt, B. A.; Neal, J. W.; Klun, J. A.; Schwarz, M.;
    Plimmer, J. R. Science 1983, 219, 314-6.
17. Brand, J. M.; Young, J. C.; Silverstein, R. M. Prog. Chem.
    Org. Nat. Prod. 1979, 37, 1-190.
18. Henricks, C. A. Tetrahedron 1977, 33, 1845-89.
19. Klun, J. A.; Leonhardt, B. A.; Schwarz, M.; Day, A.;
    Raina, A. K. (In preparation.)
20. Raina, A. K.; Klun, J. A.; Schwarz, M.; Day, A.;
    Leonhardt, B. A. (In preparation.)
21. Silverstein, R. M. Pure and Appl. Chem. 1982, 54, 2479-88.
22. Plimmer, J. R.; Inscoe, M. N.; McGovern, T. P. Ann. Rev.
    Pharmacol. Toxicol. 1982, 22, 297-320.

23. Cardé, R. T. In "Chemical Ecology of Insects"; Bell, W. J.;
    Cardé, R. T., Eds. Sinauer Associates, Inc. Sunderland,
    Massachusetts, 1984; Chap. 5.
24. Cardé, R. T.; Baker, T. C. Ibid; Chap. 13.
25. Bjostad, L. B.; Roelofs, W. L. Science 1983, 220, 1387-9.
26. Bjostad, L. B.; Roelofs, W. L. J. Chem. Ecol. 1984, 10,
    681-91.
27. Raina, A. K.; Klun, J. A. Science 1984, 225, 531-3.
28. Zacharuk, R. Y. Ann. Rev. Entomol. 1980, 25, 27-47.
29. Cornford, M. E.; Rowley, W. A.; Klun, J. A. Ann. Entomol.
    Soc. Amer. 1973, 66, 1079-88.
30. Mustaparta, H. In "Chemical Ecology of Insects"; Bell, W. J.;
    Cardé, R. T., Eds. Sinauer Associates, Inc. Sunderland,
    Massachusetts, 1984, Chap. 2.
31. O'Connell, R. J.; Grant, A. J.; Mayer, M. S.; Mankin, R. W.
    Science 1983, 220, 1408-10.
32. Boeckh, J. Proc. 15th Internat. Congress of Entomol., 1976,
    pp. 308-22.
33. Bestmann, H. J.; Vostrowsky, O. In "Olfaction and Endocrine
    Regulation"; Breipohl, W., Ed.; IRL Press Limited: Falconberg
    Court, London, 1982; Section 7, pp. 253-65.
34. Vogt, R. G.; Riddiford, L. M. XVII International Congress of
    Entomol., Abstract Volume, 1984, p. 465.
35. Krassner, M. B. Chem. Eng. News 1983, 61 (35), 22-33.
36. Campbell, W. C.; Fisher, M. H.; Stapley, E. O.;
    Albers-Schönberg, G.; Jacob, T. A. Science 1983, 221, 823-8.
37. In "Neurosciences"; Abelson, P. H.; Butz, E., Eds. Science
    1984, 225, 1253-70.

RECEIVED November 8, 1984

Alkaloidal Ant Venoms: Chemistry and Biological Activities

MURRAY S. BLUM

Department of Entomology, University of Georgia, Athens, GA 30602

By producing a series of unique chemical compounds which are employed as defensive agents, ants have prospered in the insect world. The alkaloidal ant venoms are biologically active natural products that produce a range of responses in insect ecosystems. Some possess bacterial and fungicidal properties; others are insecticidal, or repellents, or dermal necrotoxins, and have diverse pharmacological activities. This report discusses the nitrogen heterocycles produced by ants in order of structural complexity (pyrrolidines, pyrrolines, pyrroles, piperidines, piperideines, pyridines, pyrrolizidines, indolizidines, pyrazines, and indoles), and allocates the corresponding biological properties.

Solomon's exhortation, "Get thee to the ants", is probably even more appropriate today than it was at the time that he uttered it. Ants are an eminently successful group of insects, and undoubtedly consitute the major group of predatory animals on earth, a development often attributed to the en masse attacks so characteristic of these social arthropods. But their formidability as adversaries is also correlated with their chemical arsenals, which, when unleashed at their omnipresent antagonists, may often produce severe biochemical

0097–6156/85/0276–0393$06.00/0

lesions. In particular, the venoms of ants are
fortified with a potpourri of compounds that often
constitute novel deterrents, and the evolution of
these poison gland products must be regarded as a
major factor in the great success of these social
Lilliputians. Perhaps Solomon foresaw that man,
locked in his unending battle with pest insects that
have frequently developed resistance to man-made
insecticides, would eventually need the powerful
toxicants present in the venoms of ants for his own
use. Indeed, the successful utilization of these
natural insecticides for countless millenia
demonstrates that they have stood the test of time
as biocides par excellence.
      In the present review, an effort has been made
to characterize the major compounds that have been
recently identified in the venoms of ants,
specifically the alkaloids. In addition, the
extraordinarily diverse biological activities of
these venom products are described. Hopefully, this
brief exposition will leave little doubt that ants'
biosynthetic virtuosity has provided them with
chemical weaponry not only for predation, but for
use in blunting the attacks of the hostile organisms
with which they share their world.

Alkalodial Chemistry of Ant Venoms

For more than 300 years (1) ant venoms were
synonymized with formic acid, a well-known
cytotoxin. However, this poison-gland product is
now known to be limited to species in the most
specialized of the ant subfamilies -- the Formicinae
-- the members of which do not possess functional
stings. In contrast, the venoms of stinging ants
have generally been demonstrated to be alkaline,
often consisting of complex mixtures of enzymes,
toxic proteins, and biogenic amines (2, 3), in
common with poison gland products of bees and wasps.
However, the venoms of some ants are dominated by a
diversity of alkaloids, compounds whose known
distribution is limited to species in a few genera.
These novel nitrogen heterocycles are now known to
possess a wide range of biological activities,
endowing their producers with truly formidable
defensive capabilities.
      Alkaloids have only been identified in the
venoms of ant species in the Myrmicinae, the largest
of the formicid subfamilies. This diverse subfamily
includes fire ants and harvester ants, the species
of which produce highly algogenic venoms. In
addition, this ant taxon includes thief ants and

Pharaoh's ant, species which, while they do not
sting, are known to utilize their venoms in
offensive contexts (4). Although the venoms of some
myrmicine species may contain up to 95% alkaloids,
minor amounts of enzymes, characteristic products of
animal venoms, are also present and appear to
constitute the allergens associated with these
venoms (5). Indeed, even the formic acid-rich
venoms of formicine ants are fortified with
nitrogen-containing constituents (i.e., peptides), a
probable biochemical legacy of the nitrogenous
metabolism that has been emphasized in the
hymenopterous poison gland.

In particular, the venom chemistry of myrmicine
species in the genera Solenopsis and Monomorium
appears to have digressed from that of species in
other myrmicine genera, being characterized by an
emphasis on the synthesis of small nitrogen
heterocycles (6) at the expense of proteinaceous
constituents. These alkaloids, which are restricted
in their animal distribution to ant species in
several genera, are represented by both mono- and
bicyclic compounds. They may possess some
chemotaxonomic value for taxa in these myrmicine
genera (7). In the present report, these nitrogen
heterocycles are organized according to their
structures, the saturated monocyclic compounds being
presented first, followed by the bicyclic alkaloids,
and finally the monocyclic compounds containing more
than one nitrogen atom, i.e., pyrazines and
heterocyclic amines, the indoles.

Pyrrolidines.

More than a dozen 2,5-dialkylpyrrolidines have been
identified in the venoms of Solenopsis and
Monomorium species (7). All of these compounds are
of the trans configuration. Whereas Solenopsis
species may produce only one compound in their
poison glands, the venoms of Monomorium species
generally consist of mixtures of these alkaloids.

The dialkylpyrrolidines (I) contain unbranched
side chains in which one alkyl group is
even-numbered whereas the other group is
odd-numbered. These compounds have been identified
in the venoms of a variety of Solenopsis species (8,
9, 10), all members of the subgenus Diplorhoptrum.
A variety of Monomorium species in the subgenus
Monomorium produce these alkaloids (I) (6, 11, 12),
as does one species in the subgenus Xeromyrmex (6).

Monomorium species also produce
dialkylidenepyrrolidines containing a terminally

unsaturated double bond in one side chain (II) (6, 11, 12) or terminally unsaturated double bonds in both side chains (III) (12). These are in admixture with pyrrolidines in which both side chains are saturated (I) (6, 11, 12).

Several N-methylpyrrolidines have been identified in the venoms of Monomorium (Monomorium) species. M. latinode, an Old World species, is distinctive in producing an alkaloidal-rich venom containing two N-methylpyrrolidines with saturated side chains (IV) (6). On the other hand, the poison gland secretions of several North American species contain N-methylated compounds in which one (V) or both side chains (VI) are terminally unsaturated (6); they are accompanied by the corresponding norpyrrolidines.

Pyrrolines.

Both possible isomers of the 2,5-dialkyl-1-pyrrolines have been identified in the venoms of Solenopsis and Monomorium species. These compounds generally accompany the corresponding dialkylpyrrolidines as venom constituents. The venom of S. punctaticeps, an African species, contains both δ^1- and δ^5-pyrrolines (VII) (8) as do the venoms of three North American Monomorium species (12). The latter dialkylpyrrolines (VIII) are distinguished by the presence of two terminally unsaturated side chains. The pyrrolines in the venom of M. latinode are distinctive in constituting the only 5-pyrrolines (IX) in ant venoms that are not accompanied by the 1-pyrrolines (6).

Pyrrole.

A trace constituent in the venoms of the myrmicines Atta texana, A. cephalotes, A. sexdens, and Acromyrmex octospinosus is methyl 4-methylpyrrole-2-carboxylate (X) (13, 14, 15). This simple pyrrole ester is a trail pheromone for some of these species (13, 14).

Piperidines.

2-Alkyl-6-methylpiperidines have only been detected in the venoms of Solenopsis workers and their queens (16, 17, 18). These compounds, which are sometimes referred to as solenopsins, are consistent poison gland products of Solenopsis species in the subgenus Solenopsis, the fire ants (17, 18). In addition, some species in the subgenus Diplorhoptrum (thief

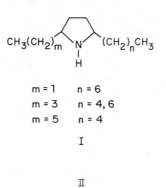

$$CH_3(CH_2)_m \qquad (CH_2)_n CH_3$$

m = 1   n = 6
m = 3   n = 4, 6
m = 5   n = 4

I

II

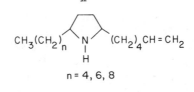

$$CH_3(CH_2)_n \qquad (CH_2)_4 CH = CH_2$$

n = 4, 6, 8

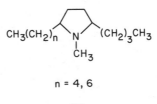

$$H_2C = HC(CH_2)_7 \qquad (CH_2)_4 CH = CH_2$$

III

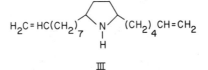

$$CH_3(CH_2)_n \qquad (CH_2)_3 CH_3$$

n = 4, 6

IV

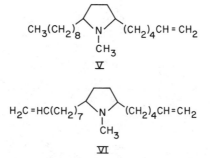

$$CH_3(CH_2)_8 \qquad (CH_2)_4 CH = CH_2$$

V

$$H_2C = HC(CH_2)_7 \qquad (CH_2)_4 CH = CH_2$$

VI

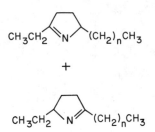

$+$

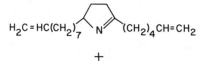

n = 4, 6

VII

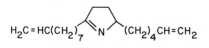

$+$

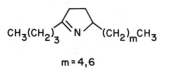

VIII

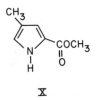

m = 4, 6

IX

CH$_3$

COCH$_3$

X

ants) and a member of the subgenus Euophthalma (6)
synthesize these compounds. Queens of some species
of Solenopsis both synthesize fewer
dialkylpiperidines than their workers and produce
different ratios of isomers as well (19).

The cis-trans isomers of four 2-alkyl-6-
methylpiperidines, in which the alkyl groups consist
of relatively long alkyl chains (C_9-C_{15}), have been
identified in fire ant venoms (XI) (16, 17); members
of each Solenopsis (Solenopsis) species group appear
to produce characteristic alkaloids. The presence
of a fifth 2-alkyl-6-methylpiperidine, 2-hepytl-6-
methylpiperidine, in the venom of queens of S.
richteri is indicated by mass spectral data (20).

Two N-methyl-2,6-dialkylpiperidines (XII) have
been identified in the venoms of two Solenopsis
(Diplorhoptrum) species (6). These alkaloids, which
are minor concomitants of the corresponding
nordialkylpiperidines, are present as trans-isomers.

The poison gland secretions of Solenopsis
(Solenopsis) species in certain taxa are enriched
with 2-alkyl-6-methylpiperidines in which the long
side chains contain a carbon-carbon double bond at
the ninth carbon from their terminal methyl groups
(XIII) (16, 24). The side chain double bonds of
these 6-alkylidene-2-methylpiperidines are all cis
(16), and both the cis- and trans-stereoisomers of
these alkaloids have been identified in a variety of
Solenopsis species (17, 18). Whereas the alkaloid
containing the 4-tridecenyl side chain (XIII, n=3)
constitutes the only unsaturated dialkylpiperidine
present in the venoms of some species (17, 18), the
compound with a 6-pentadecenyl side chain (XIII,
n=5) is always accompanied by the
tridecenyl-containing alkaloid (XIII, n=3) (18).
The corresponding saturated 2,6-dialkylpiperidines
(XI, n=12, 14) always accompany their unsaturated
counterparts (XIII, n=3, 5). On the other hand,
2-(cis-8-heptadecenyl)-6- methylpiperidine (XIII,
n=7), a minor constituent in the venom of S. invicta
(21), occurs in the absence of its saturated
counterpart.

Piperideines.

Two piperideines have been identified in Solenopsis
venoms, and in neither case do these compounds
appear to be typical products of this genus.
2-Methyl-6-undecyl-1-piperideine (XIV) is a minor
constituent in the venom of S. xyloni (17), but it
has not been detected in the venoms of closely
related species of fire ants in the subgenus

Diplorhoptrum, produces 2-(4-penten-1-yl)-
1-piperideine (XV) (16), which is particularly
distinctive because it lacks a 6-alkyl group.

Pyridines.

Anabasine (XVI) is a poison gland product of species
of Aphaenogaster (22), a North American genus in the
subfamily Myrmicinae.

Pyrrolizidines.

In contrast to the dialkylpyrrolidine theme that
characterizes the venoms of many Solenopsis
(Diplorhoptrum) species (6), that of S. xenovenenum
is distinctive in containing a pyrrolizidine as the
sole alkaloid. ^{1}H and ^{13}C NMR spectroscopy show
that this compound has a cis-fused ring junction and
is the (5Z,8E) isomer of 3-heptyl-5-methyl-
pyrrolizidine (XVII) (23).

Indolizidines.

Indolizidines have been detected as poison gland
products of both Monomorium and Solenopsis species.
Monomorine-1, 3-butyl-5-methyloctahydroindolizidine
(XVIII), is a major constituent in the venom of M.
pharaonis and is accompanied by its congener,
3-(3-hexen-1-yl)-5-methylindolizidine (XIX) (11).
Monomorine-1, which is produced on the poison gland
as the all cis-isomer, is accompanied by
pyrrolidines, characteristic Monomorium alkaloids
(12). On the other hand, 3-ethyl-5-
methylindolizidine (XX), a major constituent in the
venom of S. conjurata, is a concomitant of
2,6-dialkylpiperidines (24). Queens of an
unidentified Costa Rican Solenopsis (Diplorhoptrum)
species produce a single poison gland product,
3-hexyl-5-methylindolizidine (XXI) (24).

Pyrazines.

Alkylpyrazines, which are commonly produced in the
mandibular glands of ants and wasps (7), appear to
have a limited distribution in the venoms of ants.
2,5-Dimethyl-3-ethylpyrazine (XXII) has been
identified as a trace constituent in the venom of
Atta sexdens (15), a myrmicine species that also
produces a pyrrole (X) as a poison gland product.

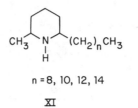

n = 8, 10, 12, 14

XI

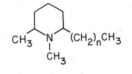

n = 8, 10

XII

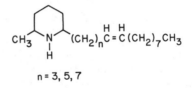

n = 3, 5, 7

XIII

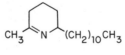

XIV

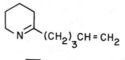

XV

Indoles.

Skatole, 3-methylindole (XXIII), is a major
constituent in the hypertrophied poison gland of
soldiers of the myrmicine Pheidole fallax (25) and
of other Pheidole species as well (24).

Biological Activities of Alkaloids

A wide variety of activities have been demonstrated
for the alkaloids identified in myrmecine ant
venoms, indicating that these small nitrogen
heterocycles have been adapted to subserve multiple
functions. Both the piperidines and pyrrolidines
possess diverse pharmacological activities (reviewed
in 7), and it seems likely that their roles in
regulating both intra- and interspecific
interactions are very significant.

Toxicology.

Studies on the modes of action of dialkylpiperidines
have shown that these compounds cause diverse
biochemical lesions. These alkaloids which are
powerful hemolysins (26) and dermal necrotoxins (27,
28), also release histamine from mast cells (29),
thus functioning as effective algogens. In
addition, these fire ant products inhibit Na^+ and K^+
ATPases (30) and at low concentrations uncouple
oxidative phosphorylation leading to reduced
mitochondrial respiration (31). 2,6-Disubstituted
piperidines also interfere with the coupling between
the ion channel and the recognition site of the
vertebrate nicotinic acetylcholine receptor (32).
Recent investigations demonstrate that these
compounds interact with the closed acetylcholine
receptor/ion channel complex at a site which is
separate from the binding site of previously known
blocking agents (33).
      These results emphasize the broad spectrum
activities of 2-alkyl-6-methylpiperidines which
offer novel probes for identifying and exposing
target sites for new classes of insect neurotoxins
(33).

Insecticidal Activities.

Fire ants rapidly immobilize insect prey by
stinging, demonstrating that the venom is strongly

XVI

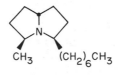

XVII

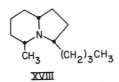

XVIII

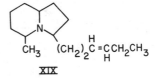

XIX

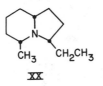

XX

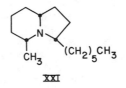

XXI

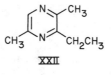

XXII

XXIII

insecticidal. Tropical applications of neat venom
to flies and beetles results in high mortality of
the target insects, which exhibit paralytic symptoms
shortly after treatment (34). On the other hand,
synthetic cis and cis-trans mixtures of
2-undecyl-6-methylpiperidine or 2-tridecyl-
6-methylpiperidine were nontoxic to adults of the
American cockroach, Periplaneta americana (33).

Workers of Monomorium minimum rapidly kill
termite workers (Reticulitermes species) with their
poison gland secretions (35). The venom of this
species is primarily fortified with 2,4-dialkyl-
pyrrolidines (12) which must function as contact
insecticides since the venom is applied topically
rather than being subdermally administered. The
same results have been reported for European species
of Monomorium attacking termites (36).

Repellent (Deterrent) Activities.

The disubstituted pyrrolidines and piperidines have
been demonstrated to be excellent repellents for
ants under field conditions. Workers of Monomorium
pharaonis effectively repel other species of ants
with venom droplets which accumulate on the tip of
the sting (37), a strategy that is used with equal
success by workers of North American Monomorium
species (38) and the European thief ant, Solenopsis
fugax (37). The venom of the latter species
contains a single alkaloid, trans-2-butyl-
5-heptylpyrrolidine (I, m=3, n=6) (10), and the
synthetic compound is as active as the neat venom in
repelling other ant species after being applied to
food or ant larvae.

Some of the dialkylpiperidines and
dialkylindolizidines are outstanding ant repellents.
A large series of these compounds were recently
evaluated as feeding repellents for four species of
ants that are considered to be major pest species
because of their aggressive habits. Hungry workers
of the fire ant Solenopsis invicta, the Argentine
ant, Iridomyrmex humilis, Tapinoma melanocephalium,
and Pharaoh's ant, Monomorium pharaonis, were
repelled by trace amounts of selected piperidines
and indolizidines added to droplets of sucrose
solution (35). These results clearly document the
abilities of the ant-derived alkaloids, at natural
concentrations, to repel feeding by ant workers at
very acceptable food sources, notwithstanding their
prolonged food deprivation.

Antimicrobial Activities.

More than 25 years ago it was demonstrated that fire
ant venom (S. invicta=saevissima) possessed
bactericidal and fungicidal activities (34).
Subsequently, the synthetic dialkylpiperidines were
shown to possess pronounced abilities to inhibit the
growth of a variety of bacteria at low
concentrations (38). Detailed studies on the
fungicidal activities of these alkaloids vis-a-vis
commercial fungicides indicated that they were
eminently capable of manifesting pronounced
fungistatic and/or fungicidal activity against a
wide variety of fungi (40).
     Both cis and trans-isomers of all the alkaloids
identified in the venom of the fire ant S. invicta
(16) were evaluated against fungi that constituted
human and plant pathogens, as well as several that
were isolated from the body surfaces of ant larvae.
In general, the stereoisomers were of equal activity
over a wide range of concentrations and rivaled
commercial fungicides (e.g., δ^2-undecenoic acid) as
fungal growth inhibitors. There is a tendency for
the alkaloids with 6-alkylidine side chains to
exhibit greater fungitoxicity than their saturated
counterparts (40).

Conclusions

Although relatively few myrmicine species have been
subjected to analytical scrutiny, the members of
this subfamily clearly demonstrate great virtuosity
in the biosynthesis of alkaloids (7). Species of
Monomorium are particularly abundant in the Old
World (41), and analyses of a few members of this
genus (6, 11) indicate that great alkaloidal
diversity may characterize their poison gland
secretions. The same is true for the genus
Solenopsis (8, 16). Furthermore, alkaloidal
surprises will probably be forthcoming when the
venoms of species in additional myrmicine genera are
examined. Indeed, a new pyrrolizidine has been
characterized in a species in the genus Chelaner
(24), an Old World taxon closely related to
Monomorium (41). In short, there are strong grounds
for concluding that myrmicine species, with their
great ecological diversity, will continue to be an
excellent source of interesting nitrogen
heterocycles for some time to come. In addition, in
terms of alkaloids, analyses of ant species in
nonmyrmicine subfamilies could also prove to be very
fruitful.

A variety of alkaloidal constituents, not
produced in poison (venom) glands, have been
identified as natural products of ant species in
four subfamilies, including the Myrmicinae (see
review in 7). This demonstrates that the biogenesis
of nitrogen heterocycles occurs widely in the
Formicidae and stresses that species in all
subfamilies should be regarded as possible alkaloid
producers. Significantly, novel nitrogen-containing
compounds (e.g., aliphatic amines and amides) have
recently been identified as venom products of two
species in the subfamily Ponerina (42), establishing
that the ponerine poison gland can synthesize
non-proteinaceous nitrogenous constituents. Thus,
the poison glands of nonmyrmicine ants can be
regarded as possible sources of nitrogenous
compounds.

Ants have prospered in a world in which they
have had to overcome a multitude of adversaries that
include pathogens as well as invertebrate and
vertebrate predators. For many ant species, it
appears that their ability to survive in such a
manifestly hostile world is in no small way due to
the alkaloidal arsenals stored in their poison gland
reservoirs. These compounds have been demonstrated
to possess a wide spectrum of toxicological
properties (7) which include pronounced neurotoxic,
antimicrobial, and insecticidal activities. Living
in moist subterranean environments, ants have been
able to prevent omnipresent pathogenic bacteria and
fungi from overwhelming them, and alkaloids have
played a major antibiotic role in this context (39,
40). The demonstrated insecticidal potencies of
alkaloidal venoms (34, 35) further documents the
biocidal properties of these compounds and augurs
well for their future consideration as viable
insecticides. Indeed, mankind can ill afford to
ignore this wondrous crop of natural fungicides and
insecticides that are available for harvesting. It
is time to give Solomon his due.

Literature Cited

1. Wray, J. Phil. Trans. Roy. Soc. London 1670,
   5, 2063.
2. Cavill, G. W. K.; Robertson, P. L.; Whitfield,
   F. B. Science 1964, 146, 79.
3. Schmidt, J. O.; Blum, M. S. Science 1978,
   200, 1064.
4. Holldobler, K. Biol. Zbl. 1928, 48, 129.

5. Baer, H.; Liu, T. Y.; Anderson, M. C.; Blum, M. S.; Schmid, W. H., James, F. J. Toxicon 1979, 17, 397.
6. Jones, T. H.; Blum, M. S.; Fales, H. M. Tetrahedron 1982, 38, 1949.
7. Jones, T. H.; Blum, M. S. In "Alkaloids: Chemical and Biological Perspectives"; Pelletier, S. W.; John Wiley and Sons, Inc., 1983; Vol. I, pp. 33-84.
8. Pedder, D. J.; Fales, H. M.; Jaouni, T.; Blum, M. S.; MacConnell, J.; Crewe, R. M. Tetrahedron 1976, 32, 2275.
9. Jones, T. H.; Blum, M. S.; Fales, H. M. Tetrahedron Letters 1979, 1031.
10. Blum, M. S.; Jones, T. H.; Holldobler, B.; Fales, H. M.; Jaouni, T. Naturwissenschaften 1980, 67, 144.
11. Ritter, F. J.; Bruggemann-Rotgans, I. E. M.; Verkuil, E.; Persoons, C. J. Proc. Symp. Pheromones and Defensive Secretions in Social Insects, 1975, pp. 99-103.
12. Jones, T. H.; Blum, M. S.; Howard, R. W.; McDaniel, C. A.; Fales, H. M.; DuBois, M. B.; Torres, J. J. Chem. Ecol. 1982, 8.
13. Tumlinson, J. H.; Silverstein, R. M.; Moser, J. C.; Brownlee, R. G.; Ruth, J. M. Nature 1971, 234, 348.
14. Riley, R. G.; Silverstein, R. M.; Carroll, B; Carroll, R. J. Insect Physiol. 1974, 20, 651.
15. Cross, J. H.; Byler, R. C.; David, U.; Silverstein, R. M.; Robinson, S. W.; Baker, P. M.; de Olveira, J. S.; Jutsum, A. R.; Cherett, J. M. J. Chem. Ecol. 1979, 5, 187.
16. MacConnell, J. G.; Blum, M. S.; Fales, H. M. Tetrahedron 1971, 26, 1129.
17. Brand, J. M.; Blum, M. S.; Fales, H. M.; MacConnell, J. G. Toxicon 1972, 10, 259.
18. MacConnell, J. G.; Blum, M. S.; Buren, W. F.; Williams, R. N.; Fales, H. M. Toxicon 1976, 14, 79.
19. Brand, J. M.; Blum, M. S.; Ross, H. H. Insect Biochem. 1973, 3, 45.
20. MacConnell, J. G.; Williams, R. N.; Brand, J. M.; Blum, M. S. Ann. Ent. Soc. Am. 1974, 67, 134.
21. Blum, M. S.; Fales, H. M.; Leadbetter, G.; Bierl, B. A. Unpublished results, 1984.
22. Wheeler, J. W.; Olubajo, O.; Storm, C. B.; Duffield, R. M. Science 1981, 211, 1051.
23. Jones, T. H.; Blum, M. S.; Fales, H. M.; Thompson, C. R. J. Org. Chem. 1980, 45, 4778.

24.   Jones, T. H.; Blum, M. S.   Unpublished results,
      1984.
25.   Law, J. H.; Wilson, E. O.; McCloskey, J. A.
      Science 1965, 149, 544.
26.   Adrouny, G. A.; Derbes, V. J.; Jung, R. C.
      Science 1959, 130, 449.
27.   Caro, M. R.; Derbes, V. J.; Jung, R.   Arch.
      Derm. 1957, 75, 475.
28.   Bufkin, D. C.; Russell, F. E.   Toxicon 1972,
      10, 526.
29.   Read, G. W.; Lind, N. K.; Oda, C. S.   Toxicon
      1978, 16, 361.
30.   Koch, R. B.; Desiah, D.; Foster, D.; Ahmed, K.
      Biochem. Pharm. 1977, 26, 983.
31.   Cheng, E. Y.; Cutkomp, L. K.; Koch, R. B.
      Biochem. Pharm. 1977, 26, 1179.
32.   Yeh, J. Z.; Narahashi, T.; Almon, R. R.
      J. Pharm. Exp. Ther. 1975, 194, 373.
33.   David, J. A.; Crowley, P. J.; Hall, S. G.;
      Battersby, M.; Sattelle, D. B.   J. Insect
      Physiol. 1984, 30, 191.
34.   Blum, M. S.; Walker, J. R.; Callahan, P. S.;
      Novak, A. F. Science 1958, 128, 306.
35.   Tomalski, M.; Blum, M. S.; Jones, T. H.
      Unpublished results, 1984.
36.   Clement, J. L.   Unpublished results, 1984.
37.   Holldobler, B.   Oecologica 1973, 11, 371.
38.   Urbani, C. B.; Kannowski, P. B.   Environ.
      Ent. 1974, 3, 755.
39.   Jouvanez, D. P.; Blum, M. S.; MacConnell,
      J. G.   Antimicrob. Ag. Chemother. 1972, 2,
      291.
40.   Cole, L. K.   Ph.D. Thesis, University of
      Georgia, 1973.
41.   Emery, C. Genera Insectorum 1922, 174, 166.
42.   Fales, H. M.; Blum, M. S.; Bian, Z.; Jones,
      T. H.; Don, A. W.   J. Chem. Ecol. 1984,
      10, 651.

RECEIVED December 14, 1984

Use of Natural Products and Their Analogues

For Combating Pests of Agricultural and Public Health Importance in Africa

D. L. WHITEHEAD

International Centre of Insect Physiology and Ecology (I.C.I.P.E.), P.O. Box 30772, Nairobi, Kenya

Ectoparasites such as ticks, mites and haematophagous flies together with certain endoparasites may be "managed" by use of avermectins (macrocyclic lactones) or hormone analogues and by immunization of the hosts with antigens derived from the vectors or parasites themselves. Recently this combined approach has been used most effectively against helminth infestations as well as against ticks and flies. Use of components of host odors and pheromones to trap, sterilize or kill tsetse flies is described as well as inexpensive methods of crop and livestock protection, such as oils and plant extracts, used traditionally in less developed countries.

There are urgent economic reasons for trying to control predation in Africa by biting, sucking flies and acarines (ticks and mites). Among them are identified the vectors of mammalian parasitic and virus diseases such as trypanosomiasis (sleeping sickness), filariasis such as onchocerciasis (river blindness), leishmaniasis, malaria, Dengue fever and East Coast fever (Theileriosis) (ECF).

One disease alone, caused by trypanosomes transmitted by the tsetse fly (Glossina spp.), is estimated (1) to limit ranching and human settlement in over 4 million square miles of Africa. Effective trypanocidal drugs have been available for 35 years, but because of their expense, toxicity and now resistance to them, the only sure way to eliminate the disease completely is considered to be by vector eradication (2). A more realistic approach adopted in West Africa (3) is aimed at introducing trypanotolerant breeds of livestock in areas where the challenge from infected flies is not unduly high.

Control of reproduction

Research carried out in this laboratory and at the Tsetse Research Laboratory near Bristol in England has focused on attempts to

0097-6156/85/0276-0409$07.00/0

disrupt reproduction or sterilize tsetse. Their specialized mode
of a adenotrophic larviparous reproduction, whereby a single larva
(stage III) is deposited every 8 - 10 days, would seem to offer an
avenue whereby population size could be effectively and selectively
controlled (4). The quantity and frequency of blood meals
available to pregnant females does, of course, constitute a limit
on the size and therefore success of the next generation.

In 1971 it was reported (5) that some substance(s) present in
the diet of the host resulted in reduced fecundity when Glossina
morsitans morsitans were fed on these animals. Subsequently Turner
and Marashi reported similar problems with G. m. morsitans fed on
rabbits which had eaten contaminated food (6). The toxin(s) were
not identified in either case, but the effects of coccidiostats
added to the diet were not realized until Jordan and Trewern
reported that sulphaquinoxaline (75 ppm) with pyrimethamine (7.5
ppm) together markedly lowered fecundity of G. austeni and G. m.
morsitans when fed to the rabbit hosts (7).

Using hormones as insect growth regulators

These observations prompted an investigation into the use of insect
growth regulators (IGR) and insect hormones - applied topically or
added to the diet of tsetse fed in vitro (8) (i.e. through a
sterile silicone membrane) - to disrupt reproduction. Further
impetus was given to these studies in Bristol when Delinger
reported from ICIPE that 5-10 μg of 20-hydroxyecdysone (20-OHE) or
juvenile hormone (JH) analogue, when injected or applied topically
(respectively) to pregnant G. m. morsitans, caused over half to
abort (9). By comparison, feeding of 20-OHE to pregnant G. m. morsitans was surprisingly more effective (Figure 1) than injecting
the hormone (10). This could mean that a metabolite of the sterol
is the active abortifacient, especially as entry from the gut into
the haemolymph of label from ^{3}H-20-OHE (NEN, Boston) was very slow
(10).

Ecdysone itself and the phytoecdysones, inokosterone and
makisterone A, were much less effective as abortifacients than
20-OHE (Table I). Cyasterone and ponasterone A were inactive; the
implication being that the hydroxyl group on C25 is required for
activity (10).

Observation of the distal tubules of the uterine gland responsible for nourishment of the larva in utero showed (Figure 2) that
adding ecdysteroids to the diet of female tsetse inhibited gland
function in some way. As a result, less "milk" from the gland was
available to nourish the larva, hence smaller pupae were produced,
or the larva was aborted. The diameter of distal tubules was also
smaller in females to which JH had been applied (Table II) but,
contrary to Denlinger's observation (9), in this experiment (where
HPLC pure JH III was used instead of his crude mixture), no abortions resulted.

Attempts to use 20-OHE as a systemic abortifacient - it was
injected into rabbits used as host - was not successful (10). The
sterol (mixed with ^{3}H-OHE) had rapidly been removed from circulation. Therefore, analogues of the sterol which will not
rapidly be metabolized (into sulphates or glucuronides) in the mammalian liver might be longer acting. The 2, 22-ditosyl derivative

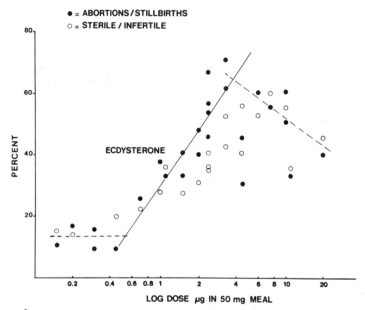

Figure 1

Dose response curve for percentage observed abortions or still births and for percentage sterile females (including unrecorded egg abortions) resulting from feeding 20-OHE <u>in vitro</u> to pregnant <u>G. m. morsitans</u> (ED_{50} = 2.5 μg for abortions alone ($Y = 37.0\overline{7} + 17.34X$)   or = 1.0 μg for all abnormalities) "Reproduced with permission from Ref. 10. Copyright 1981, 'Insect Sci. Applic. Pergamon Press, Oxford'."

Table 1. The effects of potential larvicides fed in vitro to pregnant G. m. morsitans. "Adapted from Table 1, reproduced with permission in part from ref. 10 Copyright 1981 Insect Science & its Application, Pergamon Press, Oxford."

Substance	No. of flies fed	Estimated dose (μg/50 mg meal)	Percentage of abnormal larvae*		
			1st cycle	2nd cycle	3rd cycle
20-Hydroxy-Ecdysone	15	0.1 - 0.2	30	50	20
	30	0.8 - 1.1	64	56	40
	15	1.5 - 3.4	70	-	-
	30	2.8 - 3.4	100	100	25
Ecdysone	30	0.8 - 1.1	37	42	-
	30	1.5 - 2.0	55	42	-
	30	3.6 - 3.8	100	90	-
Makisterone A	15	5	7	-	-
	30	10	15	-	-
	15	35	100	-	-
Inokosterone	15	1.8	33	18	11
	15	3.7	51	14	8
	15	8.7	100	35	12
Cyasterone	15	1.9	20	23	14
	15	2.5	21	30	14
Fonasterone A	15	5.0	7	0	-
	30	15.0	5	0	-

*Abortions and flies that did not deposit larvae

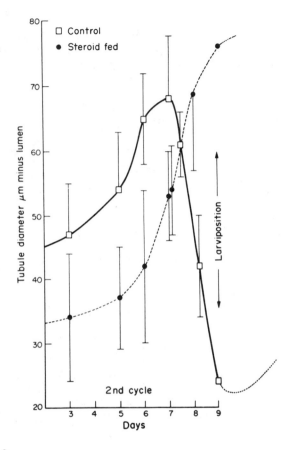

Figure 2
Variation in diameter of distal tubules of uterine gland during
second pregnancy cycle of G. m. morsitans fed in vitro on day 1
with 20-OHE (4 μg/50 mg blood) in defibrinated pig blood compared with control (analysis of variance gives P<.001 for
treatment and time on day 5 – 7). (Vertical bars = ± S.D.)
"Reproduced with permission from Ref. 10. Copyright 1981
'Insect Sci. Applic. Pergamon Press, Oxford'."

was synthesized (10) with a view to exchanging the labyl groups
with naked fluorine ions but yields were extremely low (<1%).
20-Fluorophytosterols have recently (11) proved to be selective
pro-insecticides because insects and not mammals dealkylate the
side chain releasing flurooacetate.

Table II. The Effect on Larval Development and Uterine Gland
           Activity of feeding 20-hydroxyecdysone (20-OHE) on
           day 1 with or without Juvenile Hormone III on day 2,
           4, and 6 of Second Pregnancy Cycle of G. m. morsitans

Added to diet	Treatment (Topical)	Diameter of distal tubules (minus lumen) (μm $\pm$ S. E. M.) (n)	Percent larvae normally developed
20-OHE[a]	Nil	53.0+1.4(25)	20
20-OHE	Acetone	41.5+2.1(10)	10
20-OHE	JH III[b]	49.6+6.2(16)	44
NIL	JH III	37.4+4.6(10)	90
NIL	Nil	68.0+2.2(25)	100

a) 5 μg in 50 mg blood,   b) 5 μg in acetone,   n) no. of files
                                                            dissected

Previously it was thought that JH from the corpora allata (CA)
regulated the activity of the uterine gland (4). Hence JH III was
used (Table II) to attempt to abolish the effect on the gland of
feeding 20-OHE; the inhibition, however, was not reversed.
    The reason for this may be that there is now evidence to show
that the corpora cardiaca (CC) secretes a peptide (12) which stimulates amino acid uptake by the gland. Furthermore, many attempts
to demonstrate that JH is present in the pregnant tsetse have so
far failed. Activity of the large sexual accessory gland of allatectomized Periplaneta americana is also inhibited in vivo by
20-OHE (13).

Ivermectin feeding studies

Recently, the present author and Langley and Roe (14) have demonstrated independently that Ivermectin (MK933) is >10^3 times more
effective than 20-OHE when this mixture of avermectins (Figure 3) is
fed to female G. m. morsitans through a membrane (Figure 4). More
importantly, it was effective for 2-3 weeks when pregnant flies were
fed on blood taken from a horse (14) treated once orally (0.4 mg/kg)
with the drug (Table III). Species of flies that lay their eggs in
the manure of cattle which were given oral capsules daily (5 μg/kg)
(Table IV) or given a single injection (200 μg/kg) of MK933, failed
to develop normally (15). Not only dipters can be controlled in this
way but also ticks (Table V) (16) and mites (17) which start to imbibe
the blood of treated animals are unable to engorge properly, therefore
metamorphosis or reproduction is incomplete or prevented.

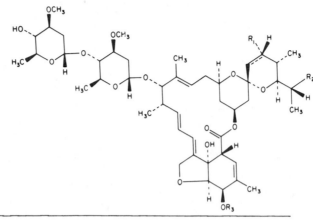

Avermectin	R_1	R_2	R_3
A_{1a}		Et	Me
A_{1b}		Me	Me
A_{2a}	OH	Et	Me
A_{2b}	OH	Me	Me
B_{1a}		Et	H
B_{1b}		Me	H
B_{2a}	OH	Et	H
B_{2b}	OH	Me	H

Where R_1 is absent, the double bond (- - - -) is present. Both sugars are α– L – oleandrose.

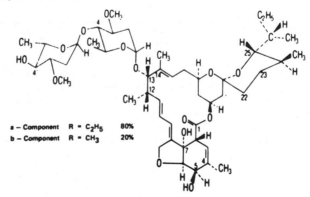

a – Component R = C_2H_5 80%
b – Component R = CH_3 20%

22,23-Dihydroavermectin B_{1a} which may also contain up to 20% 22,23 dihydroavermectin B_{1b}

<u>Figure 3</u>
Structural formulae for the avermectins showing (bottom)
Ivermectin – MK933.

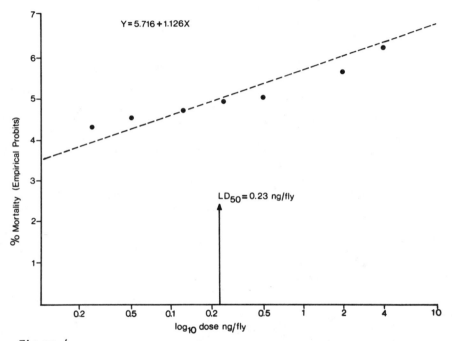

Figure 4
Maximum likelihood probit analysis (65) of dose response data
for larval mortality obtained following addition of Ivermectin
to defibrinated pig blood fed once in vitro to pregnant G. m.
morsitans (χ_2 (7df) = 8.82 ns)
"Reproduced with permission adapted from Ref. 14
Copyright 1984. 'Ent. Exp. Applic. Dr. W. Junk, The Hague'."

Table III.   Fecundity of G. m. morsitans females fed once in vitro on blood taken from a horse (236 kg) at intervals after oral administration of 95 mg of Ivermectin paste "Reproduced with permission. Ref. 14. Copyright 1984 'Ent exp. Applic-Junk'."

Days Elapsed	Percent Larval mortality	Number of pregnant flies engorged
0	0	20
1	100	30
2	80	10
3	80	10
4	70	10
5	20	10
7	20	10
8	33	30
9	40	10
10	40	10
11	11	9
12	30	10
13	10	10
14	20	10
15 – 18	5	20

Table IV. Mortality of the larvae of Horn fly, Face fly, Stable fly and House fly placed in manure from two steers (ca. 200 kg) treated with oral capsules containing MK 933 "Reproduced with permission Ref. 15 Copyright 1981. 'J. Econ. Entomol., Entomological Society of America'."

Corrected % mortality at indicated dosages (μg/kg body wt.) for indicated species

Day of treatment	Horn fly		Face fly		Stable fly		House fly	
	1	5	1	5	1	5	1	5
1	100	100	100	100	0	0	–	–
2	100	100	–	–	0	45	–	–
3	100	100	–	–	0	0	86	72
6	100	100	100	100	64	68	0	100
7	100	100	100	100	0	88	56	100
8	100	100	100	100	75	98	9	86
9	100	100	100	100	41	59	13	100
10	100	100	83	100	96	42	13	91
11	–	–	84	100	–	–	30	86
13	100	100	–	–	33	93	–	–
14	100	100	–	–	57	93	–	–

TABLE V.   Control of Adults and Nymphs (test 1 only) of Five Species of Three-host Ticks (Amblyomma, Dermacentor, Rhipicephalus spp. on Cattle Given Daily Oral or Subcutaneous Treatment with MK-933 "Reproduced with permission in part from ref. 16. Copyright 1981, 'J. Econ. Entomol., Entomological Society of America'."

% CONTROL OF EL[a] OF ADULTS

TEST	TREATMENT (µg/kg/day)	A. americanum	A. cajennense	D. andersoni	D. variabilis	R. sanguineus
Adults	50	94.5	100	100	100	100
	200	100	100	100	100	100
Nymphs[b]	50	5	71	0	–	–
	200	84	100	0	–	–
First infestation adults	10	45.8	59.8	0	0	78.8
	20	71.4	50.0	54.9	43.9	94.8
	50	100	78.4	100	100	100
	100	100	99.6	100	100	100
Second infestation adults	10	0	0	0	–	86.2
	20	40.0	0	0	60.7	100
	50	100	0	100	18.3	96.6
	100	100	100	100	57.3	100

a   EL (estimated larvae)=g, of eggs x estimated % hatch x 20,000 see Drummond et al J. Econ. Entomol. 1972, 65, 1641
b   Percent control of nymphs corrected by Abbott's formula

Because the avermectins, a family of lactones found in the fermentation broth of a unique soil bacterium from Japan, Streptomyces avermitilis (18), are reputed to act by blocking signal transmission from interneurons to excitatory motoneurons (i.e. they are gamma-amino-butyric acid (GABA) agonists). They have been tested and found effective for immobilizing the root knot nematode (19) and for inhibiting feeding of the larvae of a beetle, the Alfafa weevil, when leaf disks had been treated with MK933 at doses of 1-10 p.p.m. or when field trials using 0.01 - 0.10 Kg/ha were conducted (20). Such application in the future for crop protection would depend on formulation of a stable form of the degradable drug which is non-toxic to mammals (18).

Mention should be made at this point that the avermectins have been used primarily to clear livestock of those infesting endoparasites (18) that utilize GABA as a transmitter between nerve cells. World-wide losses of production in livestock infested with helminths cannot be overemphasized.

With regard to the possible use of ivermectin for tsetse and tick control in Africa, a light-stable formulation of MK933 , or MK936, the less expensive precursor, might find application for the treatment of domestic and even wild animals if the drug were to be offered in combination with a coccidiostat like Pancoxin (sulphaquinoxaline and pyrimethamine) in a salt lick block (7). Any form of salt, which in nature is in short supply, is highly attractive and desirable to the tsetse's wild hosts, many species of which harbour the trypanosomes responsible for causing sleeping sickness in man and nagana in livestock (1).

Unfortunately, protozoan flagellates like trypanosomes are unaffected by ivermectin (21) but, at least, the animals that take up the ivermectin (with coccidiostat) will be rid of most gastrointestinal nematodes and ectoparasites. Extraintestinal filariasis can also be treated: for example, microfilarial dermatitis in horses (22) and even onchocerciasis in humans (23), which is transmitted by the bite of infected black flies (Simulium spp.). However, the doses required (30 - 50 µg/kg) in the latter case were not without short term side effects.

Safety

The most remarkable property of these derivatives of microbial natural products is the wide margin of safety accorded to treated mammals - the reason being that the drug does not readily cross the blood-brain barrier to any extent (18). Fortunately in mammals, GABA mediated nerves only occur in the central nervous system. Doses as high as 6 mg/kg (30 times the effective dose) were injected into cattle with impunity and no foetal abnormalities resulted when pregnant animals received doses higher than those recommended (18). Furthermore, ivermectin residues proved to be extremely low in muscle, kidney and urine at any time after livestock were treated orally or by injection. More than 98% of the residue of the drug was excreted in the faeces, regardless of the method used during treatment of livestock. Obviously culturing of the actinomycete Streptomyces avermitilis by the Kitasato Institute in Japan has led to the most fortuitous breakthrough in

combating "pests", both endo- and ectoparasites, since the discovery of DDT. Unfortunately, the cost of the chemically derived
ivermectin makes it unlikely that third world countries can afford
it. However, the avermectins, although less active, should be
cost-effective if cultures of S. avermitilis mutants were to be set
up in Africa.

The need for integrated control programs

Research at ICIPE and in Zimbabwe is aimed at finding inexpensive
practical ways of monitoring and eventually controlling tsetse
populations thus reducing the challenge to man and his livestock by
infected flies. The use of salt blocks impregnated with coccidiostats and avermectins, formulated to stabilize them, could be
used as part of an integrated program in areas where tsetse are
isolated such as in the Lambwe Valley in Western Kenya, a focus of
human and livestock trypanosomiasis. Aerial spraying of the valley
with Endosulphan in 1981 followed by natural pyrethrum extracts in
1983 failed to control tsetse because other methods were not
employed to prevent resurgence of the fly population. Deployment of
insecticide impregnated screens and odor baited biconical traps,
after the costly spraying operation, would have been more effective. Use of host odors markedly increased trap catches of G.
pallidipes in the Nkuruman escarpment area of Kenya's Rift Valley
(24) (Table VI).

Table VI - Comparison[1] of Means of transformed figures of catch
           size of G. pallidipes in Biconical Traps baited with
           Odors from a Buffalo.
           "Reproduced with permission. Ref. 24 Copyright 1984.
           'Insect Sci. Applic., Pergamon Press, Oxford'."

Trap	Females		Males	
	Means	Significant*	Means	Significant*
1. Control	2.17(165.6)	*	1.93(93.0)	*
2. with 4 black strips	2.23(243.6)	*	1.90(117.2)	*
3. with body wash	2.10(74.4)	*	1.65(47.2)	*
4. with faeces	2.10(166.4)	*	1.93(104.2)	*
5. with urine	3.00(1177.2)		2.67(625.0)	

* P<0.01   1 Using Duncan's multiple comparison
                              test.

Likely, components of the odors such as acetone, methylethylketone
(MEK), and methylvinylketone (MVK), have also been tested separa-

tely and in conjunction (25) with some success (Table VII). The
attractiveness of tsetse to ox breath, acetone, CO_2, acetone and
CO_2 (26) and 1-octen-3-ol (one active component of ox odor) had
previously been established in Zimbabwe (24).

Use of the most attractive odors such as buffalo urine (which
increased catches of females 11.6 fold) to entrap and then sterilize tsetse before releasing them again would contribute greatly to
an integrated control program. Since the attractant is a combination of many natural products, the effort to mimic buffalo
urine with laboratory chemicals might not in the long run be worth
the expense and effort for field application. However, the use at
ICIPE, of electroantennagram (EAG) techniques (27) has shown the
order of responsiveness to be: acetone>1-octen-3-ol>MEK>pentanol.

Auto-sterilization

Attempts to sterilize tsetse caught in the wild have been made
using aerosol sprays within biconical traps (28). In order that
male G. m. morsitans might be autosterilized, Bisazir (a mutagenic
azaridine) was applied to decoys, attached to cloth or netting
screens, treated with the sex stimulant "Morsilure" (29) (15, 19,
23-trimethyl heptatriacontane). Successful results in the laboratory had encouraged these field trials (30).

The sex-stimulant for male G. pallidipes, the (13R, 23S) - 13,
23 - dimethylpentatriacontrane (Figure 5) has been synthesized (31)
and tested (Figure 6) (32). It is problematical whether this
short-range "Pallidilure" pheromone can be used to immobilise male
flies for long enough to allow sufficient chemosterilant to be
absorbed through the tarsal and abdominal cuticle. The search is
continuing for a chemosterilant safe enough for humans to use (33).
More reactive analogues might totally disrupt mating behaviour.

Derivatives of natural products (the benzyl phenols and
benzyl-1, 3-benzodioxoles) (Figure 5) effectively sterilize
screwworm flies (34), insecticide-resistant female house flies (35)
and female G. m. morsitans (36). In the latter case, 2 µg/fly of
benzyl-1, 3-benzodioxole was effective. This group of natural products was found to be nonmutagenic in standard Ames bacterial tests
and the compounds were relatively non-toxic to mammals (36).
Therefore, incorporation of these compounds in salt blocks might be
envisaged, or trapped tsetse might be exposed to them before
release, as part of an integrated control programme. Sterilizing
insects reduces populations far more quickly than trying to kill
them.

Immunological methods

The systemic route remains the most desirable method for
controlling haematophagous pests or ectoparasites, especially if
they happen to be vectors of virus, protozoan or metazoan inflicted
diseases. It has been understood for quite some time that the bite
of the tick transfers substances, like an anticoagulant for
instance, which somehow eventually leads to acquired resistance to
infestation by that tick (37). Therefore, use of proteins from the
ectoparasite, as antigen to immunize livestock likely to be exposed

Table VII. The effect of various Ketones and 1-Octen-3-ol on
the size of 6h catches of G. pallidipes in Biconical
Traps set under Nkuruman Escarpment (25)

Treatment		Dose rate (mg/h)	Detransformed mean catches	Index of Increase	F ratio
Acetone	Control	0	10.2	–	
	Low dose	139	18.1	x1.8	5.9*
	Medium dose	463	18.2	x1.8	
	High dose	2391	27.9	x2.7	
Methylethylketone					
(MEK)	Control	0	5.9	–	
	Low dose	77	13.4	x2.7	4.3*
	Medium dose	190	10.3	x1.7	
	High dose	1546	9.8	x1.7	
Methylvinylketone					
(MVK)	Control	0	30.0	–	
	Low dose	64	25.2	x0.8	5.4*
	Medium dose	145	29.6	x1.0	
	High dose	703	37.7	x1.2	
High dose of 3 ketones					
	Control	0	44.9	–	
	Acetone	2216	81.9	x1.8	2.9ns
	MEK	1694	69.4	x1.5	
	MVK	936	48.0	x1.1	
Acetone+1-Octen-3-ol					
	Control	0	26.3	–	
	Acetone	2216	62.2	x2.4	12.5***
	1-Octen-3-ol[1]	50	30.0	x1.1	
	Acetone+1-Octen-3-ol[1]	2216/50	53.7	x1.0	
MEK+1-Octen-3-ol					
	Control	0	47.0	–	
	MEK	85	73.9	x1.6	10.3***
	Octenol	50	88.0	x1.9	
	MEK+Octenol	85/50	125.9	x2.7	

*P <0.05. ***P <0.001. ns = nonsignificant. [1]Octenol discolored.

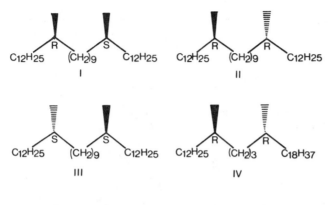

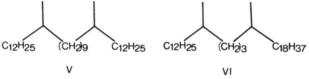

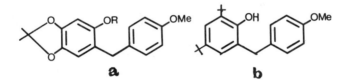

Figure 5

●Pheromones (I-IV)
Diastereoisomers (1-111) of the sex-stimulant 'Pallidilure'
13,23-dimethylpentatriacontane (V) and 13,17-isomers (IV, VI)
synthesized for testing on decoys to attract G. pallidipes
males.

●New Chemosterilants (a, b) modified biologically active
allelochemicals
a) Benzyl-1,3-benzodioxoles (R = ethyl (A 13-53024) or 2-pro-
     penyl (A 13-70714))
b) Ortho Benzylphenol
     2,4-bis (1, 1-dimethylethyl)-6-(4-methoxy phenylmethyl)
     phenol (A 13-70691).

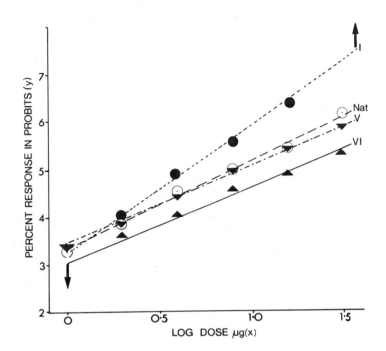

Figure 6

Maximum likelihood probit analyses (65) of figures for male G. pallidipes sexual activity responding to various doses of three compounds shown in figure 5 or the natural 'Pallidilure' extracted from females:

● I; (13R, 23S) 'palidilure' ED_{50} = 4.47 ± 1.09 µg
$Y = 3.256 + 2.681 X$; χ_2 (4df) = 1.02 ns

O Nat; natural 'pallidilure' from female cuticle
ED_{50} = 8.07 ± 1.17 µg; $Y = 3.35 + 1.82 x$;
χ_2 (4df) = 0.487 ns

▼ V; racemic synthetic 'pallidilure' ED_{50} = 8.88 ± 1.15 µg;
$Y = 3.437 + 1.648 x$; χ_2 3.437 + 1.648 x; χ_2 (4df)
= 0.205 ns

▲ VI; racemic 13, 17-dimethylpentatriacontane
ED_{50} = 17.88 ± 1.25 µg; $Y = 3.048 + 1.555 x$; χ_2 (4df)
= 1.10 ns.

on the range, should slowly reduce the population of the pest in
question to a level where the threat of disease is lowered. (38).
     Midgut cell proteins have been used as antigens in guinea pigs
to induce resistance to the tick Dermacentor variabilis, the vector
of Rickettsial disease, commonly called Rocky Mountain Spotted
Fever (39). Similar trials with calves and rabbits have been conducted (40, 41). With some success, we at ICIPE have followed this
approach - immunizing rabbits with the homogenates of various
tissues from two tick species and the tsetse G. m. morsitans. In
the latter case, fecundity was reduced in females if guts, uterii
or salivary glands were used (42). Proteins from the gut also
decreased digestion of the bloodmeal (43) in males as well as females as does feeding of trypsin inhibitors.
     Partial host resistance to engorgement, development, and repro-
duction in Rhipicephalus appendiculatus (ECF vector) appears to
result if whole tick homogenate (44), gut or egg preparations are
used. Currently we are targeting on the use of individual purified
egg proteins (45, 46) to immunize livestock. For instance, the
vitellin (500,000 daltons) (45) and the lighter glycolipoproteins
(106,000 daltons) (47) which weakly bind and carry (46, 48) the
morphogenic 20-OHE to the developing embryo, are both antigenic
(38, 49).
     The immunological approach does not prevent spread of diseases,
like Theileriosis, because the vectors will still attempt to feed
on potential hosts, thus endangering them if they are not resistant
to the pathogen. However, in the long run, as part of an
integrated approach to control vector population, the result must
benefit the rancher or pastoralist. Obviously, greater immediate
benefit accrues when livestock can be immunized against the disease
itself (50) but, in the case of trypanosomiasis, success has not
been achieved yet because the parasites can rapidly acquire a new
surface coat thus protecting themselves against the reaction caused
by circulating antibodies (51, 52). Tolerance to trypanosomes
which evolved in game animals is unfortunately more difficult to
acquire for livestock in the course of a few generations. Even
among trypanotolerant breeds of cattle, meat and milk production
are restricted once the challenge from infected tsetse rises above
a certain threshold (3).
     An interesting discovery has been reported concerning tremato-
des (Digenea) which are unaffected by avermectins - they do not
have GABA receptors. Injecting antibodies, raised to a protein
covalently bound to ecdysone, into rats infested with Schistosoma
mansoni reduced the burden of worms significantly (53). Cestodes,
filarial nematodes and trematodes all appear to produce, utilize
and secrete ecdysteroids (54). This prompted the suggestion that
not only antibodies (designed for R.I.A.) but ecdysteroids themselves and anti-ecdysones (55) might be used to interfere with the
reproduction and growth of these parasitic helminths.

Use of oils and allelochemicals

On a small scale, there is possibly no less expensive way of protecting stored agricultural produce like grain against predation by
pests than using a thin coating of vegetable oil. It must be

remembered that the cost of pesticides like MK933 or conventional
insecticides, whatever their use, usually places them beyond the
reach of the farmer in most of Africa, Asia and S. America (56).
Therefore, traditional methods which utilize readily available
plant materials need to be encouraged in this day and age. The
Giriama, a coastal tribe living in Kenya, use smoke from open wood
fires to protect their granaries from attack by weevils and
beetles. For centuries sunflower oil has been used in Asia to
deter predators of stored rice (57). Peanut, coconut, safflower
and even mineral oil or polyethylene glycol has been shown to be
effective for cowpeas (58). Elsewhere, the oil from seeds which
cost less like Indian Neem (Azadirachta indica) has been
demonstrated (59) to be equally effective (Table VIII). We have
shown that oils from turmeric contain insecticidal sesquiterpenoids
especially active against the brown plant hopper, a rice pest (60).
Such herbs have been used for centuries to preserve grain in Asia.
     Neem oil applications indirectly protect rice crops against
Rice Tungro Virus (RTC) by repelling or even killing (55) the vector Nephotettix virescens, a leaf hopper (Figure 7). The active
allelochemical in Neem seeds is Azadirachtin, a potent antifeedant
against many agricultural pests (56, 57), reputed to be an antiec-
dysone (55), which, in some extraordinary way, can be taken up
through the roots of crops (58) like maize or rice to protect the
plant from insect pests (59). Synthesis of such molecules, apart
from the sheer difficulty, would put their cost beyond the reach of
agriculture even in N. America. Here, therefore, is an excellent
example of the wisdom of reverting to traditional methods in the
less developed Third World instead of having to rely on costly
conventional pesticides (51).
     However, no one will deny that new remedies like those offered
by the broad spectrum ivermectin or by prophylactics, in the form
of specific vaccines, must eventually become available to boost
food production and fight famine and disease in Africa.
Governments, therefore must explore ways of subsidizing farmers,
ranchers and health services to afford the new vaccines, drugs, or
chemicals which research workers are developing. Traditional
methods must also be better understood by modern science in order
that they may be even more fully exploited.

Legend of Symbols

ECF = East Coast Fever
EL (estimated larvae) = g. of eggs x % hatch x 20^3
R.I.A. = radioimmunoassay
20-OHE = 20-hydroxyecdysone

Table VIII. Average repellency of plant extracts and curcumin to Tribolium castaneum adults "Reproduced with permission from Ref.59 Copyright 1983. 'J. Econ. Entomol., Entomological Society of America'."

Treatment	Solvent extract	Rate of Application ($\mu g/cm^2$)	Avg. % repellency[a] at indicated wk after treatment:				Overall avg[b]
			1	2	4	8	
Turmeric	Acetone	680	66.2	47.0	32.5	30.0	43.9
		170	54.4	42.5	15.5	15.0	31.9
	Ethanol	680	61.8	42.5	26.5	25.0	39.0
		170	49.4	42.0	15.5	11.8	29.7
	Petroleum ether	680	92.6	78.5	67.5	47.5	71.5
		170	74.4	57.5	48.0	42.5	55.6
Neem	Acetone	680	33.5	28.5	23.0	12.0	24.3
		170	31.5	22.0	19.0	7.5	20.0
	Ethanol	680	38.0	29.5	32.0	17.0	29.1
		170	29.5	19.0	12.0	6.5	16.8
	Petroleum ether	680	81.5	54.5	49.5	42.5	57.0
		170	63.0	45.5	36.0	34.5	44.8
Fenugreek	Acetone	680	65.6	53.0	24.0	13.5	39.0
		170	55.0	37.0	13.0	8.5	28.4
	Ethanol	680	65.0	50.5	25.0	17.5	39.5
		170	53.8	40.5	12.0	6.5	28.2
	Petroleum ether	680	78.8	59.0	36.0	31.0	51.2
		170	60.6	47.5	27.0	17.5	38.2
Curcumin	-	200	4.0	-2.0	-1.2	6.5	1.8
		50	-5.0	-3.5	-6.3	-1.5	-6.6

a Average of data from four replicates, 10 insects per replicate
b Average data of weeks 1,2,4, and 8.

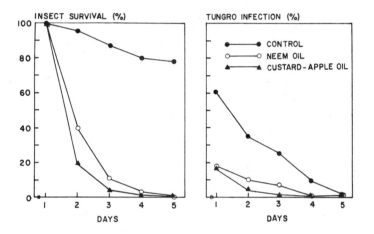

Figure 7
Average survival of and TRV transmission by N. virescens
exposed for different periods on TN1 rice seedlings treated
with 50% emulsified oils.
"Reproduced with permission, Ref: 61 Copyright 1983, 'J. Econ.
Entomol. Entomological Society of America'."

Acknowledgements

The author thanks Drs. Jack R. Plimmer, Horace G. Cutler, Paul
Hedin, Peter Ashton, Ian Sutherland and all my colleagues at ICIPE
for their support and encouragement. Mrs. Rosemary Okoth provided
secretarial expertise while Dr. Robert Dransfield kindly computed
the data (from ref. 14) shown in Figure 3.

Literature cited

1.   Nash, T.A.M., "African's Bane - the Tsetse fly"; Collins;
     London. 1969; p. 17.
2.   Jordan, A.M. Nature, Lond. 1978, 273, 607.
3.   ILCA Monograph 2 Trypanotolerant Livestock in W. & C. Africa;
     Volume 1, ILCA, Addis Ababa; 1979; p. 87.
4.   Tobe, S. A., Langley, P.A. Ann. Rev. Entomol. 1978, 23, 283.
5.   Saunders, D.A. Bull. ent. Res. 1971, 60, 431.
6.   Turner, D.A., Marashi, M.H. Trans. R. Soc. trop. Med. Hyg.
     1973, 65, 24.
7.   Jordan, A.M., Trewern, M.A. Ent. exp. Applic. 1976, 19, 115.
8.   Mews, A.R., Langley, P.A., Pimley, R.W., Flood, M.E.T. Bull.
     ent. Res. 1977, 67, 119.
9.   Denlinger, D. Nature, Lond. 1975, 253, 347.
10. Whitehead, D.L. Insect Sci. Applic. 1981, 1, 281.
11. Prestwich, G.D., Gayen, A.K., Phirwa, S., Kline, T.B.
     Bio/Technol. 1983. 1.
12. Pimley, R.W., Physiol. Entomol. 1983, 8, 429.
13. Whitehead, D.L. Gen. Comp. Endocrinol. 1973, 74, 412.
14. Langley, P. A., Roe, J. M. Ent. exp. Applic. 1984, In Press.
15. Miller, J.A., Kunz, S.E., Ochler, D.A., Miller, R. W. J. Econ.
     Entomol. 1981, 74, 608.
16. Drummond, R. O., Whetstone, T.M., Miller, J.A. J. Econ.
     Entomol. 1981, 74, 432.
17. Lee, R.P., Dodge, D.J.D., Preston, J.M. Vet. Rec. 1980, 107,
     503.
18. Campbell, W.C., Fisher, M.H. Stapley, E.O., Albers-Schonberg,
     G., Jacob, T.A. Science 1983, 221, 823.
19. Wright, D.J., Birtle, A.J., Roberts, I.T.J. Parasitology,
     1984, 88, 375.
20. Pienkowski, R.L., Mehring, P.R. J. Econ. Entomol. 1983, 76,
     1167.
21. Golder, T.K., Whitehead, D.L. unpublished data.
22. Herd, R.P., Donham, J.C. Am. J. Vet. Res. 1983, 44, 1102.
23. Aziz, M.A., Diallo, S., Diop, I.M., Lariviere, M., Porta, M.
     Lancet, 1982, 2, 171 and 1456.
24. Owaga, M.L.A. Insect Sci. Applic. 1984, 5, 87.
25. Dransfield, R.D. Chaudhury, M.F.B., Tarimo, R.S. Golder, T.K.
     unpublished data.
26. Vale, G.A. Bull. ent. Res. 1980, 70, 563.
27. Saini, R.K., Den Otter, C.J. unpublished data.
28. Bursell, E., Vale, G.A., unpublished data.
29. Langley, P.A. Coates, T.W., Carlson, D.A. Proc. O.A.U./
     S.T.R.C. 17th Meeting 1981, 463.
30. Langley, P.A., Huyton, P.M., Carlson, D.A. Swarz, M. Bull.
     ent. Res. 1981, 71, 57.
31. Kuwahara, S., Mori, K. Agric. Biol. Chem. 1983, 47, 2599.

32. McDowell, P.G., Hassanali, A., Dransfield, R.D. Physiol. Entomol. in press.
33. Borkovec. A.B. personal communication.
34. Rawlins, S.C., Jurd, L. J. Econ. Entomol. 1981, 74, 215.
35. Rawlins, S.C., Woodard, D.B. Coppedge, J.R. Jurd, L. Ibid. 1982, 75, 728.
36. Langley, P.A., Pimley, R.W., Leedham, M.P. Tsetse Research Laboratory, Annual Report, University of Bristol, U.K. 1982.
37. Nelson, F.R., Hooseintehrani, B. J. Econ. Entomol. 1982, 75, 877.
38. Mongi, A.O., Cunningham, M.P. Eleventh Annual Report, ICIPE Nairobi, 1983, p. 25-9.
39. Ackerman, S., Floyd, M., Sonenshine, D.E. J. Med. Entomol. 1980, 17, 391.
40. Schlein, Y., Lewis, C. T. Physiol. Entomol. 1970, 1, 55.
41. Allen, J.R. Humphreys, S.J. Nature, Lond. 1979, 280, 491.
42. Kaaya, G.P., Alemu, P. Insect Sci. Applic. 1984, 5, In Press.
43. Otieno, L.H. Vundla, R.M.W., Mongi, A.O. Ibid. 1984, 5, 297.
44. Mongi, A.O. Ph.D. Thesis, University of Nairobi, Nairobi, 1983.
45. Dhadialla, T. unpublished data.
46. Whitehead, D.L. Ninth Annual Report, ICIPE, 1981, p. 79.
47. Vundla, R.M.W., Whitehead, D.L. Obenchain, F.D., Osir, E.W. Proc. Acarology Congress, Edinburgh. September 1982.
48. Whitehead, D.L., Osir, E.W., Thomas, L.S., Vundla, R.M.W. Regulation of Insect Reproduction III, Czech, Acad. Sci., Prague, June 1982.
49. Vundla, R.M.W. Tenth Annual Report, ICIPE, Nairobi, 1982, p. 14.
50. Bittle, J.I., Houghten, R.A., Alexander, H., Skinnick, T.M. Sutcliffe, J.G., Lerner, R.A. Nature, Lond. 1982, 298, 30.
51. de Gee, A.L.W. Vet. Quart, 1982, 4, 32.
52. Goddeeris, B.M., Katende, J.M., Irvin, A.D. Chumo, R.S.C. Res. in Vet. Sci. 1982, 33, 360.
53. Nirde, P., Torpier, G., Capron, A., Delaage, M. In "Biosynthesis, Metabolism, and Mode of Action of Invertebrate Hormones"; Hoffmann, J. A., Porchet, M., Eds.; Springer-Verlag: Berlin, 1984; Section II, 14, p. 331.
54. Whitehead, D.L. Nature, Lond. 1984, 306, 540.
55. Koolman, J. personal communication.
56. Whitehead, D.L. Proc. 2nd Int. Chem. Conf. Africa. Hassanali, A. Ed., University of Nairobi, Nairobi, 1985, Chap. 1.
57. Qi. Y. T. Burkholder, W.E. J. Econ. Entomol. 1981, 74, 502.
58. Messina, F.J. Renwick, J.A.A. Ibid. 1983, 76, 634.
59. Jilani, G., Su, H.C.F. Ibid. 1983, 76, 154.
60. Hassanali, A., McDowell, P.G., Saxena, R.C. Tenth Annual Report, ICIPE, Nairobi, 1982, p. 44.
61. Mariappan, V., Saxena, R.C. J. Econ. Entomol. 1983, 76, 573.
62. Schmutterer, H., Ascher, K.R.S., Rembold, H. Proc. 1st Int. Neem. Conf. Rottach-Egern, 1980.

63. Bernays, E.A. "Natural Products for Innovative Pest Management"; Whitehead, D.L.; Bowers, W.S. Eds.; CURRENT THEMES IN TROPICAL SCIENCE Volume 2, Pergamon Press: Oxford, 1983; p. 261-2.
64. Gill, J.S., Lewis, C.T., Nature, Lond. 1971, 232, 402.
65. Finney, D.J. "Probit Analysis" 4th ed. Cambridge University Press. U.K. 1981.

RECEIVED November 15, 1984

Insect Antifeedant Terpenoids in Wild Sunflower

A Possible Source of Resistance to the Sunflower Moth

JONATHAN GERSHENZON[1,4], MARYCAROL ROSSITER[1,5], TOM J. MABRY[1], CHARLIE E. ROGERS[2,6], MICHAEL H. BLUST[3], and THEODORE L. HOPKINS[3]

[1] Departments of Botany and Zoology, University of Texas, Austin, TX 78713
[2] Conservation and Production Research Laboratory, Agricultural Research Service, U.S. Department of Agriculture, Bushland, TX 79012
[3] Department of Entomology, Kansas State University, Manhattan, KS 66506

Sunflower (<u>Helianthus</u> <u>annuus</u>) cultivation in the United States is frequently limited by the severity of insect damage. However, many wild <u>Helianthus</u> species are resistant to the major insect pests of cultivated sunflower. This resistance has been suggested to have a chemical basis. We found high concentrations of sesquiterpene lactones and diterpenes in glandular hairs on several resistant wild species of <u>Helianthus</u> and demonstrated that these compounds were toxins and antifeedants towards some major sunflower insect pests. Experiments were conducted on the southern armyworm (<u>Spodoptera</u> <u>eridania</u>), the migratory grasshopper (<u>Melanoplus</u> <u>sanguinipes</u>) and the sunflower moth (<u>Homoeosoma</u> <u>electellum</u>). Of particular interest was the presence of terpenoids on the portions of the flower immediately adjacent to the pollen. The early larval stages of the sunflower moth, the most destructive insect pest of cultivated sunflower in the United States, feed principally on pollen. Examination of several cultivated lines of sunflower showed that these had lower densities of glandular hairs than the wild species. Thus, increased resistance to sunflower moth predation might be achieved simply by breeding for an increased density of glandular hairs on floral parts surrounding the pollen.

[4] Current address: Institute of Biological Chemistry, Washington State University, Pullman, WA 99164.
[5] Current address: Department of Entomology, Pennsylvania State University, University Park, PA 16802.
[6] Current address: Insect Biology and Population Management Research Laboratory, Agricultural Research Service, U.S. Department of Agriculture, Tifton, GA 31793.

0097–6156/85/0276–0433$06.00/0
© 1985 American Chemical Society

The environmental and economic drawbacks to the large scale use of
synthetic insecticides have focused attention on ways of using plant
natural products to control insect pests of crop plants. Perhaps the
best-studied approach of this type is simply to breed for greater
levels of toxic or antifeedant natural products in plant parts
subject to insect attack to increase their resistance to insect
damage (1). Our investigations with sunflower (Helianthus annuus)
suggest that plant-produced natural products may make a significant
contribution towards increasing resistance to insect damage in this
crop.

        Sunflower has become one of the world's major vegetable oil
crops in recent years because of its adaptability to a wide range of
soil, temperature and water conditions (2) and because of the
development of new hybrid varieties with very high concentrations of
oil in their achenes (3). In the United States, sunflower cultivation has increased over a hundred-fold in the last 20 years (3).
Extensive insect damage is one of the principal factors limiting
further cultivation of sunflower and depressing the yield of oil (4).
Insect problems are particularly serious in the United States because
the genus Helianthus is native to North America (5) and therefore
many insect taxa in this country have had the opportunity to adapt to
feed on sunflowers (4). However, a number of wild sunflowers
(species of Helianthus) have been shown to be resistant to some of
the major insect pests of the cultivated varieties (6-12). Because
there is no obvious morphological feature in these resistant plants
which might account for their relative immunity to insect predation,
it was suggested that this resistance has a chemical basis (12).

        In this chapter, we first review the terpenoid constituents
isolated from resistant wild species of Helianthus and present the
results of some insect bioassays with these compounds. Then we
discuss the evidence which implicates terpenoids in the resistance to
the sunflower moth (Homoeosoma electellum), the most destructive
insect pest of cultivated sunflower in the United States (4), and
describe how this information might be exploited in reducing insect
damage to cultivated sunflower.

Terpenoids from wild species of Helianthus

Approximately half of the fifty species of Helianthus have shown some
resistance to the major insect pests of cultivated sunflower in
greenhouse and field trials (6-12). We began our studies by
chemically investigating a small group of Helianthus species which
were reported to be especially insect resistant. The major
lipophilic secondary metabolites of the aerial parts of these plants
proved to be terpenoid compound of two types: sesquiterpene lactones
and diterpenes. We have now studied 17 species of Helianthus
chemically and have found high levels of sesquiterpene lactones and
diterpenes in most of these (13-25). Both sesquiterpene lactones and
diterpenes have been isolated from species of Helianthus by other
workers (26-44).

        Sesquiterpene lactones are typical secondary metabolites of
many members of the Asteraceae, the composite family (45, 46). The
lactone moiety in these compounds is usually present as an α -
methylene- γ -lactone function with the lactone bridge joining carbon
atoms 12 and 6 or 12 and 8 on the sesquiterpene skeleton (see Figure

1 for numbering scheme). Most Helianthus sesquiterpene lactones are characterized by the fusion of the lactone ring to carbon 6 and the presence of a five carbon angelate or angelate-derived acid side chain esterified to position 8. Four structural types of sesquiterpene lactones have been isolated from Helianthus: germacrolides, heliangolides, eudesmanolides and guaianolides (Figure 1). The biosynthetically-simpler germacrolide and heliangolide types are the most common in the genus.

Four structural types of diterpenes are known from Helianthus: labdanes, kauranes, atisiranes and trachylobanes (Figure 2), with the tetracyclic kaurane type being the most widespread in the genus. The majority of Helianthus diterpenes have an α-oriented carboxylic acid function attached to carbon 4.

Toxicity and antifeedant activity of Helianthus sesquiterpene lactones to sunflower insects

Terpenoids isolated in large quantities from resistant species were tested with several sunflower insect pests. Since sesquiterpene lactones were present in much higher concentrations than diterpenes in the most resistant species studied, our investigations emphasized sesquiter-pene lactones. Three species of insects were used in these studies: Spodoptera eridania, Melanoplus sanguinipes and Homoeosoma electellum.

Spodoptera eridania (Lepidoptera: Noctuidae) - the southern armyworm. This species was chosen for study because it is closely related to S. exigua, the beet armyworm, an insect which has periodically damaged fields of cultivated sunflower in the southern Great Plains (47). A laboratory colony of S. exigua could not be successfully maintained. Both S. eridania and S. exigua are termed generalist feeders because they can feed on various organs of a variety of taxonomically-unrelated plants (48, 49). Experiments on S. eridania were performed in collaboration with Dr. K. Nakanishi's laboratory at Columbia University.

The growth of S. eridania larvae was significantly reduced by sesquiterpene lactones added to their diet. Two sesquiterpene lactones were used in these tests: 8 β-sarracinoyloxycumambranolide (8β SC) from Helianthus maximiliani (18) and desacetyleupasserin from H. mollis (13) (Figure 3). These compounds were added to an artificial diet at concentrations of 0.1% and 1.0% and fed to fifth instar larvae of S. eridania. At a concentration of 1.0%, both compounds caused significant growth inhibition (Table 1). Both compounds are present in Helianthus leaves at levels of 1-5%.

Larvae of S. eridania were also subjected to preference tests to see if they would avoid ingesting sesquiterpene lactones if given the choice of feeding on treated or untreated food. Starved fifth instar larvae were simultaneously presented with bean leaves (Phaseolus vulgaris) coated with a 1% acetone solution in which a sesquiterpene lactone had been dissolved and bean leaves coated with solvent only. After 24 hours, leaves coated with 8 β SC showed significantly less feeding than the solvent-coated controls, but larvae showed no preference between controls and leaves coated with desacetyleupasserin (Table 2).

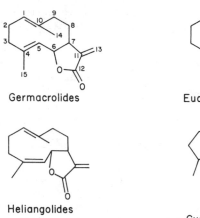

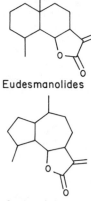

Germacrolides   Eudesmanolides

Heliangolides

Guaianolides

Figure 1. Structural types of sesquiterpene lactones in Helianthus.

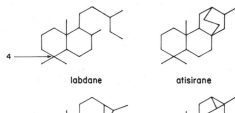

labdane   atisirane

kaurane   trachylobane

Figure 2. Structural types of diterpenes in Helianthus.

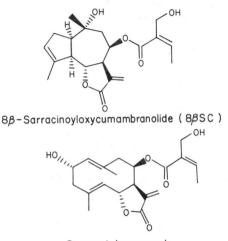

8β-Sarracinoyloxycumambranolide (8βSC)

Desacetyleupasserin

Figure 3. Sesquiterpene lactones used in the insect bioassays.

Table 1. Effect of two <u>Helianthus</u> sesquiterpene lactones on the growth of <u>S. eridania</u> larvae.*

| compound and | average daily weight gain, mg | | | |
concentration	1st day	2nd day	3rd day	4th day
8 β SC				
control	122	97	143	82
0.1%	128	102	118	71
1.0%	-28**	-14**	-9**	-11**
desacetyleupasserin				
control	97	83	76	45
0.1%	76	67	64	40
1.0%	44**	56**	45**	50

*Sesquiterpene lactones were added to artificial diets at concentrations indicated. Ten early fifth instar larvae were raised on each concentration. Gain in weight was measured daily for each individual. Negative numbers indicate average weight loss for larvae in that group.
**Significantly different from control at 1% level (t-test).

Melanoplus _sanguinipes_ (Orthoptera: Acrididae) - the migratory grasshopper.

Grasshoppers occasionally damage cultivated sunflower in the United States (4). _Melanoplus_ _sanguinipes_ is a generalist feeder and a major crop pest species widely distributed in North America (50). 8 β SC added to the diet of _M._ _sanguinipes_ at a concentration of 1% had no effect on the growth or survivorship of this insect or its rate of development to the adult stage (Table 3). However, 8 β SC was shown to be a significant feeding deterrent in preference tests with _M._ _sanguinipes_. Grasshoppers that had been starved for 48 hours were given the choice of feeding on sucrose-treated nitrocellulose membrane filter disks with and without added 8β SC. 8 β SC was dried onto the disks in chloroform solutions at levels of 0.25%, 2.5% and 25% of the dry weight of the disks. Control disks were treated with chloroform only. At all levels of 8 β SC tested, _M._ _sanguinipes_ consumed more of the control disks than the test disks (Table 4).

Homoeosoma _electellum_ (Lepidoptera: Pyralidae) - the sunflower moth.

In contrast to _S._ _eridania_ and _M._ _sanguinipes_, _H._ _electellum_ is a specialist which feeds on the inflorescences of a few species of the Asteraceae (4, 51). Since sesquiterpene lactones are found in several parts of the infloresences of _Helianthus_ species which are considered to be resistant to the sunflower moth (52), it was thought that these compounds might serve to limit the damage caused by this insect.

Larvae of _H._ _electellum_ were raised on artificial wheat germ-based diets to which varying concentrations of 8 β SC had been added. This compound was dissolved in acetone, coated on cellulose powder under vacuum and then mixed into the diet at concentrations of 0.01%, 0.1% and 1%. The control diet contained cellulose powder which had been soaked in acetone and dried under vacuum. At levels of 0.1% and 1%, 8 β SC significantly reduced pupal weight (Table 5). However, larval survival and development time were not affected.

For feeding preference tests, 8 β SC was added to squares of artificial diet at a concentration of 5%. Starved _H._ _electellum_ larvae were placed in the center of a dish containing both treated and control diet squares and allowed to feed. After two hours, significantly more first and second instar larvae were found feeding on the control diet (73%) than on the treated diet (12%) (Table 6). However, the third, fourth and fifth instars did not show a significant preference in this experiment.

The results of these bioassays indicate that sesquiterpene lactones isolated from species of _Helianthus_ resistant to major insect pests of cultivated sunflower are toxins and feeding deterrents to some of these insects. Previous investigations have shown that sesquiterpene lactones have toxic and antifeedant activities towards a variety of phytophagous insects (53-62). Evolutionarily, sesquiterpene lactones may have come to function as feeding deterrents because insects have been selected for their ability to recognize the presence of these toxic compounds in potential foodstuffs and avoid ingesting them. Sesquiterpene lactones exhibit a number of toxic effects at the cellular level, including the inhibition of protein synthesis (63, 64), nucleic acid synthesis (63-66) and respiratory enzyme activity (67). The toxicity of these compounds is thought to be due to their ability to alkylate nucleophilic sites on proteins (63, 68-70) or DNA (64, 66).

Table 2. Feeding deterrence of two <u>Helianthus</u> sesquiterpene
lactones to <u>S. eridania</u> larvae.*

compound	average amount of leaves eaten	
	control	treated
8 β SC	54.2%	21.0%**
desacetyleupasserin	30.0%	36.7%

*1% Acetone solutions of each sesquiterpene lactone were coated on
  bean leaves. Control leaves were coated with solvent only.
  Starved fifth instar larvae were given a choice of feeding on
  control or treated leaves. Tests of each compound against a
  control were repeated with ten larvae.
**Significantly different from control at 5% level (t-test for paired
  comparisons).

Table 3. Effect of 8 β SC on the growth and development of
<u>Melanoplus</u> <u>sanguinipes</u>.*

	diet	
	control	control with added 8 β SC
average adult weight, g ± s.d.		
female	0.224 ± .021	0.210 ± .015
male	0.187 ± .033	0.198 ± .024
average time of development from first instar to adult, days ± s.d.	40.3 ± 1.9	41.7 ± 3.6

*Diet consisted of finely-ground freeze-dried rye grass. 8 βSC in
  dichloromethane solution was coated on the diet at a level of 1%
  of the dry weight of the diet and the solvent allowed to
  evaporate. Control diet was treated with solvent only. Twenty
  insects were individually reared on each treatment. There were
  no significant differences between the treatments.

Table 4. Effect of 8 β SC on the feeding preference of M.
          sanguinipes.

percentage of 8 βSC per dry wgt. of disk	average amount of control disks eaten in excess of test disks, cm^2 ± s.d.
0.25	0.30 ± 0.46**
2.5	1.19 ± 0.99**
25	0.97 ± 1.25**

*Grasshoppers that had been starved for 48 hours were presented with
  5 cm^2 nitrocellulose membrane filter disks to which 5% sucrose
  solution had been added. Test disks were treated with 8 β SC in
  chloroform solution while control disks received chloroform
  only. All disks were air-dried. Area of disk eaten was determined with an area meter.
**Significant preference at the 1% level (t-test for paired compari-
  sons).

Table 5. Effect of 8 β SC on the survival, development and pupal
          weight of H. electellum larvae.*

| | concentration in diet | | | |
	control	0.01%	0.1%	1.0%
survival to adult, %	82	83	89	77
development time (from hatch to adult eclosion), days	25.7	25.6	25.7	25.7
pupal weight, g	34.4	33.3	32.4**	31.3**

*8 β SC was dissolved in acetone and coated on cellulose powder which
  was added to the diet to give the concentrations indicated.
  Control diet contained cellulose powder which had been soaked in
  acetone. Each concentration was tested on at least 100 larvae.
**Significantly different from controls at the 5% level (analysis of
  variance). Data on development time and survival did not show
  any significant effect of 8 β SC dose. Development time data
  were subjected to an analysis of variance and survival data to a
  test of independence using the G-statistic.

Possible role of sesquiterpene lactones in resistance to the sunflower moth.

Homoeosoma electellum, the sunflower moth, is currently the most serious pest of cultivated sunflower in the United States (4). In the previous section, we showed that sesquiterpene lactones from species of Helianthus resistant to H. electellum are toxic and antifeedant to H. electellum larvae when added to artificial diets. To evaluate the possible involvement of sesquiterpene lactones in resistance to H. electellum in the intact plant, it is necessary to consider the feeding habits of this insect in relation to the location of sesquiterpene lactones in the infloresences of resistant species of Helianthus. Sesquiterpene lactones are found in glandular trichomes on the terminal anther appendages, immediately adjacent to the pollen (52) (Figure 4). Young H. electellum larvae (first and second instars) eat principally pollen, while later instars feed on a variety of floral parts: corollas, styles, ovaries, developing achenes and parts of the receptacle (4, 51, 71).

How might young H. electellum larvae obtain pollen? Before the florets open, the pollen is produced and stored in the anthers. To reach the pollen in an unopened floret, larvae must crawl or eat their way into the floret and then eat into the anthers. The sesquiterpene lactone-containing glandular trichomes on the anther tips seem to be situated in just the right location to prevent the larvae from reaching the pollen (Figure 4).

After a floret opens, pollen is shed into the center of a cylinder formed by the fused anthers and the style branch pushes up through the cylinder raising the pollen above the anthers for presentation to potential pollinators (72). To reach the pollen at this stage, young larvae would have to crawl up and over the anthers (Figure 4). The sesquiterpene lactone-containing glandular trichomes on the anther appendages could also deter larvae from the anthers (or stigmas) in this manner after the florets are open. However, there is as yet no evidence to support this possibility.

The proximity of sesquiterpene lactones to the pollen in resistant species of Helianthus, in conjunction with the deleterious effects of these compounds on H. electellum when incorporated into artificial diets, suggests that the presence of sesquiterpene lactones may have been selected for because these compounds can reduce sunflower moth predation by preventing the young larvae from reaching their principal food source. If young H. electellum larvae cannot obtain sufficient pollen, they may not survive to cause further damage to the infloresence. Observations on the cultivated lines of sunflower support the role of sesquiterpene lactones in resistance to the sunflower moth. Cultivars which are frequently heavily damaged by H. electellum larvae were found to have much lower densities of glandular trichomes on their anthers than the resistant species of Helianthus (52).

Therefore, it may be possible to increase the resistance of cultivated sunflower to H. electellum by breeding for increased concentrations of sesquiterpene lactone-containing glandular trichomes on the anther tips. The development of new cultivated lines with high densities of glandular trichomes may prove to be an

Table 6.   Effect of 8 βSC on the feeding preference of <u>H. electellum</u>
           larvae.*

larval instar	feeding on control	feeding on treated diet	not feeding
1st and 2nd	73%**	12%	15%
3rd, 4th and 5th	45%	32%	23%

*Larvae were allowed to choose between feeding on a diet containing
 5% 8 βSC and a control diet. Twenty larvae of each instar were
 tested. Results for first and second instars, on the one hand,
 and for third, fourth and fifth instars, on the other, were
 similar and therefore combined.
**Significant preference at the 5% level (chi-square test).

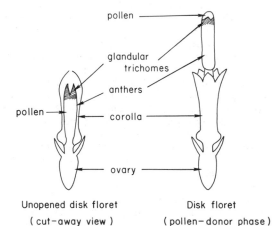

Figure 4.   Location of sesquiterpene lactone-containing glandular
trichomes on sunflower disk florets.

efficient and practical way of controlling the sunflower moth at an early stage in its development. Glandular trichomes act to put the deterrent compounds when and where they are needed for defensive purposes without the hazards and expense of applying synthetic insecticides.

Glandular trichome constituents have been associated with insect resistance in a number of other crops, including cotton, tomato, potato, tobacco and alfalfa (73). However, this is the first time that glandular trichomes or any plant natural product has been implicated in the defense of pollen against predation.

Sesquiterpene lactone-containing glandular trichomes may not serve as effective defenses against older H. electellum larvae since, in our preference bioassays, later instars (third, fourth and fifth instars) were not significantly deterred by 8 β SC (Table 6). In addition, older larvae feeding on cultivars have been observed to chew through the base of the corollas of unopened florets to reach the pollen (71). This behavior allows the larvae to avoid any glandular trichomes which might be present on the anther tips. Other chemical and morphological features might be important in the resistance to older H. electellum larvae, such as the presence of a so-called "phytomelanin" layer in the walls of developing achenes which may form a barrier to larval feeding (74). The protection of sunflower from damage by all stages of H. electellum larvae may thus require the development of cultivars with a combination of defensive traits.

Acknowledgments

We thank V. Leskinen and K. Nakanishi for the S. eridania test data, G. Gradowitz for technical assistance with the M. sanguinipes studies and M. Conoley for help with the H. electellum experiments. This work was supported by grants to T. J. Mabry from the National Institutes of Health (HDO-4488) and the Robert A. Welch Foundation (F-130) and a grant to J. Gershenzon from the Sunflower Association of America. This is contribution No. 84-13-A from the Department of Entomology, Kansas Agricultural Experiment Station. We acknowledge the use of the University of Texas Brackenridge Field Laboratory.

Literature Cited

1. Waiss, A. C., Jr.; Chan, B. G.; Elliger, C. A. In "Host Plant Resistance to Pests"; Hedin, P. A., Ed.; ACS SYMPOSIUM SERIES NO. 62, American Chemical Society: Washington, D. C, 1977; pp 115-128.

2. Robinson, R. G. In "Sunflower Science and Technology"; Carter, J. F., Ed.; American Society of Agronomy: Madison, Wisconsin, 1978; pp 89-143.

3. Putt, E. D. In "Sunflower Science and Technology"; Carter, J. F., Ed.; American Society of Agronomy: Madison, Wisconsin, 1978; pp 1-29.

4. Schulz, J. T. In "Sunflower Science and Technology"; Carter, J. F., Ed.; American Society of Agronomy: Madison, Wisconsin, 1978; pp 169-223.

5. Heiser, C. B., Jr.; Smith, D. M.; Clevenger, S. B.; Martin, W. C., Jr. Mem. Torr. Bot. Club 1969, 22, 1-218.

6.   Rogers, C. E.; Thompson, T. E. J. Econ. Entomol. 1978, 71, 221-
     222.
7.   Rogers, C. E.; Thompson, T. E. J. Econ. Entomol. 1978, 71, 622-
     623.
8.   Rogers, C. E.; Thompson, T. E. J. Econ. Entomol. 1978, 71, 760-
     761.
9.   Rogers, C. E.; Thompson, T. E. Southwestern Entomol. 1979, 4,
     321-324.
10. Rogers, C. E.; Thompson, T. E. J. Kansas Entomol. Soc. 1980,
     53, 727-730.
11. Rogers, C. E.; Thompson, T. E. Environ. Entomol. 1981, 10, 697-
     700.
12. Rogers, C. E. In "Biology and Breeding for Resistance to
     Arthropods and Pathogens in Agricultural Plants"; Harris, M.
     K., Ed.; Texas Agric. Expt. Sta. Misc. Publ. 1451, 1980; pp
     359-389.
13. Ohno, N.; Mabry, T. J. Phytochemistry 1979, 18, 1003-1006.
14. Ohno, N.; Mabry, T. J.; Zabel, V.; Watson, W. H. Phytochemistry
     1979, 18, 1687-1689.
15. Ohno, N.; Mabry, T. J. Phytochemistry 1980, 19, 609-614.
16. Ohno, N.; Gershenzon, J.; Neuman, P.; Mabry, T. J.
     Phytochemistry 1981, 20, 2393-2396.
17. Watanabe, K.; Ohno, N.; Yoshioka, H.; Gershenzon, J.; Mabry, T.
     J. Phytochemistry 1982, 21, 709-713.
18. Gershenzon, J.; Mabry, T. J. Phytochemistry 1984, 23, 1959-
     1966.
19. Gershenzon, J.; Mabry, T. J. Phytochemistry 1984, 23, 2557-2559.
20. Gershenzon, J.; Mabry, T. J. Phytochemistry 1984, 23, 2561-2571.
21. Lee, E.; Gershenzon, J.; Mabry, T. J. J. Nat. Prod. 1984, 47,
     in press.
22. Melek, F.; Ahmed, A. A.; Gershenzon, J.; Mabry, T. J.
     Phytochemistry 1984, 23, 2573-2575.
23. Melek, F.; Gershenzon, J.; Lee, E.; Mabry, T. J. Phytochemistry
     1984, 23, 2277-2279.
24. Stewart, E. S.; Gershenzon, J.; Mabry, T. J. J. Nat. Prod.
     1984, 47, 748-750.
25. Gage, D. A.; Gershenzon, J.; Mabry, T. J.; Melek, F.; Pearce,
     J.; Whittemore, A. unpublished results.
26. Beale, M. H.; Bearder, J. R.; MacMillan, J.; Matsuo, A.;
     Phinney, B. O. Phytochemistry 1983, 22, 875-881.
27. Bjeldanes, L. F.; Geissman, T. A. Phytochemistry 1972, 11, 327-
     332.
28. Bohlmann, F.; Jakupovic, J.; King, M. M.; Robinson, H.
     Phytochemistry 1980, 19, 863-868.
29. Ferguson, G.; McCrindle, R.; Murphy, S. T.; Parvez, M. J. Chem.
     Res. (S) 1982, 200-201.
30. Herz, W.; De Groote, R. Phytochemistry 1977, 16, 1307-1308.
31. Herz, W.; Govindan, S. V.; Watanabe, K. Phytochemistry 1982,
     21, 946-947.
32. Herz, W.; Kulanthaivel, P. Phytochemistry 1983, 22, 2543-2546.
33. Herz, W.; Kulanthaivel, P.; Watanabe, K. Phytochemistry 1983,
     22, 2021-2025.
34. Herz, W.; Kumar, N. Phytochemistry 1981, 20, 93-98.
35. Herz, W.; Kumar, N. Phytochemistry 1981, 20, 99-104.
36. Herz, W.; Kumar, N. Phytochemistry 1981, 20, 1339-1341.

37.   Martin Panizo, F.; Rodriguez, B. Ann. Quim. 1979, 75, 428-430.
38.   Morimoto, H.; Oshio, H. J. Nat. Prod. 1981, 44, 748-749.
39.   Ortega, A.; Ayala, A.; Guerrero, C.; Romo de Vivar, A. Rev.
      Soc. Quim. Mex. 1972, 16, 191.
40.   Ortega, A.; Romo de Vivar, A.; Diaz, E.; Romo, J. Rev.
      Latinoamer. Quim. 1970, 1, 81-85.
41.   Pyrek, J. S. Tetrahedron 1970, 26, 5029-5032.
42.   Spring, O.; Albert, K.; Gradmann, W. Phytochemistry 1981, 20,
      1883-1885.
43.   Spring, O.; Albert, K.; Hager, A. Phytochemistry 1982, 21,
      2551-2553.
44.   Stipanovic, R. D.; O'Brien, D. H.; Rogers, C. E.; Thompson, T.
      E. J. Agric. Food Chem. 1979, 27, 458-459.
45.   Seaman, F. C. Bot. Rev. 1982, 48, 121-592.
46.   Fischer, N. H.; Olivier, E. J.; Fischer, H. D. Prog. Chem. Org.
      Nat. Prod. 1979, 38, 47-390.
47.   Rogers, C. E. unpublished results.
48.   Soo Hoo, C. F.; Fraenkel, G. J. Insect Physiol. 1966, 12, 693-
      709.
49.   Soo Hoo, C. F.; Fraenkel, G. J. Insect Physiol. 1966, 12, 711-
      730.
50.   Mulkern, G. B.; Pruess, K. P.; Knutson, H.; Hagen, A. F.;
      Campbell, J. B.; Lambley, J. D. North Dakota Agric. Expt.
      Sta. Bull. 148, 1969; pp 25-26.
51.   Teetes, G. L.; Randolph, N. M. J. Econ. Entomol. 1969, 62, 264-
      265.
52.   Gershenzon, J. Ph.D. Thesis, University of Texas, Austin, 1984.
53.   Burnett, W. C., Jr.; Jones, S. B., Jr.; Mabry, T. J.; Padolina,
      W. G. Biochem. Syst. Ecol. 1974, 2, 25-29.
54.   Burnett, W. C.; Jones, S. B.; Mabry, T. J. In "Biochemical
      Aspects of Plant and Animal Coevolution"; Harborne, J. B.,
      Ed.; Academic Press: London 1978, pp 233-257.
55.   Doskotch, R. W.; Fairchild, E. H.; Huang, C.; Wilton, J. H.;
      Beno, M. A.; Christoph, G. G. J. Org. Chem. 1980, 45, 1441-
      1446.
56.   Ganjian, I.; Kubo, I.; Fludzinski, P. Phytochemistry 1983, 22,
      2525-2526.
57.   Nawrot, J.; Smitalova, Z.; Holub, M. Biochem. Syst. Ecol. 1983,
      11, 243-245.
58.   Pettei, M. M.; Miura, I.; Kubo, I.; Nakanishi, K. Heterocycles
      1978, 11, 471-480.
59.   Picman, A. K.; Elliot, R. H.; Towers, G. H. N. Biochem. Syst.
      Ecol. 1978, 6, 333-335.
60.   Wisdom, C. S.; Smiley, J. T.; Rodriguez, E. J. Econ. Entomol.
      1983, 76, 993-998.
61.   Jones, S. B.; Burnett, W. C.; Coile, N. C.; Mabry, T. J.;
      Betkouski, M. F. Oecologia 1979, 39, 71-77.
62.   Nakajima, S.; Kawazu, K. Heterocycles 1978, 10, 117-121.
63.   Lee, K.-H., et al. Science 1977, 196, 533-536.
64.   Hladon, B.; Twardowski, T. Pol. J. Pharmacol. Pharm. 1979, 31,
      35-43.
65.   Spring, O.; Kupka, J.; Maier, B.; Hager, A. Z. Naturforsch.
      1982, 37c, 1087-1091.
66.   Woynarowski, J. M.; Konopa, J. Mol. Pharmacol. 1981, 19, 97-
      102.

67.   Van Aswegen, C. H.; Potgieter, D. J. J.; Vermeulen, N. M. J. S. African J. Sci. 1982, 78, 125–127.

68.   Hanson, R. L.; Lardy, H. A.; Kupchan, S. M. Science 1976, 168, 378–380.

69.   Kupchan, S. M.; Fessler, D. C.; Eakin, M. A.; Giacobbe, T. J. Science 1976, 168, 376–378.

70.   Picman, A. K.; Rodriguez, E.; Towers, G. H. N. Chem.-Biol. Interactions 1979, 28, 83–89.

71.   Rogers, C. E. Environ. Entomol. 1978, 7, 763–765.

72.   Knowles, P. F. In "Sunflower Science and Technology"; Carter, J. F., Ed.; American Society of Agronomy: Madison, Wisconsin, 1978; pp 55–87.

73.   Stipanovic, R. D. In "Plant Resistance to Insects"; Hedin, P. A., Ed.; ACS SYMPOSIUM SERIES NO. 208, American Chemical Society: Washington, D. C. 1983; pp 69–100.

74.   Rogers, C. E.; Kreitner, G. L. Environ. Entomol. 1983, 12, 277–285.

RECEIVED November 23, 1984

Insect Feeding Deterrents from Semiarid and Arid Land Plants

ELOY RODRIGUEZ

Phytochemical Laboratory, Department of Ecology and Biology, University of California, Irvine, CA 92717

Phytochemical investigations of plant species of Parthenium, Encelia and Dicoria (Asteraceae) from semi-arid and arid zones of the United States and Mexico has resulted in the isolation and identification of numerous sesquiterpene lactones and benzopyrans that are active insect feeding deterrents. The chemistry, biological effects and possible mode of action of bioactive pseudoguaianolides and chromenes are reviewed.

During the course of evolution, arid land plants, like many other species of plants from diverse environmental zones, have evolved an array of natural organic chemicals that function primarily in regulating insect growth and reproduction. Natural constituents like prenylated quinones and benzofurans from desert plants are known to be extremely toxic when topically applied to lepidopteran larvae and pathogenic microorganisms (1). In many cases of desert plant-insect interactions, it is becoming apparent that secondary chemicals primarily function in repelling a large percentage of herbivorous insects, with the plant suffering a certain degree of leaf, flower and seed loss. In surveying the literature for repellents and antifeedants from semi-arid and arid land plants, it is surprising to find that very little research has been done on the chemistry, mode of action and distribution of desert plant antifeedants and repellents. For example, in a recent survey of naturally occurring insecticides, insect repellents and attractants in higher plants, Jacobson (2) lists only one species that is from an arid zone, while the other species examined for bioactivity are from sub-tropical and tropical areas. This is indeed very surprising when one considers that in North America, about 25% of the United States is semi-arid to arid, while in Mexico, approximately 46% is dryland (3).

0097–6156/85/0276–0447$06.00/0

Desert regions which are of great importance to the
United States and Mexico are the arid lands of the
Southwest and Baja California, Mexico. It is in these
arid regions that we have concentrated our research
efforts in isolating and elucidating the structures of
new phytochemicals that inhibit insect and fungal growth
(4).

     In this brief communication we summarize some of
our recent findings on sesquiterpenoids from species of
Parthenium, Encelia and Dicoria (Asteraceae) from the
Chihuahuan and Sonoran Deserts, that exhibit insect
repellent and antifeedant activities.

Sesquiterpene Lactones as Insect Feeding Deterrents

Sesquiterpene lactones are structurally diverse and
bitter constituents of the Asteraceae that are known to
exhibit numerous biological activities. These mevalonic
acid derived constituents have been demonstrated to have
cytotoxic and antitumor activities, are potent elicitors
of allergic contact dermatitis in humans and have been
implicated in livestock poisonings (5). Several studies
also indicate that sesquiterpene lactones are important
defensive constituents against known insect pests,
because of their feeding deterrent and insect growth-
inhibiting properties (6). For example, glaucolide A
(1), a germacranolide present in various species of
Vernonia (Asteraceae), is a feeding deterrent and
ovipositional deterrent to the fall, south and yellow-
striped army worms (7-9). Parthenin (2), a
pseudoguaianolide present in species of Parthenium,
Hymenoclea and Ambrosia, is physiogiclly active
against Melanoplus sanguinipes (10). Bakkenolide A (3),
a sesquiterpene spirolactone from Hymogyne alpina (L.)
Cass. was recently shown to be an effective repellent
against the granary weevil adult beetle (Sitophilus
granarius) the flour beetle larvae and adults (Tribolium
confusum) and the Khapra beetle larvae (Trogoderma
granarium) (12).

     Recent studies in our laboratory have also
demonstrated that C_{14} and C_{15} oxygenated
pseudoguaianolides (parthenolides) from the semi-arid
and arid genus Parthenium (Asteraceae) are effective
insect larvae feeding repellents (12). We have
investigated the efficacy of several sesquiterpene
lactones as inhibitors of early larval growth of two
species of phytophagous pests, the bollworm Heliothus
zea and the beet army worm Spodoptera exigua, using a
standard laboratory chronic feeding bioassay (12). In
particular, we have compared the inhibitory effects of a
series of sesquiterpene lactones differing both in
skeletal type and degree of substitution (13).

     The sesquiterpene phenolic esters, guayulin A (4)
and -B (5) from the desert rubber plant guayule
(Parthenium argentatum) were found to be relatively

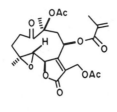

(1)

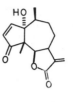

(2)

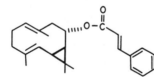

(3)

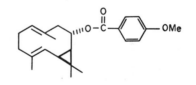

(4)   (5)

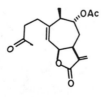

(6)

non-inhibitory (13). This was surprising, since previous studies had established that guayulin A was a potent elicitor a of delayed hypersensitivity reactions (allergic skin dermatitis) in guinea pigs (14). Twelve sesquiterpene lactones from other arid and semi-arid species of Parthenium were found to be significantly inhibitory to Heliothus zea. The xanthanolide, ivalbatin acetate (6), isolated from P. fruticosum var. trilobatum (15) and Dicoria canescens (16), was the least inhibitory; whereas the tetraneurin A parthenolide, tetraneurin A (7) from P. fruticosum and P. alpinum var. tetraneuris, was found to be equally inhibitory to H. zea and S. exigua (13). At a dietary concentration of 3.0 mM/kg fr. wt., tetraneurin A reduced levels of growth of H. zea by 88% relative to controls in chronic feeding bioassay. Dose-response relationships for four pseudoguaianolides fed to H. zea were relatively linear over the dose range tested (13). Pseudoguaianolides that are oxygenated at the C-14 and C-15 positions tend to be more inhibitory than the non-oxygenated ambrosanolides. For example, tetraneurin A (7) is more inhibitory than coronopilin (8) and parthenin (3). Similarly, ligulatin B (9) is more inhibitory than confertin (10), the major constituent of P. tomentosum from Mexico.

Similar feeding deterrent effects against Heliothus zea were also observed with dehydroleucodin (11) from Artemisia tridentata var. vaseyana (17). On the other hand, farinosin (12), the major eudesmanolide in Encelia farinosa does not seem to affect the relative growth rate of H. zea. The chloroform extract of E. farinosa, which contains both sesquiterpene lactones and chromenes, is deterrent to the fifth instar of H. zea (17). The deterrent was identified as encecalin (13), a benzopyran present in the brittle bush (E. farinosa). Comparison of encecalin with precocene I (14) and precocene II (15), suggests that the precocenes are more effective in repelling insect larvae, with the insects starving to death. Also, no larvae appeared to grow past the second instar (14). At concentrations of 0.6% wwt, encecalin and the precocenes were effective feeding deterrents. This is noteworthy, since the natural concentrations of encecalin are 8-10 times this amount.

Although a number of new sesquiterpene lactones have been shown to inhibit insect growth, few studies have distinguished between behavioural effects and actual post-ingestive physiological inhibition (Isman, personal communicartion). Isman (18), has recently demonstrated that parthenin (1) is extremely toxic to the grasshopper Melanopus sanguinipes when injected into the hemocoel at doses greater than 0.25 μmole per 300 mg insect. Dose-dependent sublethal symptoms range from a reduction in normal locomotary ability to paralysis and eventual death. The toxic symptoms observed by Isman are consistent with the grasshopper heart as a major target site in vivo. Furthermore, the toxicity of parthenin and

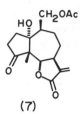

(7)

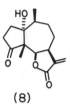

(8)

(9)

(10)

(11)

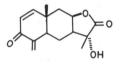

(12)

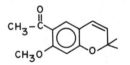

(13)

(14)

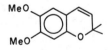

(15)

other active sesquiterpene lactones is dependent on the presence of the exocyclic methylene function and the cyclopentone ring. Further studies are still needed to clearly understand the role of different skeletal substituents on sesquiterpene lactones that effectively deter feeding by phytophagous insects.

Concluding Remarks

One reason that dominant plants are successful in semi-arid and arid ecosystems is due to the high concentration of terpenoids and prenylated phenolics that repel and/or deter feeding by herbivorous insects. Compounds like sesquiterpene lactones and benzopyrans are present in high quantities in the leaves of species of Parthenium, Encelia and Dicoria and have been demonstrated to be effective feeding deterrents against known economic insect pests. In vivo experiments are still needed in the field with bioactive constituents to better understand their effects on native phytophagous insects.

Acknowledgements

This research has been supported by NIH (RCDA-100472) and NSF (PCM-83172240). I am greatly indebted to my research associates cited in the references, in particular, Dr. M. Isman (University of British Columbia, Vancouver) who provided me with unpublished data on the physiological toxicity of sesquiterpene lactones to insects.

Literature Cited

1.   Rodriguez, E. In "Plant Resistance to Insects"; Hedin, P. Ed., ACS SYMPOSIUM SERIES No. 208, American Chemical Society: Washington D.C., 1983; p. 291.
2.   Jacobson, M. In "Plants - The Potential for Extracting Protein, Medicines and Other Useful Chemicals", Office of Technology, Washington D. C., 1983; p. 138.
3.   Becker, R., Sayre, R.N., Saunders, R.M. JOACS 1984, 62, 931-938.
4.   Rodriguez, E. In "Economical Plants from Arid Zones", Wickens, G. Ed., Kew Garden Publications, London, 1984 (in press).
5.   Rodriguez, E., ·Towers, G.H.N., Mitchell, J. Phytochemistry, 1976, 15, 1573-1580.
6.   Burnett, W.C., Jones, S.B., Mabry, T.J., Padolina, W.G. Biochem. System. Ecol., 1974, 2, 25-29.
7.   Burnett, W.C., Jones, S.B., Mabry, T.J. Plant System. and Evol., 1977, 128, 277-286.
8.   Burnett, W.C., Jones, S.B., Mabry, T.J. Amer. Midland Nat., 1978, 100, 242-246.

9.   Jones, S.B., Burnett, W.C., Coile, N.C., Mabry, T.J., Betkouski, M.F. Oecologia, 1979, 39, 71-77.
10.   Picman, A.K., Elliot, R.H. and Towers, G.H.N. Can. J. Zoology, 1981, 59, 285-292.

11.   Harmatha, J., Nawrot, J. Biochem. System. Ecol., 1984, 12, 95-98.
12.   Isman, M.B., Rodriguez, E. Environ. Entomol., 1984, 13, 539-542.
13.   Isman, M.B., Rodriguez, E. Phytochemistry, 1983, 22, 2709-2713.
14.   Rodriguez, E., Thompson, G., Reynolds, G. Science, 1980, 211, 1444-1445.
15.   Rodriguez, E. Biochem. System. Ecol., 1977, 5, 207-218.
16.   Fang, S.D., Rodriguez, E. Phytochemistry, 1984 (in press).
17.   Wisdom, C.S., Smiley, J.T., Rodriguez, E. J. Econ. Entomol., 1983, 76, 993-998.
18.   Isman, M.B. unpublished data.

RECEIVED February 11, 1985

Secondary Metabolites from Plants and Their Allelochemic Effects

HORACE G. CUTLER

Plant Physiology Research Unit, Richard B. Russell Center, Agricultural Research Service, U.S. Department of Agriculture, Athens, GA 30613

Lower and higher plants are sources of secondary metabolites that have allelochemic effects in plants, fungi, bacteria and vertebrates. Examples of two biologically active natural products from a fungus and one from a higher plant are presented. Ophiobolin G and H are new metabolites from Aspergillus ustus whose stereochemistry differs from other fungal ophiobolins but resembles those of insect origin. Both inhibit growth of etiolated wheat coleoptiles and Bacillus subtilis. Ophiobolin H is more active than ophiobolin G, induces phytotoxicity in corn plants and causes hyperacusia in chicks. The higher plant product is carboxyatractyloside from Xanthium strumarium. It is a hypoglycemic agent and has selective phytotoxic properties. It inhibits etiolated wheat coleoptiles, stunts corn plants, and produces slightly malformed leaves in tobacco plants. It has fungistatic properties. Both sets of compounds are discussed relative to their potential as biological agents, markers for elucidating biochemical pathways, and in synthesis of novel products.

The isolation and identification of natural products for potential use in agriculture has increased sharply during the past decade. With few exceptions the synthesis and alteration of these compounds for further use has been slow.

Perhaps plans have not been formulated in several cases, or results disseminated. The avenues of research that the identification of a natural product may open are many, so that in the true sense these compounds may be considered allelochemic (1)as opposed to strictly allelopathic. For example, the discovery of a novel biologically active natural product logically presupposes that the metabolic pathway will eventually be elucidated, that possible synthesis, or partial synthesis may be attempted, that homologs and analogs will be described and that other areas may be considered for eventual employment of these materials, not only in more obvious uses such as pesticides, but also medicinal and non-biological areas. An example of a non-biological use is that of orlandin (2) which is denatured in acetone to produce an acid and base resistant product which is ceramic-like and is not soluble in organic solvents.

Biochemical pathways may be followed with relative ease in one organism because of facility of culture and the availability of large quantities of a specific metabolite,yet with another organism metabolic pathways are extremely difficult to elucidate because the organism is fastidious and the yield of an interesting metabolite is minute. The occurrence of a novel metabolite common to both discrete organisms then becomes a rational starting point for the study of the metabolic sequences in the organism that is simple to culture and gives high yields of the metabolite under study. Upon clarification of the biochemical pathways it is then generally easier to chart and experimentally discern the nature of the biochemical sequences in the more fastidious organism.

We now examine two organisms, one a fungus, the other a higher plant, with respect to the production of allelochemicals and their utility, both direct and indirect. This includes their potential use as agricultural chemicals, or medicinal agents, their use in model systems for studying biochemical pathways, or as templates for further synthesis work. As each of the examples is explored it will become apparent that the study of natural products generally takes place as isolated pieces of research wherein their roles as biologically useful tools may not be linked until some time has elapsed. Therefore, their full potential as allelochemicals may not be obvious upon initial isolation and, upon first consideration, they may only be regarded as novel structures.

In 1965, Arigoni reported the isolation of gascardic acid, a sesterterpene from the secretions of the scale insect Gascardia madagascariensis (3) (Figure 1). Later, other sesterterpenes based on the ophiobolane skeleton (Figure 2) were isolated from another scale insect, Ceroplastes albolineatus, and were shown to be ceroplastol, ceroplasteric acid (4) (Figure 3) and albolic acid (5). All three compounds have similar configurations, with the only

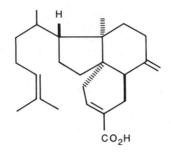

GASCARDIC ACID

Figure 1.   Gascardic acid.

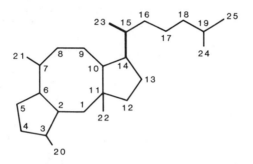

OPHIOBOLANE SKELETON

Figure 2.   Opiobolane skeleton and numbering system.

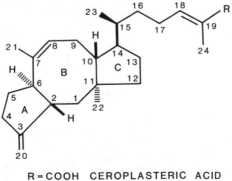

R=COOH CEROPLASTERIC ACID
R=CH₂OH CEROPLASTOL

Figure 3.   Ceroplastol and ceroplasteric acid.

difference between ceroplasteric acid and albolic acid being
a C2-C3 double bond in the latter. The unique common feature
of these three insect derivatives is the trans configuration
between the A and B rings. Other insect ophiobolins, if they
exist, have not been elucidated.

Earlier, cochliobolin was isolated from Helminthosporium
oryzae (Cochliobolus miyabeanus, perfect stage) by Orsenigo
in 1957 (6), and it was shown to be toxic to rice seedlings.
Even at low concentrations the metabolite was toxic to root
and coleoptile growth. No structural data were reported even
though the material was a white crystalline powder and had a
melting point of 180-182°C. Ophiobolin A, and other
ophiobolones, were isolated from H. oryzae and other
Helminthosporium species (7,8,9,10), and there followed a
description of the structure and stereochemistry of
ophiobolin A (11,12) (Figure 4), which was found to be
identical to cochliobolin. Reports of the isolations of
other ophiobolins followed with ophiolobolin B (12,13);
ophiolobin C (11,13) from Helminthosporium sp, ophiobolin D
(cephalonic acid) (14) from Cephalosporium caerulens; and
ophiolobin F (15) from Cochliobolus heterostrophus.
Ophiobolin E remains an enigma and one can only speculate
that the substance was identical to a previous product, was
unstable upon purification, or was not available in
sufficient quantities for physical, chemical and biological
studies. Whatever, the literature is blank for ophiobolin
E. All of these fungally derived ophiobolins have one
structural feature in common: the junction between rings A
and B is cis.

We have recently isolated two new ophiobolins, G and H,
from a novel source, Aspergillus ustus (ATCC No. 38849) (16)
(Figures 5 and 6). Structurally, these two compounds are
interesting in that the ring fusion between rings A and B is
trans, as opposed to cis in the ophiobolins previously
described from other fungal sources. In this respect these
new metabolites more closely resemble the ophiobolins
extracted from insects. Structural features common to both
the insect and fungal ophiobolins are the double bonds at
C7-C8 and C18-C19. Ophiobolins G and H are unique in the
possession of a C16-C17 cis double bond.

There is only slight structural modification between
ophiobolins G and H. Ophiobolin G is a ketoaldehyde with the
ketone function present at C5 in the A ring and the aldehyde
group at C7 in the B ring. Ophiobolin H is an hemiacetal
having the A-B rings joined by an oxygen bridge between C5
and C21 (C21 joins to C7) and OH groups is located at C5, and
C3. It is precisely this difference that seems to account
for the diverse allelochemic effects of the two molecules.
In plant bioassays, etiolated wheat coleoptiles (Triticum
aestivum L.,) were significantly inhibited (P<0.01) by both
compounds at concentrations ranging from 10^{-3} to 10^{-5} M in
buffered aqueous solutions containing sucrose. Ophiobolin G

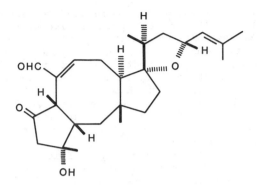

OPHIOBOLIN A

Figure 4.   Opiobolin A.

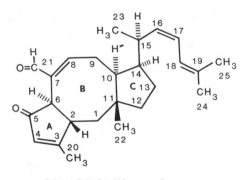

OPHIOBOLIN   G

Figure 5.   Opiobolin G.   Reproduced from Ref. 16.   Copyright 1984
American Chemical Society.

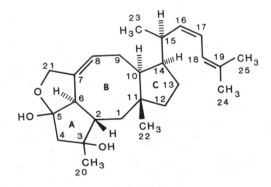

OPHIOBOLIN   H

Figure 6.   Opiobolin H.   Reproduced from Ref. 16.   Copyright 1984
American Chemical Society.

inhibited coleoptiles 81, 53 and 23% while H inhibited 99, 70 and 58% at 10^{-3}, 10^{-4} and 10^{-5} M respectively, relative to controls, indicating that ophiobolin H is the more active of the metabolites in this bioassay. When intact, greenhouse-grown plants were treated with the two ophiobolins a high degree of phytotoxic specificity was observed. Bean and tobacco plants sprayed with solutions that ranged from 10^{-2} to 10^{-4} M were generally unaffected by either metabolite. Corn plants (Zea mays L.,) each treated with 100 ul droplets of each metabolite placed into uppermost leaf sheaths were unaffected by ophiobolin G, but effects with ophiobolin H were quite pronounced(16). That is, ophiobolin H caused complete necrosis of the innermost sheath of leaves within 5 days following treatment at 10^{-2} M. The initial response was marked by a wilting of the sensitive leaves followed by necrosis indicating a possible effect on the stomates. Since treatment consisted of droplets being placed in the base of the leaf sheath, the whole leaf was not treated; but the response observed was over the entire surface of the treated leaves indicating translocation within the individual leaf. However, the wilting was uniform throughout the treated leaf though it would have been anticipated that severe necrosis would have been obvious at the site of application. By contrast, the remainder of the leaves on the plants showed no evidence of a phytotoxic response. Further studies are needed to determine whether the transport and, therefore, the induction of phytotoxicity is unidirectional along the leaf or not. Another member of the same class, ophiobolin A, has been shown to be the probable allelochemic agent responsible for brown spot symptoms in rice (17), another member of the Graminiae. A recent study (18) using spinach leaf slices or Chlorella ellipsoidea C-27 and an oxygen electrode reports that ophiobolin A and the newly discovered 6-epiophiobolin A, anhydroophiobolin A and anhydro-6-epiophiobolin A, isolated from the culture filtrate of a phytopathogenic species of Helminthosporium, were inhibitory to photosynthesis. The I_{50} values were between 10^{-3} and 10^{-4} M for all those ophiobolins that inhibited photosynthesis. In comparison, diuron, bromacil and ioxynil, known inhibitors of the Hill reaction, were active between 10^{-5} and 10^{-7} M in the same biological systems. Additionally, ophiobolins B, C and F are phytotoxins (19) while the fusicoccins, which are close structural relatives of the ophiobolins (even though there is some dispute as to whether the fusicoccins are sesterterpenes, or diterpenes [19]), are also reported to have phytotoxic properties (20,21). New evidence (22) obtained from bioassays incorporating 6-epiophiobolin A and 3-anhydro-6-epiophiobolin A indicate that they are host specific inhibitors of Texas-male-sterile corn mitochondria, and the mitochondria are one hundred times less sensitive to ophiobolin A than to 6-epiophiobolin A. It would appear that the ophiobolins may have selective phytotoxic properties and may be useful templates for synthesizing products for agricultural use against pests.

Of further interest is the antibacterial activity of ophiobolin G and H (16). Both compounds, prepared on assay disks, were challenged with Bacillus subtilis and Escherichia coli in petri dish bioassays, with concentrations ranging from 25 to 1000 ug per disk. Following incubation for 18 hours at 37°C it was noted that neither compound inhibited E. coli development, but both compounds inhibited B. subtilis. Further, ophiobolin G moderately inhibited B. subtilis at concentrations of 50 to 1000 ug/disk and the inhibition zones were uniformly circular. In contrast, ophiobolin H inhibited from 250 to 1000 ug/disk and the zones of inhibition were greater than doubled, but the inhibition zones were not circular. A pH or solubility factor appeared to be responsible for the lack of uniformity. Again, the activity of ophiolobin H was greater than G. The potential use of ophiobolins as control agents against microbiological populations needs further study and these preliminary results sugest that they may have selective properties.

Testing of the biological activity of ophiolobin G and H has not been limited to plants and bacteria, but has been expanded to include preliminary studies with vertebrates (16). Day-old-chicks were dosed via crop intubation with corn oil containing each metabolite at concentrations up to 375 mg/kg. Results were dramatic with ophiobolin H. Within an hour of dosing at 375 mg/kg the chicks were extremely nervous, refused feed and had symptoms of hyperacusia, indicating increased irritability of the sensory neural mechanism. The threshold dose for hyperacusial induction was ca. 250 mg/kg. Chicks had totally recovered within 24 hours and there were no apparent long term ill effects. The lack of mortality and need of relatively high dosages to induce hyperacusia suggest the use of ophiobolin H as a tool for studying the mechanisms of hyperacusial induction of epileptic seizure in clinical studies.

The biochemical linkages between the ophiobolins isolated from the insects, Gascardia madigascariensis and Ceroplastes albolineatus, and the fungus, Aspergillus ustus, remain to be evaluated. The fact that both the insect and fungal ophiobolins have a similar conformation between the A and B rings indicates that similar biochemical pathways may exist. In the other fungi, in which the A to B ring fusion is cis, it is necessary for a double bond in geranylfarnesyl pyrophosphate to isomerize to the cis configuration (23). Apparently this does not take place in either the aforementioned insects or A. ustus. If, then, the biosynthetic systems are identical in both Aspergillus and the scale insects, the fungus, because of the availability of large quantities of metabolites and ease of culture, becomes an ideal model for studying methods for blocking, or

shunting, pathways that give rise to ophiobolin production in
those insects. The nature of the scale insects is such that
they are difficult to control for many reasons. First, they
tend to congregate as masses on stems and under leaves;
second, they produce waxy secretions which cover the body of
the insects and act as physical barriers against desiccation
(4). It is probable that the waxy secretions may also act as
a barrier against chemical sprays. Whatever, the ophiobolins,
ceroplasteric acid and ceroplastol are an integral part of
these secretions. Agents, which are bioregulatory in
function, are also produced by the insects and these induce
loss of vigor in infected plants, rosetting of new shoot
growth, and pitting of the bark on stems (24). It is not
known whether these bioregulating agents are ophiobolins or
not. It should also be pointed out that the genus Gascardia
is considered to be a junior synonym of Ceroplastes by some
entomologists (24). Other fungal ophiobolins are presently
under investigation.

We now turn our attention to carboxyatractyloside, a
compound from the higher plant Xanthium strumarium that has
interesting allelochemic properties in both plants and
animals. The genus, Xanthium, produces burrs (fruiting
bodies) that are biloculate, each of which contains a seed.
One seed is located just above and to one side of the other.
The superior seed is commonly referred to as the "upper" and
the inferior seed as the "lower". The lower seed is known to
germinate rapidly upon dispersal but the upper one may take
months or even years to imbibe and germinate, even when
environmental conditions are perfect (25). It has long been
recognized that cocklebur seed contain a self inhibitor, and
research to elucidate the mechanisms and substances that
control dormancy has been conducted for almost ninety years
(26,27,28,29). In 1957, Wareing and Foda (28) showed the
presence of two plant growth inhibitors in cocklebur seed.
They made water extracts of seed and these proved to be
inhibitory to etiolated wheat coleoptile (Triticum aestivum
L.,) growth. The extracts were placed on paper chromatograms
which were developed in isopropanol – 1% NH_4OH in water (4:1,
v/v) by the ascending method. Two inhibitory zones were
observed, one at Rf 0.0–0.3 designated inhibitor A, the other
at Rf 0.4–0.5, designated inhibitor B. No further
publications followed the 1957 report and further work on the
isolation and characterization of the metabolites ceased.
The presence of abscisic acid, a potent plant growth
inhibiting substance since found in many plants and plant
parts including ash, birch, sycamore, willow, young cotton
bolls and seeds of ash and pear (30), was not detected in the
seed at that time. Furthermore, inhibitors A and B could not
have been abscisic acid because they were both much more
polar (P. F. Wareing, personal communication, 1982).

An apparently unrelated but significant event occurred
in 1972 when Danieli and co-workers isolated two chemically

interesting glycosides from the Mediterranean thistle, Atractylis gummifera L.(31). These were atractyloside and carboxyatractyloside (Figure 7). The latter was unstable in the free state and rapidly degraded when kept in solution, except in the salt form (a characteristic also noted by Cole et al., (32)). Four years later (33) carboxyatractyloside was isolated from Xanthium strumarium because of its hypoglycemic properties. Then in 1980, a toxic agent implicated in the death of pigs in South Georgia was isolated from cocklebur burrs containing seed. The toxic principle was found to be carboxyatractyloside (32) and not, as had been previously claimed (34), hydroquinone. Because of the striking chemical features of carboxyatractyloside, its unique behaviour in some early experiments with paper chromatograms and wheat coleoptile bioassays, we became interested in its biological properties in plants (35).

It became apparent to us that carboxyatractyloside was probably identical to the inhibitor A previously described (28). When dried, sprayed with anisaldehyde reagent and gently heated, paper chromatograms revealed a bright magenta area at RF 0.00–0.28, and when a complementary part of the chromatogram was cut into three equal areas, eluted with buffer–sucrose solution and bioassayed in the wheat coleoptile assay, coleoptiles were significantly inhibited 88, 76 and 86% respectively, relative to controls. Another area at Rf 0.46–0.56 also reacted positively to anisaldehyde reagent to give a salmon pink color, but this was not inhibitory to coleoptiles. Additionally, ($\pm$) abscisic acid was chromatographed in the identical solvent, and upon treatment with anisaldehyde and heat it produced a brown color at Rf 0.60–0.90. Equivalent areas on chromatograms induced 100% inhibition in coleoptile bioassays, as expected. Professor Wareing's observation that abscisic acid was less polar was correct, though the absolute proof for either the presence or absence of abscisic acid in cocklebur seed remains to be tested. Also, the choice of solvent by those earlier pioneers was serendipitous because, as has been earlier stated, purified carboxyatractyloside is particularly unstable in solution as the free acid. In later studies we have also discovered that the dipotassium salt loses biological activity if it becomes hydrated in a humid atmosphere, even at 4° C.

Insofar as the allelochemic effects on higher plants are concerned the most dramatic effects were obtained with intact week-old corn plants (Zea mays L.) treated with concentrations of carboxyatractyloside at 10^{-2}, 10^{-3} and 10^{-4} M (100 ul placed inside the uppermost leaf sheath) (35). Responses were visible at 72 hours and consisted of necrotic lesions on leaves at 10^{-2} M and chlorosis within leaf sheaths at 10^{-3} and 10^{-4} M. One week following treatment there was massive necrosis of the leaves at 10^{-2} M and a 50% inhibition of growth, while 10^{-3} M treated plants were inhibited 25%

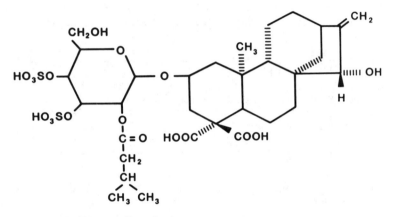

Figure 7.   Carboxyatractyloside.

compared to controls. By comparison there were no apparent
effects on week-old bean plants with applications of the
metabolite at 10^{-2} to 10^{-4} M. Effects on six-week-old
tobacco seedlings were quite subtle but there was no
phytotoxicity. Treatments with carboxyatractyloside at all
concentrations induced cupping of leaves within 24 hours so
that the tobacco plants had the appearance of having been
treated with cold night temperatures (2-7° C), and these
effects lasted for more than 72 hours.

The somewhat selective action of carboxyatractyloside
against the monocotyledon corn suggests that it may be
developed as an herbicide to control grasses in crops.
Indeed the chemistry of the molecule is such that several
synthetic adaptations are suggested. One, which is readily
available, is the compound atractyloside, which differs from
carboxyatractyloside by lacking a carboxyl group (31). It is
biologically inactive in plants (35) and in animals (R. J.
Cole, B. P. Stuart, and H. S. Gosser, unpublished data).
Another compound, the aglycone carboxyatractyligenin, is
readily obtained by refluxing carboxyatractyloside with 20%
KOH followed by acidification (31). Surprisingly, this
compound has not been tested for activity in plant systems,
but it should be biologically active. Because the methyl
ester of carboxyatractyligenin can be simply prepared with
diazomethane, this in turn can be oxidized with chromic acid
to yield diketocarboxyatractyligenin methyl ester (31).
Partial, or total reduction of this molecule should yield
interesting starting material for further testing as
biologicals, including the possibility of phosphate esters to
aid in translocation. The biological activity of
carboxyatractyloside is attributed to its role in inhibiting
translocation of adenine nucleotides across the mitochondrial
membrane (36).

Preliminary bioassays with carboxyatractyloside indicate
that it has selective fungistatic properties (Cutler,
unpublished data). In repeated disk tests various
concentrations of the metabolite have consistently inhibited
(>15 mm zone of inhibition) cultures of Chaetomium
cochlioides 189 grown in petri dishes containing nutrient
agar. However, carboxyatractyloside did not inhibit
Chaetomium cochlioides 195, Aspergillus flavus, or Curvularia
lunata 49.

Carboxyatractyloside has not yet been tested for either
insecticidal or insect repelling properties. We have
observed that cocklebur does not appear to be the subject of
massive infestations by insects, and we anticipate that the
plant contains substances that repel, or are toxic to
insects. And, as we shall see shortly, the effects of
carboxyatractyloside on enzyme systems are noteworthy and
further suggest the possible use of carboxyatractyloside as
an insecticide.

Any discussion of the allelochemic effects of

carboxyatractyloside must include the work carried out with
vertebrates. In 1981, Stuart and co-workers (37) redefined
the cause of swine intoxication induced by cocklebur
(Xanthium strumarium L.). Using cotyledonary cocklebur
seedlings, ground burrs containing seed, aqueous extracts of
burrs, and authentic carboxytractyloside administered either
orally or intravenously at various concentrations, it was
demonstrated that carboxyatractyloside was the causal agent
of cocklebur intoxication. The clinical signs observed were
similar to those described in earlier reports and included
depression, nausea, lack of coordination, convulsions with
paddling of the feet, coma and death. There was
hypoglycemia, as determined from serum chemistry profiles,
with increases in serum oxaloacetic transaminase and
isocitric dehydrogenase. Hypoglycemia appeared to be related
to the uncoupling of oxidative phosphorylation (36) and
death was due to hypoglycemia. Blood sugar levels were
recorded at zero (R. J. Cole, personal communications).
Additionally, the gross pathology was such that there was
acute hepatic necrosis, centrilobular accentuation and
disruption of the lobular hepatocytes (37). Although it was
originally postulated that hydroquinone was the origin of
swine intoxication (34) it could not be found in cocklebur.
Additionally, authentic hydroquinone did not induce lesions
(serofibrinous ascites, edema of the gallbladder wall and
lobular accentuation of the liver) but it did induce
hyperglycemia (37), the exact opposite response to
carboxyatractyloside. Interestingly, hydroquinone has been
shown to regulate auxin-oxidase activity and is,therefore,
considered to be a plant growth regulator(38). Further, it is
an inhibitor of wheat coleoptile growth and root formation in
Phaseolus cuttings (39).
    While the relative mammalian toxicity of
carboxyatractyloside would seem to preclude its use as an
agricultural chemical, there is some exciting chemistry to be
carried out relative to biological activity and further
testing in plant and other systems. All indications are that
carboxyatratyloside is not persistent and is easily degraded
so that even though initial application may pose problems to
the applicator there are no long term effects on the
environment. In the case of medicinal drugs, hypoglycemic
agents are especially useful when it is necessary to control
high blood sugar levels in patients; for example, when
insulin production is lacking or insufficient to equilibrate
the biological system.
    Ophiobolins G and H and carboxyatractyloside exemplify
the rich diversity of natural product structures that possess
biological activity and have the potential for use in
agricultural and other areas. Linkages, which are slow in
developing, need to be established across the various fields
of research and include their use for elucidating biochemical
pathways in unrelated organisms and synthetic structural
changes to give novel products.

Literature Cited

1. Whittaker, R. H. In "Chemical Ecology", ed. E. Sondheimer and J. B. Simeone. Academic Press: New York, 1970; pp 33-70.
2. Cutler, H. G.; Crumley, F. G.; Cox, R.H.; Hernandez, O.; Cole, R.J.; Dorner, J. W. J. Agric. Food Chem. 1979, 27, 592.
3. Arigoni, D. Chemical Society Autumn Meeting, Nottingham, England, 1965.
4. Iitaka, Y.; Watanabe, I., Harrison, I. T.; Harrison, S. J. Am. Chem. Soc. 1968, 90, 1092.
5. Rios, T.; Gomez, G. F. Tetrahedron Lett. 1969, 2929.
6. Orsenigo, M. Phytopathol. Z. 1957, 29, 189.
7. Ishibashi, K.; Nakamura, N. J. Agric. Chem. Soc. Jpn. 1958, 32, 739.
8. Ishibashi, K. J. Agric. Chem. Soc. Jpn. 1961, 35, 323.
9. Ishibashi, K. J. Agric. Chem. Soc. Jpn. 1962a, 36, 226.
10. Ishibashi, K. J. Antibiot. 1962b, 15A, 88.
11. Nozoe, S.; Morisaki, M.; Tsuda, K.; Iitaka, Y.; Takahashi, N.; Tamura, S., Ishibashi, K.; Shirasaka, M. J. Am. Chem. Soc. 1965, 87, 4968.
12. Canonica, L.; Fiecchi, A.; Kienle, M. G.; Scala, A. Tetrahedron Lett. 1966, 1329.
13. Nozoe, S.; Hirai, K.; Tsuda, K. Tetrahedron Lett. 1966, 2211.
14. Itai, A.; Nozoe, S.; Tsuda, K.; Okuda, S. Tetrahedron Lett. 1967, 4111.
15. Nozoe, S.; Morisaki, M.; Fukushima, K.; Okuda, S. Tetrahedron Lett. 1968, 4457.
16. Cutler, H. G.; Crumley, F. G.; Cox, R.H.; Springer, J. P.; Arrendale, R. F.; Cole, R. J.; Cole, P. D. J. Agric. Food Chem. 1984, 32, 778.
17. Chattopadhyay, A. K.; Samaddar, K. R.; Phytopath. Z. 1980, 98, 118.
18. Kim, J-M.; Hyeon, S-B.; Isogai, A.; Suzuki, A. Agric. Biol. Chem. 1984, 48, 803.
19. Turner, W. B. In "Fungal Metabolites"; Academic Press: New York, 1971; p. 250.
20. Graniti, A. Phytopathol. Mediterr. 1966, 5, 146.
21. Ballio, A. In "Regulation of Cell Membrane Activities in Plants," ed. E. Marré and O. Cifferi. North Holland:Amsterdam, 1977; p. 217.
22. Canales, M. W. Ph.D. Thesis. University of Minnesota, 1983 [University Microfilms International; Ann Arbor, MI 48106; 1984).
23. Canonica, L.; Fiecchi, A.; Kienle, M. G.; Ranzi, B. M.; Scala, A. Tetrahedron Lett. 1968, 275.
24. Gimpel, W. F.; Miller, D. R.; Davidson, J. A. Ag. Expt. Sta. Univ. Maryland Misc. Publ. 841 June, 1974.

25. King, L. J. In, "Weeds of the World, Biology and Control," Leonard Hill Books: London, 1966.
26. Arthur, J. C. Proc. Soc. Prom. Agric. Sci. 1895, 16,70.
27. Thornton, N. C. Contrib. Boyce Thompson Inst. 1935, 7, 477.
28. Wareing, P. F.; Foda, H. A. Physiologia Plant. 1957, 10, 266.
29. Esashi, Y.; Komatsu, H.; Ishihara, N.; Ishizawa, K. Plant Cell Physiol. 1982, 23, 41.
30. Wareing, P. F.; Phillips, I. D. J. In "The Control of Growth and Differentiation in Plants". Pergamon Press:Oxford, 1970. Chapter II.
31. Danieli, B.; Bombardelli, E.; Bonati, A.; Gabetta, B. Phytochemistry 1972, 11, 3501.
32. Cole, R. J., Stuart, B. P.; Lansden, J. A.; Cox, R. H. J. Agric. Food Chem. 1980, 28, 1330.
33. Craig, J. C.; Mole, M. L.; Billets, S., El-Feraly, F. Phytochemistry 1976, 15, 1178.
34. Kusel, N. R.; Miller, C. E. J. Am. Pharm. Assoc. Sci. Ed. 1950, 39, 202.
35. Cutler, H. G.; Cole, R. J. J. Natural Products 1983, 46, 609.
36. Luciani, S.; Martini, N.; Santi, R. Life Sci. 1971, 10, 961.
37. Stuart, B. P.; Cole, R. J.; Gosser, H. S. Vet. Pathol. 1981, 18, 368.
38. Wada, S.; Nagao, M. Sci. Repts. Tohoku Univ: IV, Ser. B. 26: 181, 1961.
39. Kefeli, V. I. In "Natural Plant Growth Inhibitors and Phytohormones", Dr. W. Junk b.v. Publishers, The Hague, 1978, p. 208.

RECEIVED February 8, 1985

Insect Antifeedants from the Peruvian Plant *Alchornea triplinervia*

D. HOWARD MILES[1], BARBARA L. HANKINSON[1], and SHIRLEY A. RANDLE[2]

[1]Department of Chemistry, Mississippi State University, Mississippi State, MS 39762
[2]Department of Entomology, Mississippi Agricultural and Forestry Experiment Station, Mississippi State, MS 39762

The boll weevil antifeedants, anthranilic acid,
gentisic acid, senecioic acid, trans-cinnamic acid,
trans-cinnamaldehyde and camphor have been isolated
from the Peruvian plant Alchornea triplinervia
(Euphorbiaceae). Anthranilic acid and camphor
also showed significant inhibition of growth of
the tobacco budworm.

Although pesticides continue to be the major approach to boll
weevil control, problems related to their use have led to a
search for alternative forms of pest control. These include
chemicals that modify behavior and/or development, biological
agents, and genetic manipulation.

Antifeedants appear to be a promising approach to agricultural
pest control. Kubo et. al. (1) define insect antifeedants
as "substances which when tasted can result in cessation of
feeding either temporarily or permanently depending upon potency."
Since antifeedants do not directly kill the insect, problems
associated with pesticides, hormones, biological agents, or
genetic manipulation may not be operative; however, other problems
may be operative. Antifeedants may be obtained by the isolation
of naturally occurring antifeedants or by synthesis.

Antifeedants have been isolated which represent many broad
classes of compounds (2). Since it is widely accepted that
tropical flora have built-in defense mechanisms (2) due to
their constant exposure to attack by many types of biological
organisms including insects, the plants chosen for this study
were selected from the Amazon River basin of Peru and the southern
portion of the United States (Mississippi). The ethanolic extracts
of more than five hundred plants from Peru and Mississippi were
evaluated for potency with a boll weevil antifeedant bioassay.
Those plants which showed total inhibition of feeding were fractionated and examined in additional bioassays using the tobacco
budworm. Three of the plants which were selected for further
examination include the Peruvian plants Alchornea triplinervia
(Euphorbiaceae) and Machaerium floribundum Bentham (Leguminosae)
and the Mississippi plant Heterotheca camporum (Compositae).

0097–6156/85/0276–0469$06.00/0
© 1985 American Chemical Society

Recent publications (3,4) in regard to the latter two plants indicate that they were also toxic to two bacteria which occur in the gut of the tobacco budworm Heliothis virescens (Fab.) and the corn earworm Heliothis zea (Boddie). The compounds, rhamnetin 3-0-glucoside and isoquercitrin, were isolated (3) from H. camporum based upon their activity against Pseudomonas maltophilia and Enterobacter cloacae. Also, a procyanidin was isolated (4) from M. floribundum on the basis of activity against P. maltophilia.

Boll Weevil Antifeedant Bioassay

The agar-plug bioassay procedure developed by Hedin et. al. (5) was used with a few modifications. The agar plugs were formed by boiling 3 grams of agar and 3 grams of freeze-dried cotton bolls in 100 ml of distilled water to effect a viscous sol. The sol was poured into 13 mm diameter hollow glass rods and gelation occurred after cooling. These gelatinous rods were cut into individual 3.5 cm plugs.

The extracts of the plant samples were applied to preweighed 4 cm squares of Whatman #1 chromatography paper by dipping the paper into a solution of the extract. After drying, the papers were weighed to determine the "concentration" of the extract being tested. A control paper was prepared by dipping a 4 cm square of paper into the solvent used for the extract and allowing the paper to air-dry. The papers were wrapped around the agar-cotton plugs and fastened with staples. The ends of the plugs were sealed with corks. The plugs were then placed staple-side down in the petri dishes so that the boll weevils could feed only by puncturing the papers.

Twenty newly emerged boll weevils were placed in 14 x 2 cm petri dishes containing the test and control plugs. The bioassay was carried out in the dark at 80°F for 4 hours, after which the papers were removed and the punctures counted.

Antifeedant activity was expressed as a %T/C value, where

$$\%T/C = \frac{\text{\#punctures of test paper (T)}}{\text{\#punctures of control paper (C)}} \times 100$$

%T/C values of zero represented total inhibition of feeding, while values greater than 100 represented feeding attractancy.

Tobacco Budworm Larval Growth Bioassay

The bioassay was performed according to the procedure of Hedin et. al. (6). A randomized complete block of eight replicates of five larvae were used for each data point. Compounds were tested at 5-6 concentrations ranging from 0.02 to 0.50% of the total diet. Larval weights were expressed as percent of control (%T/C). For estimation of ED_{50} values, the larval weight expressed as %T/C was regressed as the percent of the compound in the diet using the linear portion of the curve. The percent of the compound that would decrease the weight gain by 50% (ED_{50}) was then estimated.

Extraction of Alchornea triplinervia Leaves

The Alchornea triplinervia used in this project was collected by
the Institute for Botanical Exploration at their research facility
in Iquitos, Peru, in the upper Amazon Valley during the first
quarter of 1980.

The plant material was extracted according to the procedure
in Figure 1. The ethanol extract [2] (Figure 1) exhibited total
inhibition of feeding against the cotton boll weevil at levels
of 10, 25, and 50 mg (Table I). Further fractionation of the
ethanol extract [2] showed no abatement of this antifeedant activity
in either the $CHCl_3$ [3] or the aqueous fraction [4] with %T/C
values of less than 12% at the 10-50 mg dose levels as indicated
in Table I. Further fractionation (Figure 1) and bioassay (Table
I) showed that the methanol fraction [6], and the $CHCl_3$/ethanol
fraction [7] demonstrated excellent antifeedant activity.

The $CHCl_3$/ethanol extract [7] was successively extracted
with 5% acetic acid, 5% hydrochloric acid, 5% sodium bicarbonate,
5% sodium carbonate, and 5% sodium hydroxide. Bioassay of these
fractions indicated that antifeedant activity resided in the
sodium bicarbonate and sodium carbonate fractions.

A crystalline material (compound 1, mp 144-145°C) was isolated
from the sodium bicarbonate fraction by crystallization from
ether. Compound 1 was shown to be anthranilic acid according
to IR, NMR, and UV comparisons with an authentic sample (Aldrich
Chemical Company).

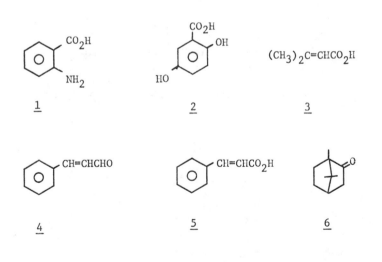

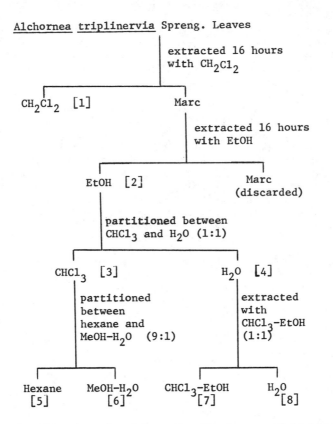

Figure 1. Fractionation Scheme for the Leaves of <u>Alchornea triplinervia</u> Spreng. (Euphorbiaceae)

Table I. Bioassay Test Results from Extraction Fractions of
Alchornea triplinervia

Layer	%T/C 10 mg	25 mg	50 mg
1 (CH_2Cl_2)	66	31	19
	161	100	219
	237	263	148
2 (Ethanol)	0	0	0
	0	0	0
	0	0	0
3 ($CHCl_3$)	3	0	0
	12	0	0
	3	0	0
4 (H_2O)	3	0	6
	0	0	1
	11	1	0
5 (Hexane)	78	0	6
	54	24	1
	51	1	7
6 (Methanol)	0	1	3
	0	0	0
	0	0	0
7 ($CHCl_3$/Ethanol)	0	0	0
	2	0	0
	0	1	0
8 (H_2O)	9	0	0
	1	4	1
	0	0	0

Column chromatography on silica gel of the sodium carbonate
fraction utilizing benzene, chloroform, and methanol as the eluting
solvents resulted in the isolation of compounds 2 (mp 199-200°C)
and 3. Compound 2 was shown to be gentisic acid according to
IR, NMR, UV and mixed melting point comparison with an authentic
sample (Aldrich Chemical Company).

Compound 3 was shown to be senecioic acid based on IR, NMR,
and mixed melting point comparison with an authentic sample (Aldrich
Chemical Company).

Investigation of the chloroform fraction [3] in Figure 1
resulted in the isolation of compounds 4, 5, and 6. This was
achieved through extensive column chromatography on silica gel.
Compounds 4 and 5 were present in the oily chromatographic fraction
which had the characteristic smell of cinnamon. Compound 4 was
isolated as an oil and identified as cinnamaldehyde by formation
of the 2,4-dinitrophenylhydrazine derivative which melted at
251-253° [lit(7) m.p. 255°C]. NMR, IR, and UV data were consistent
with this assignment. Further proof was provided by air oxidation
to form a white crystalline material which was. shown to be identical
with compound 5 which upon IR, NMR, and UV comparison with an
authentic sample (Aldrich Chemical Company) was shown to be cinnamic
acid.

Compound 6 was isolated via repetitive column chromatography
of the $CHCl_3$ fraction [3]. Compound 6 was shown to be identical
upon IR, NMR, and UV comparison with an authentic sample of camphor.

Boll Weevil Antifeedant Bioassay Results

The results of the boll weevil antifeedant bioassays with anthranilic acid, gentisic acid, senecioic acid, cinnamaldehyde and
cinnamic acid are presented in Table II. Camphor was not bioassayed due to its extreme volatility. However camphor is a
known moth repellent (8) and is probably a boll weevil antifeedant
since the fraction from which it was isolated was active.

The results with cinnamaldehyde are perhaps the most significant since total inhibition of feeding was demonstrated at a
dose of 30 mg for an entire agar plug (area = 16 cm^2). Moreover
the activity remained at a high level (96% inhibition) as the
dose was reduced to 20 mg and 10 mg. A dose of 10 mg for the
entire plug is equivalent to a dose of 0.62 mg per cm^2. Tests
will be performed at even lower doses in the future.

Tobacco Budworm Growth Study

Anthranilic acid, gentisic acid, senecioic acid, cinnamic acid,
and camphor were tested as inhibitors of tobacco budworm larval
growth. The results of this study are presented in Table III.
This table shows that gentisic acid and senecioic acid did not
show significant inhibition of larval growth of the tobacco
budworm. However anthranilic acid, cinnamic acid, and camphor
showed significant inhibition of growth with ED_{50} levels of
0.68%, 0.38%, and 0.50% respectively. This compares fairly
well with tannins, gossypol, and anthocyanin which have ED_{50}
values of 0.05-0.07% as reported by Hedin (9).

Table II. Boll Weevil Antifeedant Activity of Compounds Isolated
from A. triplinervia

Compound	Dose Level (mg)	%T/C			Average %T/C
Anthranilic Acid	10	12	11	13	12
	20	14	8	10	11
	30	4	7	8	6
Gentisic Acid	5	48	95	51	65
	10	8	9	6	8
	12	17	11	14	14
Senecioic Acid	10	6	16	24	15
	15	15	14	21	17
	40	1	4	2	2
Cinnamaldehyde	10	3	1	9	4
	20	0	10	3	4
	30	0	0	1	0
Cinnamic Acid	10	42	26	12	27
	20	31	24	24	26
	30	9	46	21	25

Table III. Tobacco Budworm Growth Inhibition by Compounds Isolated
from A. triplinervia

Compound	ED_{50}[a]
Anthranilic Acid	0.68
Gentisic Acid	NS[b]
Senecioic Acid	NS[b]
Cinnamic Acid	0.38
Camphor	0.50

[a]ED_{50} = percent of compound required to reduce weight gain by 50%.
[b]NS = not significant.

Conclusion

This study describes the isolation and bioassay against insects of
six compounds from the Peruvian plant Alchornea triplinervia. The
compounds anthranilic acid, cinnamic acid, and camphor showed
significant inhibition of the growth of the tobacco budworm.
Anthranilic acid, gentisic acid, senecioic acid, cinnamic acid, and
cinnamaldehyde demonstrated low to moderate activity in the boll
weevil antifeedant bioassay. Cinnamaldehyde, a constituent of
the spice cinnamon, showed the highest level of inhibition to
boll weevil feeding.

Acknowledgements

This research has been supported by the Mississippi Agricultural and Forestry Experiment Station. We are greatly indebted to our colleague, Dr. Paul A. Hedin, for his assistance throughout this study and in the preparation of this manuscript. We also acknowledge Dr. Sidney McDaniel of the Institute for Botanical Exploration for plant collection.

Literature Cited

1. I. Kubo, Y.W. Lee, M. Pettei, F. Pilkiewicz, and K. Nakanishi, Chem. Comm., 1013 (1976).
2. I. Kubo and K. Nakanishi. In "Host Plant Resistance to Pests"; Hedin, P.A., Ed; ACS SYMPOSIUM SERIES No. 62, American Chemical Society: Washington D.C. 1977; pp. 165-178.
3. S.K. Waage and Paul A. Hedin, Phytochemistry, in press, 1984.
4. S.K. Waage and P.A. Hedin, and E. Grimley, Phytochemistry, in press, 1984.
5. P.A. Hedin, A.C. Thompson, and J.P. Minyard, J. Econ. Entomol. 59, 181 (1966).
6. P.A. Hedin, D.H. Collum, W.H. White, W.L. Parrott, H.C. Lane, and J.N. Jenkins, Scientific Papers of the Institute of Organic and Physical Chemistry of Wroclaw Technical University, No. 22, Conferences 7, 1971 (1980).
7. C.D. Hodgman, ed., Tables for Identification of Organic Compounds, Chemical Rubber Publishing Company, 1960, 72.
8. M. Windhoz, ed., Merck Index, 9th Edition, Merck and Co., Inc., 1976, p. 1735.
9. P.A Hedin, J.N. Jenkins, D.H. Collum, W.H. White, W.L. Parrott, and M.W. MacGown, Experientia 39, 799 (1983).

RECEIVED November 8, 1984

Biotechnology in Crop Improvement

JOHN T. MARVEL

Monsanto Agricultural Products Company, St. Louis, MO 63167

Biotechnology promises to play a significant role
in crop improvement and productivity in the 1990's
and beyond. Early advances will probably be in
the development of selective and safe microbial
pesticides and the transfer of one to three gene
traits to agronomic crops. While microbial pesticides are technically fairly straightforward,
genetically improving crop plants, using recombinant techniques, will require the solution of numerous technical problems. Of initial importance is
the development of transformation vectors and regeneration technology in key crop plants, particularly legumes and cereals. Once these hurdles have
been overcome, the key emphasis will shift to the
discovery of genes to be transferred.

This paper reviews the status of regeneration and
transformation technology in the major crop plants
and highlights recent progress in plant biochemistry which may serve as a source of important
traits for genetic engineering.

Agricultural biotechnology has been in the public eye a good
deal recently. However, the basic thrust of biotechnology in agriculture is actually mundane. In fact, genetic engineering of plants
will be just another tool for plant breeders to use in their continuing efforts to improve plant productivity.
Classical breeding has been the mainstay of crop improvement
since the rediscovery of Mendelian genetics at the beginning of this
century. The improvements have been significant, e.g., the development of hybrid corn resulted in a steady 1-2% increase in yield per
year. Other crop breeding programs led to the development of strains
that would sustain food production in previously sterile environments.
Cell biology, in conjunction with genetic engineering, promises new ways to improve this record by enhancing yield potential,
improving pest tolerance, decreasing stresses due to the environment
and to agricultural chemicals, and improving overall agronomic

0097-6156/85/0276-0477$09.50/0
© 1985 American Chemical Society

acceptability. In order to comprehend how these new improvements
will occur, an understanding of classical breeding methods is
essential.
     Classical breeding consists of four distinct activities:
(1) screening for desirable traits and transferring those traits to
adapted lines by sexual crosses (2-4 years); (2) selecting progeny
with the desired combinations of traits (3-4 years); (3) field
evaluation of the selected varieties for yield and performance under
several environments (3-4 years); and finally (4) seed increase for
sale (2-3 years). In all, the process requires 10-15 years to produce a new variety which will typically have a lifetime of only 6-8
years.
     In addition, this process is limited by the sexual compati-
bility between the lines used for a cross. Typically, only lines
from the same species or very closely related ones can be used as a
source of new traits. Biotechnology can address these bottlenecks
of time and gene sources by genetically engineering plant cells and
then regenerating them into whole plants with the new traits. This
is possible because plants, alone among higher organisms, can be
regenerated into whole plants from somatic cells. This is a
phenomenon called "totipotency".
     The regeneration cycle is illustrated for alfalfa in Figure
1. A cutting, or explant, taken from the parent plant, is put onto
a medium containing plant hormones and nutrients. Soon the tissue
begins to proliferate cells in a rather disorganized mass to form a
callus. Upon treatment with appropriate plant nutrients and hormones, the callus will form structures which develop into shoots, a
process referred to as "organogenesis". These shoots may be removed from the callus, rooted, and grown into normal fertile
plants (1).
     This process, outlined in Figure 2, could be useful in a
breeding program. As in classical breeding, first a quality cultivar is chosen. Established tissue culture techniques are then
utilized. The material is placed into culture which allows selection by classical methods or the insertion of new genes. After the
tissue with a new trait has been produced, it can be regenerated
into a quality cultivar containing the new desired trait.
     The potential benefits of this scheme are two-fold. First,
the time-lines to develop new cultivars may be dramatically shortened. In cell biology-facilitated breeding, the identification,
isolation and cloning of a gene requires 1-3 years. To culture a
plant tissue, transform it and regenerate it takes approximately
6 months. Field evaluation and seed increase are unchanged by this
technology, so the total time is 6-10 years, a considerable savings
in time over conventional breeding.
     Genetic engineering allows the introduction of genes from any
source into plants; hence, a crop's germplasm base becomes all living organisms rather than just closely related, sexually compatible
plants. This means that genes (Figure 3A) become available from
bacteria which produce insecticidal proteins, from bacteria or
fungi which produce antibiotics active against plant pathogens, or
from stress-tolerant wild plants which would normally be sexually
incompatible. The promise of this technology is in its ability to
"teach" plants to produce their own insecticides, fungicides and
growth regulators.

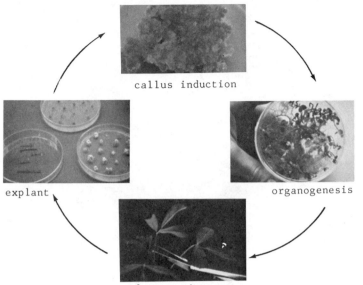

callus induction

explant

organogenesis

plant totipotency

Figure 1. Alfalfa regeneration (clockwise). A cutting is taken from a petiole (explant), placed on medium to induce the formation of a callus, then transferred to an altered medium causing shoots to form.

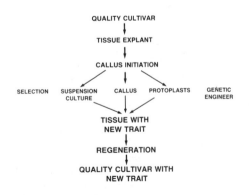

Figure 2. Tissue culture crop improvement. Sequence shows the integration of cell biology techniques into crop improvement. Hurdles to using the scheme include callus initiation, protoplast preparation, selection in culture, and plant regeneration.

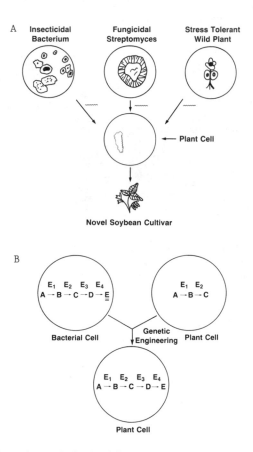

Figure 3. Examples of desirable genes to be inserted into crop plants. A. Unrelated organisms may have genes beneficial to crop plants. B. Suggested extension of a biochemical pathway in plants. Bacteria produces desired molecule $\underline{E}$ by enzymatic steps E_1 to E_4; plant pathway stops at intermediate $\underline{C}$. Introduction of bacterial genes for steps E_3 and E_4 causes plant cell to produce $\underline{E}$.

Even more sophisticated improvements may be possible
(Figure 3B). For example, if a microbe produces a molecule $\underline{E}$
which is nematocidal, and the plant has the biosynthetic machinery
to make a key intermediate of this molecule, $\underline{C}$, then perhaps genes,
coding for the enzymes necessary to complete the biosynthetic pathway, could be moved into the plant, causing the plant to produce
its own nematocide. The result is literally chemical synthesis in
living tissues. Plant genetic engineering could be competitive
with the chemical pesticide business.

However, enough must be understood about plant metabolism to
support the useful manipulation of the plants' biosynthetic apparatus to respond better to stresses or disease. Genetic manipulation
of complex pathways, such as the shikimate pathway (Figure 4), will
be a large task requiring considerably more biochemical knowledge
about plants than is currently available.

Having considered the kinds of advances that tissue culture
and genetic engineering can make in crop improvements, it is necessary to explore the technical limitations to making those kinds of
changes (Figure 2). For the major crops one must first have the
ability to culture and regenerate plants from various explants.
Unfortunately, not all crops respond to current tissue culture
techniques and regenerate in vitro. It is also necessary to have
the technology to either select new cell lines in culture or to
genetically engineer new traits into those tissues that are in culture, and then regenerate plants that will express the new trait.
Finally, these new cultivars must be extensively evaluated in the
field to assure that the desired trait has been inserted and is
expressed at the proper time and in the proper plant tissue.

Much progress has been made in the regeneration of plants and
in understanding the regeneration process. This includes the development of selection technology with particular emphasis on
resistance and the development of genetic engineering technology
using the Agrobacterium tumefaciens vector system.

The process of plant regeneration begins with the selection
of the proper explant which, when placed in the appropriate culture
media, will form a callus. In the case of alfalfa, somatic embryos
will form on the callus surface (Figure 5A) after the calli have
been exposed to the appropriate ratio of cytokinins and auxins.
Eventually these embryos will precociously germinate and form
shoots. The developing embryos can be excised and placed onto a
rooting media to develop a root system (Figure 5B). A final transfer of the rooted plants to gravel tubs and gradual exposure to
greenhouse conditions result in the development of normal plants
(Figure 5C). In time they flower and produce seeds. When these
seeds are planted, they produce normal fertile plants. Some variability in phenotype has been observed in plants arising from the
tissue culture process. These somaclonal variations may be present
in 2-20% of the opoulation, and may depend upon the stresses encountered during the tissue culture process.

This kind of media manipulation is indispensable in tissue
culture for maximizing the frequency of regeneration. When the
regeneration system for a new species is in the early stages of
development, experimental modifications are necessary in order to

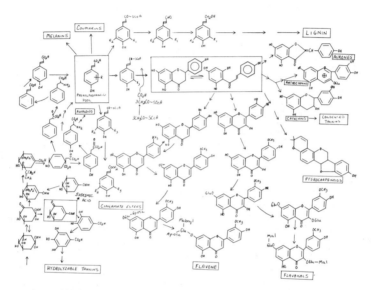

Figure 4.   Shikimate-derived metabolism in plants.   A complicated biosynthetic pathway is a possible genetic engineering
target.

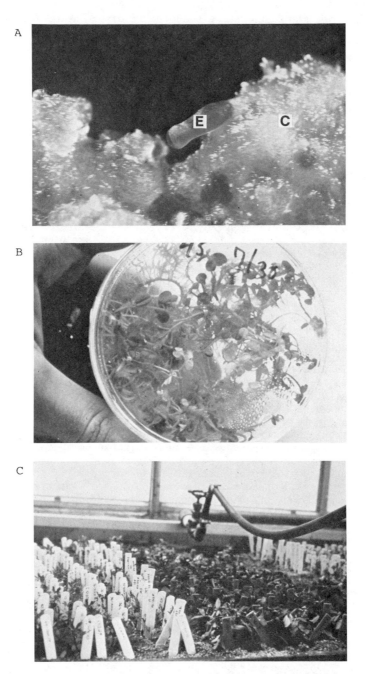

Figure 5.   Alfalfa embryogenesis.   A.   An alfalfa somatic embryo, E, about to germinate, which is surrounded by callus C. B.   Plantlets rooting.   C.   Regenerated plants in gravel in the greenhouse.

achieve optimal results. Figure 6 shows the effect of manipulating
the amino acid composition of the media on the frequency of regeneration (2). It is evident that when either Shenk and Hildebrandt's
or Blaydes basal media are used, the frequency of embryo formation
is low. However, if either media is supplemented with alanine or
proline, the frequency of embryogenesis is greatly enhanced.

Research of a more fundamental nature is also necessary in
order to understand and effectively manipulate the regeneration
process. This is particularly true in crops such as soybean and
cereals which are recalcitrant to regeneration. For example, histological and histochemical studies can be conducted during regeneration in order to understand the growth and development process. The
formation of alfalfa somatic embryos (Figure 5A) is well suited for
such a basic investigation. Very early in regeneration, alfalfa
callus (as visualized in a cross-section stained with safranin and
fast green in Figure 7A) has already differentiated into distinct
tissues. The darkly staining purple tissue highlights the embryo
in an early stage of formation. The lighter blue staining cells
below the embryo have been named the "proregenerative mass" and
appear to function as cellular progenitors to the embryo (3). From
this proregenerative mass, the incipient embryo develops from a
single cell. This is an interesting and significant finding because
it clarifies an early organizational event which occurs in regeneration. In Figure 7B the proregenerative mass remains as the embryo
grows larger; in fact, when the embryo is approximately at the stage
of the whole embryo shown in Figure 5A, the proregenerative mass
remains attached to the embryo (Figure 7C).

It is speculated that this mass substitutes for the suspensor,
an organ that normally aids in feeding developing embryos in planta.
Recalcitrant soybean and cereal tissue culture systems are currently
being investigated for evidence that these kinds of proregenerative
cell masses are formed (4).

Another approach to studying the developmental process in
regeneration is to observe the histochemical changes in cells.
Figure 8A illustrates alfalfa cells which have been stained with
aniline blue-black for total protein before induction of the regeneration process. These normal callus cells are elongated and
not densely staining. Figure 8B shows a cell mass which has been
induced to regenerate. The cells are very compact and tightly
associated, and the cytoplasm is darkly staining with aniline blue,
indicating a high concentration of proteins which may be necessary
for the regeneration process. Studies of this type will aid in the
design of biochemical experiments designed to better understand the
molecular basis for regeneration in plants.

The development of selection technology is necessary in order
to derive cell lines with specific traits, such as herbicide resistance, from tissue culture. Figure 9 depicts the somatic cell
selection process using cell culture techniques. At the bottom
left is a flask containing a suspension of alfalfa cells. These
cells have been cultured for 4-8 weeks and then sieved to yield
very small cell clumps which will be used for selection in vitro.
In the example shown here, selection was made for herbicide resistance. In the upper left-hand corner, glyphosate, the active
ingredient in Roundup, is added to the growth media. After several

1
SHO   SHA   SHP

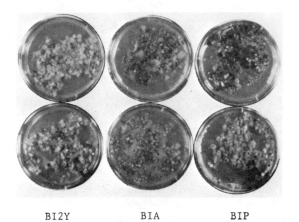

BI2Y   BIA   BIP

Figure 6. Effect on alfalfa regeneration of amino acid addition to media. The top row shows petri plates of Shenk and Hildebrandt medium with no amino acid addition, SHO; with the addition of L-alanine, SHA; or with L-proline, SHP. The bottom row shows Bladyes medium, with no amino acid addition, BI2Y; with the addition of L-alanine, BIA; or with L-proline, BIP. These amino acid additions enhance the frequency of embryogenesis.

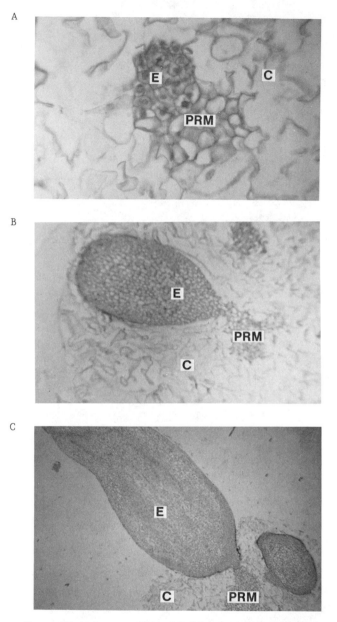

Figure 7. Histology of alfalfa somatic embryogenesis.
A. Cross-section of callus after induction of embryogenesis.
Lightly staining cells are the proregenerative mass (PRM)
which gives rise to the darkly staining cells, the embryo (E),
surrounded by very lightly staining non-regenerating callus (C)
B. A somatic embryo at a later stage of development. C.
Somatic embryo beginning to germinate, comparable to the whole
embryo shown in Figure 5A.

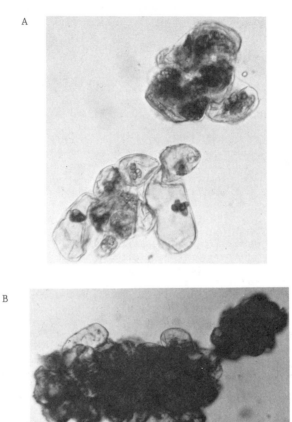

Figure 8. Histochemical studies of alfalfa callus cells. A. Normal callus cells stained with aniline blue black for total protein before induction of regeneration. B. Callus cells stained with aniline blue black after induction of regeneration; very darkly staining material is protein.

days' exposure to glyphosate, the cells are plated onto solid media
as shown in the center picture. From these plated cells only a few
will develop into calli.

Once these calli are formed, the hormone levels can be manipulated to induce shoot formation. This is followed by the rooting, hardening and transfer-to-greenhouse processes. During this
sequence the selected lines can be screened at the cellular level,
at the regenerated plantlet level, at the whole plant level in the
greenhouse, and finally in the field which is the "acid test".

Figure 10 shows, in considerably more detail, the sequence of
mutagenesis and selection which has been actually used to develop
herbicide resistant lines. Calli were initiated from a healthy alfalfa plant. After 4-8 weeks, these calli were broken up into suspensions, and either treated with a mutagenizing agent or screened
simply by selection for spontaneous mutations. After either procedure, the selected lines, i.e., the lines which survived exposure
to the herbicide, were then regenerated, and the plants were evaluated in a number of schemes. In addition, plants selected at the
cellular level for resistance were recycled through the entire system of mutagenesis and selection to enhance the desired resistance
trait.

A number of cell lines identified in cell culture were resistant to 10 millimolar glyphosate. These lines were regenerated
and the plantlets were placed on media containing 10 or 100 millimolar glyphosate. For comparison, regenerated but not selected
control plantlets were placed on similar media. Some of the selected lines grew and developed on the glyphosate-containing media.
In contrast, unselected control plantlets failed to survive. Survivors were rooted and transferred to the greenhouse where they were
sprayed with Roundup at rates equivalent to 2 or 4 pounds per acre.
Survivors of this test were subsequently evaluated for their resistance to Roundup in the field.

The data for this field experiment are summarized in Table 1.
Data are presented here for 13 lines which were derived from culture
and field-evaluated for resistance to Roundup herbicide. Each of
these lines was significantly more tolerant to Roundup than the regenerated non-selected control B74. However, the level of resistance
in these 13 cell lines was not commercially significant. Nevertheless, this does indicate that resistant plants can be derived by
selections at the cellular level.

Eleven-thousand cell lines were mutagenized and screened to
identify these 13 lines which were resistant at the whole plant
level in the field. This is a frequency of approximately one-tenth
of one percent; undoubtedly this frequency can be improved by
genetic engineering.

herbicide
exposure
↑

selection

regeneration

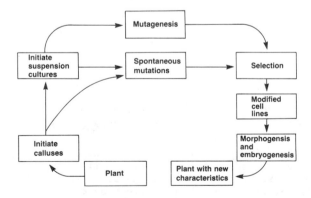

Figure 9.   Somatic cell selection for herbicide resistance.
Bottom left, a flask of alfalfa cells in suspension. Top left,
addition of herbicide to the cells. Center, cells plated
onto solid medium containing herbicide; a resistant callus
growing on herbicide-containing medium. Top right, resistant
plantlets regenerating. Bottom right, tolerant plants selected
from tissue culture growing in the field after being sprayed
with the herbicide.

Figure 10.   Generalized scheme for cellular selection involving
spontaneous or induced mutagenesis.

Table 1.   Superior Alfalfa Clones from the Glyphosate Field Test
Based on Survival 21 Days Post-Application.

Clone ID	Mutagenesis and Selection Conditions	Percent Survival and One-Tailed P Value[6]			
		2 Lbs/Acre		4 Lbs/Acre	
B74	None	5%		0%	
64s-8	EMS-AMP[2]	(not tested)		30%	.0288
65s-1[1]		57%	.0012	30%	.0288
65s-2		31%	.0386		
65s-8[1]		57%	.0004		
141s-3		29%	.0586		
141s-6[1]		70%	.0000		
80s-3	PF-GLY[3]	63%	.0001	27%	.0425
1s-1	5FU-GLY[4]	44%	.0014		
1s-2		38%	.0139		
B62-3-10[1]	NQO-GLY[5]	63%	.0001		
B62-3-13		47%	.0031		
B62-3-26		42%	.0076		
B89-4		31%	.0490		

[1]Also superior to the 20% survival level of RA3-24 ($p \leq 0.05$) in the
2 lb/acre treatment.

[2]Ethylmethanesulfonate - aminomethylphosphonic acid.

[3]Proflavin-glyphosate.

[4]5-Fluorouracil-glyphosate.

[5]Nitroquinoline oxide-glyphosate.

[6]P values were obtained from a binary T-test comparison with B74,
an unselected, non-mutagenized regenerated clone.

A number of systems and plant species have been used to study
the mode of action of glyphosate. Figure 11 illustrates the effect
of treating soybean leaves with glyphosate. The soluble organic
extracts from those leaves were analyzed using high pressure liquid
chromatography, HPLC. The chromatogram of the extract from the
untreated leaf is shown at the top. The profile from the glypho-
sate-treated leaf is shown at the bottom. There are some differences between the chromatograms. Especially significant is a large

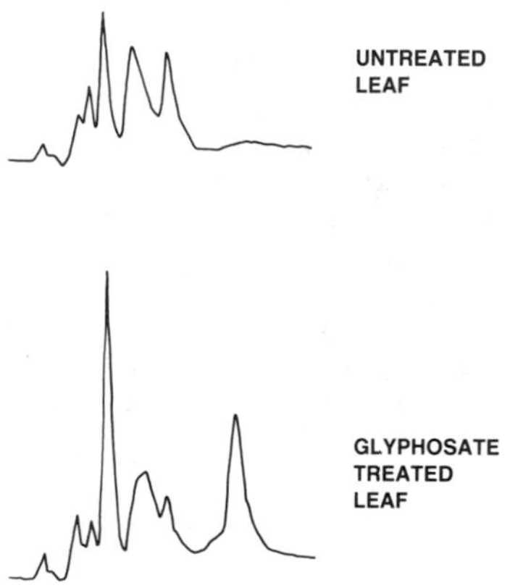

Figure 11. HPLC chromatograms comparing organic extracts of glyphosate treated soybean leaves to extracts of non-treated leaves. Origins of the chromatograms are on the left. Right-most peak from treated leaf extract is shikimate-3-biphosphate.

peak not present in the untreated control, which appears in the
extract of the treated plant (rightmost peak in the lower chromatogram). On further analysis it was found that this peak is
shikimate-3-biphosphate, the substrate of enolpyruvyl shikimate
phosphate (EPSP) synthase, the enzyme which glyphosate is thought
to inhibit.

Alfalfa plants, derived from tissue culture and tolerant to
Roundup, were evaluated to investigate their mechanism of resistance. No difference was found in the level of shikimate-3-
biphosphate which accumulated in both resistant and susceptible
lines upon treatment with glyphosate.

This result indicates that the structural gene coding for
EPSP synthase has not been modified to code for a glyphosate-
resistant enzyme because these tissues still accumulate high levels
of shikimate-3-biphosphate. Radio-labeled glyphosate has been used
to study uptake of the chemical in both susceptible and tolerant
lines. The data tends to support the conclusion that the selected
regenerated alfalfa lines take up considerably less glyphosate than
their non-selected counterparts, suggesting that impaired glyphosate
uptake is probably the mechanism of this resistance. This type of
resistance is not dissimilar from resistance to some anti-cancer
drugs in cultured cancer cell lines. In those systems the constitutive level of membrane glycoproteins increases greatly with continuous exposure to the drug. It can be hypothesized that exposure of
plant cells to glyphosate might also induce glycoprotein synthesis
which make the membranes of the cells less permeable to the chemical.

Other problems, besides obtaining commercially insignificant
levels of resistance and the requirement to screen large numbers of
lines, were encountered using tissue culture selection. Figure 12A
shows one of the glyphosate tolerant regenerates, HG-2. It is no
longer a normal trifoliate alfalfa plant. The effect of long-term
tissue culture, mutagenesis and selection for glyphosate resistance
or tolerance was definitely deleterious and increased the somaclonal
variation. Figure 12B shows the flower of this same line; it is
evident that the flower morphology more closely resembles a hya-
cinth than an alfalfa flower. Because of results such as these
observed from tissue culture selection, the advent of genetic
engineering technology in plants was greeted with great enthusiasm.

A natural transformation system that has been discovered and
exploited for plants (Figure 13) depends on a natural soil bacterium which is a pathogen of dicotyledonous plants, Agrobacterium
tumefaciens. A. tumefaciens contains a plasmid (circular extra-
chromosomal DNA) called the "Ti-plasmid". This plasmid is so named
because it carries the tumor inducing DNA which is transferred into
the plant cells. During the course of infection, a 16 kilobase
piece of DNA called the "T-DNA" is transferred from the Ti-plasmid
into the chromosomal DNA of the plant cell. In the wild, this
results in the formation of a crown gall, a disorganized tumorous
growth of plant cells which can be seen on trees and other plants
in the field. These tumors can be removed from the plant and grown
in culture. There are several points which have been observed about
these cultures.

First, the tumor cultures produce their own plant hormones;
thus, unlike other callus, they do not require hormone supplemented

Figure 12. HG-2 glyphosate tolerant alfalfa derived from tissue culture selection. A. Abnormal multifoliate leaf structure. B. Abnormal flower morphology.

media. In addition, they produce unusual amino acid metabolites
called "opines" which are not present in the untransformed plant
tissue. These metabolites are derivatives of arginine and provide
a nutrient source for the bacteria. It was noted that, after the
bacteria were eliminated by treatment of the callus tissue with
appropriate antibiotics, the plant cells still produced opines and
still grew independent of plant hormones.

Researchers perceived that in tumor formation A. tumefaciens
T-DNA was transferred into the plant cells. An organism with a
natural DNA transformation ability could be very useful for genetically engineering or transferring traits into plant cells. Of
course, there were a number of obstacles to be overcome in using
A. tumefaciens for plant genetic engineering.

First, the tissue that came from infection of plants with
A. tumefaciens was a chimeric tissue containing a mixture of transformed and non-transformed cells. Second, vector technology was
needed for actually putting foreign genes into the T-DNA of
A. tumefaciens. Third, a method was needed for selection of transformants. Fourth, it had been found that cells, which were able to
regenerate normally before transformation, were inhibited from regeneration after transformation due to the hormone genes (tumor
genes) that were transferred into the plant cells. Therefore, techniques were needed to remove the tumor genes to allow regeneration
of transformed cells.

The chimeric tissue problem has been solved by co-cultivating
plant protoplasts with A. tumefaciens (Figure 14), then treating
these protoplast suspensions with an antibiotic that selectively
kills the A. tumefaciens. The protoplasts are plated onto a nurse
layer of tobacco cells which feed the individual protoplasts and aid
them in regenerating into pure colonies. The colonies ultimately
form callus, and this allows for the production of pure cultures of
transformed cells. These cultures can then be regenerated into
plants.

The technology for shuttling the desired foreign genes into
the A. tumefaciens T-DNA is shown in Figure 15. The salient features are: first, the intermediate plasmid (pMON120) has homology
with the A. tumefaciens Ti-plasmid, allowing co-integration into the
Ti-plasmid and the insertion of desired genes into the T-DNA of
A. tumefaciens. Second, the plasmid contains the gene coding for
nopaline synthase which serves as a convenient, rapidly assayable
marker for transformation in plants because this metabolite can be
easily identified by chromatography. In addition, the intermediate
plasmid carries the gene for spectinomycin and streptomycin resistance which allows its selection and manipulation in E. coli and
A. tumefaciens. It also contains a unique Eco RI restriction site
which allows the opening or cutting of this plasmid for the insertion of desired genes.

When foreign genes were originally put into plants using the
A. tumefaciens Ti-plasmid co-integration technology, they were not
expressed. It was hypothesized that the problem was that bacterial
promoters are not recognized in plants. Hence, three chimeric genes
were constructed as shown in Figure 15. One gene was composed of
the nopaline synthase (NOS) promoter, which is known to function in

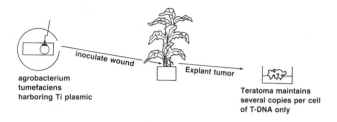

Figure 13. <u>Agrobacterium</u> <u>tumefaciens</u>, the natural transformation system for plants. The bacterium contains a circular
piece of DNA, the Ti-plasmid. The bacterium infects a
dicotyledonous plant and transfers the T-DNA from the Ti-
plasmid to the plant chromosomal DNA causing a tumor to form.
The tumor is called a "crown gall". The gall, or teratoma, can
be removed from the plant and placed into culture on medium
without exogenous hormones.

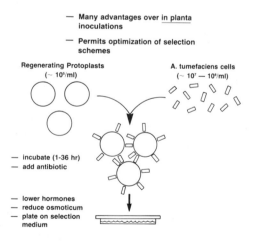

Figure 14. Co-cultivation procedure for transformation of
plant protoplasts by <u>Agrobacterium</u> <u>tumefaciens</u>.

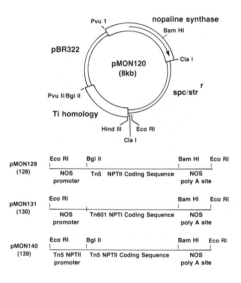

Figure 15. Plasmid vector system for shuttling chimeric genes into the T-DNA. Upper, the intermediate vector, pMON120, consists of a segment of pBR322, a gene for resistance to spectinomycin and streptomycin, a nopaline systhase gene (NOS) and a region of homology (LIH) with the Ti-plasmid to allow cointegration and insertion of chimeric genes into the T-DNA. Lower, the three chimeric kanamycin resistance genes shown were inserted into the unique Eco RI site of pMON120 to create the new intermediate plasmids, pMON129, pMON131 and pMON140.

plants, coupled to a gene coding for kanamycin resistance (the Tn5 neomycin phosphotransferase II coding sequence). The Tn 601 neomycin phosphotransferase I coding sequence was also fused with the NOS promoter. Finally, Tn5–NPTII with its own promoter was used to test whether or not bacterial promoters would, in fact, function in plants. These three genes were inserted into the Eco RI site of pMON120 to produce the intermediate plasmids designated pMON129, pMON131 and pMON140.

In Figure 16 the recombination, which shuttles the vector with the chimeric gene coding for kanamycin resistance (Tn5–NPTII) into the Ti-plasmid, is demonstrated. After co-integration occurs, a new Ti-plasmid is formed which carries a left- and a right-hand border. The border sequences form the ends of the T-DNA segment which integrates into the plant chromosome. The new Ti-plasmid also contains the chimeric kanamycin resistance gene and the nopaline synthase gene, as well as the tumor genes which were on the original Ti-plasmid.

When A. tumefaciens infects the protoplast, it will transfer all these genes, including the kanamycin resistance gene, into the plant genome. Selection for transformants can then be accomplished on kanamycin-containing media. Because the pMON vector system, which co-integrates into the Ti-plasmid, contains a right-hand border next to the NOS gene, the T-DNA actually has two right-hand borders. Either of these right borders can function as one end of the transferred T-DNA. This occasionally allows for integration of a shortened segment of DNA. When this "short transfer" occurs, the tumor genes are deleted and plant regeneration can proceed.

Information to support the transfer of these selectable markers is shown in Figure 17. Figure 17A shows the results of a paper electrophoresis separation of petunia calli which were selected on the basis of nopaline production. Comparison of the spots observed, after staining with phenanthrene quinone reagent, demonstrates that opines are being produced by these plant cells.

Figure 17B shows the actual selected callus. Calli on the left were derived from transformed protoplasts which did not contain the gene coding for kanamycin resistance, while those on the right were derived from transformed protoplasts which contained the resistance gene. In a media containing 50 μg /ml of kanamycin, those cells containing the resistance gene have regenerated to form a healthy green callus, while the transformed control cells have not.

The kill curve in Figure 17C graphically illustrates the necessity of the nopaline promoter in the chimeric gene rather than depending upon the bacterial promoter for gene expression. The top curve is the kill curve for the neomycin phosphotransferase II system connected to the nopaline synthase promoter, while the middle curve is the same promoter now fused to the neomycin phosphotransferase I gene. The bottom curve, showing the same resistance as a non-transformed control, is the neomycin phosphotransferase II with its own bacterial promoter. The chimeric gene with the NOS promoter and the NPT-II coding sequence gives an LD50 for kanamycin of approximately 150 μg /ml.

Southern hybridization, shown in Figure 17D, confirmed that the DNA from A. tumefaciens had been transferred into the plant

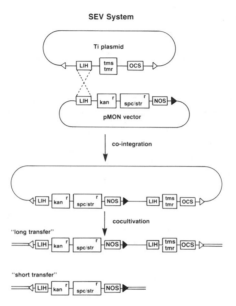

Figure 16. Using the Ti-plasmid vector system for plant
transformation. The arrows represent the T-DNA border sequences. LIH is a region of homologous DNA for recombination
between Ti and the pMON vector. The tumor genes are denoted
tms and tmr. OCS and NOS are octopine and nopaline synthase
genes, respectively. The chimeric kanamycin resistance gene is
designated Kan^r. The resistance gene for spectinomycin and
streptomycin is designated spc/str^r. Recombination between the
LIH regions on the Ti-plasmid and pMON vector produces the
cointegrate plasmid with two right borders. After co-
cultivation and selection for kanamycin resistance either the
entire T-DNA including the tumor genes ("long transfer") or a
truncated T-DNA without tumor genes ("short transfer") is
integrated into the plant genome. Only plant cells receiving
the short transfer can regenerate.

cells. These blots were prepared from the total plant chromosomal
DNA.

The entire tissue culture sequence, used to obtain the
transformed petunia plants, is shown in Figure 18. Two types of
callus are growing in the petri plate containing kanamycin media.
One received the "long" transfer including the tumor genes which
prevent regeneration. The other is a callus which was derived
from a "short" transfer of the T-DNA, i.e., one that did not contain the tumor genes. As a consequence, this callus is beginning
to send out shoots which can be rooted and grown into whole plants.
Leaves from these regenerated transformed plants still express the
kanamycin resistance trait when put back into culture. This trait
is inherited in a dominant Mendelian fashion (3:1) (6).

The entire technology has been reduced to practice. A foreign gene has been inserted into plant cells, those cells have
been regenerated, and the regenerated plants expressed the trait
(kanamycin resistance) which was inserted. Although many problems
remain, the technology is available to start determining whether
genetic engineering can impact the plant breeder in a useful
fashion.

The next issue to confront is the question of the source of
genes. One potentially useful gene is the Bacillus thuringiensis
(B.t.) protein toxin gene. This protein is not toxic to mammals,
but it is toxic to lepidopterous insects at nanogram levels (7).
This protein has a billion-fold safety factor for humans and is
acceptable for engineering into plants for insect control. The
Farm Chemicals Handbook contains this statement about the B.t.
toxin: "Harmless to humans, animals and useful insects. Safe for
the environment. Exempt from requirements for a tolerance on all
raw agricultural commodities when applied to growing crops, for both
preharvest and postharvest uses" (8).

When the bacterium sporulates at the end of its growth cycle,
it produces a proteinaceous crystal which is the active ingredient
in the B.t. toxin. It is thought that this crystal is degraded in
the alkaline gut of lepidopteran pests to form a protoxin protein
of 134,000 dalton molecular weight, and that these protoxins are
further degraded to lower molecular weight toxins in the range of
60,000-70,000 daltons (9).

The trigonal protein crystals, which are visible in crude
B. thuringiensis cell preparations, can be isolated by centrifugation, purified and treated with a pH-10 buffer to mimic the conditions found in the insect gut. The manipulations allow the isolation and purification of the 134,000 molecular weight protoxin,
Figure 19. The amino acid sequence from the amino terminus of this
B.t. protein was determined, the possible DNA coding sequences were
deduced, and a set of degenerate DNA probes was synthesized. The
amino acid sequence and the probes are shown below:

AA Sequence: MET - ASP - ASN - ASN - PRO - ASN - ILE

DNA Probes: ATG - GAT - AAT - AAT - CCA - AAT - ATT
                      C   C   C   G   C   C
                                        T   A
                                        C

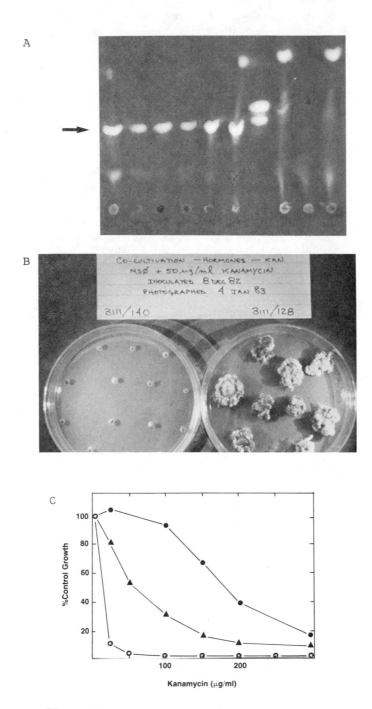

Figure 17. See caption on next page.

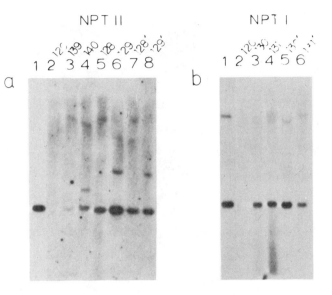

Figure 17. Verification of plant transformation. A. Nopaline
production in transformed petunia. Arrow indicates nopaline.
Lanes, left to right: 1, callus; 2, pedicel; 3, sepal; 4,
petal; 5, stigma; 6, anther; 7, nopaline (bottom) and octopine
(top) standard; 8-10, not relevant. B. Growth of transformed
petunia calli on kanamycin-containing media. Calli in petri
plate on the right were derived from transformed protoplasts
containing a functional chimeric kanamycin resistance gene.
Calli on left were derived from transformed protoplasts which
did not contain the resistance gene. C. Comparison of the
activity of chimeric kanamycin-resistance genes. Curves show
relative growth of transformed petunia calli containing different chimeric genes over a range of kanamycin concentrations.
Genes tested were: NOS promoter-NPTII from pMON129 (closed
circles), NOS promoter-NPTI from pMON131 (open triangles),
NPTII with its bacterial promoter from pMON140 (open circles),
untransformed control (dots). D. Detection of foreign DNA in
transformed petunia calli. Total plant DNA from transformants
selected for kanamycin resistance or hormone independent growth
(for pMON120 transformants) was digested with Eco RI, separated
by electrophoresis and hybridized with probes specific for
NPTII (panel a) or NPTI (panel b). Numbers above each lane
refer to pMON vectors shown in Figure 14. Lane 1 in each panel
contains a plasmid DNA marker.

Figure 18. Tissue culture sequence to obtain transformed petunia plants expressing a foreign gene, kanamycin resistance. The petri plate at the bottom contains two calli. The callus not forming shoots received the "long transfer", and the shoot-forming callus, the "short transfer". The "short trans-fer" shoots are removed from the callus and rooted in the container in the center. The rooted plant is transferred to the greenhouse. The leaves of the regenerated plant express the foreign gene.

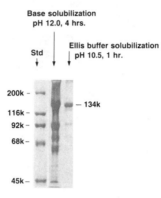

Figure 19. <u>Bacillus</u> <u>thuringiensis</u> crystal toxin protein. Isolated protein crystals were solubilized as indicated and analyzed by denaturing polyacrylamide gel electrophoresis. Solubilization in Ellis buffer yields the 134,000 dalton protoxin.

The probes were hybridized to blots of total plasmid DNA prepared from B. thuringiensis. This is shown schematically in Figure 20. Plasmid DNA from toxin-producing B. thuringiensis is first cut with restriction enzymes, resulting in a collection of restriction fragments which can be sized by gel electrophoresis. Hybridizing the gel with the DNA probe identifies the fragments that correspond to the DNA coding for the toxin.

The actual Southern hybridization is shown in Figure 21. In the left panel is an agarose gel showing restriction enzyme analysis of plasmid DNAs containing the B.t. toxin gene. In the right panel is the blot of the same gel hybridized with the radio-labeled DNA probe. The leftmost lane is plasmid DNA isolated from B. thuringiensis itself. The other lanes are E. coli plasmids containing cloned segments of B. thuringiensis DNA. In the B. thuringiensis plasmid only a single Bam HI fragment hybridizes with the probe. Two of the E. coli plasmids, pMAP1 and pMAP2, contain cloned copies of this hybridizing fragment which contains the B.t. toxin coding sequence.

Once this gene has been identified by Southern blot, it can be cloned and expressed in E. coli. Using an appropriate plasmid cloning vector, such as pBR322, this gene can be inserted into E. coli for multiple copy production. Southern hybridization can be used to verify that the gene is being cloned (Figure 21), and Western blots using antibodies to the B.t. toxin can be used to confirm that the protein is being produced. Additionally, whole cells or protein extracts of the cells can be fed to susceptible insects to verify that the protein toxin is being made and is active. Presently, this gene should be put into the appropriate A. tumefaciens vectors and tested for transfer and expression in plant cells.

Although this is a graphic example of what can be done with a single gene which has been thoroughly studied, many useful traits are multigenic. Response to stress is almost certainly controlled by multiple genes. Stress is one of the most important areas for investigation in agricultural biology. Figure 22 is a graphic representation of the yield losses incurred by environmental and biological stresses. It is estimated that some 60% of the genetic yield potential is lost to heat, cold, water and disease stress.

Recently progress has been made in studying various aspects of one particular stress, i.e., the process by which plants protect themselves from heat. Under normal conditions, it has been found that plants produce enzymes required for normal cellular metabolism (Figure 23A). Figure 23B illustrates the situation when the plant senses an increase in temperature. New genes, called the "heat shock genes", are turned on, with the subsequent production of heat shock proteins and expression of normal metabolism genes is reduced.

This phenomenon is shown graphically in Figure 24. Under a normal range of temperatures, normal cellular protein synthesis proceeds. However, as the temperature rises, normal cellular metabolism decreases and heat stress metabolism increases until finally normal protein synthesis is stopped. Perhaps by understanding how to control the expression of normal cellular proteins, it may be possible to engineer plants that continue to express

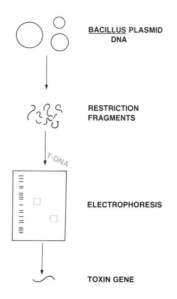

Figure 20.   Schematic diagram of the isolation of the Bacillus
thuringiensis toxin gene.

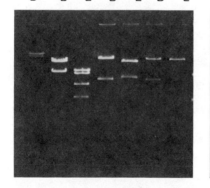
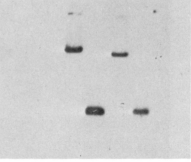

Figure 21.   Cloning the Bacillus thuringiensis protein toxin
gene.   Left panel, agarose gel restriction enzyme analysis of
plasmid DNAs.   Right panel, Southern blot of the same gel after
hybridization with a synthetic DNA probe 21 nucleotides in
length (21-mer) corresponding to the B.t. toxin amino acid
sequence.   Labels B and E denote plasmid DNAs digested with
restriction enzymes Bam HI or Eco RI, respectively.   B.t.
denotes plasmid DNA isolated from Bacillus thuringiensis
itself.   Clone #20, pMAP1, and pMAP2 are E. coli plasmids
containing cloned segments of Bacillus DNA.

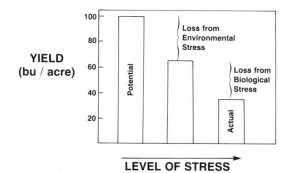

Figure 22.   Theoretical graph of the genetic yield potential of an agronomic crop compared to actual yield after losses due to environmental and biological stresses.

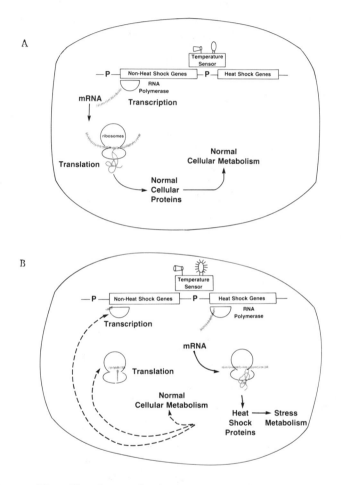

Figure 23. The heat shock response. A. At normal temperature, non-heat shock genes required for normal cellular metabolism are expressed, while heat shock genes are turned off. B.
At high temperatures, heat shock genes are expressed, while
expression of normal genes is reduced.

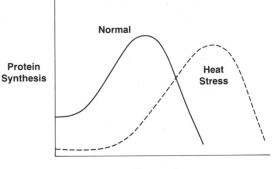

Figure 24. Schematic representation of changes in the profile of protein synthesis at normal and elevated temperatures.

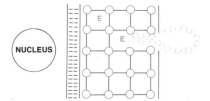

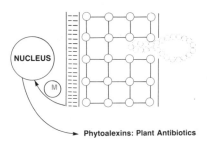

Figure 25. Schematic representation of phytoalexin production following exposure of the plant cell wall to a pathogen. E represents plant cell wall and pathogen enzymes; M, plant cell messengers.

their normal metabolism at higher temperatures while simultaneously
producing the needed heat shock proteins which protect the plant
against heat stress. Hypothetically then, this could result in an
increased level of protein synthesis at elevated temperatures which
may relate to an increase in yield. Research towards this end is
on-going in many institutions.

Another system, which has been well studied at the biochemical level, is the induced resistance to plant disease. For many
years it has been known that susceptible plants can be treated with
heat-killed or otherwise attenuated pathogens and have immunity
induced. Figure 25 shows schematically what is thought to happen
during this process. First, a pathogen, such as a fungal spore,
contacts a plant cell and begins to germinate an infection tube.
As this tube develops, it secretes enzymes which begin to dissolve
the plant cell walls. This enzymatic action in turn releases plant
enzymes which are cemented in the wall. These plant enzymes in turn
begin to dissolve the cell wall of the fungus with resulting "biochemical warfare" between these two organisms.

As a consequence of this enzymatic activity, fragments of
both the fungal cell wall and the plant cell wall are released and
recognized by the plant cells. This stimulates as yet unknown
messengers to initiate de novo protein synthesis, ultimately resulting in the synthesis of compounds called "phytoalexins". These
plant antibiotics help to inhibit the growth of the fungus in the
plant tissues. Understanding these mechanisms of resistance at a
molecular level should allow us to genetically engineer plants with
improved disease resistance in the future.

Acknowledgments

Alfalfa Tissue Culture: M. G. Carnes, M. S. Wright, K. A. Walker,
D. H. Mitten, T. K. Skokut, D. V. Connor, M. A. Hinchee,
S. M. Colburn and S. J. Sato.

Bacillus thuringiensis Cloning: T. L. Graham, F. J. Perlak,
L. S. Watrud, K. K. Budd and E. Mayer.

Protein Isolation: P. B. Lavrik.

Synthesis of DNA Probe: E. T. Sun.

Insect Testing: P. Marrone, T. Mosley.

Agrobacterium Transformation: R. Fraley, R. Horsch, and S. Rogers.

Literature Cited

1. Walker, Keith A., Poli C. Yu, Shirley J. Sato, and
    E. G. Jaworski. (1978) "The Hormonal Control of Organ
    Formation in Callus of Medicago sativa L. Cultured In Vitro."
    Am. J. Bot. 65:654-659.
2. Walker, K. A., M. L. Wendeln and E. G. Jaworski. (1979)
    "Organogenesis in Callus Tissue of Medicago sativa. The
    Temporal Separation of Induction Processes from
    Differentiation Processes." Plant Sci. Lett. 16:23-30.
3. Colburn, S. M. and M. A. W. Hinchee. (1984) "Development
    of Alfalfa Somatic Embryos in Tissue Culture." Am. J. Bot.
    71:22.
4. Pierson, P. E. and M. S. Wright. (1984) "Histology of
    Structures from a Soybean (Glycine max) Somatic Embryogenesis
    Protocol." Symposium on Propagation of Higher Plants through
    Tissue Culture: III Development and Variation, Knoxville.
5. Fraley, R. T., S. G. Rogers, R. B. Horsch, P. R. Sanders,
    J. S. Flick, S. P. Adams, M. L. Bittner, L. A. Brand,
    C. L. Fink, J. E. Fry, G. R. Galluppi, S. B. Goldberg,
    N. L. Hoffman and S. C. Woo. (1983) "Expression of Bacterial
    Genes in Plant Cells." Proc. Nat. Acad. Sci. USA, 80:4803-4897.
6. Horsch, R. B., R. T. Fraley, S. G. Rogers, P. R. Sanders,
    A. Lloyd and N. Hoffman. "Inheritance of Functional Foreign
    Genes in Plants." (1984) Science, 223:496-498.
7. Bulla, L. A. Jr., R. A. Rhodes and G. St. Julian. "Bacteria
    as Insect Pathogens." (1975) In Annual Review of Microbiology,
    eds. M. P. Starr, J. L. Ingraham and S. Raffel, 29:163-190.
8. Farm Chemicals Handbook (1984), Meister Publishing Co.,
    Willoughby, Ohio, p.C22.
9. Bulla, L. A. Jr., K. J. Kramer, D. J. Cox, B. L. Jones,
    L. I. Davidson and G. L. Lookhart. "Purification and
    Characterization of the Entomocidal Protoxin of Bacillus
    Thuringiensis." J. Biol. Chem. 256:3000-3004.

RECEIVED January 7, 1985

Why Are Green Caterpillars Green?

JOHN K. KAWOOYA, PAMELA S. KEIM, JOHN H. LAW, CLARK T. RILEY, ROBERT O. RYAN, and JEFFREY P. SHAPIRO

Department of Biochemistry, University of Arizona, Tucson, AZ 85721

Insects use camouflage coloration as a means of avoiding predation. The green color of the tobacco hornworm larvae, (Manduca sexta) can be separated into constituent blue and yellow components. The water soluble blue component is the biliprotein, insecticyanin. The yellow color is derived from lipoprotein bound carotenes. This lipoprotein, lipophorin, is the major lipid transport vehicle in insect hemolymph. In addition to transporting dietary lipid, lipophorin is also involved in the transport of lipophilic insecticides. Nearly all the recovered radioactivity in hemolymph from topically applied [^{14}C]-DDT is associated with lipophorin. Lipophorin of adult M. sexta is larger, less dense and is associated with small amounts of a third, adult specific, apoprotein. Alterations in adult lipophorin density, lipid content and apoprotein stoichiometry can be caused by injection of the decapeptide, adipokinetic hormone.

Green is a popular color for insects. Green caterpillars abound. Who has grown tomatoes without encountering the ubiquitous tomato (and sometimes tobacco) hornworm? Its color blends so successfully with that of the host plant that finding the cause of the damage can be frustrating in the extreme. Even some adult insects are green and the green insect egg evades detection when placed on a host plant. Clearly, green coloration imparts a protective advantage to the phytophagous insect.

Insecticyanin

The remarkable matching of insect pigmentation to that of the host plant led to the idea that the pigment of host and pest were related, possibly identical. Early workers compared the green

0097–6156/85/0276–0511$06.00/0

pigments of plants and larval insects and concluded that the insects retained chlorophyll from the diet and sequestered it in their tissues for camouflage coloration (1).

The chlorophyll hypothesis was thoroughly destroyed by Lederer and Przibram in 1933, when they reported that the green color of some orthopterans could be resolved into a water soluble blue pigment and a fat soluble yellow carotene fraction (2). Subsequent work on a number of species showed that the blue color results from protein-bound bile pigments, usually biliverdin IXγ or IXα (3) (Figure 1). The yellow color results from protein-bound carotenes, which in the lepidoptera are usually lutein and β-carotene (4) (Figure 1). It is interesting to note that the green coloration of the eggs of a tropical tree frog, Agalychnis dacnicolor, has recently been shown to result from a mixture of biliverdin IXα and lutein (5).

Recent investigations have been focused on the identification of the protein-pigment complexes of insects. For example, in the tobacco hornworm, Manduca sexta, a blue biliprotein, insecticyanin, has been found in the hemolymph, epidermal cells and eggs (6). This protein was purified to homogeneity and crystallized by Cherbas (6). It was shown to be an oligomeric protein composed of 22,000 dalton subunits. The chromophore was tentatively identified as biliverdin IXγ, associated by non-covalent bonds to the apoproteins.

The structure of the insecticyanin apoprotein has recently been determined (7) (Figure 2). In comparing this protein with others that bind bile pigments, short regions of homology have emerged. Those segments may thus represent the sites of bile pigment binding to insecticyanin. The holoprotein appears to be a tetramer, as indicated by cross-linking experiments (7).

We do not know how insects produce bile pigments. Some evidence points to de novo synthesis (8), but it is possible that some dietary component is involved. In mammals, protoporphyrin IX is cleaved to biliverdin IXα (Figure 1), but we do not know if an analogous process leads to insect biliverdins. Mammals process large amounts of protoporphyrin IX resulting from heme degradation. As insects do not make hemoglobin for oxygen transport, their supply of protoporphyrin IX is much more limited. If one could understand the route of bile pigment synthesis in insects and disrupt it, interference with camouflage coloration might be an attainable goal.

Hemolymph Lipoprotein

The yellow carotene binding protein of M. sexta hemolymph is a more complicated case. Carotenes are extremely water-insoluble materials. They share this property with several other natural products including sterols, fats and hydrocarbons, all of which are important to insects. This property is also shared by many xenobiotics, including pesticides. Transport of hydrophobic materials within the aqueous compartments of living organisms, e.g. blood or hemolymph, is accomplished by lipoproteins. Extensive

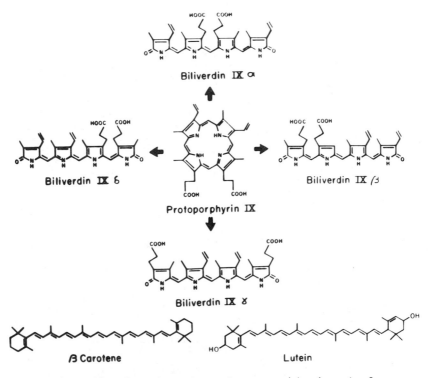

Figure 1. Bile pigments and carotenes used by insects for coloration. The conversion of protoporphyrin IX to the various bile pigments is indicated.

```
                                       10                                    20
GLY-ASP-ILE-PHE-TYR-PRO-GLY-TYR-CYS-PRO-ASP-VAL-LYS-PRO-VAL-ASN-ASP-PHE-ASP-LEU-

                                       30                                    40
SER-ALA-PHE-ALA-GLY-ALA-TRP-HIS-GLU-ILE-ALA-LYS-LEU-PRO-LEU-GLU-ASN-GLU-ASN-GLN-

                                       50                                    60
GLY-LYS-CYS-THR-ILE-ALA-GLU-TYR-LYS-TYR-ASP-GLY-LYS-LYS-ALA-SER-VAL-TYR-ASN-SER-

                                       70                                    80
PHE-VAL-SER-ASN-GLY-VAL-LYS-GLU-TYR-MET-GLU-GLY-ASP-LEU-GLU-ILE-ALA-PRO-ASP-ALA-

                                       90                                   100
LYS-TYR-THR-LYS-GLN-GLY-LYS-TYR-VAL-MET-THR-PHE-LYS-PHE-GLY-GLN-ARG-VAL-VAL-ASN-

                                      110                                   120
LEU-VAL-PRO-TRP-VAL-LEU-ALA-THR-ASP-TYR-LYS-ASN-TYR-ALA-ILE-ASN-TYR-ASN-CYS-ASP-

                                      130                                   140
TYR-HIS-PRO-ASP-LYS-LYS-ALA-HIS-SER-ILE-HIS-ALA-TRP-ILE-LEU-SER-LYS-SER-LYS-VAL-

                                      150                                   160
LEU-GLU-GLY-ASN-THR-LYS-GLU-VAL-VAL-ASP-ASN-VAL-LEU-LYS-THR-PHE-SER-HIS-LEU-ILE-

                                      170                                   180
ASP-ALA-SER-LYS-PHE-ILE-SER-ASN-ASP-PHE-SER-GLU-ALA-ALA-CYS-GLN-TYR-SER-THR-THR-

                                      189
TYR-SER-LEU-THR-GLY-PRO-ASP-ARG-HIS
```

Figure 2. The covalent amino acid structure of the blue
biliprotein of M. sexta. Residues 60–90, 102–108 and 147–154
show some homology with other bile pigment binding proteins and
may represent regions involved in chromophore binding.

studies on human serum lipoproteins have provided a model in which extremely hydrophobic materials (e.g. triacylglycerol, sterol esters, hydrocarbons) form a spherical core (9). The acyl chains of a phospholipid monolayer as well as the apolar ring system and side chain of free sterols associate with the hydrophobic core, while the polar phospholipid head groups and sterol hydroxyl groups face the aqueous exterior. Interspersed near the aqueous interface are the apoproteins, which have well defined, segregated, hydrophobic and hydrophilic regions on their surfaces. The hydrophilic regions of the apoproteins interface with the aqueous environment while the hydrophobic regions associate with the phospholipids. Based on this structural organization, the apoproteins are said to be amphiphilic. This arrangement provides for the packaging of hydrophobic material within a water compatible envelope that may be efficiently transported through the blood.

Human lipoproteins exist in several sizes and densities with differing lipid to protein ratios. These various lipoproteins have different origins in the body, different destinations and different functions (10). Thus, chylomicrons are extremely large low density particles formed in the intestine and designed to deliver dietary fat to adipose tissue. Very low density lipoproteins (VLDL), on the other hand, are smaller, more dense particles designed to deliver lipids from the liver to adipose and other tissues. Low density lipoproteins (LDL), formed from VLDL or produced in the liver or intestine deliver cholesterol to peripheral tissue, while high density lipoproteins (HDL) function to return cholesterol from peripheral tissues to the liver for catabolism. There is a complex exchange of lipids and apoproteins between the lipoprotein classes.

If one draws hemolymph from the green larva of Manduca sexta and mixes it with potassium bromide to a concentration of 44 percent, places this solution in an ultracentrifuge tube, overlayers with saline and subjects the mixture to centrifugation at 200,000 x g for 4 hours in a vertical rotor, one resolves the green color into a lower blue phase, a clear zone and a bright yellow band in the middle of the tube (Figure 3). This is the result of a density gradient of KBr set up in the centrifugal field. Most proteins, including the blue insecticyanin, have a density greater than 1.30 g/ml and thus are sedimented to the bottom of the tube. The yellow carotene is associated with the lipoprotein of larval hemolymph, which has a density of 1.15 g/ml, and thus floats above the remainder of the hemolymph proteins.

It is thought that dietary carotene is transferred to the hemolymph lipoprotein, which is called lipophorin (11), at the midgut during digestion of food. It is transported to epidermal cells, where it probably associates with a different protein inside the cells. Unlike the blue component of green coloration, insects appear to be completely dependent upon dietary carotenes for the yellow component (4). M. sexta larvae, raised on a standard laboratory diet, are distinctly blue in color, rather than green.

What do we know about the structure and multiple functions of insect lipophorin? Larval lipophorin from M. sexta (12,13), with a density of 1.15 g/ml, is comparable to the high density lipoprotein

of human serum. Table I compares the composition of mammalian HDL
with M. sexta larval lipophorin. The differences are in the
content of diacylglycerol, a major component of lipophorin, and
sterol esters, which are present in only small amounts in
lipophorin. The polypeptide components are also different.
Mammalian HDL contains several copies of relatively small
apoproteins, the apoA series, while lipophorin contains an
extremely large apoprotein, apoLp-I, of about 240,000 daltons, and
a moderate sized apoLp-II of about 80,0000 daltons (13). Each
lipophorin particle has only one copy of each apoprotein.

Table I. Composition of High Density Lipoproteins
in Insects and Man

	Percent Total Weight	
Lipid Component	Insect*	Mammals**
Triacylglycerol	0.9	2
Diacylglycerol	14.9	0.4
Sterol (% as esters)	2.0 (0%)	18 (84%)
Phospholipid	14.9	18
Hydrocarbon	4.7	--
Protein	62.7	58
Density (g/ml)	1.15	1.06-1.21

*Taken from Prasad (26)
**Taken from Chapman (27)

Xenobiotic transport by lipophorin. We believe that the function
of the larval lipophorin is to transport water-insoluble materials
consumed by the larva from the site of digestion and absorption in
the midgut to the tissues of storage or utilization. Among these
are fats, sterols and carotenes. We also know that hydrocarbons,
produced in blood cells, are transported by lipophorin to epidermal
cells, where they are exported to the surface of the exoskeleton
(14). Lipophorins may also transport hydrophobic xenobiotics.
Figure 4 shows the result of an experiment in which a sublethal
dose of ^{14}C-DDT was applied to the cuticle of a fifth instar
larva of M. sexta. After 19 hours, hemolymph was drawn and
subjected to KBr density gradient centrifugation. The
radioactivity in each fraction was determined, and the result can
be seen. The amount of radioactive DDT in the hemolymph
represented only about 0.5-1 percent of the applied dose, but
virtually all of the radioactive material was associated with the
lipophorin. While the mode of transport of insecticides in insects
is a matter of some controversy (15) it can be clearly stated that
if DDT gets into the hemolymph, it associates with lipophorin.
Earlier workers (16-18) have mixed labeled insecticides with
hemolymph and shown that they become associated with lipoproteins,
but none of these reports identify the lipoproteins in the
hemolymph.

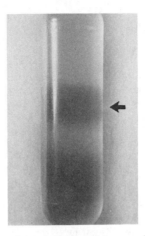

Figure 3. Potassium bromide density gradient
ultracentrifugation of M. sexta hemolymph. The less dense
yellow colored lipophorin floats above the layer of more dense
ordinary proteins, including the blue insecticyanin. Arrow
designates position of lipophorin in the gradient.

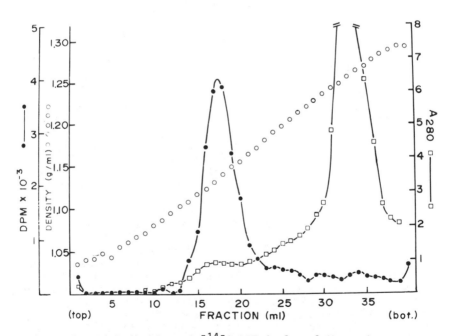

Figure 4. Distribution of ^{14}C-DDT in larval M. sexta
hemolymph. 19 h after topical application, hemolymph was
subjected to density gradient ultracentrifugation as shown in
Figure 3. Following centrifugation the tube was fractionated
and the radioactivity in each fraction determined. Most of the
labeled pesticide was found in the lipophorin fraction.

Adult Lipophorin

When the larva undergoes metamorphosis to an adult moth, a somewhat
different lipophorin is found in the blood. The density of this
adult form is 1.08 g/ml, and a new apoprotein, apoLp-III, 17,000
daltons (19), associates with the lipophorin, in addition to
apoLp-I and apoLp-II. Analysis shows that the particle has twice
the lipid content as the larval form and an apoprotein ratio of 1
apoLp-I to 1 apoLp-II to 2 apoLp-III.

ApoLp-III is one of the most abundant proteins of adult
hemolymph, reaching a concentration of 17 mg/ml. In hemolymph
apoLp-III can be found free or associated with lipophorin. Only a
small part of the total apoLp-III is associated with lipophorin
when the animal is resting.

A major difference between metabolism of the larva and that of
the adult is that connected with flight in the latter. Some adult
insects (e.g. flies and honeybees) use sugar to fuel flight, while
others, particularly those that fly long distances, use fat for
flight (e.g. butterflies, moths and locusts). A few insects (e.g.
tsetse flies and Colorado potato beetles) have an unusual reliance
on proline as a flight fuel (20). In the case of fat utilization,
it is necessary to transport large amounts of fat from the
reservoirs in the fat body to the flight muscle. It is well
established in the locust, Locusta migratoria, that fat is
transported as diacylglycerol associated with lipophorin (20).

The process of fat mobilization for use in flight metabolism
in L. migratoria is initiated by the release of a decapeptide, the
adipokinetic hormone (AKH) from the corpus cardiacum (20), a gland
posterior to the brain and a part of the neuroendocrine system.
Similar polypeptide hormones are probably found in all adult
insects and are involved in preparing the animal for flight. In
the cockroach, AKH causes mobilization of carbohydrate in the form
of the disaccharide, trehalose, which is produced in the fat body
from the glycogen reserves and transported through the hemolymph
(20). In the Colorado potato beetle Leptinotarsa decemlineata, AKH
stimulates production of proline by the fat body (21). Since
synthetic locust AKH causes these effects in all of these animals,
it is likely that each has a similar and homologous polypeptide
hormone that signals the onset of flight metabolism.

In the locust, AKH is thought to act upon receptors in the fat
body cell membrane to activate adenyl cyclase, which then activates
the enzymatic machinery to convert triacylglycerol to
diacylglycerol. The exact nature of that enzymatic machinery is
unclear, but there is an obvious parallel to the action of glucagon
on mammalian adipose tissue. When diacylglycerol leaves the locust
fat body, it is accepted by locust lipophorin in the hemolymph, and
at the same time, a small polypeptide, called "C protein"
associates with the diacylglycerol loaded lipophorin (22). We
believe that M. sexta apoLp-III is analogous to locust C protein.

To test this hypothesis, we carried out experiments in which
we injected synthetic locust AKH into adult M. sexta. We observed
a dramatic shift in the density of the adult lipophorin from

1.08 g/ml to 1.03 g/ml, or into the LDL class of lipoprotein (23). This change was accompanied by a large increase in size, diacylglycerol content and apoLp-III content (Figure 5). It can be seen that when most of the lipophorin has been loaded to capacity, the free apoLp-III in the hemolymph is greatly depleted.

ApoLp-III has been isolated and characterized (19). It is devoid of cysteine and tryptophan, and contains only one tyrosine residue. It is poor in glycine and proline, which tend to destroy helical conformation, and rich in leucine, glutamate and lysine, which are good helix formers. In keeping with this composition, the circular dichroism spectrum indicates a high content of helix. In addition, viscosity experiments, as well as studies on monolayers of apoLp-III at the air-buffer interface suggest that apoLp-III is a compact molecule, a characteristic of proteins with a high content of α-helix. The N-terminal sequence can be arranged into a perfect amphiphilic helix and would provide an excellent lipid binding site. Indeed, current experiments show apoLp-III to be an excellent lipid binding protein that binds either to phospholipid or diacylglycerol coated surfaces with high affinity.

Larvae lack the ability to load lipophorin with diacylglycerol, even when apoLp-III and AKH are supplied. On the other hand, larval lipophorin is readily converted to the adult form and loaded in the adult when AKH is supplied. If a foreign larval lipophorin, that from the honeybee Apis mellifera, is injected into adult M. sexta along with AKH, the foreign lipophorin is partially loaded. However, immunoprecipitation experiments with anti-apoLp-III antibodies indicate that apoLp-III did not associate with the honeybee lipophorin (24). This suggests that apoLp-III may recognize some feature of the apoproteins for binding and is not simply associating with exposed lipid.

Lipophorins from several orders of adult insects have been examined, and so far only in L. migratoria (22), M. sexta and a hemipteran, Leptoglossus zonatus (25), has apoLp-III been observed. Efforts are now underway to determine if apoLp-III is invariably associated with flight metabolism fueled by fat.

As one can see the question posed by the title can be answered at several levels. Pursuit of the question leads into the basic biochemistry and physiology of the insect and reveals fundamental facets of the transport of vital hydrophobic materials throughout the insect system. An understanding of the structure and functions of the lipoprotein transport vehicle may lead to a better understanding of normal physiology, as well as the mechanism for distribution of hydrophobic xenobiotics.

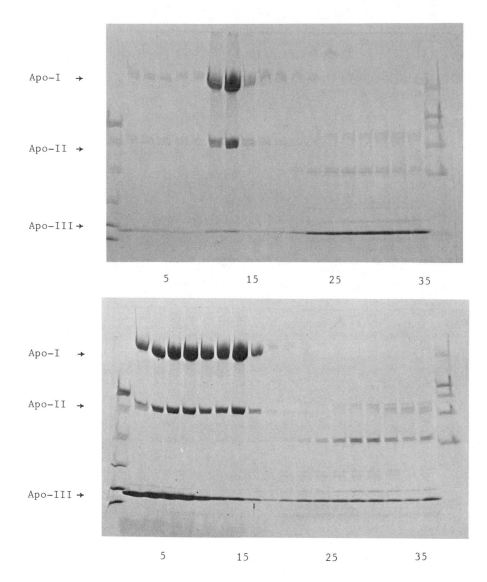

Apo-I →

Apo-II →

Apo-III →

5   15   25   35

Apo-I →

Apo-II →

Apo-III →

5   15   25   35

Figure 5. Effect of adipokinetic hormone on lipophorin.
Sodium dodecyl sulfate polyacrylamide gel electrophoresis (4-15
percent acrylamide gradient slab) of M. sexta adult hemolymph
following density gradient ultracentrifugation. Centrifuge
tubes were fractionated and aliquots applied to the gel.
Above, saline injected control animals; below, adipokinetic
hormone (200 pmoles/animal) injected. Reproduced with
permission from Ref. 23. Copyright 1983 Academic Press.

Acknowledgments

Research in the authors' laboratories was supported by grants from the National Science Foundation, No. 83-02670, the National Institute of General Medical Sciences, No. 29238 and a fellowship from the National Institutes of General Medical Science, No. GM09760.

Literature Cited

1. Poulton, E. B. Proc. Roy. Soc. 1885, 38, 269.
2. Przibram, H.; Lederer, E. Anz. Akad. Wiss. Wien. 1933, 70, 163.
3. Rudiger, W. Angew. Chem. Int. Ed. 1970, 9, 433.
4. Feltwell, J.; Rothschild, M. J. Zool. Lond. 1974, 174, 441.
5. Marinetti, G. V.; Bagnara, J. T. Biochemistry 1983, 22, 5651.
6. Cherbas, P. T. Ph.D. Thesis, Harvard University, 1973.
7. Riley, C. T.; Barbeau, B. K.; Keim, P. S.; Kezdy, F. J.; Heinrikson, R. L; Law, J. H. J. Biol. Chem., in press.
8. Rudiger, W.; Klose, W.; Vuillaume, M.; Barbier, M. Experientia 1969, 5, 487.
9. Shen, B. W.; Scanu, A. M.; Kezdy, F. J. Proc. Natl. Acad. Sci. USA, 1977, 74, 837.
10. Kostner, G. M. Adv. Lipid Res. 1983, 20, 1.
11. Chino, H.; Downer, R. G. H.; Wyatt, G. R.; Gilbert, L. I. Insect Biochem. 1981, 11, 491.
12. Pattnaik, N. M.; Mundall, E. C.; Trambusti, B. G.; Law, J. H.; Kezdy, F. J. Comp. Biochem. Physiol. 1979, 63B, 469.
13. Shapiro, J. P.; Keim, P. S.; Law, J. H. J. Biol. Chem. 1984, 259, 3680.
14. Katase, H.; Chino, H. Biochim. Biophys. Acta 1982, 710, 341.
15. Gerolt, P. Biol. Rev. 1983, 58, 233.
16. Skalsky, H. L.; Guthrie, F. E. Pest. Biochem. Physiol. 1975, 5, 27.
17. Fell, D.; Giannotti, O.; Holzhacker, E. C. Arq. Inst. Biol. Sao Paulo 1976, 43, 135.
18. Maliwal, B. P.; Guthrie, F. E. J. Lipid Res. 1982, 23, 474.
19. Kawooya, J. K.; Keim, P. S.; Ryan, R. O.; Shapiro, J. P.; Samaraweera, P.; Law, J. H. J. Biol. Chem. 1984, 259, 10733.
20. Beenakkers, A. M. Th.; Van der Horst, D. J.; Marrewijk, W. J. A. Insect Biochem. 1984, 14, 243.
21. Weeda, E. J. Insect Physiol. 1981, 27, 411.
22. Wheeler, C. H.; Goldsworthy, G. J. J. Insect Physiol. 1983, 29, 349.
23. Shapiro, J. P.; Law, J. H. Biochem. Biophys. Research Communs. 1983, 115, 924.
24. Ryan, R. O.; Prasad, S. V.; Schmidt, J. O.; Wang, X.-Y.; Wells, M. A.; Law, J. H. 1984, in preparation.
25. Ryan, R. O.; Schmidt, J. O.; Law, J. H. Arch. Insect Biochem. Physiol. 1984, 1, 375.
26. Prasad, S. V.; Ryan, R. O.; Wells, M. A.; Law, J. H. J. Biol. Chem., 1984, submitted.
27. Chapman, M. J. J. Lipid Res. 1980, 21, 789.

RECEIVED January 23, 1985

INDEXES

Author Index

Subject Index

A

ABA--See Abscisic acid
Abscisic acid
  mechanism, 90
  physiological effects, 90, 91
  qualitative effect on protein
    synthesis, 90
  role in flowering, 92
Abyssinin
  absolute configuration, 193, 195
  antifeedant activity, 200
  ^{13}C-NMR data, 185, 188f
  contour plot, 190, 191f
  groups, 185, 189f
  growth inhibitory activity, 200
  physical constants, 185, 190
  ^{1}H-NMR data, 185, 188f
  structure, 185, 190, 194
  two-dimensional COSY
    spectrum, 190, 192f
  two-dimensional NMR
    spectrum, 185, 189f
Abyssinols
  antifeedant activity, 200
  growth inhibitory activity, 200
Abyssinol A
  physical constants, 193
  ^{1}H-NMR data, 193, 194t
  structure, 193
Abyssinol B
  physical constants, 196
  ^{1}H-NMR data, 193, 194t
  structure, 193
Abyssinol C
  physical constants, 196
  ^{1}H-NMR data, 193, 194t
  structure, 184, 193, 196, 198
Adipokinetic hormone,
    function, 518, 519
Adult lipophorin
  differences with larval
    lipophorin, 518
  properties, 518, 519
Agrichemical transport
  accumulation, 15
  characteristics, 15
  determining factors, 14
AJH--See Anti juvenile hormone

AKH--See Adipokinetic hormone
Alchornea triplinervia leaves
  antifeedant activity, 471
  bioassay test results, 471, 473t
  extract structures, 471
  extraction, 471, 474
  fractionation scheme, 471, 472f
Alfalfa somatic embryogenesis,
    histology, 484, 486f
Alkaloids
  See also Nitrogen heterocycles
  biological activity, 402
  definition, 394
  identification, 394
Alkaloid toxicology, modes of
    action, 402
2-Alkyl-6-methylpiperidines
  cis-trans isomers, 399
  identification, 397, 399
Alkylpyrazines, examples, 400
Allelochemicals
  applications, 120
  categories, 111
  description, 109
  examples, 111, 456, 458, 460, 462
  indirect modes of action, 112
  interactions, 112
  schematic of field
    interrelationships, 110f
  toxicity, 111
Allelochemical effects
  effects on higher
    plants, 463, 465, 466
  examples, 113-15
  indirect, 113
Allelochemical interactions
  See also Allelochemical interference
  additive and synergistic
    interactions, 112
  environmental factors, 115-19
  herbicidal interactions, 112
Allelochemical interference
  See also Allelochemical interactions
  allelopathic weeds, 115-17
  crop inhibition of weeds, 119, 120
  crop-crop interactions, 117-19
Allelochemics--See Allelochemicals
Allelopathic agents--See
    Allelochemicals
Allelopathic effects, visible
    symptoms, 114

526

Indexing and production by Deborah H. Steiner
Jacket design by Pamela Lewis

Elements typeset by Hot Type Ltd., Washington, D.C.
Printed and bound by Maple Press Co., York, Pa.